Inorganic Reactions
and Methods

Volume 2

The Formation of the Bond
to Hydrogen (Part 2)

Editor

J.J. Zuckerman

Subject Index Editor

A.P. Hagen

Library of Congress Cataloging in Publication Data

Inorganic reactions and methods.

Includes bibliographies and indexes.
Contents: v. 1. The formation of bonds to hydrogen—
pt. 2, v. 2. The formation of the bond to hydrogen—
—v. 15. Electron-transfer and electrochemical
reactions; photochemical and other energized reactions.
1. Chemical reaction, Conditions and laws of—
Collected works. 2. Chemistry, Inorganic—Synthesis—
Collected works. I. Zuckerman, Jerry J.
QD501.I623   1987   541.3'9   85-15627
ISBN 0-89573-250-5 (set)

Printed in the United States of America.

ISBN 0-89573-252-1 VCH Publishers
ISBN 3-527-26260-1 VCH Verlagsgesellschaft

Inorganic Reactions
and Methods

Volume 2

Inorganic Reactions and Methods

Editor

Professor J.J. Zuckerman
Department of Chemistry
University of Oklahoma
Norman, Oklahoma 73019

Editorial Advisory Board

Contents of Volume 2

Formation of Bonds between Hydrogen and Elements of Group VB (N, P, As, Sb, Bi)

1.6. Formation of Bonds between Hydrogen and Elements of Group IVB (C, Si, Ge, Sn, Pb) 64

Formation of Bonds between Hydrogen and Elements of Group IVB (C, Si, Ge, Sn, Pb)

1.7.   Formation of Bonds between Hydrogen and Elements of Group IIIB (B, Al, Ga, In, Tl)   124

1.8.   Formation of Bonds between Hydrogen and Metals of Group IA (Li, Na, K, Rb, Cs, Fr) or IIA (Be, Mg, Ca, Sn, Ba, Ra)   151

Formation of Bonds between Hydrogen and Metals of Group IA (Li, Na, K, Rb, Cs, Fr) or IIA (Be, Mg, Ca, Sn, Ba, Ra)

1.9.   Formation of Bonds between Hydrogen and Metals of Group IB (Cu, Ag, Au) or IIB (Zn, Cd, Hg)   160

1.10.   Formation of Bonds between Hydrogen and Transition and Inner-Transition Metals   173

**Formation of Bonds between
Hydrogen and Transition and
Inner-Transition Metals**

1.11.   **Formation of Bonds between
Hydrogen and Elements of Group 0   236**

1.12.   **Formation of Reversible Metal
Hydrides by Direct Reaction of
Hydrogen   237**

How to Use this Book

1. Organization of Subject Matter.

1.1. Logic of Subdivision and Add-On Chapters.

This volume is part of a series that describes all of inorganic reaction chemistry. The contents are subdivided systematically and so are the contents of the entire series: Using the periodic system as a correlative device, it is shown how bonds between pairs of elements can be made. Treatment begins with hydrogen making a bond to itself in H_2 and proceeds according to the periodic table with the bonds formed by hydrogen to the halogens, the groups headed by oxygen, nitrogen, carbon, boron, beryllium and lithium, to the transition and inner-transition metals and to the members of group zero. Next it is considered how the halogens form bonds among themselves and then to the elements of the main groups VI to I, the transition and inner-transition metals and the zero-group gases. The process repeats itself with descriptions of the members of each successive periodic group making bonds to all the remaining elements not yet treated until group zero is reached. At this point all actual as well as possible combinations have been covered.

The focus is on the primary formation of bonds, not on subsequent reactions of the products to form other bonds. These latter reactions are covered at the places where the formation of those bonds is described. Reactions in which atoms merely change their oxidation states are not included, nor are reactions in which the same pairs of elements come together again in the product (for example, in metatheses or redistributions). Physical and spectroscopic properties or structural details of the products are not covered by the reaction volumes which are concerned with synthetic utility based on yield, economy of ingredients, purity of product, specificity, etc. The preparation of short-lived transient species is not described.

While in principle the systematization described above could suffice to deal with all the relevant material, there are other topics that inorganic chemists customarily identify as being useful in organizing reaction information and that do not fit into the scheme. These topics are the subject of eight additional chapters constituting the last four books of the series. These chapters are systematic only within their own confines. Their inclusion is based on the best judgment of the Editorial Advisory Board as to what would be most useful currently as well as effective in guiding the future of inorganic reaction chemistry.

1.2. Use of Decimal Section Numbers

The organization of the material is readily apparent through the use of numbers and headings. Chapters are broken down into divisions, sections and subsections, which have short descriptive headings and are numbered according to the following scheme:

1. Major Heading
1.1. Chapter Heading
1.1.1. Division Heading
1.1.1.1. Section Heading
1.1.1.1.1. Subsection Heading

Further subdivision of a five-digit "slice" utilizes lower-case Roman numerals in parentheses: (i), (ii), (iii), etc. It is often found that as a consequence of the organization, cognate material is located in different chapters but in similarly numbered pieces, i.e., in parallel sections. Section numbers, rather than page numbers, are the key by which the material is accessed through the various indexes.

1.3. Building of Headings

1.3.1. Headings Forming Part of a Sentence

Most headings are sentence-fragment phrases which constitute sentences when combined. Usually a period signifies the end of a combined sentence. In order to reconstitute the context in which a heading is to be read, superior-rank titles are printed as running heads on each page. When the sentences are put together from their constituent parts, they describe the contents of the piece at hand. For an example, see 2.3 below.

1.3.2. Headings Forming Part of an Enumeration

For some material it is not useful to construct title sentences as described above. In these cases hierarchical lists, in which the topics are enumerated, are more appropriate. To inform the reader fully about the nature of the material being described, the headings of connected sections that are superior in hierarchy always occur as running heads at the top of each page.

2. Access and Reference Tools

2.1. Plan of the Entire Series (Front Endpaper)

Printed on the inside of the front cover is a list, compiled from all 18 reaction volumes, of the major and chapter headings, that is, all headings that are preceded by a one- or two-digit decimal section number. This list shows in which volumes the headings occur and highlights the contents of the volume that is at hand by means of a gray tint.

2.2. Contents of the Volume at Hand

All the headings, down to the title of the smallest decimal-numbered sub-section, are listed in the detailed table of contents of each volume. For each heading the table of contents shows the decimal section number by which it is preceded and the number of the page on which it is found. Beside the decimal section numbers, successive indentations reveal the hierarchy of the sections and thereby facilitate the comprehension of the phrase (or of the enumerative sequence) to which the headings of hierarchically successive sections combine. To reconstitute the context in which the heading of a section must be read to become meaningful, relevant headings of sections superior in hierarchy are repeated at the top of every page of the table of contents. The repetitive occurrences of these headings is indicated by the fact that position and page numbers are omitted.

2.3. Running Heads

In order to indicate the hierarchical position of a section, the top of every page of text shows the headings of up to three connected sections that are superior in hierarchy. These running heads provide the context within which the title of the section under discussion becomes meaningful. As an example, the page of Volume 1 on which section 1.4.9.1.3 "in the Production of Methanol" starts, carries the running heads:

1.4. The Formation of Bonds between Hydrogen and O,S,Se,Te,Po
1.4.9. by Industrial Processes
1.4.9.1. Involving Oxygen Compounds

whereby the phrase "in the Production of Methanol" is put into its proper perspective.

2.4. List of Abbreviations

Preceding the indexes there is a list of those abbreviations that are frequently used in the text of the volume at hand or in companion volumes. This list varies somewhat in length from volume to volume; that is, it becomes more comprehensive as new volumes are published.

Abbreviations that are used incidentally or have no general applicability are not included in the list but are explained at the place of occurrence in the text.

2.5. Author Index

The author index is compiled by computer from the lists of references. Thus it tells whose publications are cited and in that respect is comprehensive. It is not a list of authors, beyond those cited in the references, whose results are reported in the text. However, as the references cited are leading ones, consulting them, along with the use of appropriate works of the secondary literature, will rapidly lead to the complete literature related to any particular subject covered.

Each entry in the author index refers the user to the appropriate section number.

2.6. Compound Index

The compound index lists individual, fully specified compositions of matter that are mentioned in the text. It is an index of empirical formulas, ordered according to the following system: the elements within a given formula occur in alphabetical sequence except for C, or C and H if present, which always come first. Thus, the empirical formula

for $Ti(SO_4)_2$ is O_8S_2Ti
$BH_3 \cdot NH_3$ BH_6N
Be_2CO_3 CBe_2O_3
$CsHBr_2$ Br_2CsH
$Al(HCO_3)_3$ $C_3H_3AlO_9$

The formulas themselves are ordered alphanumerically without exception; that is, the formulas listed above follow each other in the sequence BH_6N, Br_2CsH, CBe_2O_3, $C_3H_3AlO_9$, O_8S_2Ti.

A compound index constructed by these principles tells whether a given compound is present. It cannot provide information about compound classes, for example, all aluminum derivatives or all compounds containing phosphorus.

In order to open this route of access as well, the compound index is augmented by successively permuted versions of all empirical formulas. Thus the number of appearances that an empirical formula makes in the compound index is equal to the number of elements it contains. As an example, $C_3H_3AlO_9$, mentioned above, will appear as such and, at the appropriate positions in the alphanumeric sequence, as $H_3AlO_9*C_3$, $AlO_9*C_3H_3$ and $O_9*C_3H_3Al$. The asterisk identifies a permuted formula and allows the original formula to be reconstructed by shifting to the front the elements that follow the asterisk.

Each nonpermuted formula is followed by linerarized structural formulas that indicate how the elements are combined in groups. They reveal the connectivity of the compounds underlying each empirical formula and serve to distinguish substances which are identical in composition but differ in the arrangement of elements (isomers). As an example, the empirical formula $C_4H_{10}O$ might be followed by the linearized structural formulas $(CH_3CH_2)_2O$, CH_3 $(CH_2)_2OCH_3$, $(CH_3)_2CHOCH_3$, $CH_3(CH_2)_3OH$, $(CH_3)_2CHCH_2OH$ and CH_3 $CH_2(CH_3)CHOH$ to identify the various ethers and alcohols that have the element count $C_4H_{10}O$.

Each linearized structural formula is followed in a third column by keywords describing the context in which it is discussed and by the number(s) of the section(s) in which it occurs.

2.7. Subject Index

The subject index provides access to the text by way of methods, techniques, reaction types, apparatus, effects and other phenomena. Also, it lists compound classes such as organotin compounds or rare-earth hydrides which cannot be expressed by the empirical formulas of the compound index.

For multiple entries, additional keywords indicate contexts and thereby avoid the retrieval of information that is irrelevant to the user's need.

Again, section numbers are used to direct the reader to those positions in the book where substantial information is to be found.

2.8. Periodic Table (Back Endpaper)

Reference to periodic groups avoids cumbersome enumerations. Section headings in the series employ the nomenclature.

Unfortunately, however, there is at the present time no general agreement on group designations. In fact, the scheme that is most widely used (combining a group number with the letters A and B) is accompanied by two mutually contradictory interpretations. Thus, titanium may be a group IVA or group IVB element depending on the school to which one adheres or the part of the world in which one resides.

In order to clarify the situation for the purposes of the series, a suitable labeled periodic table is printed on the inside back cover of each volume. All references to periodic group designations in the series refer to this scheme.

Preface to the Series

Inorganic Reactions and Methods constitutes a closed-end series of books designed to present the state of the art of synthetic inorganic chemistry in an unprecedented manner. So far, access to knowledge in inorganic chemistry has been provided almost exclusively using the elements or classes of compounds as starting points. In the first 18 volumes of **Inorganic Reactions and Methods**, it is bond formation and type of reaction that form the basis of classification.

This new route of access has required new approaches. Rather than sewing together a collection of review articles, a framework has had to be designed that reflects the creative potential of the science and is hoped to stimulate its further development by identifying areas of research that are most likely to be fruitful.

The reaction volumes describe methods by which bonds between the elements can be formed. The work opens with hydrogen making a bond to itself in H_2 and proceeds through the formation of bonds between hydrogen and the halogens, the groups headed by oxygen, nitrogen, carbon, boron, beryllium and lithium to the formation of bonds between hydrogen and the transition and inner-transition metals and elements of group zero. This pattern is repeated across the periodic system until all possible combinations of the elements have been treated. This plan allows most reaction topics to be included in the sequence where appropriate. Reaction types that do not arise from the systematics of the plan are brought together in the concluding chapters on oxidative addition and reductive elimination, insertions and their reverse, electron transfer and electrochemistry, photochemical and other energized reactions, oligomerization and polymerization, inorganic and bioinorganic catalysis and the formation of intercalation compounds and ceramics.

The project has engaged a large number of the most able inorganic chemists as Editorial Advisors creating overall policy, as Editorial Consultants designing detailed plans for the subsections of the work, and as authors whose expertise has been crucial for the quality of the treatment. The conception of the series and the details of its technical realization were the subject of careful planning for several years. The distinguished chemists who form the Editorial Advisory Board have devoted themselves to this exercise, reflecting the great importance of the project.

It was a consequence of the systematics of the overall plan that publication of a volume had to await delivery of its very last contribution. Thus was the defect side of the genius of the system revealed, as the excruciating process of extracting the rate-limiting manuscripts began. Intense editorial effort was required in order to bring forth the work in a timely way. The production process had to be designed so that the insertion of new material was possible up to the very last stage, enabling authors to update their pieces with the latest developments. The publisher supported the cost of a computerized bibliographic search of the literature and a second one for updating.

Each contribution has been subjected to an intensive process of scientific and linguistic editing in order to homogenize the numerous individual pieces, as well as to provide the highest practicable density of information. This had several important consequences. First, virtually all semblances of the authors' individual styles have been excised. Second, it was learned during the editorial process that greater economy of language could be achieved by dropping conventionally employed modifiers (such as very) and eliminating italics used for emphasis, quotation marks around nonquoted words, or parentheses around phrases, the result being a gain in clarity and readability. Because the series focuses on the chemistry rather than the chemical literature, the need to tell who has reported what, how and when can be considered of secondary importance. This has made it possible to bring all sentences describing experiments into the present tense. Information on who published what is still to be found in the reference lists. A further consequence is that authors have been burdened neither with identifying leading practitioners, nor with attributing priority for discovery, a job that taxes even the talents of professional historians of science. The authors' task then devolved to one of describing inorganic chemical reactions, with emphasis on synthetic utility, yield, economy, availability of starting materials, purity of product, specificity, side reactions, etc.

The elimination of the names of people from the text is by far the most controversial feature. Chemistry is plagued by the use of nondescriptive names in place of more expository terms. We have everything from Abegg's rule, Adkin's catalyst, Admiralty brass, Alfven number, the Amadori rearrangement and Adurssov oxidation to the Zdanovskii law, Zeeman effect, Zincke cleavage and Zinin reduction. Even well-practiced chemists cannot define these terms precisely except for their own areas of specialty, and no single source exists to serve as a guide. Despite these arguments, the attempt to replace names of people by more descriptive phrases was met in many cases by a warmly negative reaction by our colleague authors, notwithstanding the obvious improvements wrought in terms of lucidity, freedom from obscurity and obfuscation and, especially, ease of access to information by the outsider or student.

Further steps toward universality are taken by the replacement of element and compound names wherever possible by symbols and formulas, and by adding to data in older units their recalculated SI equivalents. The usefulness of the reference sections has been increased by giving journal-title abbreviations according to the *Chemical Abstracts Service Source Index*, by listing in each reference all of its authors and by accompanying references to patents and journals that may be difficult to access by their *Chemical Abstracts* citations. Mathematical signs and common abbreviations are employed to help condense prose and a glossary of the latter is provided in each volume. Dangerous or potentially dangerous procedures are highlighted in safety notes printed in boldface type.

The organization of the material should become readily apparent from an examination of the headings listed in the table of contents. Combining the words constituting the headings, starting with the major heading (one digit) and continuing through the major chapter heading (two digits), division heading (three digits), section heading (four digits) to the subsection heading (five digits), reveals at once the subject of a "slice" of the plan. Each slice is a self-contained unit. It includes its own list of references and provides definitions of unusual terms that may be used in it. The reader, therefore, through the table of contents alone, can in most instances quickly reach the desired material and derive the information wanted.

In addition there is for each volume an author index (derived from the lists of references) and a subject index that lists compound classes, methods, techniques, apparatus, effects and other phenomena. An index of empirical formulas is also provided. Here in each formula the element symbols are arranged in alphabetical order except that C, or C and H if present, always come first. Moreover, each empirical formula is permuted successively. Each permuted formula is placed in its alphabetical position and cross referenced to the original formula. Therefore, the number of appearances that an empirical formula makes in the index equals the number of its elements. By this procedure all compounds containing a given element come together in one place in the index. Each original empirical formula is followed by a linearized structural formula and keywords describing the context in which the compound is discussed. All indexes refer the user to subsection rather than page number.

Because the choice of designations of groups in the periodic table is currently in a state of flux, it was decided to conform to the practice of several leading inorganic texts. To avoid confusion an appropriately labeled periodic table is printed on the back endpaper.

From the nature of the work it is obvious that probably not more than two persons will ever read it entire: myself and the publisher's copy editor, Dr. Lindsay S. Ardwin. She, as well as Ms. Mary C. Stradner, Production Manager of VCH Publishers, are to be thanked for their unflagging devotion to the highest editorial standards. The original conception for this series was the brainchild of Dr. Hans F. Ebel, Director of the Editorial Department of VCH Verlagsgesellschaft in Weinheim, Federal Republic of Germany, who also played midwife at the birth of the plan of these reaction volumes with my former mentor, Professor Alan G. MacDiarmid of the University of Pennsylvania, and me in attendance, during the Anaheim, California, American Chemical Society Meeting in the Spring of 1978. Much of what has finally emerged is the product of the inventiveness and imagination of Professor Helmut Grünewald, President of VCH Verlagsgesellschaft. It is a pleasure to acknowledge that I have learned much from him during the course of our association. Ms. Nancy L. Burnett is to be thanked for typing everything that had to do with the series from its inception to this time. Directing an operation of this magnitude without her help would have been unimaginable. My wife Rose stood by with good cheer while two rooms of our home filled up with 10,000 manuscript pages, their copies and attendant correspondence.

Finally, and most important, an enormous debt of gratitude toward all our authors is to be recorded. These experts were asked to prepare brief summaries of their knowledge, ordered in logical sequence by our plan. In addition, they often involved themselves in improving the original conception by recommending further refinements and elaborations. The plan of the work as it is being published can truly be said to be the product of the labors of the advisors and consultants on the editorial side as well as the many, many authors who were able to augment more general knowledge with their own detailed information and ideas. Because of the unusually strict requirements of the series, authors had not only to compose their pieces to fit within narrowly constrained limits of space, format and scope, but after delivery to a short deadline were expected to stand by while an intrusive editorial process homogenized their own prose styles out of existence and shrank the length of their expositions. These long-suffering colleagues had then to endure the wait for the very last manuscript scheduled for their volume to be delivered so that their work could be published, often after a further diligent search of the literature to insure that the latest discoveries were being cited and that claims for facts now proved false were eliminated. To these co-workers (270 for the reaction volumes alone), from whom so much was demanded but who continued to place their knowledge and talents unstintingly at the disposal of the project, we dedicate this series.

J. J. ZUCKERMAN
Norman, Oklahoma
July 4, 1985

Editorial Consultants to the Series

Contributors to Volume 2

Professor N. Bartlett
Department of Chemistry
University of California at Berkeley
Berkeley, California 94720
  (*Section 1.11*)

Professor L. Barton
Department of Chemistry
University of Missouri at St. Louis
8001 Natural Bridge Road
St. Louis, Missouri 61321
  (*Section 1.7*)

Professor F. Glockling
Inorganic Chemistry Laboratory
University of Oxford
South Parks Road
Oxford, OX1 3QR
England
  (*Section 1.9*)

Professor A.P. Hagen
Department of Chemistry
University of Oklahoma
Norman, Oklahoma 73019
  (*Subject Index*)

Professor A. Herold
Laboratoire de Chimie Minérale
Appliqué
L.A. 158–Université de Nancy I
B.P. 239
F-54506 Vandœuvre les Nancy,
Cedex France
  (*Section 1.8.2*)

Dr. G.G. Libowitz
Corporate Technology
Allied Corporation
P.O. Box 1021R
Morristown, New Jersey 07960
  (*Section 1.8.3, 1.12*)

Professor T.J. Lynch
Department of Chemistry
University of Nevada
Reno, Nevada 89557
  (*Sections 1.10.2, 1.10.3, 1.10.5*)

Dr. A.J. Maeland
Corporate Technology
Allied Corporation
P.O. Box 1021R
Morristown, New Jersey 07960
  (*Section 1.12*)

Dr. J.F. Mareche
Laboratoire de Chimie Minérale
Appliquée
L.A. 158—Université de Nancy I
B.P. 239
F-54506 Vandœvre les Nancy,
Cedex France
  (*Section 1.8.2*)

Professor A.D. Norman
Department of Chemistry
University of Colorado
Campus Box 215
Boulder, Colorado 80309
  (*Sections 1.5, 1.6*)

Professor J.R. Norton
Department of Chemistry
Colorado State University
Fort Collins, Colorado 80523
  (*Sections 1.10.6 through 1.10.8*)

Professor J. Topich
Department of Chemistry
Virginia Commonwealth University
1001 West Main Street
Richmond, Virginia 23284
  (*Section 1.10.4*)

1. The Formation of Bonds to Hydrogen (Part 2)

1.5. Formation of Bonds between Hydrogen and Elements of Group VB (N, P, As, Sb, Bi)

1.5.1. Introduction

This chapter covers reactions by which bonds between hydrogen and N, P, As, Sb and Bi are formed. A large area of synthetic chemistry is represented, because N—H, and to a lesser extent P—H, bond formation occurs in the context of bio-, organic, or-ganometallic, and inorganic chemistry. Reactions of principal importance to inorganic and organometallic chemists, including those from organophosphorus chemistry, re-ceive major attention. Nitrogen–hydrogen bond formation of principal interest to bio-logical or organic chemistry is presented in less detail.

The group VB element–hydrogen bond formation reactions are presented in broad classes developed according to reagent or reaction type, occasionally requiring arbitrary characterization of systems necessary to allow their classification. Reagents of type X—H are classed as protonic reagents if they participate in reactions as protonic acids, as is the case for molecules in which the electronegativity of X is greater than that of H. Other X—H species are classed as simple hydrides if they contain hydrogen and one element (binary) or hydrogen and two main elements (ternary, e.g., SiH_3PH_2) and as complex hydrides if they contain hydride ions (H^-) coordinated to a central element to form complex anionic species, e.g., $LiAlH_4$.

(A.D. NORMAN)

1.5.2. by Reaction of Hydrogen

1.5.2.1. with Nitrogen

1.5.2.1.1. from the Elements.

Ammonia is obtained in bulk quantities from the exothermic reaction of gaseous N_2 and H_2 in the well-known process[1]:

$$N_2 + 3\ H_2 \rightleftharpoons 2\ NH_3 \qquad \Delta H° = -92\ kJ\ mol^{-1} \qquad (a)$$

The equilibrium favors NH_3 formation at low T and high P; e.g., the vol % of NH_3 in a 3:1 H_2:N_2 mixture are (T and P in parenthesis)[2]: 14.73 (300°C, 1.01×10^3 Pa), 52.04 (300°C, 1.01×10^4 Pa), 92.55 (300°C, 1.01×10^5 Pa), 57.47 (500°C, 1.01×10^5 Pa). To achieve acceptable reaction rates and conversion to NH_3, reactions are carried out over an Fe or Fe–Fe_2O_3 catalyst[3] at 500°C and ca. 10^5 Pa. Hydrazine forms also in low quantities; too low to provide a useful synthesis[1]:

$$N_2 + 2\ H_2 \rightarrow N_2H_4 \qquad (b)$$

Photolysis of N_2–H_2 mixtures in the presence of TiO_2 catalyst produces[4] NH_3 in low yield.

3

4   1.5. Formation of Bonds between Hydrogen and N, P, As, Sb, Bi
         1.5.2. by Reaction of Hydrogen
         1.5.2.1. with Nitrogen

Hydrogen and N_2 react in the presence of $V(OH)_2$–$Mg(OH)_2$ or $V(OH)_2$–ZrO_2 catalysts in aqueous base to form NH_3 and N_2H_4. Ammonia production is favored at high catalyst concentration, whereas N_2H_4 is favored at high dilution. Yields of N_2H_4 $\leq 89\%$ are claimed[5]. Hydrogen reacts with N_2 at 50°C in H_2O (pH 9–10) containing Mg^{2+}–$Ti(OH)_3$–$Mo(OH)_3$ catalyst to form[6] both NH_3 and N_2H_4.

Atomic nitrogen (^4S) reacts with H_2 or H atoms to yield NH_3 and lesser quantities of N_2H_4[7,8]. Reactions in an electrical or microwave discharge, which lead to NH_3, occur as:

$$N + H_2 + M \rightarrow NH_2 + M \tag{c}$$

$$NH_2 + H \rightarrow NH_3 \tag{d}$$

where M is a third body for collisional deactivation.

(A.D. NORMAN)

1. F. A. Cotton, G. Wilkinson, *Advanced Inorganic Chemistry*, 4th ed., Wiley-Interscience, New York, 1980.
2. W. L. Jolly, *The Inorganic Chemistry of Nitrogen*, Benjamin, New York, 1964.
3. A. Ozaki, *Acc. Chem. Res., 14*, 16 (1981).
4. G. N. Schrauzer, T. D. Guth, *J. Am. Chem. Soc., 100*, 7189 (1978).
5. G. N. Schrauzer, N. Strampoch, L. A. Hughes, *Inorg. Chem., 21*, 2184 (1982).
6. V. V. Abalyaeva, N. T. Denisov, L. M. Khidekel, A. E. Shilov, *Bull. Acad. Sci. USSR, Div. Chem. Sci.*, 2638 (1975).
7. A. N. Wright, C. A. Winkler, *Active Nitrogen*, Academic Press, New York, 1968.
8. M. F. A. Dove, D. W. Sowerby, *Coord. Chem. Rev, 34*, 262 (1981).

1.5.2.1.2. from Compounds.

Inorganic nitrogen compounds react with H_2 yielding N–H containing products. Germanium nitride reacts with H_2 at 600°C:

$$Ge_3N_2 + 3 H_2 \rightarrow 3 Ge + 2 NH_3 \tag{a}^1$$

Nitric oxide in H_2O is reduced by H_2 over Pt catalysts to hydroxylamine[2]. Hydrazine in the presence of catalyst is cleaved by H_2 to NH_3; however, decomposition of NH_3 to N_2 and H_2 occurs. The N_2 and H_2 are favored at equilibrium except at high T and P. Calcium cyanide reacts with H_2:

$$Ca(CN)_2 + 8 H_2 \rightarrow CaH_2 + 2 CH_4 + 2 NH_3 \tag{b}^3$$

Reaction of an Os-coordinated isocyanate with H_2 (49.5×10^3 Pa) at 140°C results in a mixture of products, including those resulting from H_2 addition to the $C \equiv N$ bond[4]:

$$(\mu_2\text{-H})Os_3(CO)_{10}[(\mu_2\text{-h}^1)NCHCF_3] + H_2 \rightarrow (\mu_2\text{-H})Os_3(CO)_{10}(\mu_2\text{-HNCH}_2CF_3) \tag{c}$$

Hydrogen reduction of organic nitrogen compounds occurs readily in the presence of homogeneous and heterogeneous catalysts,[5-7] e.g., nitro, nitroso, oxime, nitrile, imine, hydrazone and azide compounds are reduced:

$$PhNO_2 + 3 H_2 \xrightarrow{Pt} 2 H_2O + PhNH_2 \tag{d}^{6,7}$$

$$PhNO + 2 H_2 \xrightarrow[EtOH]{Pt} H_2O + PhNH_2 \tag{e}^{6,7}$$

$$Ph(PhCH_2CH_2)NNO + H_2 \xrightarrow[-HNO]{finely\ divided\ Ni} Ph(PhCH_2CH_2)NH \tag{f}^6$$

$$Me(Ph)CNOH + 2 H_2 \xrightarrow[HOAc]{Pd} H_2O + Me(Ph)CHNH_2 \qquad (g)^{6,7}$$

$$RCN + 2 H_2 \xrightarrow{RhH(PPh_3)_3} RCH_2NH_2 \qquad (h)^{7,8}$$

$$PhCHNPh + H_2 \xrightarrow{Pd} PhCH_2NHPh \qquad (i)^{6,7}$$

$$R_2CNNHR' + H_2 \xrightarrow[EtOH]{Pd} R_2CHNHNHR' \qquad (j)^{9}$$

$$PhC_6H_4N_3 + H_2 \xrightarrow{finely\ divided\ Ni} N_2 + PhC_6H_4NH_2 \qquad (k)^{8,9}$$

A novel dehydrogenation–reduction of nitrocyclohexane at 420°C over PdO–Al$_2$O$_3$ catalysts yields aniline[10]:

$$C_6H_{11}NO_2 \rightarrow 2 H_2O + C_6H_5NH_2 \qquad (l)$$

(A.D. NORMAN)

1. F. Glockling, *The Chemistry of Germanium*, Academic Press, New York, 1968.
2. W. L. Jolly, *The Inorganic Chemistry of Nitrogen*, Benjamin, New York, 1964.
3. E. Wiberg, E. Amberger, *Hydrides of the Elements of Main Groups I-IV*, Elsevier, Amsterdam, 1971.
4. J. Banford, Z. Dawoodi, K. Hendrick, M. J. Mays, *J. Chem. Soc., Chem. Commun.*, 554 (1982).
5. H. O. House, *Modern Synthetic Reactions*, Benjamin, Menlo Park, California, 1972.
6. C. A. Buehler, D. E. Pearson, in *Survey of Organic Synthesis*, Vol. 2, Wiley-Interscience, New York, 1977, p. 391.
7. L. F. Fieser, M. Fieser, *Advanced Organic Chemistry*, Rheinhold, New York, 1961.
8. L. G. Wade Jr., in *Compendium of Organic Synthetic Methods*, Vol. 4, Wiley-Interscience, New York, 1980, p. 146.
9. J. R. Malpass, in *Comprehensive Organic Chemistry*, D. Barton, W. D. Ollis, eds., Vol. 2, I. O. Sutherland, ed., Pergamon Press, New York, 1979, p. 3.
10. R. J. Lindsay, in *Comprehensive Organic Chemistry*, D. Barton, W. D. Ollis, ed., Vol. 2, I. O. Sutherland, ed., Pergamon Press, New York, 1979, p. 131.

1.5.2.2. with Phosphorus.

Atomic hydrogen obtained (i) from H$_2$ dissociation in an electrical discharge, (ii) in a radio-frequency plasma discharge, (iii) photochemically in Hg-sensitized processes, or (iv) at high T reacts with phosphorus or its compounds to form phosphines. Elemental P or P$_4$O$_{10}$ and atomic H react[1,2]:

$$P_{(red)} + 3 H \rightarrow PH_3 \qquad (a)$$

$$P_4O_{10} + 32 H \rightarrow 10 H_2O + 4 PH_3 \qquad (b)$$

In the P$_{(red)}$ reaction, small quantities of P$_2$H$_4$ also form. Atomic hydrogen bombardment of InP surfaces generates PH$_3$ in small quantities[3].

Reduction of thio- or selenophosphinic acids by H$_2$ over finely divided Ni produces phosphine oxides[4]. Reduction of the chiral (R) acids results in inversion of configuration and formation of the (S) oxides[5]:

$$(R)\text{-}t\text{-}BuPhP(X)OH + H_2 \rightarrow (S)\text{-}t\text{-}BuPhP(O)H + H_2X \qquad (c)$$

where X = S, Se. Hydrogenolysis of compounds containing activated C—P bonds can result in P—H bond formation[6]:

$$(RO)_2PCH(R')NHNCHR'' + H_2 \xrightarrow{\text{finely divided Ni}} CH_2(R')NHNCHR''$$
$$+ (RO)_2P(O)H \qquad (d)$$

where R, R', R'' = alkyl.

(A.D. NORMAN)

1. W. L. Jolly, A. D. Norman, *Prep. Inorg. React.*, *4*, 1 (1968).
2. D. T. Hurd, *Chemistry of the Hydrides*, Wiley, New York, 1952.
3. D. T. Clark, T. Fok, *Thin Solid Films*, *78*, 271 (1981); *Chem. Abstr.*, *94*, 201,425 (1981).
4. J. Emsley, D. Hall, *The Chemistry of Phosphorus*, Harper and Row, London, 1976.
5. J. Michalski, Z. Skrzpzynski, *J. Organomet. Chem.*, *97*, C31 (1975).
6. J. Roction, C. Wasielewski, *Rozn. Chem.*, *50*, 477 (1976); *Chem. Abstr.*, *85*, 108,708 (1976).

1.5.2.3. with Arsenic.

Hydrogen atoms produced from H_2 in an electrical discharge react with As or As_4O_{10} to produce AsH_3 in low yields[1,2]:

$$As + 3 H \rightarrow AsH_3 \qquad (a)$$

$$As_4O_{10} + 32 H \rightarrow 10 H_2O + 4 AsH_3 \qquad (b)$$

These are not practical syntheses for AsH_3.

(A.D. NORMAN)

1. W. L. Jolly, A. D. Norman, *Preparative Inorganic Reactions*, Vol. 4, W. L. Jolly, ed., Interscience, New York, 1968, p. 1.
2. S. Miyamato, *J. Chem. Soc. Jpn.*, *53*, 724 (1932).

1.5.2.4. with Antimony.

Hydrogen atoms generated from H_2 in an electrical discharge, by thermolysis of H_2 or in a Hg-sensitized photochemical process react with Sb targets to form SbH_3 in low yields[1,2]:

$$3 H + Sb \rightarrow SbH_3 \qquad (a)$$

(A.D. NORMAN)

1. W. L. Jolly, A. D. Norman, *Preparative Inorganic Chemistry*, Vol. 4, W. L. Jolly, ed., Interscience, New York, 1968, p. 1.
2. D. T. Hurd, *Chemistry of the Hydrides*, Wiley, New York, 1952.

1.5.2.5. with Bismuth.

Electrolytic reduction of Bi(III) in H_2O yields BiH_3 through H atoms generated at the cathode:

$$Bi^{3+} + 3 H^+ + 6 e^- \rightarrow BiH_3 \qquad (a)$$

A discussion of active metal reductions is contained in §1.5.3.5.

(A.D. NORMAN)

1. A. G. Barikov, V. P. Gladyshev, *Sov. Electrochem. (Engl. Transl.)*, *8*, 795 (1972); *Chem. Abstr.*, *77*, 42,368 (1972).

1.5.3. by Protonation

1.5.3.1. of Nitrogen and Nitrogen Compounds

1.5.3.1.1. in Aqueous Systems.

Protonation of amines in H_2O results in N–H bond formation in an equilibrium:

$$\gtrdot N + H_2O \underset{}{\overset{H_2O}{\rightleftharpoons}} OH^- + [\gtrdot NH]^+ \qquad (a)$$

The extent of protonation in pure H_2O depends on temperature and the amine basicity. Equilibrium constants (K_b) for several selected amines at 25°C are[1,2]: NH_3, 8.6×10^{-5}; N_2H_4, 8.5×10^{-7}; NH_2OH, 6.6×10^{-9}; $MeNH_2$, 4.5×10^{-4}; Me_2NH, 5.4×10^{-4}; $PhNH_2$, 4.2×10^{-10} and C_5H_5N, 2.3×10^{-9}. For even the strongest bases, e.g., Me_2NH and Et_2NH, at equilibrium in 1 M solutions, $< 1\%$ of the amine is present in the protonated (ammonium) form. Addition of protonic acids to aq amine results in further protonation to ammonium species, depending on the acid and concentrations of reacting species[3,4]:

$$NH_{3(aq)} + [H_3O]^+_{(aq)} + X^-_{(aq)} \xrightarrow{H_2O} H_2O + X^-_{(aq)} + [NH_4]^+_{(aq)} \qquad (b)$$

where $X^- =$ e.g., Cl^-, Br^-, $[HSO_4]^-$.

Hydrolysis of ionic nitrides yields[1,5] NH_3:

$$M_3N_2 + 6\ H_2O \rightarrow 3\ M(OH)_2 + 2\ NH_3 \qquad (c)$$

where M = Be, Mg, Ca, Sr, Ba. Replacement of H_2O by D_2O provides[6,7] a synthesis for ND_3. Likewise, ionic amides are rapidly converted to amines in high yield[1,8–10]:

$$MNH_nR_{4-n} + H_2O \rightarrow MOH + R_{3-n}NH_n \qquad (d)$$

where M = Li, Na, K, Rb, Cs; R = alkyl, aryl, Me_3Si; n = 0, 1, 2; or:

$$M(NH_2)_2 + 2\ H_2O \rightarrow M(OH)_2 + 2\ NH_3 \qquad (e)$$

where M = Ca, Ba, Mg, Zn, Cd.

Nitrogen compounds activated or made more basic by metalation or metal coordination undergo protolysis to form N—H bonds. Hydrolyses of Li hydrazide or thioarylamide produce free amines[10]:

$$R_2(Me)CN(Li)NR'_2 + H_2O \rightarrow LiOH + R_2(Me)CNHNR'_2 \qquad (f)$$

$$R_2(Me)CN(Li)SPh + H_2O \rightarrow LiOH + PhSOH + R_2(Me)CNH_2 \qquad (g)$$

Hydrolysis of magnesium amides from organomagnesium-halide reagent addition to unsaturated nitrogen compounds proceeds similarly[11,12]:

$$R_2(R')CN(R'')MgBr + H_2O \rightarrow Mg(OH)Br + R_2(R')CNHR'' \qquad (h)$$

Hydrolysis of the methyldiazenidomolybdenum complex yields the methylhydrazido complex[13]:

$$MoI(N_2Me)(dppe)_2 + HBF_4 \xrightarrow{H_2O} [MoI(N_2HMe)(dppe)_2]BF_4 \qquad (i)$$

where dppe $= Ph_2PCH_2CH_2PPh_2$. Protonolysis of a disubstituted hydrazidorhenium complex[14]:

8   1.5. Formation of Bonds between Hydrogen and N, P, As, Sb, Bi
     1.5.3. by Protonation
     1.5.3.1. of Nitrogen and Nitrogen Compounds

$$h^5\text{-CpRe(CO)}_2NNMe(p\text{-MeC}_6H_4) + HBF_4 \xrightarrow{\ H_2O\ }$$
$$[h^5\text{-CpRe(CO)}_2NHNMe(p\text{-MeC}_6H_4)]BF_4 \qquad (j)$$

or a coordinated alkylisocyanide[15] yields N—H bond-containing products:

$$Fe_2(h^4\text{-diene})_2(CO)_{4-n}(CNR)_n + n\ H^+ \xrightarrow{\ H_2O\ } [Fe_2(h^4\text{-diene})_2(CO)_{4-n}(CNHR)]^{n+} \qquad (k)$$

where R = alkyl; n = 1, 2; diene = C_5H_5. Reaction of N_2 with $TiCl_3$ or VCl_2 and Mg in aq THF, followed by CO_2, produces the complexes $MMg_2Cl_2(NCO)(O)(THF)_3$ (M = Ti or V), which when hydrolyzed produces[16] CO and NH_3.

Covalently bonded inorganic amides hydrolyze easily to amines:

$$HN(SO_3H)_2 + H_2O \xrightarrow{\ HCl_{(aq)}\ } H_2SO_4 + HSO_3NH_2 \qquad (l)^{17}$$

$$(NH_2)_3PO + KOH \xrightarrow{\ KOH_{(aq)}\ } [(NH_2)_2P(O)O]K + NH_3 \qquad (m)^{18}$$

$$Ph(Me)NBCl_2 + 3\ H_2O \xrightarrow{\ HCl_{(aq)}\ } B(OH)_3 + 2\ HCl + Ph(Me)NH \qquad (n)^{11}$$

$$HONHSO_3H + H_2O \xrightarrow{\ H_2O\ } H_2SO_4 + NH_2OH \qquad (o)$$

Amido transition-metal complexes are hydrolyzed similarly[10]. Complex species, such as the amido-bridged dicobalt complex, undergo reversible hydrolysis in aq acid[19]:

$$[(NH_3)_4CoO_2(NH_2)Co(NH_3)_4]^{3+} \xrightarrow{\ NH_{3(aq)}\text{-}H^+\ } [(NH_3)_5CoO_2Co(NH_3)_5]^{4+} \qquad (p)$$

Calcium cyanamide:

$$CaCN_2 + 2\ H^+ \xrightarrow{\ H^+_{(aq)}\ } Ca^{2+} + H_2NCN \qquad (q)$$

and azodiphosphate:

$$2\ K_4[O_3PN{=}NPO_3] + 4\ H_2O \xrightarrow{\ H^+_{(aq)}\ } 4\ K_2HPO_4 + N_2 + N_2H_4 \qquad (r)$$

hydrolyses produce H_2NCN[1] and N_2H_4[18], respectively. Passage of $[B_{10}H_{13}CN]^{2-}$ through an acidic ion-exchange column yields the N—H-containing product[19]:

$$[B_{10}H_{13}CN]^{2-} + 2\ H^+ \xrightarrow{\ H^+_{(aq)}\ } B_{10}H_{12}NH_3 \qquad (s)$$

Aqueous reductions are useful for amine synthesis. Organic reducing agents (N_2H_4 or formic acid) reduce amides or imines[2,9]. Active metals in acid reduce nitro groups[2,9]. Reduction of Ph_3PNBr with I^-:

$$Ph_3P{=}NBr + 2\ HI + H_2O \xrightarrow{\ H_2O\ } I_2 + Ph_3PO + NH_4Br \qquad (t)^{18}$$

or HNO_2 with $[HSO_3]^-$ produces $[NH_4]^+$ or $[NH_3OH]^+$, respectively:

$$H^+ + HNO_2 + 2\ [HSO_3]^- + H_2O \xrightarrow{\ H^+_{(aq)}\ } 2\ [HSO_4]^- + [NH_3OH]^+ \qquad (u)^1$$

Chloramine:

1.5.3. by Protonation
1.5.3.1. of Nitrogen and Nitrogen Compounds
1.5.3.1.1. in Aqueous Systems.

9

$$3\ NH_2Cl + 3\ [OH]^- \xrightarrow{[OH]^-_{(aq)}} N_2 + 3\ Cl^- + 3\ H_2O + NH_3 \qquad (v)^1$$

or $[NH_2]^- - [NO_3]^-$ soln reactions yield[1] NH_3:

$$3\ [NH_2]^- + 3\ [NO_3]^- \xrightarrow{[OH]^-_{(aq)}} 3\ [OH]^- + N_2 + 3\ [NO_2]^- + NH_3 \qquad (w)^1$$

Reduction of dinitrogen to N—H bond products (N_2 fixation) occurs in H_2O-containing inorganic catalysts and reducing agents, perhaps not unlike the biological enzyme nitrogenase[16,19,20]. Basic solutions (pH = 9–10) containing initially Ti(III), Cr(II) or Zn as reducing agents along with V(II) and Mo(V) species and Mg^{2+} ions react with N_2 to form NH_3 and/or N_2H_4. The primary reactions are V(II) reductions of N_2:

$$N_2 + 4\ V(OH)_2 + 4\ H_2O \rightarrow 4\ V(OH)_3 + N_2H_4 \qquad (x)$$

$$N_2 + 6\ V(OH)_2 + 6\ H_2O \rightarrow 6\ V(OH)_3 + 2\ NH_3 \qquad (y)$$

Vanadium(III) reduction by the reducing agent present (e.g., Zn)[20] regenerates the necessary V(II) species.

Electrolytic reduction of HNO_3 in aq HCl or H_2SO_4 at Pb electrodes produces NH_2OH in 69% yield[1]:

$$HNO_3 + 6\ e^- + 6\ H^+ \xrightarrow{H_2O} 2\ H_2O + NH_2OH \qquad (z)$$

Low yields of N_2H_4 are obtained from N_2 reduction at a Hg cathode in the presence of Mo complexes[16]. Electrolysis of N_2 at 40 V for 11 days at a Ni–Cr cathode in $(i\text{-}PrO)_4Ti–(i\text{-}PrO)_3Al$ solvent, followed by hydrolysis produces low yields[21] of NH_3:

$$N_2 + 4\ H^+ + 4\ e^- \xrightarrow{H_2O\,-Mo\ complex} N_2H_4 \qquad (aa)$$

(A.D. NORMAN)

1. W. L. Jolly, *The Inorganic Chemistry of Nitrogen,* Benjamin, New York, 1964.
2. R. T. Morrison, R. N. Boyd, *Organic Chemistry,* 4th ed., Allyn and Bacon, Boston, 1983.
3. T. Moeller, *Inorganic Chemistry,* Wiley-Interscience, New York, 1982.
4. R. P. Bell, *The Proton in Chemistry,* 2nd ed., Cornell Univ. Press, Ithaca, NY, 1969.
5. D. T. Hurd, *Chemistry of the Hydrides,* Wiley, New York, 1952.
6. G. Brauer, in *Handbook of Preparative Inorganic Chemistry,* 2nd ed., Vol. 1, G. Brauer, ed., Academic Press, New York, 1963, p. 137.
7. L. K. Krannich, U. Thewalt, W. J. Cook, S. R. Jain, H. H. Sisler, *Inorg. Chem., 12,* 2304 (1973).
8. T. L. Gilchrist, in *Comprehensive Organic Chemistry,* D. Barton, W. D. Ollis, eds. Vol. 2, I. O. Sutherland, ed., Pergamon Press, New York, 1979, p. 273.
9. L. F. Fieser, M. Fieser, *Advanced Organic Chemistry,* Rheinhold, New York, 1961.
10. D. C. Bradley, *Adv. Inorg. Chem. Radiochem., 15,* 259 (1972).
11. J. R. Malpass, in *Comprehensive Organic Chemistry,* D. Barton, W. D. Ollis, eds., Vol. 2, I. O. Sutherland, ed., Pergammon Press, New York, 1979, p. 3.
12. C. A. Buehler, D. E. Pearson, *Survey of Organic Syntheses,* Vol. 2, Wiley-Interscience, New York, 1977.
13. D. C. Busby, T. A. George, S. D. A. Iske, Jr., S. W. Wagner, *Inorg. Chem., 20,* 22 (1981).
14. C. F. Barrientos-Pennos, F. W. B. Einstein, T. Jones, D. Sutton, *Inorg. Chem., 21,* 2578 (1982).
15. S. Willis, A. R. Manning, *J. Chem. Soc., Dalton Trans.,* 23 (1979).
16. J. Chatt, J. R. Dilworth, R. L. Richards, *Chem. Rev., 78,* 589 (1978).

10   1.5. Formation of Bonds between Hydrogen and N, P, As, Sb, Bi
         1.5.3. by Protonation
         1.5.3.1. of Nitrogen and Nitrogen Compounds

17. M. Becke-Goehring, E. Fluck, *Developments in the Inorganic Chemistry of Nitrogen*, C. B. Colburn, ed., Elsevier, Amsterdam, 1966, p. 150.
18. E. Fluck, in *Topics in Phosphorus Chemistry*, Vol. 4, M. Grayson, E. J. Griffith, eds., Wiley-Interscience, New York, 1967, p. 291.
19. F. A. Cotton, G. Wilkinson, *Advanced Inorganic Chemistry*, 4th ed., Wiley-Interscience, New York, 1980.
20. D. V. Sokol'skii, Ya. A. Dorfman, Yu. M. Shindler, S. S. Stroganov, A. N. Sharopin, *J. Gen. Chem. USSR (Engl. Transl.)*, 43, 252 (1973).
21. E. E. Van Tamelen, D. A. Sieley, *J. Am. Chem. Soc.*, 91, 5194 (1969).

1.5.3.1.2. in Other Protonic Solvents.

Protonation of nitrogen compounds possessing a lone pair of electrons in a sufficiently acidic solvent yields N—H bonds. Ionic species (e.g., N^{3-}) and neutral compounds (e.g., R_3N) react[1,2]:

$$N^{3-} \xrightarrow{H^+} [NH]^{2-} \xrightarrow{H^+} [NH_2]^- \xrightarrow{H^+} NH_3 \qquad (a)$$

$$R_3N + HX \rightleftharpoons [R_3NH]^+ + X^- \qquad (b)$$

where X = e.g., OR, SR, CO_2Me, OSO_3H, NH_2, F, Cl, Br or I. The order of relative acidities of common protonic solvent acids is $H_2SO_4 > HX$ (X = halogen) $> RCO_2H > RSH$, $ROH > NH_3$. The order of nitrogen species basicity is $N^{3-} > [RN]^{2-} > [R_2N]^- > R_3N$. Treatment of the anions with moderately strong acids (assuming no other reactions at functional R groups), such as ROH, RSH or RCO_2H, yields NH_3, RNH_2, or R_2NH, quantitatively[3,4]. Ammonia dissociates to a small extent:

$$2\ NH_3 \underset{NH_{3(l)}}{\rightleftharpoons} [NH_4]^+ + [NH_2]^- \qquad (c)$$

where $K(25°C) = 10^{-27}$ and $K(-50°C) = 10^{-33}$, and substituted amines to a lesser extent form N—H bonds through self-association[5].

Liquid H_2SO_4[6] and hydrogen halides[7,8] are acids capable of protonating tricoordinated nitrogen. The order of hydrogen halide acidity is[8] $HI > HBr > HCl > HF$. Weakly basic amines, e.g., Ph_3N, react with HX[7,8] or H_2SO_4[6]:

$$Ph_3N + H_2SO_4 \xrightarrow{H_2SO_{4(l)}} [HSO_4]^- + [Ph_3NH]^+ \qquad (d)$$

Difluoramine in $HF–AsF_5$ is protonated[9] at $-78°C$:

$$NHF_2 + HF + AsF_5 \xrightarrow[-78°C]{HF-AsF_5} (NH_2F_2)AsF_6 \qquad (e)$$

Difluoramine is potentially explosive and should be handled with care[5,9].

Solvolysis of nonmetal amides, through cleavage of the nonmetal–nitrogen bond, can form amines, although these reactions are of interest for the reaction–product other than the amine:

$$P(NMe_2)_3 + 3\ EtOH \xrightarrow{EtOH} (EtO)_3P + 3\ Me_2NH \qquad (f)[10]$$

$$(Me_3Si)_2NPh + 2\ MeOH \xrightarrow{MeOH} 2\ Me_3SiOMe + PhNH_2 \qquad (g)[11]$$

Metal-amide alcoholysis proceeds similarly[12], e.g., the dimethylamidoditungsten complexes react with ROH (R = Me, Et) forming the alkoxide complexes and amine[13]:

1.5.3. by Protonation
1.5.3.1. of Nitrogen and Nitrogen Compounds
1.5.3.1.2. in Other Protonic Solvents.

11

$$W_2(NMe_2)_6 + 6\ ROH \xrightarrow{ROH} W_2(OR)_6 + 6\ Me_2NH \tag{h}$$

Methanolysis of metal–carbon bonded amides yields[14] NH_3:

$$Mn(CO)_3(PPh_3)_2CO_2NH_2 + MeOH \xrightarrow{MeOH-[MeO]^-} Mn(CO)_3(PPh_3)_2CO_2Me + NH_3 \tag{i}$$

whereas solvolysis of the Pt-coordinated isocyanide produces a coordinated imine[15]:

$$(Ph_3P)Pt(Cl)_2CNPh + EtOH \xrightarrow{EtOH} (Ph_3P)Pt(Cl)_2C(OEt)NHPh \tag{j}$$

Strong-acid protonation of metal-coordinated N_2 yields products containing $N-H$ bonds. Hydrogen chloride in MeOH reacts with $(h^5\text{-}Cp_2Ti)_2N_2MgCl$, forming N_2H_4 in 80% yield[16]. Protonation of Mo and W complexes depends on the ligands, the complex geometry, the acid and the reaction conditions[17]. The cis-$[M(N_2)_2(PMe_2Ph)_4]$ complexes react with HBr in MeOH to form azenido complexes[17]:

$$cis\text{-}[M(N_2)_2(PMe_2Ph)_4] + 3\ HBr \xrightarrow{MeOH} [PMe_2PhH]Br + N_2$$
$$+ MBr_2(NNH_2)(PMe_2Ph)_3 \tag{k}$$

where M = Mo, W. Similar reactions with H_2SO_4 occur[2,17]:

$$Mo(N_2)_2(dppe)_2 + 2\ H_2SO_4 \xrightarrow{MeOH} N_2 + [Mo(NNH_2)(dppe)HSO_4]HSO_4 \tag{l}$$

where dppe = $Ph_2PCH_2CH_2PPh_2$, although such reactions proceed further to produce NH_3 and small quantities of N_2H_4. Ammonia is produced upon strong-acid protonolysis of both cis- and trans-$M(N_2)_2(phosphine)_4$ complexes:

$$M(N_2)_2(PMe_nPh_{3-n})_4 \xrightarrow[MeOH]{H_2SO_4} [PMe_nPh_{3-n}H][HSO_4]$$
$$+ Mo(VI)\ product + N_2 + 2\ NH_3 \tag{m}$$

where M = Mo, W, and n = 1,2. From the reaction of trans-$W(N_2)_2(PMe_2Ph)_4$ with H_2SO_4 in MeOH or EtOH near quantitative yields of NH_3 arise[17-19]. Characterization of the Mo product(s) is better accomplished in nonprotonic solvent systems (see §1.5.3.1.3).

Oxidation of NH_3 in liq NH_3 produces[5] $[NH_4]^+$:

$$2\ NH_3 + X_2 \xrightarrow{NH_{3(l)}} NH_2X + X^- + [NH_4]^+ \tag{n}$$

where X = Cl, Br, I. Nitrate oxidation of amide ion yields NH_3 and azide[5]. Hydroxylamine reacts with ClO_3F in EtOH to form $[NH_3OH]^+F^-$, $[NH_3OH][ClO_3]$, O_2, N_2 and H_2O:

$$3\ K^+ + 3\ [NH_2]^- + [NO_3]^- \xrightarrow{NH_{3(l)}} N_3^- + 3\ KOH + NH_3 \tag{o}[20]$$

Isolation of products in this reaction presents a potential explosion hazard.

(A.D. NORMAN)

1. W. J. Jolly, *Inorganic Chemistry of Nitrogen,* Benjamin, Inc. New York, 1964.
2. F. A. Cotton, G. Wilkinson, *Advanced Inorganic Chemistry,* 4th ed., Wiley-Interscience, New York, 1980.
3. T. Moeller, *Inorganic Chemistry,* Wiley-Interscience, New York, 1982.

12   1.5. Formation of Bonds between Hydrogen and N, P, As, Sb, Bi
          1.5.3. by Protonation
          1.5.3.1. of Nitrogen and Nitrogen Compounds

4. T. L. Gilchrist, in *Comprehensive Organic Chemistry*, D. Barton, W. D. Ollis, eds., Vol. 2, I. O. Sutherland, ed., Pergamon Press, 1979, p. 273.
5. W. L. Jolly, C. J. Hallada, in *Non-Aqueous Solvent Systems*, T. C. Waddington, ed., Academic Press, London, 1965, p. 1.
6. R. J. Gillespie, E. A. Robinson, in *Non-Aqueous Solvent Systems*, T. C. Waddington, ed., Academic Press, London, 1965, p. 117.
7. H. H. Hyman, J. J. Katz, in *Non-Aqueous Solvent Systems*, T. C. Waddington, ed., Academic Press, London, 1965, p. 47.
8. M. E. Peach, T. C. Waddington, in *Non-Aqueous Solvent Systems*, T. C. Waddington, ed., Academic Press, London, 1965, p. 83.
9. K. O. Christie, *Inorg. Chem.*, 14, 2821 (1975).
10. E. Fluck, in *Topics in Phosphorus Chemistry*, M. Grayson, E. J. Griffith, eds., Vol. 4, Wiley-Interscience, 1967, p. 291.
11. F. D. King, D. R. M. Walton, *J. Chem. Soc., Chem. Commun.*, 256 (1974).
12. D. C. Bradley, *Adv. Inorg. Chem. Radiochem.*, 15, 259 (1972).
13. M. A. Chisholm, I. P. Rothwell, *Prog. Inorg. Chem.*, 29, 1 (1982).
14. T. Kruck, M. Noack, *Chem. Ber.*, 97, 1693 (1964).
15. E. O. Fischer, *Adv. Organomet. Chem.*, 14, 1 (1976).
16. Y. G. Borodko, I. N. Ivleva, L. M. Kachapina, E. F. Kvashina, A. K. Shilova, A. E. Shilov, *J. Chem. Soc., Chem. Commun.*, 169 (1973).
17. J. Chatt, J. R. Dilworth, R. L. Richards, *Chem. Rev.*, 78, 589 (1978).
18. Z. Dori, *Prog. Inorg. Chem.*, 28, 239 (1981).
19. T. Takahashi, Y. Mizobe, M. Sato, Y. Uchida, M. Hidai, *J. Am. Chem. Soc.*, 102, 7461 (1980).
20. K. V. Titana, E. I. Kolmakova, V. Ya. Rasolovskii, *Russ. J. Inorg. Chem. (Engl. Transl.)*, 23, 634 (1978).

1.5.3.1.3. with Protonic Acids in Nonprotonic Solvents.

Ionic metal nitrides[1] and amides[2,3] react with protonic acids (HX):

$$H_{3-n}NM_n + n\ HX \xrightarrow{Et_2O} n\ MX + NH_3 \tag{a}$$

where M = Li, Na, K, Cs; n = 1, 2, 3; X = Cl, Br, I, HSO_4, H_2PO_4, etc.

$$M(NH_2)_2 + 2\ H_2O \xrightarrow{Et_2O} M(OH)_2 + 2\ NH_3 \tag{b}$$

where M = Zn, Cd, Ca, Ba, Sr. Azide[2], isocyanate[2], and organic or nonmetal moiety-substituted amides[3,4] react similarly:

$$NaN_3 + HX \rightarrow NaX + HN_3 \tag{c}$$

$$NaNCO + HX \rightarrow NaX + HNCO \tag{d}$$

$$[R_2N]M + HX \rightarrow MX + R_2NH \tag{e}$$

where M = Li, Na, K; R = alkyl, aryl, R_3Si, R_3Ge. Amide ions in the gas phase react with proton sources[6]:

$$[NH_2]^-_{(g)} + H_2O_{(g)} \rightarrow [OH]^-_{(g)} + NH_{3(g)} \tag{f}$$

Reactions (a), (b) and (e) are of limited value as routes to N—H bonds because they only regenerate the amine from which the amide is prepared. However, they are useful for preparation of deuterated products, e.g.:

$$Na[Me_2N] + DCl \xrightarrow{Et_2O} NaCl + Me_2ND \tag{g}$$

1.5.3. by Protonation
1.5.3.1. of Nitrogen and Nitrogen Compounds
1.5.3.1.3. with Protonic Acids in Nonprotonic Solvents.

13

$$2 \, Na[(Me_3Si)_2N] + 3 \, D_2O \xrightarrow{Et_2O} 2 \, NaOD + (Me_3Si)_2O + 2 \, Me_3SiND_2 \qquad (h)^5$$

In contrast to ionic phosphides which upon protonolysis yield P_2H_4 along with PH_3 (see §1.5.3.2.1), ionic nitride protonolyses do not yield N_2H_4, except for Ba_3N_2 which, when heated with H_2O at 380°C, produces[7] N_2H_4.

Amines react with protonic acids in nonprotonic solvents or in the gas phase to form ammonium ions. From studies of gas-phase equilibrium reactions between amines and protonic reference acids:

$$[NH_4]^+ + B \rightleftharpoons [BH]^+ + NH_3 \qquad (i)$$

relative basicities of amines, e.g., $Me_3N > Me_2NH > MeNH_2 > NH_3$, are known[8,9]. In solvents the order of relative basicities may differ from that in the gas phase, e.g., $Me_2NH > MeNH_2 > Me_3N > NH_3$[8]. Strong acids, such as hydrogen halides, react with amines in nonprotonic solvents:

$$R_{3-n}NH_n + HX \rightarrow [R_{3-n}NH_{n+1}]X \qquad (j)$$

Because the ammonium salts have low solubilities in ether or hydrocarbon solvents, they precipitate, allowing quantitative synthesis[4,8].

Haloamines react with hydrogen halides in exchange–oxidation and form amine products. Trifluoromethyldichloroamine reacts with HCl at $-78°C$:

$$CF_3NCl_2 + 2 \, HCl \rightarrow 2 \, Cl_2 + CF_3NH_2 \qquad (k)^9$$

N-Bromosuccinimide with HBr forms succinimide and Br_2 in high yield[4]:

$$\overline{C(O)CH_2CH_2C(O)}NBr + HBr \rightarrow Br_2 + \overline{C(O)CH_2CH_2C(O)}NH \qquad (l)$$

Addition of HF to Me_3NBH_3 produces the tetrafluoroborate-ammonium salt[10]:

$$Me_3HBH_3 + 4 \, HF \rightarrow 2 \, H_2 + [Me_3NH]BF_4 \qquad (m)$$

Addition of HX to unsaturated nitrogen moieties yields amines or ammonium salts. Photolysis of a silylazide in the presence of t-BuOH yields amine through t-BuOH reaction with a $R_2Si=NR$ intermediate[11]:

$$R_3SiN_3 + t\text{-BuOH} \xrightarrow{h\nu} N_2 + t\text{-BuOSiR}_2NHR \qquad (n)$$

Phosphazenes react with hydrogen halides (HX) to form quaternary salts:

$$R_3P=NR' + HX \rightarrow [R_3PNHR']X \qquad (o)^{12,13}$$

where R = alkyl, aryl; R' = H, alkyl, aryl, SiR_3, GeR_3, although HF adds[14] to $(CCl_3)_2P(Cl)=NH$:

$$(CCl_3)_2P(Cl)=NH + HF \rightarrow (CCl_3)_2P(F)ClNH_2 \qquad (p)$$

and HCO_2H adds[13] to $(Cl_3PN)_2$:

$$Cl_3P=NN=PCl_3 + 2 \, HCO_2H \xrightarrow{C_2Cl_4H_2} 2 \, HCl + 2 \, CO + [Cl_2P(O)NH]_2 \qquad (q)$$

to form aminophosphorus(V) products.

Alcohol addition to the $P=N$ bond of a tricoordinated phosphazene occurs:

14   1.5. Formation of Bonds between Hydrogen and N, P, As, Sb, Bi
        1.5.3. by Protonation
        1.5.3.1. of Nitrogen and Nitrogen Compounds

$$Me_3SiNRPS=NBu\text{-}t + MeOH \rightarrow Me_3SiNRPSOMeNHBu\text{-}t \qquad (r)^{15}$$

where $R = Me_3Si$, t-Bu. From reaction of catechol and substituted chlorophosphazene an amidophosphorane is obtained[16]:

Alcoholysis of the phosphoryl chloride produces the bis(phosphoryl)amine:

$$Cl_3P=NPOCl_2 + 5\ ROH \rightarrow 4\ HCl + RCl + [(RO)_2PO]_2NH \qquad (t)$$

where R = Me, Et, n-Pr.

Oxidation of phosphazenes by halogens, interhalogens, or acid chlorides yields amidophosphonium salts[12], e.g.:

$$2\ Ph_3P=NH + X_2 \rightarrow Ph_3P=NX + X[Ph_3PNH_2] \qquad (u)$$

where X = Cl, Br.

Isocyanates, thioisocyanates, azobenzene and nitriles react with protonic acids to form secondary amines[4,8].

Protonic acid cleavage of compounds containing nonmetal–N or metal–N bonds results in N—H bonds. These reactions produce amines that are more readily prepared by other reactions; hence, they are of limited synthetic utility. Groups IIIB, IVB (except C), and VB (except N) element–N bond cleavage reactions are:

$$R_3ENH_3 + HCl \rightarrow R_3E + NH_4Cl \qquad (v)^{18,19}$$

$$(R_2N)_3E + 6\ HCl \rightarrow ECl_3 + 3\ R_2NH_3Cl \qquad (w)^{18,19}$$

where E = B, Al, Ga and R = alkyl, aryl. Group IVB elements, in both 4+ and 2+ oxidation states, are cleaved:

$$R_3MNR'_2 + HX \rightarrow R_3MX + R'_2NH \qquad (x)^{21\text{-}24}$$

where M = Si, Ge, Sn, Pb; R and R' = alkyl, aryl; X = Cl, Br, I, OH, etc.;

$$(R_3Sn)_2NN(SnR_3)_2 + 4\ HX \rightarrow 4\ R_3SnX + N_2H_4 \qquad (y)^{25}$$

$$Me_3MN_3 + H_2O \rightarrow (Me_3M)_2O + HN_3 \qquad (z)^{21}$$

$$(RR'N)_2Sn + 2\ HX \rightarrow SnX_2 + 2\ RR'NH \qquad (aa)^{20}$$

where R,R' = Me_3Si, Me_3Si; Me_3Si, t-Bu; and X = Cl, $MeCO_2$, C_5H_5. Nitrogen—P, As, or Sb bonds react:

$$(R_2N)_3E + 6\ HCl \rightarrow ECl_3 + 3\ R_2NH_2Cl \qquad (ab)^{13,26\text{-}29}$$

where E = P, As, Sb; and phosphazenes hydrolyze:

1.5.3. by Protonation   15
1.5.3.1. of Nitrogen and Nitrogen Compounds
1.5.3.1.3. with Protonic Acids in Nonprotonic Solvents.

$$Ph_3P = NR + H_2O \rightarrow Ph_3PO + RNH_2 \qquad (ac)[13]$$

where R = Me, Et, Ph. In a compound containing both Si—N and P—N bonds, H_2O selectively cleaves the Si—N bond[31]:

$$2\ Me_2P(S)N(SiMe_3)_2 + H_2O \rightarrow (Me_3Si)_2O + 2\ Me_2P(S)NH(SiMe_3) \qquad (ad)$$

Metal amides react with protonic reagents to produce simple amines:

$$2\ Mo(NMe_2)_4 + 8\ i\text{-}PrOH \rightarrow Mo_2(O\text{-}i\text{-}Pr)_8 + 8\ Me_2NH \qquad (ae)[1,3]$$

$$(h^5\text{-}Cp)_2Ti(NR_2)_2 + 2\ HCl \rightarrow (h^5\text{-}Cp)_2TiCl_2 + 2\ R_2NH \qquad (af)[1,3]$$

$$U(NEt_2)_4 + 2\ C_5H_6 \rightarrow U(C_5H_5\text{-}h^5)_2(NEt_2)_2 + 2\ Et_2NH \qquad (ag)[32]$$

Element (metal or nonmetal)–N protonic acid cleavage yields nitrogen compounds. Reaction of $(Me_3Si)_4N_4$ with CF_3CO_2H yields N_4H_4:

$$(Me_3Si)_2NN = NN(SiMe_3)_2 + 4\ CF_3CO_2H \xrightarrow[CH_2Cl_2]{-78°C}$$
$$4\ CF_3CO_2SiMe_3 + H_2NNNNH_2 \qquad (ah)[33]$$

Hydrolyses of the carbohydrazamide:

$$Et_3SnN(CO_2Et)N(CO_2Et)SnEt_3 + 2\ H_2O \rightarrow 2\ Et_3SnOH + (EtCO_2N)_2H_2 \qquad (ai)[23]$$

or $B_{10}H_{13}CN(SiMe_3)_2$ yield novel products:

$$B_{10}H_{13}CN(SiMe_3)_2 + H_2O \rightarrow (Me_3Si)_2O + B_{10}H_{12}CNH_3 \qquad (aj)[34]$$

Transamination, a type of protonic–acid cleavage, yields amine upon exchange with the amide moiety of a nonmetal or metal amide. These reactions are used to prepare amides or imides. The amine formed is of incidental interest, e.g.:

$$B(NR_2)_3 + 3\ R'_2NH \rightarrow B(NR'_2)_3 + 3\ R_2NH \qquad (ak)[19]$$

where R,R′ = alkyl, aryl, etc.;

$$(Et_2N)_3P + 3\ PhNH_2 \rightarrow (PhNH)_3P + 3\ Et_2NH \qquad (al)[35]$$

$$(n\text{-}Bu_3Ge)_2NH + RNH_2 \rightarrow (n\text{-}Bu_3Ge)_2NR + NH_3 \qquad (am)[21]$$

$$Me_2AsNMe_2 + 2\ RNH_2 \rightarrow Me_2AsNR_2 + Me_2NH \qquad (an)[36]$$

where R = Et, n-Pr, i-Bu. Product formation often requires shifting an equilibrium by removal of the amine formed.

Reactions of transition-metal complex-coordinated ligands with protonic acids can yield N—H bond-containing products. Coordinated nitrosyl groups react with HCl:

$$Os(NO)Cl(CO)(PPh_3)_2 + HCl \rightarrow OsCl_2(NHO)(CO)(PPh_3)_2 \qquad (ao)[1]$$

$$Os(NO)_2(PPh_3)_2 + 2\ HCl \rightarrow OsCl_2NHOHNO(PPh_3)_2 \qquad (ap)[12]$$

$$IrNO(PPh_3)_3 + 3\ HCl \rightarrow PPh_3 + IrCl_3NH_2OH(PPh_3)_2 \qquad (aq)[37]$$

Protonation of an N-coordinated nitrile:

$$[Ru(NH_3)_5NCR]^{3+} + H_2O \rightarrow [Ru(NH_3)_5NH_2C(O)R]^{3+} \qquad (ar)$$

or a cyanoalkyl ion moiety proceed similarly[1]:

16   1.5. Formation of Bonds between Hydrogen and N, P, As, Sb, Bi
        1.5.3. by Protonation
        1.5.3.1. of Nitrogen and Nitrogen Compounds

$$h^5\text{-Cp(CO)}_2\text{FeCH(Me)CN} + H^+ \rightarrow [h^5\text{-Cp(CO)}_2\text{Fe}[(\mu_2\text{-Me(H)C}=\text{CNH})]]^+ \quad \text{(as)}$$

Reactions of dinitrogen (N_2) titanium- or group VIA-(Mo, W) coordinated species with protonic acids can yield N-protonated complexes, or free hydrazines or amines. The $(h^5\text{-Me}_5\text{Cp})_2\text{TiN}_2$ is protonated by HCl to form N_2H_4 in 55% yields[38]:

$$2\,(h^5\text{-Me}_5\text{Cp})_2\text{TiN}_2 + 4\,\text{HCl} \rightarrow 2\,(h^5\text{-Me}_5\text{Cp})_2\text{TiCl}_2 + N_2 + N_2H_4 \quad \text{(at)}$$

The bis(titanium)hydrazido complex formed by HCl reaction with $(h^5\text{-Cp})_2\text{TiN}_2\text{MgCl}$ in MeOH reacts further with HCl to produce[39] N_2H_4:

$$[(h^5\text{-Cp})_2\text{TiCl}]_2 N_2H_2 + 2\,\text{HCl} \rightarrow 2\,(h^5\text{-Cp})_2\text{TiCl}_2 + N_2H_4 \quad \text{(au)}$$

Molybdenum and W complexes undergo protonation to (i) NH intermediates, (ii) NH intermediates which can be further protonated to NH_3 or N_2H_4, and (iii) N_2H_4 and/or NH_3 directly. Trans-$M(N_2)(\text{dppe})_2$ (M = Mo, W) reacts with a deficiency[40] of HBr:

$$\text{trans-}M(N_2)_2(\text{dppe})_2 + 2\,\text{HBr} \rightarrow \text{trans-}MBr_2(N_2H_2)(\text{dppe})_2 \quad \text{(av)}$$

where dppe = $Ph_2PCH_2CH_2PPh_2$. The WBr_2 complex containing PMe_2Ph ligands, in monoglyme reacts with HCl to form a N_2H_3 complex:

$$WBr_2(NNH_2)(PMe_2Ph)_3 + \text{HCl} \xrightarrow{\text{monoglyme}} WClBr_2(N_2H_3)(PMe_2Ph)_3 \quad \text{(aw)[41]}$$

Reactions of metal–N_2R complexes with protonic acids yield hydrazido complexes[42]:

$$\text{trans-}WBr(N_2R)(\text{dppe})_2 + \text{HCl} \rightarrow [\text{trans-}WBr(N_2HR)(\text{dppe})_2]\text{Cl} \quad \text{(ax)}$$

where R = Me, Et, t-Bu, etc. Dialkylhydrazo–Mo complexes are cleaved by HBr:

$$\text{Mo}(N_2R_2)(\text{dppe})_2 + \text{HBr} \rightarrow [\text{MoBr(NH)(dppe)}_2]\text{Br} + R_2\text{NH} \quad \text{(ay)[43]}$$

where R = $CH_2(CH_2)_3CH_2$. A Pt–N_2Ph complex undergoes protonation to the coordinated hydrazido product[39]:

$$\text{PtCl}(N_2\text{Ph})(PEt_3)_2 + 2\,\text{HCl} \xrightarrow{\text{THF}} [\text{PtCl(NHNPh)}(PEt_3)_2]\text{Cl} \quad \text{(az)}$$

With excess protonic acid, under more vigorous conditions, conversion of N_2 complexes to NH_3 and N_2H_4 occurs. Reaction of cis-$M(N_2)_2(PMe_2Ph)_4$ (M = Mo, W) with xs HCl yields N_2H_4 and NH_3 in ca. 50 and 25 mol % yields[44]. Treatment of these complexes with H_2SO_4 yields NH_3 and N_2 in a 2:1 mol ratio; NH_3 is produced nearly quantitatively:

$$2\,\text{Mo}(N_2)_2(\text{triphos})\text{PPh}_3 + 8\,\text{HBr} \xrightarrow{\text{THF}} 2\,\text{PPh}_3 + 3\,N_2$$
$$+ 2\,\text{MoBr}_3(\text{triphos}) + 2\,\text{NH}_4\text{Br} \quad \text{(ba)[40]}$$

where triphos = $P(CH_2CH_2PPh_2)_2$.

Nitrides, formed in N_2–metal complex reactions, with protonic acids yield N—H bond-containing products. Sodium napthalide reduction of Et_2TiCl_2 under N_2 in ether[45] or $(h^5\text{-Me}_5\text{Cp})_2\text{TiN}_2$ in THF[39], followed by H_2O treatment, results in 65% conversion of 1 mol of N_2 to NH_3 per mol initial Ti complex:

$$(h^5\text{-Me}_5\text{Cp})_2\text{TiN}_2 \xrightarrow{\text{NaNp}} \xrightarrow{H_2O} (h^5\text{-Me}_5\text{Cp})_2\text{Ti(OH)}_2 + 2\,\text{NH}_3 \quad \text{(bb)}$$

1.5.3. by Protonation   17
1.5.3.1. of Nitrogen and Nitrogen Compounds
1.5.3.1.3. with Protonic Acids in Nonprotonic Solvents.

Titanumin ethoxide absorbs N_2 to form a complex that after reduction with 4 or 6 equiv Na napthalide in THF and treatment with H_2O produces[45] mainly N_2H_4 or NH_3:

$$\frac{1}{x} [Ti(OEt)_2N_2]_x \xrightarrow{\ 4\,NaNp\ } \frac{1}{x} [Ti(OEt)_2N_2]_x^{4-} \xrightarrow{\ H^+\ } \frac{1}{x} [Ti(OEt)_2]_x + N_2H_4 \qquad (bc)$$

$$\xrightarrow{\ 6\,NaNp\ } \frac{1}{x} [Ti(OEt)_2N_2]_x^{6-} \xrightarrow{\ H^+\ } \frac{1}{x} [Ti(OEt)_2]_x + 2\,NH_3 \qquad (bd)$$

Reaction of $FeCl_3$ with Mg under N_2 yields a complex formulated as $[FeMgCl_3\text{-}(THF)_{1.5}]_2N_2$, which produces N_2H_4 on hydrolysis[39]. Reaction of trans-$Mo(N_2)_2(dppe)_2$ with a reduced ferrodoxin, $[EtSFeS]_4^{3-}$, in monoglyme followed by treatment with HCl yields[46] NH_3.

Intramolecular H-exchange reactions, common in organic N-containing molecules, can yield N—H bonds in inorganic or organometallic systems. Metal coordination of the PH phosphazene:

$$(Ph_2P)_2N_3PMeH + COAuCl \xrightarrow{\ -CO\ } (Ph_2P)_2N_2(NH)PMeAuCl \qquad (be)^{47}$$

or phosphorane transfers H from P to N:

$$h^5\text{-}CpMo(CO)_3Cl + \;\; \overset{\displaystyle O}{\underset{\displaystyle O}{\overset{Ph}{\underset{H}{\Large\diagdown}}}}\!\!P\!-\!N \xrightarrow{\ -CO\ } h^5\text{-}CpMo(CO)_2Ph(OC_2H_4)_2NH \qquad (bf)^{48}$$

Structural rearrangement accompanies H-transfer with silylaminophosphines.

(A.D. NORMAN)

1. F. A. Cotton, G. Wilkinson, *Advanced Inorganic Chemistry*, 4th ed., Wiley-Interscience, New York, 1980.
2. W. L. Jolly, *The Inorganic Chemistry of Nitrogen*, Benjamin, New York, 1964.
3. D. C. Bradley, *Adv. Inorg. Chem. Radiochem., 15*, 259 (1972).
4. L. F. Fieser, M. Fieser, *Advanced Organic Chemistry*, Reinhold, New York, 1961.
5. H. Bürger, *Inorg. Nucl. Chem. Lett., 1*, 11 (1965).
6. L. B. Brewster, E. Lee-Ruff, D. K. Bohme, *J. Chem. Soc., Chem. Commun.*, 35 (1973).
7. K.-H. Linke, K. Schroedter, *Z. Anorg. Allg. Chem., 413*, 165 (1975).
8. J. R. Malpass, in *Comprehensive Organic Chemistry*, D. Barton, W. D. Ollis, eds., Vol. 2, I. O. Sutherland, ed., Pergamon Press, New York, 1979, p. 3.
9. M. F. A. Dove, D. B. Sowerby, *Coord. Chem. Rev., 34*, 262 (1981).
10. J. M. Van Paasschen, R. A. Geanangel, *J. Am. Chem. Soc., 94*, 2680 (1972).
11. D. R. Parker, L. H. Sommer, *J. Am. Chem. Soc., 98*, 618 (1976).
12. E. W. Abel, S. A. Mucklejohn, *Phosphorus Sulfur, 9*, 235 (1981).
13. E. Fluck, in *Topics in Phosphorus Chemistry*, M. Grayson, E. J. Griffith, eds., Vol. 4, Wiley-Interscience, New York, 1967, p. 291.
14. E. S. Kozlov, L. G. Dubenko, M. I. Povalotskii, *J. Gen. Chem. USSR (Engl. Transl.), 48*, 1734 (1978).
15. O. J. Scherer, N.-T. Kulbach, W. Glässel, *Z. Naturforsch., Teil B, 33*, 652 (1978).

16. V. P. Kukhar, E. V. Grishkun, V. P. Radovskii, *J. Gen. Chem. USSR (Engl. Transl.), 48,* 1308 (1978).
17. L. Riesel, G. Pich, C. Ruby, *Z. Anorg. Allg. Chem., 430,* 227 (1977).
18. E. Wiberg, E. Amberger, *Hydrides of the Elements of Main Groups I–IV,* Elsevier, Amsterdam, 1971.
19. K. Niedenzu, J. W. Dawson, *Boron-Nitrogen Compounds,* Academic Press, New York, 1965.
20. D. H. Harris, M. F. Lappert, *J. Chem. Soc., Chem. Commun.,* 895 (1974).
21. F. A. Glockling, *The Chemistry of Germanium,* Academic Press, London, 1969.
22. D. A. Armitage, ed., *Organometallic Chemistry, Specialist Periodical Reports,* Vol. 10, Royal Society of Chemistry, London, 1980, p. 86.
23. J. G. A. Luyten, F. Rykens, G. J. M. Van der Kerk, *Adv. Organomet. Chem., 3,* 397 (1965).
24. K. Jones, M. F. Lappert, in *Organotin Compounds,* A. K. Sawyer, ed., Marcel Dekker, New York, 1971, Vol. 2, p. 509.
25. N. Wiberg, M. Veith, *Chem. Ber., 104,* 3176 (1971).
26. J. E. Emsley, D. Hall, *The Chemistry of Phosphorus,* Harper and Row, New York, 1976.
27. H. R. Allcock, *Phosphorus-Nitrogen Compounds; Cyclic, Linear, and High-Polymeric Systems,* Academic Press, New York, 1972.
28. F. Kober, W. J. Rühl, *Z. Anorg. Allg. Chem., 406,* 52 (1974).
29. H. A. Meinemar, J. G. Noltes, *J. Organomet. Chem., 25,* 139 (1970).
30. H. J. Kleiner, *Justus Leibigs Ann. Chem.,* 751 (1974).
31. J. C. Wilburn, R. H. Neilson, *Inorg. Chem., 16,* 2519 (1977).
32. T. Marks, *Prog. Inorg. Chem., 25,* 223 (1979).
33. N. Wiberg, H. Bayer, H. Bachlihuber, *Angew. Chem., Int. Ed. Engl., 14,* 177 (1975).
34. F. R. Scholer, L. J. Todd, *J. Organomet. Chem., 14,* 261 (1968).
35. A. Tarassoli, R. C. Haltiwanger, A. D. Norman, *Inorg. Chem., 21,* 2684 (1982).
36. F. Kober, *Z. Anorg. Allg. Chem., 400,* 285 (1973).
37. J. A. McCleverty, *Chem. Rev., 79,* 1 (1979).
38. J. E. Bercaw, *J. Am. Chem. Soc., 96,* 5087 (1974).
39. J. Chatt, J. R. Dilworth, R. L. Richards, *Chem. Rev., 78,* 589 (1978).
40. J. A. Baumann, T. A. George, *J. Am. Chem. Soc., 102,* 6153 (1980).
41. T. Takahashi, Y. Mizobe, M. Sato, Y. Uchida, M. Hidai, *J. Am. Chem. Soc., 101,* 3405 (1979).
43. W. Hussain, G. J. Leigh, C. J. Pickett, *J. Chem. Soc., Chem. Commun.,* 747 (1982).
44. T. Takahashi, Y. Mizobe, M. Sato, Y. Uchida, M. Hidai, *J. Am. Chem. Soc., 102,* 7461 (1980).
45. E. E. Van Tamelen, *Acc. Chem. Res., 3,* 360 (1970).
46. E. E. Van Tamelen, J. A. Gladyscz, C. D. Brûlet, *J. Am. Chem. Soc., 96,* 3020 (1974).
47. K. C. Dash, A. Schmidpeter, H. Schmidbaur, *Z. Naturforsch, Teil B, 35,* 1286 (1980).
48. J. Wachler, F. Jeanneaux, J. G. Riess, *Inorg. Chem., 19,* 2169 (1980).

1.5.3.2. of Phosphorus and Phosphorous Compounds

1.5.3.2.1. in Aqueous Systems.

Treatment of P_4 with hot alkali or alkaline-earth hydroxide solutions occurs principally in two ways to form PH_3 and the phosphite and hypophosphite ions[1-3]:

$$P_4 + 4\,[OH]^- + 4\,H_2O \rightarrow 4\,[H_2PO_2]^- + 2\,H_2 \qquad (a)$$

$$P_4 + 4\,[OH]^- + 2\,H_2O \rightarrow 2\,[HPO_3]^{2-} + 2\,PH_3 \qquad (b)$$

Disproportionation of P_4 in $Ba(OH)_2$ soln proceeds primarily as:

$$2\,P_4 + 3\,Ba(OH)_2 + 6\,H_2O \rightarrow 3\,Ba(H_2PO_2)_2 + 2\,PH_3 \qquad (c)[3]$$

The $Ba(H_2PO_2)_2$ formed is a major source of the hypophosphite ion and the source of H_3PO_2 which is obtained upon H_2SO_4 acidification of the salt. Moderate yields of PH_3 are reported from reaction[4] with SeO_2 in H_2O at 19°C:

$$P_4 + 3 SeO_2 + 6 H_2O \rightarrow 3 H_3PO_4 + 3 Se + PH_3 \tag{d}$$

Direct reaction of red phosphorus with $Mg(OH)_2$, $Ca(OH)_2$ or $Ba(OH)_2$ at 225–300°C yields PH_3.

Protonation of -ide salts (e.g., Ca_3P_2, Mg_3P_2, Zn_3P_2) in H_2O or strong aq acids yields[5] PH_3:

$$M_3P_2 + 6 H^+ \rightarrow 3 M^{2+} + 2 PH_3 \tag{e}$$

where $M = Ca$, Mg, Sr, Zn. From reactions of AlP with H_2SO_4 at $-70°C$, laboratory quantities of PH_3 (ca. 5 g) can be obtained[6]:

$$MP + 3 H^+ \rightarrow M^{3+} + PH_3 \tag{f}$$

where $M = Al$, Ga, In. Substantial quantities of P_2H_4 and higher phosphine formation accompanies PH_3 production in Ca_3P_2 hydrolysis. This reaction constitutes the most convenient laboratory synthesis[3,5,7] of P_2H_4. Diphosphine is formed directly during hydrolysis, whereas the higher hydrides, e.g., P_nH_{n+2} ($n = 3–9$), P_nH_n ($n = 3–10$), P_nH_{n-2} ($n = 4–12$), etc., result secondarily through P_2H_4 disproportionation (see §1.5.4.2.4).

Small quantities of GeH_3PH_2 or AsH_2PH_2 can be obtained through aqueous acid hydrolysis of ternary alloys or compressed $CaGe–Ca_3P_2$ or $Ca_3As_2–Ca_3P_2$, respectively[8]. In addition to the ternary hydrides, the expected binary hydrides, e.g., P_2H_4, GeH_4, As_2H_4 or P_2H_4, are obtained.

Cleavage of P—P bond-containing compounds, as in the hydrolysis of $(CF_3P)_4$, yields P—H products[9]. At RT reaction occurs as:

$$(CF_3P)_4 + 4 H_2O \xrightarrow{25°C} 2 CF_3P(O)(OH)H + 2 CF_3PH_2 \tag{g}$$

Under more forcing conditions, hydrolysis of the F_3C-P bond occurs to form CF_3H and H_3PO_3. Tetraalkyldiphosphine disulfides in aqueous base are cleaved:

$$R_2P(S)P(S)R_2 + NaOH \rightarrow R_2P(S)ONa + R_2P(S)H \tag{h}$$

where $R = Me$, Et, n-Bu. Although the reaction does not work well when $R = Ph$[10], excellent yields of $Me_2P(S)H$ are obtained[11]. Subsequent conversion of this product in NaOH soln to $Me_2P(O)H$ occurs:

$$Me_2P(S)H + NaOH \rightarrow Me_2P(O)H + NaSH \tag{i}$$

and if conditions are not controlled carefully this hydrolysis product can dominate. Diphosphorus tetraiodide in aq $Ba(OH)_2$ yields hypodiphosphite as $Ba(H_2P_2O_4)$ along with a complex mixture of other phosphorus acid materials[12]:

$$P_2I_4 + 3 Ba(OH)_2 \rightarrow 2 BaI_2 + 2 H_2O + Ba(H_2P_2O_4) \tag{j}$$

Group IVB (Si, Ge, Sn) element–phosphorous compound hydrolysis in H_2O yields phosphines, although reactions with protonic acids in nonprotic solvents are more useful (see §1.5.3.2.3).

Hydrolyses of P(III) compounds, e.g., P_4O_6, yields H_3PO_3 quantitatively[2]:

$$P_4O_6 + 6 H_2O \rightarrow 4 H_3PO_3 \tag{k}$$

Phosphorus trihalide (PCl_3, PBr_3, PI_3) products depend on T, reagent ratios and solution pH[2,13]. In hydrochloric acid at 25°C, PCl_3, PBr_3 and PI_3 react primarily as:

$$PX_3 + 3 H_2O \rightarrow 5 H^+ + 3 X^- + [HPO_3]^{2-} \tag{l}$$

Evaporation of volatiles from PCl_3–H_2O yields crystalline H_3PO_3 quantitatively. Hydrolysis of phosphorus trihalides at 0°C in an $NaHCO_3$ buffer forms mixtures, including $[HPO_3]^{2-}$, $[H_2PO_3]^-$, $[H_2PO_5]^{2-}$, $[P_2O_6]^{4-}$, $[PO_4]^{3-}$, and $[P_2O_7]^{4-}$. From the PBr_3 reaction, upon addition of HBr and alcohol, crystalline $Na_3[HP_2O_5]\cdot12\ H_2O$ is obtained.

Aqueous acid-solution hydrolysis of R_2PX or RPX_2 species forms phosphine oxides and phosphinic acids, respectively[2,14]:

$$RPX_2 + 2 H_2O \rightarrow 2 HX + RP(O)(OH)H \tag{m}$$

$$R_2PX + H_2O \rightarrow 2 HX + R_2P(O)H \tag{n}$$

where R = alkyl, aryl; X = halide, alkoxy, aryloxy. The $Me_2P(O)H$ can be isolated as a hydrochloride adduct, which at 80°C decomposes as:

$$2\ Me_2P(O)H\cdot HCl \rightarrow Me_2P(O)OH + [Me_2PH_2]Cl \tag{o}[15]$$

Phosphine is formed when P_4 in aqueous acid is treated with Zn amalgam[5]. Electrolysis of H_3PO_3 or H_3PO_2, at Pd or Hg cathodes, produces[5,16] some PH_3:

$$P_4 + 12\ e^- + 12\ H^+ \xrightarrow{H_2O} 4\ PH_3 \tag{p}$$

At black phosphorous[17] or InP cathodes[18], electrolytic reduction of the electrode to PH_3 occurs. Zinc in aqueous acid reduces $[H_2PO_4]^-$ or $[HPO_3]^{2-}$ to PH_3[19] and Ph_2PCl to Ph_2PH[20]:

$$[H_2PO_2]^- + 5 H^+ + 2 Zn \rightarrow 2 H_2O + 2 Zn^{2+} + PH_3 \tag{q}$$

$$[HPO_3]^{2-} + 8 H^+ + 3 Zn \rightarrow 3 H_2O + 3 Zn^{2+} + PH_3 \tag{r}$$

$$2\ Ph_2PCl + 2 Zn + 2 H^+ \rightarrow 2 Zn^{2+} + 2 Ph_2PH + 2 Cl^- \tag{s}$$

(A.D. NORMAN)

1. F. A. Cotton, G. Wilkinson, *Advanced Inorganic Chemistry*, 4th ed., Wiley-Interscience, New York, 1980.
2. J. Emsley, D. Hall, *The Chemistry of Phosphorus*, Harper and Row, New York, 1976.
3. G. Brauer, *Handbook of Preparative Inorganic Chemistry*, 2nd ed., Vol. 1, Academic Press, New York, 1963.
4. E. Montignie, *Z Anorg. Allg. Chem., 306*, 235 (1960).
5. E. Fluck, *Top. Curr. Chem. 35*, 3 1973.
6. R. C. Mariott, J. D. Odom, C. T. Sears, Jr., *Inorg. Synth., 14*, 1 (1973).
7. M. Baulder, *Angew. Chem., Int. Ed. Engl., 21*, 492 (1982).
8. W. L. Jolly, A. D. Norman, *Prep. Inorg. React. 4*, 1 (1968).
9. A. H. Cowley, in *Topics in Phosphorus Chemistry*, Vol. 4, M. Grayson, E. J. Griffith, eds., Interscience, New York, 1967, p. 1.
10. L. Maier, *Prog. Inorg. Chem. 5*, 27 (1963).
11. R. A. Malevannaya, E. N. Tsvetkov, M. I. Kabachnik, *Bull. Acad. Sci., USSR Div. Chem. Sci.*, 936 (1976).
12. M. Baudler, M. Mengel, *Z. Anorg. Allg. Chem., 374*, 159 (1970).
13. D. S. Payne, in *Topics in Phosphorus Chemistry*, Vol. 4, M. Grayson, E. J. Griffith, eds., Interscience, New York, 1967, p. 35.
14. A. W. Frank, *Chem. Rev., 61*, 389 (1961).
15. H. J. Kleiner, *Justus Liebigs Ann. Chem.*, 751 (1974).
16. V. P. Gladyshev, *Tr. Inst. Khim. Nauk Akad. Nauk, Kaz. SSR, 35*, 74 (1973); *Chem. Abstr., 80*, 43,447 (1974).

1.5. Formation of Bonds between Hydrogen and N, P, As, Sb, Bi   21
1.5.3. by Protonation
1.5.3.2. of Phosphorus and Phosphorous Compounds

17. I. Chernykh, E. V. Zubova, V. V. Savranskii, A. P. Tomilov, *Sov. Electrochem. (Engl. Transl.)*, *16*, 1797 (1980).
18. G. I. Erusalimchik, D. M. Levin, *Sov. Electrochem. (Engl. Transl.)*, *16*, 1854 (1980).
19. L. Maier, *Prog. Inorg. Chem.*, *5*, 27 (1963).
20. D. T. Burns, A. Townshend, A. H. Carter, in *Inorganic Reaction Chemistry*, Vol. 2, Part B, Ellis Harwood, Chichester, England, 1981, p. 327.

1.5.3.2.2. in Other Protonic Solvents.

Reactions of alkali-metal phosphides in liq NH_3 with protonic acids, such as NH_4Br, readily yield phosphine or substituted derivatives[1,2]:

$$R_{3-n}PM_n + NH_4Br \xrightarrow{NH_{3(l)}} n\ MBr + R_{3-n}PH_n \tag{a}$$

e.g., PH_3[3] and Ph_2PH[4] are obtained upon protonation of Ph_2PNa and KPH_2, respectively, in over 80% yield. Similarly, NH_4Br or H_2O protonation of Na_4P_2 in liq NH_3 produces[5] P_2H_4:

$$Na_4P_2 + 4\ NH_4Br \xrightarrow{NH_{3(l)}} 4\ NH_3 + 4\ NaBr + P_2H_4 \tag{b}$$

White phosphorus reacts in n-BuOH–CCl_4 mixed solvent with [n-BuO]$^-$ to form $(n\text{-}BuO)_2P(O)H$, in yields[6] $\leq 82\%$:

$$P_4 \xrightarrow[H_2O,\ RT]{n\text{-}BuOH,\ [n\text{-}BuO]^-} (n\text{-}BuO)_3P,\ (n\text{-}BuO)_3PO,\ (n\text{-}BuO)_2P(O)H \tag{c}$$

With 2 equiv of [n-BuO]$^-$, no $(n\text{-}BuO)_2P(O)H$ is obtained.

Halophosphorus-(III) species undergo[2] alcoholysis to alkyl halides and H_3PO_3:

$$3\ ROH + PX_3 \rightarrow 3\ RX + (HO)_2P(O)H \tag{d}$$

Under controlled conditions, phosphonate products (e.g., R = Me, Et, Ph, C_6H_{11}) can be obtained[7]:

$$3\ ROH + PX_3 \rightarrow RX + (RO)_2P(O)H + 2\ HX \tag{e}$$

Alkyldichlorophosphines react in alcohol to form phosphinate esters, e.g.:

$$RPCl_2 + 2\ R'OH \rightarrow HCl + R'Cl + RP(O)(OR')H \tag{f}$$

Yields of $t\text{-}BuP(O)(OEt)H$[8] and $MeP(O)(OR')H$ (R' = Et, n-Pr, i-Bu, n-Bu)[9] $\leq 98\%$ can be obtained.

Phosphaethene reacts in refluxing MeOH to form a secondary phosphine:

$$Me_3SiN(Ph)C(t\text{-}Bu){=}PPh \xrightarrow[-Me_3SiOMe]{MeOH} PhN{=}C(t\text{-}Bu)P(Ph)H \tag{g}$$

In contrast, treatment of the vinylphosphonate with alcoholic acid cleaves the phosphoryl moiety with no addition to the vinyl double bond[10]:

$$MePhCHCHC(OSiMe_3)P(O)(OEt_2) \xrightarrow[ROH]{[H_3O]^+} MePhCHCH_2CO_2R + (EtO)_2P(O)H \tag{h}$$

Protonation of phosphine, phosphites or phosphorous halides in strong acid forms P—H bonds. Phosphine in conc H_2SO_4 or HCl is converted[11,12] to $[PH_4]^+$:

$$PH_3 + H_2SO_4 \xrightarrow{H_2SO_4} [HSO_4]^- + [PH_4]^+ \tag{i}$$

22   1.5. Formation of Bonds between Hydrogen and N, P, As, Sb, Bi
        1.5.3. by Protonation
        1.5.3.2. of Phosphorus and Phosphorous Compounds

Protonation of phosphinic acid, phosphites,[13] or halophosphines[14] by FSO_3H or $FSO_3H–SbF_5$ in SO_2 at low T yields phosphonium ions, as observed in ^1H and ^{31}P NMR spectra, e.g.:

$$(RO)_3P + H^+ \xrightarrow{\text{strong acid}} [(RO)_3PH]^+ \qquad (j)$$

where R = H, Me, Et, i-Pr, n-Bu, Ph;

$$PF_{3-n}Cl_n + H^+ \xrightarrow{\text{strong acid}} [F_{3-n}Cl_nPH]^+ \qquad (k)$$

where n = 1–3.

(A.D. NORMAN)

1. F. A. Cotton, G. Wilkinson, *Advanced Inorganic Chemistry*, 4th ed., Wiley-Interscience, New York, 1980.
2. J. Emsley, D. Hall, *The Chemistry of Phosphorus*, Harper and Row, New York, 1976.
3. R. G. Hayter, F. S. Humiec, *Inorg. Chem.*, 2, 306 (1963).
4. R. I. Wagner, A. B. Burg, *J. Am. Chem. Soc.*, 75, 3869 (1953).
5. E. C. Evers, E. H. Street, Jr., S. L. Lung, *J. Am. Chem. Soc.*, 73, 5088 (1951).
6. C. Brown, R. F. Hudson, G. A. Warten, H. Coates, *Phosphorus Sulfur*, 6, 481 (1979).
7. D. S. Payne, in *Topics in Phosphorus Chemistry*, M. Grayson, E. J. Griffith, eds., Vol. 4, Wiley-Interscience, New York, 1967, p. 85.
8. D. C. Crofts, D. M. Parker, *J. Chem. Soc., C*, 332 (1970).
9. M. Fink, H. J. Kleiner, *Justus Leibigs Ann. Chem.*, 741 (1974).
10. T. Hata, M. Nakajima, M. Sekine, *Tetrahedron Lett.*, 2047 (1979).
11. M. E. Peach, T. C. Waddington, in *Non-Aqueous Solvent Systems*, T. C. Waddington, ed., Academic Press, London,, 1965, p. 83.
12. R. J. Gillespie, E. A. Robinson, in *Non-Aqueous Solvent Systems*, T. C. Waddington, ed., Academic Press, London, 1965, p. 117.
13. G. A. Olah, C. W. McFarland, *J. Org. Chem.*, 36, 1374 (1971).
14. L. J. Vande Griend, J. G. Verkade, *J. Am. Chem. Soc.*, 97, 5958 (1975).

1.5.3.2.3. with Protonic Acids in Nonprotonic Solvents.

Protonation of phosphides results in direct formation of P—H bonds.[1,2] Reactions of alkali-metal mono- and disubstituted phosphides with H_2O in ethers leads to:

$$R_{3-n}PM_n + n H_2O \rightarrow n MOH + R_{3-n}PH_n \qquad (a)$$

where n = 1, 2, 3; R = alkyl, aryl; M = Li, Na, K; e.g., PH_3, $PhPH_2$, Et_2PH, Ph_2PH, EtPhPH, in 30–80% yields. Metal salts of di- and tetraphosphines react to form phosphines and cyclopolyphosphines[3]:

$$M_2(RP)_4 \xrightarrow[H_2O]{THF} MOH + (RP)_4 + RPH_2 \qquad (b)$$

where M = Li, K; R = Me, Et, C_6H_{11}. t-Butyl-substituted lithio di- and triphosphines upon hydrolysis generate the parent phosphines[3,4]:

$$(t\text{-BuP})_2K_2 + 2 H_2O \rightarrow 2 KOH + (t\text{-BuPH})_2 \qquad (c)$$

$$(t\text{-BuP})_2PK + H_2O \rightarrow KOH + (t\text{-BuP})_2PH \qquad (d)$$

The h^5-$Cp(CO)_2Mn$-coordinated $(PhPH)_2$ forms in a reaction involving an -ide salt in which protons are abstracted from the THF solvent[5]:

$$h^5\text{-CpMn(CO)}_2PPhLi_2 + h^5\text{-CpMn(CO)}_2PPhCl_2 \xrightarrow{THF}$$
$$2 LiCl + h^5\text{-CpMn(CO)}_2P(H)PhP(H)PhMn(CO)_2Cp\text{-}h^5 \qquad (e)$$

1.5.3. by Protonation
1.5.3.2. of Phosphorus and Phosphorous Compounds
1.5.3.2.3. with Protonic Acids in Nonprotonic Solvents.

23

Complex phosphido- or alkylphosphidoaluminates quantitatively hydrolyze to form[6] PH_3 or primary and secondary phosphines, e.g.:

$$LiAl(PHR)_4 + 2 H_2O \xrightarrow{Et_2O} LiAl(OH)_4 + 4 RPH_2 \qquad (f)^7$$

The strongly basic Et_2PLi deprotonates $(Et_2P)_3SiH$:

$$Et_2PLi + (Et_2P)_3SiH \rightarrow (Et_2P)_3SiLi + Et_2PH \qquad (g)$$

however, no reaction occurs[8] with $(Et_2P)_2SiH_2$. Reaction of HCl with a spiro-phosphorus(V) anion and treatment of the $(NPCl_2)_3$ alkylmagnesium halide reagent product with i-PrOH proceed:

$$(h)^9$$

$$(NPCl_2)_3 \xrightarrow[\text{2. n-PrOH}]{\text{1. RMgBr, (Bu}_3\text{PCuI)}_4 \text{ catal.}} [(NPCl_2)_2NP(H)R] \qquad (i)^{10}$$

where R = Me, Et, i-Pr, n-Pr, n-Bu. Deuterium-labeling experiments show the P—H hydrogen comes from the n-PrOH.

Phosphonium salts containing a P—H bond form in condensed phases upon strong-acid protonation[1,11] or in the gas phase by hydrogen-ion protonation[7,12] of phosphines:

$$R_{3-n}PH_n + HX \rightarrow [R_{3-n}PH_{n+1}]X \qquad (j)$$

where R = H, alkyl, aryl; X = Cl, Br, I and n = 0–3. The more strongly basic alkylphosphines react quantitatively. Phosphonium chloride or bromide is highly dissociated at RT; PH_4I forms a solid stable adduct[1]. The perfluoroalkylphosphines are cleaved by HX to CF_3H and the respective P(III) halide. Phosphorus(III) halides and phosphites are protonated in strong acid media, e.g. HSO_3F–SbF_5 in SO_2 at $-60°C$ to $-70°C$[13]:

$$PF_{3-n}X_n + H^+ \xrightarrow{SO_{2(l)}, \ -70°C} [F_{3-n}X_nPH]^+ \qquad (k)$$

where X = Cl, n = X = Br, n = 1–3;

$$(RO)_3P + H^+ \xrightarrow{SO_{2(l)}, \ -60°C} [(RO)_3PH]^+ \qquad (l)$$

where R = Me, Et, i-Pr, n-Bu, Ph. Tetramethyldiphosphine is protonated[11] by HCl in $CHCl_3$. Reaction of the P-containing cage, $P(OCH_2CH_2)_3N$, with $[R_3O][BF_4]$ in CH_3CN gives the intramolecularly P—N-coordinated cage in 80% yield[14]:

24   1.5. Formation of Bonds between Hydrogen and N, P, As, Sb, Bi
        1.5.3. by Protonation
        1.5.3.2. of Phosphorus and Phosphorous Compounds

$$P(OCH_2CH_2)_3N + [R_3O][BF_4] \xrightarrow{CH_3CN} \left[H-\underset{\underset{O}{|}}{\overset{\overset{O}{|}}{P}}-N \right] BF_4 \qquad (m)$$

Protonic acid cleavage of bonds between P and group IVB elements (E = Si, Ge, Sn) yields P—H bonds[15-17]:

$$>P-E + HA \rightarrow P-H + E-A \qquad (n)$$

Reactions involving the acids H_2O, H_2S, ROH, RSH, NH_3, RNH_2, RCCH, and RCO_2H in the gas phase or in nonprotic solvents, form, e.g., $R_{3-n}PH_n$ (n = 1–3), PH_3[15-20], RPH_2 (R = Me, Et, t-Bu, Ph)[15,21] and R_2PH (R = Me, Et, Ph)[15,16,21,22]:

$$2 SiH_3PH_2 + H_2O \rightarrow (SiH_3)_2O + 2 PH_3 \qquad (o)$$

$$(Me_3Si)_2PPh + 2 MeOH \xrightarrow{THF} 2 Me_3SiOMe + PhPH_2 \qquad (p)$$

$$Et_3GePEt_2 + PhNH_2 \rightarrow Et_3GeNHPh + Et_2PH \qquad (q)$$

The order of acid reactivity is hydrogen halides > alcohol, thiols > amines. It is noteworthy that $(GeH_3)_3P$ is inert to H_2O cleavage but reacts rapidly[16] with H_2S:

$$2 (GeH_3)_3P + 3 H_2S \rightarrow 3 (GeH_3)_2S + 2 PH_3 \qquad (r)$$

Cleavage of Si—P bonds is fast; Ge—P and Sn—P bonds are less reactive; Ph_3SnPPh_2 is inert[15] to hot H_2O. The stepwise nature of multiple cleavage reactions can be used to prepare intermediate cleavage products, e.g., under controlled conditions both $Me_3SiP(H)Ph$:

$$(Me_3Si)_2PPh + MeOH \xrightarrow{triglyme} MeOSiMe_3 + Me_3SiP(H)Ph \qquad (s)$$

and $PhPH_2$:

$$Me_3SiP(H)Ph + MeOH \xrightarrow{triglyme} MeOSiMe_3 + PhPH_2 \qquad (t)$$

can be obtained[21] from the alcoholysis of $(Me_3Si)_2PPh$.

Protonic acid cleavage of group IVB element (Si, Ge)-P bonds is most valuable synthetically to prepare P—H bond-containing higher phosphines or metal complexes:

$$(t-BuP)_2PSiMe_3 + MeOH \rightarrow MeOSiMe_3 + (t-BuP)_2PH \qquad (u)^3$$

$$5 (Me_3SiP)_4 + 20 MeOH \rightarrow 20 MeOSiMe_3 + 4 P_5H_5 \qquad (v)^3$$

$$Me_3Si(PPh)_4SiMe_3 + 2 MeOH \rightarrow 2 MeOSiMe_3 + (PPh)_4H_2 \qquad (w)^3$$

Methanolysis of silylated metal phosphido complexes yields phosphido complexes,[23] e.g.:

$$3 [h^5\text{-}CpNiP(SiMe_3)_2]_2 \xrightarrow[-12\ MeOSiMe_3]{12\ MeOH} 2 (h^5\text{-}CpNiPH_2)_3 \qquad (x)$$

Acid cleavage of the cyclic germylphosphines yields acyclic secondary phosphines[24]:

$$R_2Ge{-\!\!\!-}PR' \xrightarrow{HX} R_2Ge(X)CH_2CH_2CH_2P(H)R' \qquad (y)$$

1.5.3. by Protonation
1.5.3.2. of Phosphorus and Phosphorous Compounds
1.5.3.2.3. with Protonic Acids in Nonprotonic Solvents.

25

where R = Et, R′ = Et, Ph; X = Cl, OH, OMe, SEt, OAc.

Alcohol cleavage of C—P bonds can form P—H bond-containing products. Methanol cleaves $(MeCO)_3P$ stepwise, forming[25] PH_3:

$$(MeCO)_3P + 3\ MeOH \rightarrow 3\ MeCO_2Me + PH_3 \qquad (z)$$

Cleavage of the P=C bond in a phosphaethene proceeds:

$$Me_3SiPC(OSiMe_3)Bu\text{-}t + 2\ ROH \rightarrow 2\ ROSiMe_3 + t\text{-}BuCOPH_2 \qquad (aa)^{26}$$

Protonic acid cleavage of the P—P bonds in P_4 yields compounds containing P—H bonds. Reaction at 25°C with HI yields PH_3 and P_2I_4 quantitatively[27]:

$$5\ P_4 + 24\ HI \xrightarrow{CS_2} 8\ PH_3 + 6\ P_2I_4 \qquad (ab)$$

Dimethylphosphine, from H_2O cleavage of $Me_2P(S)P(S)Me_2$ in the presence of n-Bu_3P at 160–170°C, is obtained in 66% yield[28]:

$$2\ Me_2P(S)P(S)Me_2 + 4\ n\text{-}Bu_3P + 2\ H_2O \rightarrow 4\ n\text{-}Bu_3PS + Me_2PO_2H + 3\ Me_2PH \quad (ac)$$

This synthesis of Me_2PH is better than the $LiAlH_4$–$Me_2P(S)P(S)Me_2$ method described in §1.5.5.2.2. The unsymmetrically substituted diphosphine $Me_2PP(CF_3)_2$ reacts with HCl:

$$Me_2PP(CF_3)_2 + HCl \rightarrow Me_2PCl + (CF_3)_2PH \qquad (ad)$$

or H_2O at 25°C, forming $(CF_3)_2PH$ in 98% and 79% yields, respectively[29]. Dry HI, H_2O, or Me_2AsH and $(CF_3)_2PP(CF_3)_2$ react:

$$(CF_3)_2PP(CF_3)_2 + HX \rightarrow (CF_3)_2PX + (CF_3)_2PH \qquad (ae)$$

where X = I, OH, $AsMe_2$. Ring opening and partial degradation of $(CF_3P)_4$ by H_2O in diglyme yields a mixture[29] of $(CF_3P)_2H_2$ and $(CF_3P)_3H_2$. Reduction–scission of Me_4P_2 by $RuCl_3 \cdot x\ H_2O$ forms metal-coordinated[30] Me_2PH:

$$RuCl_3 \cdot x\ H_2O + 2\ Me_4P_2 \xrightarrow{THF} RuCl_2(Me_2PH)_4 \qquad (af)$$

Trimethylstannane cleavage of metal-coordinated diphosphines yields metal-coordinated[31] $(CF_3)_2PH$:

$$Me_2PPR_2M(CO)_5 + Me_3SnH \rightarrow Me_3SnPMe_2 + (CO)_5MPHR_2 \qquad (ag)$$

where R = Me, CF_3; M = Cr, Mo.

Hydrolysis or alcoholysis of P(III) halides, especially PCl_3 and PBr_3, yields H_3PO_3; however, these reactions are better carried out in H_2O or alcoholic media[1,7] (see §1.5.3.2.1 and §1.5.3.2.2). Hydrolysis of trialkyl or triarylphosphites occurs readily, e.g., reaction[7] of $(i\text{-}PrO)_3P$ with H_2O in CH_3CN:

$$(i\text{-}PrO)_3P + H_2O \rightarrow (i\text{-}PrO)_2P(O)H + i\text{-}PrOH \qquad (ah)$$

Dry HCl reacts with $Ph_2P(O)Ac$, yielding $Ph_2P(O)H$[33]:

$$Ph_2P(O)OAc + HCl \xrightarrow{-10°C} Ph_2P(O)H + AcOCl \qquad (ai)$$

Controlled reactions of Me_2PCl with H_2O or MeOH and $MePCl_2$ with MeOH[34] yield $Me_2P(O)H$ and MeP(O)HOMe, respectively:

26   1.5. Formation of Bonds between Hydrogen and N, P, As, Sb, Bi
       1.5.3. by Protonation
       1.5.3.2. of Phosphorus and Phosphorous Compounds

$$Me_2PCl + ROH \rightarrow RCl + Me_2P(O)H \tag{aj}$$

$$MePCl_2 + 2\ MeOH \rightarrow MeP(O)(OR)H \tag{ak}$$

where R = H, Me. Boron-trichloride coordination of tertiary Ph_2PX phosphines, followed by addition of H_2O, produces $Ph_2P(O)H$ in 68–87% yield[35]:

$$Ph_2PX \cdot BCl_3 \xrightarrow[-HX]{H_2O} Ph_2P(O)H \tag{al}$$

where X = CH_2OMe, CH_2SMe, $C(O)Ph$. Hydrolysis of cyclic secondary phosphine chlorides results in the cyclic phosphine oxide in high yield[36].

Aminophosphines can react with protonic acids (HX) to form P—H bonds. Generally, such reactions produce NH and PX products, owing to the polarity of the P—N bond[1,37]. Hydrogen sulfide and dialkylaminodialkylphosphines in benzene at 80°C yield dialkylphosphine sulfides quantitatively[37]:

$$R_2PNR_2' + 2\ H_2S \xrightarrow{C_6H_6} [R_2'NH_2]SH + R_2P(S)H \tag{am}$$

where R = Et; R' = Me, Et, t-Bu, Ph. Tris(dimethylamino)phosphine reacts with H_2S:

$$(Me_2N)_3P + 4\ H_2S \rightarrow [Me_2NH_2][HP(S)S_2] + [Me_2NH_2][SH] \tag{an}$$

Conversion of thiophosphite to its Na salt, followed by thermolysis, yields PH_3 quantitatively[38]:

$$3\ [Na]_2[HP(S)S_2] \rightarrow 2\ [Na]_3[P(S)S_3] + PH_3 + S \tag{ao}$$

Ammonia and $(Me_2N)_2PCl$ in diethyl ether form the hydridocyclophosphazene[39]:

$$3\ (Me_2N)_2PCl + 3\ NH_3 \xrightarrow[-Me_2NH_2Cl]{Et_2O} [NP(H)NMe_2]_3 \tag{ap}$$

An aminophosphine intermediate forms, which tautomerizes to the final product.

Oxidative addition of protonic acids to P(III) compounds yields P—H-phosphoranes. Alkyldichlorophosphines react at -30 to $-20°C$ with HF in $FCCl_3$ solvent to form equatorially substituted products[40]:

$$RPCl_2 + 3\ HF \rightarrow RP(F)_3H + 2\ HCl \tag{aq}$$

where R = Me, Et, n-Bu, t-Bu, Ph. In the presence of H_2O, the phosphonous acid fluorides forms[41]:

$$RPCl_2 + HF + H_2O \xrightarrow[-2\ HCl]{} RP(O)HF \tag{ar}$$

Gaseous HBr and allyldifluorophosphine undergo exchange and addition[42]:

$$3\ CH_2{=}CHCH_2PF_2 + 2\ HBr \xrightarrow{25°C} 2\ CH_2{=}CHCH_2PF_3H$$
$$+ CH_2{=}CHCH_2PBr_2 \tag{as}$$

Gaseous NH_3 and PF_2NH_2 or PF_2Cl react[43] at 25°C. Alcohols and thiols, with PF_2H at 0°C, form[44] PH_3F_2:

$$PF_2H + 2\ REH \rightarrow (RE)_2 + PF_2H_3 \tag{at}$$

1.5.3. by Protonation
1.5.3.2. of Phosphorus and Phosphorous Compounds
1.5.3.2.3. with Protonic Acids in Nonprotonic Solvents.

27

where R = Me, Et; E = O, S;

$$PF_2NH_2 + 2 NH_3$$ (au)

$$PF_2Cl + 3 NH_3 \xrightarrow[-NH_4Cl]{} P(NH_2)_2F_2H$$ (av)

Alcohols react with $PhPF_2$ in benzene to form P—H-fluorophosphoranes[45]:

$$PhPF_2 + ROH \rightarrow PhP(OR)(F)_2H$$ (aw)

where R = Me, Et, CF_3CH_2, $(CF_3)_2CH$, $Ph(CF_3)CH$, t-Bu, whereas H_2O reacts in a two-step process to produce the final $PhPF_3H$:

$$PhPF_2 + H_2O \rightarrow PhP(O)FH + HF$$ (ax)

$$PhPF_2 + HF \rightarrow PhPF_3H$$ (ay)

Six-coordinated P—H-phosphoranes form[46] P(III) compounds and from $[HF_2]^-$:

$$(CF_3)_nPF_{3-n} + KHF_2 \xrightarrow[CH_3CN]{60-100°C} K[(CF_3)_nP(H)F_{5-n}]$$ (az)

where n = 0,1,2; or from:

$$Me_2NPF_2 + 3 KHF_2 \xrightarrow{CH_3CN} [Me_2NH_2][PF_5H] + 3 KF$$ (ba)

and from pentacoordinated P—H-phosphoranes[47] with F^-:

$$CsF + H_2PF_3 \rightarrow Cs[PF_4H_2]$$ (bb)

In contrast, PF_2I with PH_3 or HI[48], or CF_3PI_2 with HI[49] in the presence of Hg yield P(III) products:

$$PF_2I + HI + 2 Hg \xrightarrow{85°C} Hg_2I_2 + PF_2H$$ (bc)

$$2 PF_2I + PH_3 + 2 Hg \xrightarrow{25°C} Hg_2I_2 + \frac{1}{n}(PH)_n + 2 PF_2H$$ (bd)

$$CF_3PI_2 + 2 HI + 2 Hg \xrightarrow{25°C} Hg_2I_2 + (CF_3PH)_2$$ (be)

Reaction (bd) is preferable for PF_2H synthesis if PH_3 is available. Both meso and d,l-forms of $(CF_3PH)_2$ are observed. With xs HI in the presence of Hg, CF_3PI_2, or $(CF_3)_2PI$ form the respective phosphines[49]:

$$CF_3PI_2 + 2 HI + 2 Hg \rightarrow Hg_2I_2 + CF_3PH_2$$ (bf)

$$2 (CF_3)_2PI + 2 HI + 2 Hg \rightarrow Hg_2I_2 + 2(CF_3)_2PH$$ (bg)

Phosphorous(V) P—H-phosphoranes, stabilized by the P atom being in at least one ring, can be synthesized; e.g., phosphonamidites react with carboxylic acids, alcohols, and thiols (HR) to form five-coordinated products[50,51]:

(bh)

28   1.5. Formation of Bonds between Hydrogen and N, P, As, Sb, Bi
      1.5.3. by Protonation
      1.5.3.2. of Phosphorus and Phosphorous Compounds

where X = Et; R = OEt, OPh, SPh, OAc, SEt, OCOCH=CH$_2$, o-NH$_2$C$_6$H$_4$. Reactions proceed in aprotic solvents (ethers, benzene, toluene, CH$_2$Cl$_2$) at reflux T. Diol reactions with phosphonamidates or Ph$_2$PCl proceed similarly[50]:

$$\text{(bi)}$$

where X = NMe$_2$, R = H; X = N(CH$_2$CH$_2$)$_2$, R = Me. Equilibrium between the P—H-phosphorane and its P(III) form is observed in solution, e.g.:

$$\text{(bj)}$$

Reactions between compounds with mixed hydroxyl–amino functionality and P(III) species yield P—H-phosphoranes also[50]:

$$HN(CH_2CHMeOH)_2 + PhOPCl_2 \xrightarrow{-2\,HCl}$$

$$\text{(bk)}$$

Using MeP(NMe$_2$)$_2$ instead of PhOPCl$_2$, the product mixture consists of one d,l- and two meso-forms[51]. Cyclic tetramines consisting of 12 (a–d = 2), 13 (a–c = 2; d = 3), 14 (a = c = 2, b = d = 3), 15 (a–c = 3; d = 2) and 16 (a–d = 3) membered rings react with (Me$_2$N)$_3$P at 120°C to form tetramine P—H-phosphoranes[52]:

$$\text{(bl)}$$

Compound stability decreases with increasing tetramine ring size.

Disproportionation of proton-acidic P species can result in P—H bond formation. Anhydrous H$_3$PO$_3$ decomposes at 220°C to PH$_3$ and H$_3$PO$_4$:

$$4\ H_3PO_3 \xrightarrow{\Delta} 3\ H_3PO_4 + PH_3 \qquad \text{(bm)[17]}$$

$$NaP(OH)_2 + RX \rightarrow NaX + RPH(O)(OH) \qquad \text{(bn)}$$

1.5.3. by Protonation
1.5.3.2. of Phosphorus and Phosphorous Compounds
1.5.3.2.3. with Protonic Acids in Nonprotonic Solvents.

29

Alkyl halide reactions with $NaP(OH)_2$ yield alkylphosphinites[7] which upon heating disproportionate[32]:

$$3\ RPH(O)OH \xrightarrow{150°C} RPH_2 + 2\ RP(O)(OH)_2 \qquad \text{(bo)}$$

Intramolecular H-rearrangement processes yield P—H products in a phosphorane oxide rearrangement to an alkenylphosphinite[53]:

$$\text{(structure)} \longrightarrow \text{t-BuCHC(t-Bu)P(O)(Bu-t)H} \qquad \text{(bp)}$$

Tautomeric exchange occurs in carboxy-substituted phosphines[54] or in hydroxyphosphaethenes[55], both tautomers existing in solution at equilibrium:

$$MePhPCH(CO_2Me)_2 \rightleftharpoons MePhP(H)=C(CO_2Me)_2 \qquad \text{(bq)}$$

$$RP=C(OH)(t\text{-Bu}) \rightleftharpoons RP(H)CO(t\text{-Bu}) \qquad \text{(br)}$$

Aminophosphines containing an NHR group on phosphorus can rearrange to the P—H-phosphinimine product[56]:

$$R_2NP(H)N(H)R \rightarrow R_2NP(H)_2NR \qquad \text{(bs)}$$

where R = alkyl, Me_3Si. When R = alkyl, the P(V) form dominates; however, with Me_3Si substitution or P coordination to a metal, the systems can be stabilized in the P(III) form. Dialkylamine addition to phosphinimines yields P—H products through aminophosphine intermediates[57]:

$$(Me_3Si)_2NPNR + HNR'R'' \rightarrow (Me_3Si)_2N(R'R''N)P(H)=NR \qquad \text{(bt)}$$

where R = Me_3Si, t-Bu; R′ = R″ = Me_3Si; Et, $-(CH_2)_5-$; R′, R″ = H, t-Bu. In this system, the latter predominates when R = t-Bu. Hydriodocyclophosphazenes form similarly:

$$H_2NP(Ph)_2NP(Ph)_2NH + RP(OPh)_2 \xrightarrow[-2\ PhOH]{} (NPPh_2)_2NP(H)R \qquad \text{(bu)}$$

where R = Me, Et, Ph.

(A.D. NORMAN)

1. F. A. Cotton, G. Wilkinson, *Advanced Inorganic Chemistry*, 4th ed., Wiley-Interscience, New York, 1980.
2. V. D. Bianco, S. Doronzo, *Inorg. Synth., 16*, 161 (1976).
3. M. Baudler, *Angew. Chem., Int. Ed. Engl., 21*, 492 (1982).
4. M. Baudler, C. Gruner, H. Tschähunn, J. Hahn, *Chem. Ber., 115*, 1739 (1982).
5. G. Huttner, H. D. Müller, V. Bejenke, O. Orama, *Z, Naturforsch, Teil B, 31*, 1166 (1976).
6. A. E. Finholt, C. Helling, V. Imhof, L. Nielson, E. Jacobson, *Inorg. Chem., 2*, 504 (1963).
7. J. Emsley, D. Hall, *The Chemistry of Phosphorus*, Harper and Row, New York, 1976. Good general coverage of the chemistry of phosphorus.
8. G. Fritz, G. Becker, *Angew. Chem., Int. Ed. Engl., 6*, 1078 (1967).
9. I. Granoth, J. C. Martin, *J. Am. Chem. Soc., 101*, 4623 (1979).
10. H. R. Allcock, P. J. Harris, *J. Am. Chem. Soc., 101*, 6221 (1979).
11. F. Seel, H. J. Bassler, *Z. Anorg. Allg. Chem., 418*, 263 (1975).
12. J. W. Long, J. L. Franklin, *J. Am. Chem. Soc., 96*, 2370 (1974).
13. L. J. Vande Griend, J. G. Verkade, *J. Am. Chem. Soc., 97*, 5958 (1975).

14. D. S. Milbrath, J. G. Verkade, *J. Am. Chem. Soc.*, 99, 6607 (1977).
15. E. W. Abel, S. M. Illingworth, *Organomet. Chem. Rev., A,* 5, 143 (1970).
16. J. E. Drake, C. Riddle, *Q. Rev. Chem. Soc.*, 263 (1970).
17. W. L. Jolly, A. D. Norman, *Prep. Inorg. React.,* 4, 1 (1968).
18. A. D. Norman, W. L. Jolly, *Inorg. Chem.,* 18, 1594 (1979).
19. K. G. Sharp, *J. Chem. Soc., Chem. Commun.*, 564 (1977).
20. A. D. Norman, *Inorg. Chem.,* 9, 870 (1970).
21. G. Becker, O. Mundt, M. Rössler, E. Schneider, *Z. Anorg. Allg. Chem.,* 443, 42 (1978).
22. R. Demuth, *Z. Anorg. Allg. Chem.,* 427, 221 (1976).
23. H. Schaeffer, *Z. Anorg. Allg. Chem.,* 459, 157 (1979).
24. C. Couret, J. Escudié, J. Satgé, G. Redoules, *Synth. React. Inorg. Metal-Org. Chem.,* 7, 99 (1977).
25. G. Becker, *Z. Anorg. Allg. Chem.,* 480, 38 (1981).
26. G. Becker, *Z. Anorg. Allg. Chem.,* 480, 21 (1981).
27. M. Schmidt, H. J. Schroeder, *Z. Anorg. Allg. Chem.,* 378, 185 (1970).
28. A. Trenkle, H. Vahrenkamp, *Inorg. Synth.,* 21, 180 (1982).
29. R. G. Cavell, R. C. Dobbie, *J. Chem. Soc.*, 1308 (1967).
30. F. A. Cotton, B. A. Franz, L. Hunter, *Inorg. Chim. Acta,* 16, 203 (1976).
31. J. Grobe, L. V. Duc, *Z. Naturforsch., Teil B,* 36, 666 (1981).
32. A. W. Frank, *Chem. Rev.,* 61, 389 (1961).
33. J. A. Miller, D. Stewart, *J. Chem. Soc. Perkin Trans. 1,* 1898 (1977).
34. K. Weissermel, H. J. Kleiner, M. Finke, U.-H. Felcht, *Angew. Chem., Int. Ed. Engl.,* 20, 223 (1981).
35. K. C. Hansen, G. B. Sollender, C. L. Holland, *J. Org. Chem.,* 39, 267 (1974).
36. S. D. Venkataramu, G. D. MacDonell, W. R. Purdum, M. El-Deek, K. D. Berlin, *Chem. Rev.,* 77, 121 (1977).
37. L. Maier, *Topics in Phosphorus Chemistry,* M. Grayson, E. J. Griffith, eds., Vol. 2, Wiley-Interscience, New York, 1965, p. 43.
38. F. Seel, G. Zindler, *Z. Anorg. Allg. Chem.,* 470, 167 (1980).
39. A. Schmidpeter, H. Rössknecht, *Chem. Ber.,* 107, 3146 (1974).
40. R. Appel, A. Gilak, *Chem. Ber.,* 108, 2693 (1975).
41. U. Ahrens, H. Falius, *Chem. Ber.,* 105, 3317 (1972).
42. E. R. Falardeau, K. W. Morse, J. G. Morse, *Inorg. Chem.,* 14, 1239 (1975).
43. D. E. J. Arnold, D. W. H. Rankin, *J. Chem. Soc., Dalton Trans.*, 1130 (1976).
44. L. F. Centofanti, R. W. Parry, *Inorg. Chem.,* 12, 1456 (1973).
45. A. F. Janjen, J. L. Kruczynski, *Can. J. Chem.,* 57, 1903 (1979).
46. J. F. Nixon, J. R. Swain, *Inorg. Nucl. Chem. Lett.,* 5, 295 (1969).
47. K. O. Christie, C. J. Schrock, E. C. Curtis, *Inorg. Chem.,* 15, 843 (1976).
48. L. Centofanti, R. W. Rudolph, *Inorg. Synth.,* 12, 281 (1970).
49. R. C. Dobbie, P. D. Gosling, *J. Chem. Soc., Chem. Commun.*, 585 (1975).
50. S. Tripett, ed., *Organophosphorus Chemistry, Specialist Periodical Reports,* The Chemical Society, London; Vol. 9, p. 33, 1978; Vol. 10, p. 33, 1979; Vol. 11, p. 30, 1980.
51. D. Houalla, J. Mouhuch, M. Sanchez, R. Wolf, *Phosphorus,* 5, 229 (1975).
52. T. J. Atkins, J. E. Richman, *Tetrahedron Lett.,* 52, 5149 (1978).
53. H. Quast, M. Heuschmann, *Justus Leibigs Ann. Chem.*, 977 (1981).
54. O. I. Kolodiazhnyi, *Tetrahedron Lett.,* 21, 2269 (1980).
55. G. Becker, M. Rosler, E. Schneider, *Z. Anorg. Allg. Chem.,* 439, 121 (1978).
56. E. Niecke, C. Ringel, *Angew. Chem., Int. Ed. Engl.,* 16, 486 (1977).
57. A. Schmidpeter, J. Ebeling, *Angew. Chem., Int. Ed. Engl.,* 7, 209 (1968).

1.5.3.3. of Arsenic and Arsenic Compounds

1.5.3.3.1. in Aqueous Systems.

Arsine is prepared by treatment of Mg or Zn arsenides with aqueous acid[1,2]:

$$M_3As_2 + 6 H^+ \xrightarrow{\ HCl_{(aq)}\ } 3 M^{2+} + 2 AsH_3 \qquad (a)$$

1.5. Formation of Bonds between Hydrogen and N, P, As, Sb, Bi   31
1.5.3. by Protonation
1.5.3.3. of Arsenic and Arsenic Compounds

Using an Mg-Al-As alloy low yields of As_2H_4 in addition to AsH_3 arise[2]. Slow addition of H_2O to powdered Na_3As at 25°C produces AsH_3 in 85–90% yield[3]:

$$Na_3As + 3 H_2O \rightarrow 3 NaOH + AsH_3 \qquad (b)$$

along with traces of As_2H_4. Hydrolysis of mixed $CaGe-Ca_3As_2$ alloys produces small amounts of GeH_3AsH_2, along with binary hydrides[2].

Reactions of As metal, As_4O_6, or As(III) halides with H_2O do not yield As—H bonds, in contrast to the P analogues (see §1.5.3.2.1).

Arsine forms upon reduction of As(III) species in acid[4]:

$$As^{3+} + 3 Zn + 3 H^+ \rightarrow 3 Zn^{2+} + AsH_3 \qquad (c)$$

Phenylarsonic acid reacts with Zn amalgam in H_2O to form $PhAsH_2$ in 83% yield[1]. Similarly, Cu-Zn alloy in acid reduces CF_3AsI_2 and $(CF_3)_2AsI$ to CF_3AsH_2 and $(CF_3)_2AsH$, respectively[5]:

$$2 (CF_3)_{3-n}AsI_n + n Zn \xrightarrow{\text{5 N HCl}} n ZnI_2 + (CF_3)_{3-n}AsH_n \qquad (d)$$

where $n = 1, 2$. Zinc dust in acid reduces Me_2AsOH:

$$Me_2AsOH + Zn + 2 H^+ \rightarrow Zn^{2+} + H_2O + Me_2AsH \qquad (e)$$

or $Me_2As(O)OH$ (cacodylic acid) to Me_2AsH nearly quantitatively[6,7]:

$$Me_2As(O)OH + 2 Zn + 4 HCl \rightarrow 2 ZnCl_2 + 2 H_2O + Me_2AsH \qquad (f)$$

Zinc amalgam reduction of $PhMeAsO_2H$ yields $PhMeAsH$. Reduction of the arsonic acid function, without attack on the substituted aryl ring, yields $p\text{-}MeC_6H_4AsH_2$:

$$p\text{-}MeOC_6H_4AsO(OH)_2 + 2 Zn + 4 HCl \rightarrow 2 ZnCl_2 + 3 H_2O$$
$$+ p\text{-}MeOC_6H_4AsH_2 \qquad (g)[8]$$

Electrolytic reduction of As(III) or As(V) species in solution at the cathode yields arsine[2,9]; e.g., AsH_3 is formed upon reduction of either As(III) or As(V) species in 0.5 N H_2SO_4. Reduction of As(V) is more cathodic than As(III).

(A.D. NORMAN)

1. F. A. Cotton, G. Wilkinson, *Advanced Inorganic Chemistry,* 4th ed., Wiley-Interscience, New York, 1980.
2. W. L. Jolly, A. D. Norman, *Prep. Inorg. React., 4,* 1 (1968).
3. J. E. Drake, C. Riddle, *Inorg. Synth., 13,* 14 (1972). A preferred synthesis of AsH_3.
4. G. O. Doak, L. D. Freedman, *Organometallic Compounds of Arsenic, Antimony and Bismuth,* Wiley-Interscience, New York, 1970.
5. H. J. Emeléus, R. N. Hazeldine, E. Walaschewski, *J. Chem. Soc.,* 1552 (1953).
6. R. D. Feltham, A. Kasenally, R. S. Nyholm , *J. Organomet. Chem., 7,* 285 (1967).
7. R. D. Feltham, W. Silverthorne, *Inorg. Synth., 10,* 159 (1967).
8. F. G. Mann, A. J. Wilkinson, *J. Chem. Soc.,* 3336 (1957).
9. V. P. Gladyshev. *Tr. Inst. Khim. Nauk Akad. Nauk, Kaz., SSR., 35,* 74 (1973); *Chem. Abstr.,* 80, 43,447 (1974).

1.5.3.3.2. in Other Protonic Solvents.

Alkali-metal arsenides M_3As in liq NH_3 react with NH_4Br to form[1,2] AsH_3:

$$M_3As + 3 NH_4Br \xrightarrow{NH_{3(l)}} 3 MBr + NH_3 + AsH_3 \qquad (a)$$

32   1.5. Formation of Bonds between Hydrogen and N, P, As, Sb, Bi
          1.5.3. by Protonation
          1.5.3.3. of Arsenic and Arsenic Compounds

where $M = Na, K$. The dipotassium salt, $[K]_2^+[\text{n-BuAsNH}]^{2-}$ is protonated by NH_4Cl in NH_3:

$$[K]_2^+[\text{n-BuAsNH}]^{2-} + 2\ NH_4Cl \xrightarrow{NH_{3(l)}} NH_3 + 2\ KCl + \text{n-BuAs(H)NH}_2 \quad \text{(b)}$$

The aminoarsine product, readily decomposes to NH_3 and a cyclopolyarsine[3].

Strong-acid protonation of t-alkyl- or arylarsines yields arsonium ions in solution[4]:

$$Ph_3As + 2\ HCl \xrightarrow{HCl_{(l)}} [HCl_2]^- + [Ph_3AsH]^+ \quad \text{(c)}$$

(A.D. NORMAN)

1. G. Brauer, ed., *Handbook of Preparative Inorganic Chemistry*, 2nd ed., Vol. 1, Academic Press, New York, 1963.
2. W. C. Johnson, A. Petchukas, *J. Am. Chem. Soc.*, *59*, 2068 (1937).
3. B. Ross, W. Marzi, W. Axmacher, *Chem. Ber.*, *113*, 292B (1980).
4. M. E. Peach, T. C. Waddington, in *Non-Aqueous Solvent Systems*, T. C. Waddington, ed., Academic Press, London, 1965, p. 83.

1.5.3.3.3. with Protonic Acids in Nonprotonic Solvents.

Protonation of metal arsenides provides high yields of AsH_3 and alkyl- or arylarsines[1-4]. These reactions, by substitution of D_2O for H_2O, are especially useful sources of deuterium-labeled products:

$$Na_3As + 3\ H_2O \rightarrow 3\ NaOH + AsH_3 \qquad \text{(a)[3]}$$

$$LiAl(AsH_2)_4 + 4\ H_2O \rightarrow LiAl(OH)_4 + 4\ AsH_3 \qquad \text{(b)[5]}$$

Hydrolysis of RR′AsM in dioxane at 25°C yields, e.g., the arsines[2] Me_2AsH, Et_2AsH, Ph_2AsH, $(C_6H_{11})_2AsH$, EtPhAsH and $(p\text{-MeC}_6H_4)_2AsH$:

$$MAsRR' + H_2O \xrightarrow{\text{dioxane}} MOH + RR'AsH \qquad \text{(c)}$$

where $M = Li, Na, K$; R and $R' =$ alkyl or aryl. Protonation of $KAsH_2$ by SiH_3AsH_2 yields AsH_3, in a proton exchange[6]:

$$KAsH_2 + SiH_3AsH_2 \rightarrow AsH_3 + SiH_3AsHK \qquad \text{(d)}$$

Protonic acid (HX) cleavage of arsenic–group IVB (Si, Ge, Sn) bonds in silyl-, germyl- or stannylarsines yields[4] AsH_3 or Me_2AsH:

$$R_3As + 3\ HX \rightarrow 3\ RX + AsH_3 \qquad \text{(e)}$$

where $R = SiH_3, GeH_3, Me_3Si$;

$$RAsR_2' + HX \rightarrow RX + R_2'AsH \qquad \text{(f)}$$

where $R = GeH_3, SiF_3, Me_3Si, Me_3Ge, MeSiH_2, MeGeH_2$; $R' = H$, alkyl. Water, H_2S, ROH or RSH also react to form arsines[3,4,6]; e.g., $(Me_3Si)_3As$ and H_2O:

$$2\ (Me_3Si)_3As + 3\ H_2O \rightarrow 3\ (Me_3Si)_2O + 2\ AsH_3 \qquad \text{(g)}$$

yield AsH_3, $Me_3SiAsMe_2$ and H_2O, or Me_2AsSiF_3 and H_2S yield Me_2AsH:

$$2\ RAsMe_2 + H_2Y \rightarrow R_2Y + 2\ Me_2AsH \qquad \text{(h)}$$

where $R = Me_3Si$, $Y = O$; $R = SiF_3$, $Y = S$. These cleavages are general except that Ge—P and Sn—P bonds are resistant to H_2O.

Hydrogen iodide reacts with CF_3AsI_2 or $(CF_3)_2AsI$ in the presence of Hg to form arsines. A deficiency of HI with CF_3AsI_2 yields the diarsine which is present both in meso- and d,l-forms[7]:

$$2\ CF_3AsI_2 + 2\ HI + 3\ Hg \rightarrow 3\ HgI_2 + CF_3AsHAsHCF_3 \tag{i}$$

With xs HI or HCl, CF_3AsX_2 and $(CF_3)_2AsX$ yield the respective arsines[4,8]:

$$(CF_3)_{3-n}AsX_n + n\ HX + \frac{n}{2}\ Hg \rightarrow \frac{n}{2}\ HgX_2 + (CF_3)_{3-n}AsH_n \tag{j}$$

where $X = Cl, I; n = 1, 2$.

Intramolecular proton transfer and C—As bond cleavage in the alkoxylmethylarsines, Ph_2AsCH_2OR, results in the equilibrium formation of Ph_2AsH:

$$Ph_2AsCH_2OR \rightleftharpoons Ph_2AsH + RCHO \tag{k}$$

where $R = Et, Ph$. At 25°C in benzene, the equilibrium constant for the reaction is[9] $K \approx 1$.

(A.D. NORMAN)

1. A. Tzschach, W. Voightlander, Z. Chem., 19, 393 (1979). A current review of the use of arsenides for arsine synthesis.
2. G. O. Doak, L. D. Freedman, Organometallic Compounds of Arsenic, Antimony and Bismuth, Wiley-Interscience, New York, 1971.
3. W. L. Jolly, A. D. Norman, Prep. Inorg. React., 4, 1 (1968).
4. J. E. Drake, C. Riddle, Q. Rev. Chem. Soc., 263 (1970).
5. A. E. Finholt, C. Helling, V. Imhof, L. Nielson, E. Jacobson, Inorg. Chem., 2, 504 (1963).
6. C. Glidwell, G. M. Sheldrick, J. Chem. Soc., A, 350 (1969).
7. R. C. Dobbie, P. D. Gosling, J. Chem. Soc. Chem. Commun., 585 (1975).
8. R. G. Cavell, R. C. Dobbie, J. Chem. Soc., A, 1308 (1967).
9. P. J. Busse, C.-P. Hrung, K. J. Irgolic, D. H. O'Brien, F. L. Kolar, J. Organomet. Chem., 185, 1 (1980).

1.5.3.4. of Antimony and Antimony Compounds.

Stibine forms from reaction of dil HCl with finely divided Mg-Sb alloy[1-3]:

$$Mg_3Sb_2 + 6\ H^+ \xrightarrow{HCl_{(aq)}} 3\ Mg^{2+} + 2\ SbH_3 \tag{a}$$

Yields are not over 10%.[3] Protonic acids react with alkali-metal stibnides, e.g.:

$$K_3Sb + 3\ H_2O \rightarrow 3\ KOH + SbH_3 \tag{b[1,2]}$$

$$R_2SbLi + H_2O \rightarrow LiOH + R_2SbH \tag{c[3]}$$

where $R = Me, Et, Ph$, etc.

Cleavage of the Si—Sb or Ge—Sb bonds of $(MH_3)_3Sb$ $(M = Si, Ge)$ with HCl results in SbH_3 in high yield[4]:

$$(H_3M)_3Sb + 3\ HCl \rightarrow 3\ H_3MCl + SbH_3 \tag{d}$$

The electrolytic reduction of Sb(III) or Sb(V) species in alkali or acid yields SbH_3 at the cathode[1,5]:

$$Sb^{3+} + 3\ H_2O + 6\ e^- \rightarrow 3\ [OH]^- + SbH_3 \tag{e}$$

Using a metallic Sb cathode, yields up to 15% of the total gas ($SbH_3 + H_2$) can be attained[5]. Active metal (e.g., Na or Zn amalgam) reduction of Sb(III) in H_2O produces SbH_3:

$$Sb^{3+} + 3 H^+ + 3 Zn \rightarrow 3 Zn^{2+} + SbH_3 \qquad (f)$$

(A.D. NORMAN)

1. W. L. Jolly, A. D. Norman, *Prep. Inorg. React.*, **4**, 1 (1968).
2. G. Brauer, ed., *Handbook of Preparative Inorganic Chemistry*, 2nd ed., Vol. 1, Academic Press, New York, 1963.
3. G. O. Doak, L. D. Freedman, *Organonmetallic Compounds of Arsenic, Antimony, and Bismuth*, Wiley-Interscience, New York, 1970.
4. E. A. V. Ebsworth, D. W. H. Rankin, G. M. Sheldrick, *J. Chem. Soc., A*, 2828 (1968).
5. V. P. Gladyshev. *Tr. Inst. Khim. Nauk Akad. Nauk, Kaz. SSR*, **35**, 74 (1973); *Chem. Abstr.*, **80**, 43,447 (1974).

1.5.3.5. of Bismuth and Bismuth Compounds.

Bismuthine forms in low yields when dil HCl is added to a Mg-Bi alloy of approximate composition[1,2] Mg_3Bi_2:

$$Mg_3Bi_2 + 6 H^+ \xrightarrow{HCl_{(aq)}} 3 Mg^{2+} + 2 BiH_3 \qquad (a)$$

Water reacts with $(Me_3Ge)_3Bi$ at 25°C without formation[3] of BiH_3[3], because of the high thermal instability of the latter.

Electrolytic reduction of Bi(III) in H_2O yields BiH_3 at the cathode[1]:

$$Bi^{3+} + 3 H^+ + 6 e^- \rightarrow BiH_3 \qquad (b)$$

Similarly, Na or Zn amalgams in 1 N H_2SO_4 or $HClO_4$ containing complexing agents such as thiourea yield BiH_3 at the electrode surface[4]:

$$Bi^{3+} + 3 Zn + 3 H^+ \rightarrow 3 Zn^{2+} + BiH_3 \qquad (c)$$

(A.D. NORMAN)

1. W. L. Jolly, A. D. Norman, *Prep. Inorg. React.*, **4**, 1 (1968).
2. F. E. Saalfeld, H. J. Svec, *Inorg. Chem.*, **2**, 46 (1963).
3. I. Schuman-Rudisch, *Z. Naturforsch., Teil B*, **22**, 1081 (1967).
4. A. Barikov, V. P. Gladyshev, *Sov. Electrochem. (Engl. Transl.)*, **8**, 795 (1972); *Chem. Abstr.*, **77**, 42,368 (1972).

1.5.4. by Reaction of Hydrides

1.5.4.1. with Compounds of Nitrogen

1.5.4.1.1. Involving Ionic Hydrides.

Cesium and Rb hydrides react with anhyd N_2 on heating to form metal-amide salts[1,2]:

$$2 RbH + 2 N_2 \rightarrow RbN_3 + RbNH_2 \qquad (a)$$

$$4 CsH + N_2 \rightarrow 2 Cs + 2 CsNH_2 \qquad (b)$$

The other alkali-metal hydrides do not react similarly; e.g., LiH with N_2 yields Li_3N

1.5. Formation of Bonds between Hydrogen and N, P, As, Sb, Bi   35
1.5.4. by Reaction of Hydrides
1.5.4.1. with Compounds of Nitrogen

and H_2. Calcium hydride in alcohol or acetic acid, with $PdCl_2$ or $PtCl_2$ catalysts, reduces $PhNO_2$ to $PhNH_2$.

(A.D. NORMAN)

1. D. T. Hurd, *Chemistry of the Hydrides,* Wiley, New York, 1952.
2. E. Wiberg, E. Amberger, *Hydrides of the Elements of Main Groups I–IV,* Elsevier, Amsterdam, 1971.

1.5.4.1.2. Involving Covalent Hydrides.

Boranes, alanes, organosilanes and organostannanes are used for reduction of organic nitrogen compounds to compounds containing N—H bonds[1-4]; however, few inorganic systems form by such reactions. The nitrogen chloride $C_2F_5NCl_2$ reacts with Me_3SiH above $-25°C$ to form the fluoroalkylimine[5]:

$$C_2F_5NCl_2 + 2 Me_3SiH \rightarrow HF + 2 Me_3SiCl + CF_3CFNH \tag{a}$$

Diborane and N_2F_4 react at 150°C, yielding a mixture[6] of NH_3BH_3, BF_3, B_5H_9 and H_2.

Transition-metal hydrides react with nitrogen compounds to form N—H bond-containing organometallic products[1]. The $[HFe_3(CO)_{11}]^-$ cluster anion reacts with nitriles to form a coordinated RCNH species, along with the RCHN-coordinated isomer[7]:

$$Me_4N[HFe_3(CO)_{11}] + RC\equiv N \xrightarrow[-2\,CO]{} [Fe_3(RCNH)(CO)_9][Me_4N] \tag{b}$$

where R = Me, Ph. Similar products form in $Re_2(CO)_6(Ph_2PCH_2PPh_2)H_2$—MeCN reactions[8]. Imines react[9] with Os:

$$H_2Os_3(CO)_{10} + PhCH{=}NMe \xrightarrow[-CO]{} Os_3(CO)_9(\mu\text{-}H_2)(\mu\text{-}NMeCH_2C_6H_5) \tag{c}$$

and Pt[10] hydrides:

$$PtHCl(PPh_3)_2 + Me_3SnN{=}C(CF_3)_2 \rightarrow Me_3SnCl + (PPh_3)_2PtNHC(CF_3)_2 \tag{d}$$

Diazo compounds and diazonium salts[5] react with Zr:

$$Ph_2CN_2 + h^5\text{-}CpZr(H)Cl \rightarrow h^5\text{-}CpZr(Cl)NHNCPh_2 \tag{e}$$

and W hydrides:

$$h^5\text{-}CpWH_2 + RN_2[PF_6] \xrightarrow{-20°C} [h^5\text{-}CpW(H_2NNR)][PF_6] \tag{f}$$

where R = Ph, p-FC_6H_4, p-MeC_6H_4. Azide reductions by $PtHX(PR_3)_2$ produce amido complexes in high yield[10]:

$$PtXH(PR_3)_2 + R'N_3 \xrightarrow[-N_2]{} PtX(NHR')(PR_3)_2 \tag{g}$$

where R = Et, Ph; X = Cl, N_3, CN; R' = Ph, PhCO.

Nitrogen atoms or radicals form N—H bonds. Nitrogen (^4S) atoms, from N_2 dissociation in an electrical discharge, react[11] with alkanes to form NH_3:

$$\cdot\overset{\cdot}{N}\cdot \xrightarrow{CH_4} \cdot\overset{\cdot}{N}H \xrightarrow{CH_4} \cdot NH_2 \xrightarrow{CH_4} NH_3 \tag{h}$$

and HCN. Nitrogen radicals, $R_2N\cdot$ or $R\overset{\cdot}{N}\cdot$ (triplet) react similarly forming N—H bonds upon H-atom abstraction:

36   1.5. Formation of Bonds between Hydrogen and N, P, As, Sb, Bi
        1.5.4. by Reaction of Hydrides
        1.5.4.1. with Compounds of Nitrogen

$$R_2N\cdot \xrightarrow{R-H} R_2NH \tag{i}$$

$$R\overset{\cdot}{N}\cdot \xrightarrow{R-H} RNH\cdot \xrightarrow{R-H} RNH_2 \tag{j}$$

Radicals from N_2F_4 dissociation react with thiols to form[12] HNF_2:

$$2\ NF_2 + 2\ RSH \rightarrow (RS)_2 + 2\ HNF_2 \tag{k}$$

(A.D. NORMAN)

1. A. Hajós, *Complex Hydrides,* Elsevier, Amsterdam, 1979. A concisely written, excellent review.
2. E. Wiberg, E. Amberger, *Hydrides of the Elements of Main Groups I–IV,* Elsevier, Amsterdam, 1971.
3. H. O. House, *Modern Synthetic Reactions,* 2nd ed., Benjamin, Menlo Park, CA, 1972.
4. H. C. Brown, *Boranes in Organic Chemistry,* Cornell Univ. Press, Ithaca, NY, 1972. A readable, authoritative review.
5. M. F. A. Dove, D. B. Sowerby, *Coord. Chem. Rev., 40,* 261 (1982).
6. L. H. Long, *Adv. Inorg. Chem. Radiochem., 16,* 201 (1974).
7. M. A. Andrews, H. D. Kaesz, *J. Am. Chem. Soc., 101,* 7238 (1979).
8. M. J. Mays, D. W. Post, P. R. Raithby, *J. Chem. Soc., Chem. Commun.,* 171 (1980).
9. R. D. Adams, J. P. Selegne, *Inorg. Chem., 19,* (1980).
10. D. M. Roundhill, *Adv. Organomet. Chem., 13,* 273 (1975).
11. J. I. G. Cadagan, *Q. Rev., Chem. Soc., 16,* 208 (1962).
12. M. Muraki, T. Mukaiyama, *Chem. Lett.,* 875 (1975).

1.5.4.1.3. Involving Exchange–Cleavage.

Reactions of amido–group IVB element (Si, Ge, Sn) bonds with group IVB hydrides result in bond cleavage and N—H bond formation[1-3], e.g.:

$$Ph_3EH + R_3SnNEt_2 \rightarrow Ph_3ESnR_3 + Et_2NH \tag{a}$$

where E = Sn, Ge; R = Et, Ph;

$$Me_3SnH + Me_3PbNEt_2 \rightarrow Me_3SnPbMe_3 + Et_2NH \tag{b}$$

Tris(diethylamino)ethylstannane reacts rapidly with Ph_3GeH cleaving two Sn–N bonds:

$$EtSn(NEt_2)_3 + 2\ Ph_3GeH \rightarrow EtSn(GePh_3)_2NEt_2 + 2\ Et_2NH \tag{c}$$

but slowly to cleave the third[2]. Cleavage of the Sn—N bond of a stannylamide:

$$Et_3SnN(Ph)COH + Ph_2SnH_2 \rightarrow Et_3SnSnPh_2H + PhNHCOH \tag{d}$$

yields the parent phenylformamide[4]. Similar cleavage–exchange reactions involving P—H or As—H and P—N[5] or Ge—N[6] bonds:

$$Me_2PNMe_2 + HPMe_2 \rightarrow Me_2PPMe_2 + Me_2NH \tag{e}$$

$$2\ Me_3GeNMe_2 + MeAsH_2 \rightarrow (Me_3Ge)_2AsMe + 2\ Me_2NH \tag{f}$$

yield Me_2NH.

(A.D. NORMAN)

1. F. Glockling, *The Chemistry of Germanium,* Academic Press, New York, 1969.
2. M. Lesbre, P. Mazerolles, J. Satgé, *The Organic Compounds of Germanium,* Interscience, London, 1971.

1.5. Formation of Bonds between Hydrogen and N, P, As, Sb, Bi   37
1.5.4. by Reaction of Hydrides
1.5.4.1. with Compounds of Nitrogen

3. J. D. Kennedy, W. McFarlane, B. Wrackmeyer, *Inorg. Chem.*, *15*, 1299 (1976).
4. J. G. A. Luijten, F. Rykens, G. J. M. Van der Kerk, *Adv. Organomet. Chem.*, *3*, 397 (1955).
5. E. Fluck, in *Topics in Phosphorus Chemistry*, Vol. 4, M. Grayson, E. J. Griffith, eds., Interscience, New York, 1967, p. 291. An excellent, extensive review.
6. J. W. Anderson, J. E. Drake, *J. Chem. Soc., Chem. Commun.*, 1372 (1971).

1.5.4.1.4. Involving Redistribution–Disproportionation.

Redistribution produces amines, although the reactions are not often used for amine synthesis. Gem-bis- and tris(amido)P(III) and (V) compounds undergo redistribution with amine elimination, e.g.[1]:

$$2\ P(NHPh)_3 \rightarrow [(PhNH)_2P]_2NPh + PhNH_2 \qquad (a)^2$$

$$2\ (PhNH)_2PNMe_2 \rightarrow [PhNPN(H)Ph]_2 + 2\ Me_2NH \qquad (b)^1$$

$$2\ (NH_2)_3PO \rightarrow [NH_2P(O)]_2NH + NH_3 \qquad (c)^1$$

$$2\ PhP(S)(NHEt)_2 \rightarrow [(PhNH)P(S)NPh]_2 + 2\ PhNH_2 \qquad (d)^1$$

Hydrazine forms from the hydrazinothiophosphinite[3] at 80°C:

$$Ph_2P(S)NHNH_2 \rightarrow Ph_2P(S)NHNHP(S)Ph_2 + N_2H_4 \qquad (e)$$

Amidosilanes,[4] germanes,[5] and stannanes[6] react similarly:

$$2\ R_3SnNHEt \rightarrow (R_3Sn)_2NR + RNH_2 \qquad (f)$$

where R = Me, Et, n-Bu. Except for NaNH$_2$, alkali or alkaline-earth amides eliminate NH$_3$ on thermolysis:

$$3\ MNH_2 \rightarrow M_3N + 2\ NH_3 \qquad (g)$$

$$3\ M(NH_2)_2 \rightarrow M_3N_2 + 4\ NH_3 \qquad (h)$$

LiNH$_2$ eliminates NH$_3$ forming[7,8] Li$_2$NH:

$$2\ LiNH_2 \rightarrow Li_2NH + NH_3 \qquad (i)$$

Disproportionation of NH$_3$ in Ag-zeolite (Ag$_{12}$-zeolite A) forms the higher hydrides[8,9], N$_3$H$_3$, N$_3$H$_5$, N$_4$H$_4$. Normally, disproportionation occurs in the reverse direction. Tetrazene and N$_2$H$_4$ disproportionate[7] to NH$_3$ and H$_2$, faster in the presence of Ni or Pt catalysts:

$$N_4H_6 \rightarrow 2\ NH_3 + N_2 \qquad (j)$$

$$3\ N_2H_4 \rightarrow N_2 + 4\ NH_3 \qquad (k)$$

Diazine disproportionates[8] above −180°C:

$$3\ N_2H_2 \rightarrow 2\ N_2 + H_2 + N_2H_4 \qquad (l)$$

Pyrolysis of the tetraazo ring produces a polymer containing N—H bonds:

$$2\ \begin{array}{c} HC \overline{\qquad} NSiMe_3 \\ \| \qquad\quad | \\ N \diagdown \quad \diagup N \\ N \end{array} \xrightarrow{-N_2} (Me_3SiN)_2C + \frac{1}{x}(H_2NCN)_x \qquad (m)^{10}$$

hydrazoic acid[11] or NH$_2$OH [12], on photolysis:

$$6 \ NH_2OH \xrightarrow{h\nu} 2 \ N_2 + 6 \ H_2O + 2 \ NH_3 \qquad (n)^{13}$$

or HN_3 upon thermolysis[13]:

$$3 \ HN_3 \rightarrow 4 \ N_2 + NH_3 \qquad (o)$$

yield NH_3 and N_2. Heating $NaNH_2$ with $NaNO_3$:

$$3 \ NaNH_2 + NaNO_3 \rightarrow NaN_3 + 3 \ NaOH + NH_3 \qquad (p)$$

or $N_2O(g)$ yields[8] NH_3:

$$2 \ NaNH_2 + N_2O \rightarrow NaN_3 + NaOH + NH_3 \qquad (q)$$

Disproportionation with CO_2 elimination can be a ready source of amines:

$$6 \ (NH_2)_2CO \xrightarrow{\Delta} C_3N_3(NH_2)_3 + 3 \ CO_2 + 6 \ NH_3 \qquad (r)^8$$

$$MeN(CHO)_2 + COF_2 \xrightarrow{\Delta} CO + CO_2 + MeNHCHF_2 \qquad (s)^{14}$$

$$NH_2CO_2NH_4 \rightarrow CO_2 + 2 \ NH_3 \qquad (t)^8$$

$$O_2S(NHCO_2H)_2 \rightarrow 2 \ CO_2 + O_2S(NH_2)_2 \qquad (u)^{15}$$

$$NH_2NHCO_2H \rightarrow CO_2 + N_2H_4 \qquad (v)^{16}$$

(A.D. NORMAN)

1. E. Fluck, in *Topics in Phosphorus Chemistry*, Vol. 4, M. Grayson, E. J. Griffith, eds., Wiley-Interscience, New York, 1967, p. 291.
2. A. Tarassoli, R. C. Haltiwanger, A. D. Norman, *Inorg. Chem., 21*, 2684 (1982).
3. H.-J. Jahns, L. Thielemann, *Z. Anorg. Allg. Chem., 397*, 47 (1973).
4. E. A. V. Ebsworth, *Volatile Silicon Compounds*, Pergamon Press, New York, 1963.
5. F. Glockling, *The Chemistry of Germanium*, Academic Press, London, 1969.
6. K. Jones, M. F. Lappert, *J. Chem. Soc.*, 1944 (1965).
7. W. L. Jolly, *The Inorganic Chemistry of Nitrogen*, Benjamin, New York, 1964. Excellent, brief coverage of basic inorganic nitrogen chemistry.
8. F. A. Cotton, G. Wilkinson, *Advanced Inorganic Chemistry*, 4th ed., Wiley-Interscience, New York, 1980.
9. Y. Kim, J. W. Gilje, K. Seff, *J. Am. Chem. Soc., 99*, 7075 (1977).
10. L. Birkofer, A. Ritter, P. Richter, *Chem. Ber., 96*, 2750 (1963).
11. B. A. Thrush, *Proc. R. Soc. London, Ser. A, 235*, 143 (1956).
12. M. Simic, E. Hayon, *J. Am. Chem. Soc., 93*, 5982 (1971).
13. E. L. Muetterties, W. J. Evans, J. C. Sauer, *J. Chem. Soc., Chem. Commun.*, 939 (1974).
14. E. Allenstein, G. Schrempf, *Z. Anorg. Allg. Chem., 474*, 7 (1981).
15. M. Becke-Goehring, E. Fluck, in *Developments in the Inorganic Chemistry of Nitrogen*, C. B. Colburn, ed., Elsevier, Amsterdam, 1966, p. 150.
16. E. Nachbaur, G. Leiseder, *Monatsh. Chem., 102*, 1719 (1971).

1.5.4.2. with Compounds of Phosphorus

1.5.4.2.1. Involving Ionic Hydrides.

Alkali or alkaline-earth metal hydride reductions of P compounds are less effective than those using complex hydrides, e.g., $LiAlH_4^{1,2}$. Phosphorus tribromide reacts with LiH in Et_2O at 0°C to form a phosphorus subhydride quantitatively[3]:

$$PBr_3 + 3 \ LiH \rightarrow 3 \ LiBr + \tfrac{1}{n} \ (PH)_n + H_2 \qquad (a)$$

1.5. Formation of Bonds between Hydrogen and N, P, As, Sb, Bi   39
1.5.4. by Reaction of Hydrides
1.5.4.2. with Compounds of Phosphorus

although with PCl_3 low yields of PH_3 are obtained[2]. Similarly, $P(O)Br_3$ or $P(O)Cl_3$ react[2] with LiH, yielding polymeric phosphorus subhydride:

$$P(O)X_3 + 4 \, LiH \rightarrow 3 \, LiX + LiOH + H_2 + \tfrac{1}{n} \, (PH)_n \qquad (b)$$

Phosphine forms when PCl_3 vapor in N_2 carrier gas is passed through a column packed with LiH in sand[4]:

$$PCl_3 + 3 \, LiH \rightarrow 3 \, LiCl + PH_3 \qquad (c)$$

Cesium hydride, when heated directly with phosphorus gives[1] $CsPH_2$:

$$2 \, CsH + P_{red} \rightarrow CsPH_2 + Cs \qquad (d)$$

Less reactive, alkali-metal hydrides[1] do not react similarly.

Primary and secondary phosphines are produced in low yield upon reaction of LiH with $MePCl_2$, and Me_2PCl and $PhPCl_2$[2].

(A.D. NORMAN)

1. D. T. Hurd, *Chemistry of the Hydrides*, Wiley, New York, 1952.
2. E. Wiberg, E. Amberger, *Hydrides of the Elements of Main Groups I–IV*, Elsevier, Amsterdam, 1971.
3. W. L. Jolly, A. D. Norman, *Prep. Inorg. React., 4*, 1 (1968).
4. E. Fluck, *Top. Curr. Chem., 35*, 1 (1973). Excellent discussion of PH_3 syntheses.

1.5.4.2.2. Involving Covalent Hydrides.

Organosilanes and $HSiCl_3$ reduce selectively halophosphines, phosphonous acids and phosphonic acids and esters[1-3]. Unlike ionic metal hydrides, organosilanes do not attack C—P bonds. The organosilane reductions usually can be performed without solvents. In addition, they occur with retention of configuration[4] at the phosphorus.

Silane reductions of phosphorus compounds are listed in Table 1. Mono- or dichlorophosphines and phosphoryl compounds react with Ph_2SiH_2, e.g.:

$$R_{3-n}PCl_n + \tfrac{n}{2} \, Ph_2SiH_2 \rightarrow \tfrac{n}{2} \, Ph_2SiCl_2 + R_{3-n}PH_n \qquad (a)$$

TABLE 1. REPRESENTATIVE SILANE REDUCTIONS

Phosphorus compound	Silane[a]	Phosphine	Yield[a]	Ref.
$PhPCl_2$	Ph_2SiH_2,$HSiCl_3$	$PhPH_2$	82,75	2
$n\text{-}BuP(O)(OEt)_2$	PMHS[b]	$n\text{-}BuPH_2$	90	2
$PhP(O)(OEt)_2$	Ph_2SiH_2	$PhPH_2$	87	2
$n\text{-}Bu_2P(O)Cl$	Ph_2SiH_2,$HSiCl_3$	$n\text{-}Bu_2PH$	95,63	5,6
$n\text{-}BuPhP(O)Cl$	$HSiCl_3$	$n\text{-}BuPhPH$	60	6
$(m\text{-}MeC_6H_4)_2P(O)Cl$	$HSiCl_3$	$m\text{-}MeC_6H_4PH_2$	31	6
Ph_2PCl	Ph_2SiH_2,$HSiCl_3$	Ph_2PH	59,75	5

[a] Yields are for the respective silanes, as listed.
[b] PMHS = polymethylhydridosilacone, $Me_3SiO[SiH(Me)O]_xOSiMe_3$.

40

1.5.4. by Reaction of Hydrides
1.5.4.2. with Compounds of Phosphorus
1.5.4.2.2. Involving Covalent Hydrides.

$$2 RP(O)X_2 + 2 Ph_2SiH_2 \rightarrow -(Ph_2SiO)_n^- + Ph_2SiX_2 + 2 RPH_2 \qquad (b)$$

$$2 R_2P(O)X + Ph_2SiH_2 \rightarrow -(Ph_2SiO)_n^- + Ph_2SiX_2 + 2 R_2PH \qquad (c)$$

where n = 1–3; X = OH, Cl, alkoxy; R = alkyl, aryl. Reactions proceed smoothly in neat silane at 200–275°C. Similarly, $HSiCl_3$ reductions are performed in refluxing benzene:

$$R_{3-n}PCl_n + n \, HSiCl_3 \rightarrow n \, SiCl_4 + R_{3-n}PH_n \qquad (d)$$

$$R_{3-n}P(O)X_n + (n+1) \, HSiCl_3 \rightarrow n \, SiCl_3X + R_{3-n}PH_n + -(SiCl_2O)_n- + HCl \quad (e)$$

where n = 1, 2; X = Cl, OH, alkoxy; R = alkyl, aryl. These reactions are run in the presence of a tertiary amine, (e.g., Et_3N, pyridine) which scavenges HCl from the reaction medium. Phosphinic acid chlorides are reduced by arylsilanes more efficiently than are the free acids or their esters. Organosilane reduction of cyclic phosphorus moieties occurs effectively with minimal ring cleavage[3]. Phosphorinane is obtained in 60% yield by reaction of Ph_2SiH_2 at 200°C with the cyclic phosphinic acid[7]:

$$(f)$$

Bicyclic and tricyclic phosphines[3], e.g., $C_6H_4(CH_2)_2PH$, are formed equally well.

Tri-n-butylstannane reacts smoothly with $(MeO)_2PCl$ at 25°C forming $(MeO)_2PH$ along with traces[8] of $P(OMe)_3$ and PH_3:

$$(MeO)_2PCl + n\text{-}Bu_3SnH \rightarrow n\text{-}Bu_3SnCl + (MeO)_2PH \qquad (g)$$

but the analogous reaction with $MeOPCl_2$ does not occur. Tetramethyldistanane reacts with P_4 at 0°C in the dark to form[9] PH_3:

$$3 \, Me_4Sn_2H_2 + P_4 \xrightarrow{Et_2O} (Me_2Sn)_6P_2 + 2 \, PH_3 \qquad (h)$$

Transition-metal hydride reactions can result in P—H bond formation. White phosphorus and $(h^5\text{-}C_5H_5)_2MoH_2$ yield the metal-coordinated[10] P_2H_2:

$$2 \, (h^5\text{-}C_5H_5)_2MoH_2 + P_4 \rightarrow 2 \, (h^5\text{-}C_5H_5)_2MoP_2H_2 \qquad (i)$$

Phosphine results from the Pt hydride complex cleavage of $(SiH_3)_3P$.

$$(SiH_3)_3P + 3 \, trans\text{-}Pt(H)I(PEt_3)_2 \rightarrow 3 \, trans\text{-}Pt(SiH_3)I(PEt_3)_2 + PH_3 \qquad (j)^{[11]}$$

Elemental phosphorus with covalent hydrides can be converted to phosphines. White phosphorus when passed through a silent electric discharge in the presence of C_2H_4, C_3H_8, CH_4, NH_3 or N_2H_4, yields traces[12] of PH_3. Reaction with $LiPH_2$ produces PH_3 in 95% yield[13]:

$$3 \, P_4 + 6 \, LiPH_2 \rightarrow Li_3P_7 + 4 \, PH_3 \qquad (k)$$

and reaction with alkenes in the presence of O_2 forms β-hydroxyphosphinic acids[14]:

$$RCH=CH_2 + P_4 \xrightarrow[H_2O]{O_2} H_3PO_4 + RCH(OH)CH_2P(H)(O)OH \qquad (l)$$

1.5. Formation of Bonds between Hydrogen and N, P, As, Sb, Bi   41
1.5.4. by Reaction of Hydrides
1.5.4.2. with Compounds of Phosphorus

Passage of PH_3, PH_3–SiH_4 or PH_3–GeH_4 through a silent electric discharge or hot-tube pyrolysis reactor results in complex product mixtures, which involve formation of new P—H bonds. Major products from PH_3, PH_3–SiH_4, or PH_3–GeH_4 reactions include P_2H_4, SiH_3PH_2 and GeH_3PH_2, in addition to Si_2H_6 and Ge_2H_6. Lesser products include[15,16] $Si_2H_5PH_2$, $(SiH_3)_2PH$, Si_3H_8, Si_4H_{10}, $(GeH_3)_2PH$, and Ge_3H_8.

(A.D. NORMAN)

1. E. Wiberg, E. Amberger, *Hydrides of the Elements of Main Groups I-IV*, Elsevier, Amsterdam, 1971.
2. A. Hajós, *Complex Hydrides*, Elsevier, Amsterdam, 1979.
3. S. D Venkataramu, G. D. MacDonell, W. R. Purdum, M. El-Deek, K. D. Berlin, *Coord. Chem. Rev.*, 77, 121 (1977).
4. K. Marsi, *J. Org. Chem.*, 39, 265 (1974).
5. H. Fritzsche, U. Hasserodt, F. Korte, *Chem. Ber.*, 98, 1681 (1965).
6. H. Fritzsche, U. Hasserodt, F. Korte, U.S. Pat. 3,261,871 (1966); *Chem. Abstr.*, 65, 18,619 (1966).
7. J. B. Lambert, W. L. Oliver, Jr., *Tetrahedron*, 27, 4245 (1971).
8. L. Centofanti, *Inorg. Chem.*, 12, 1131 (1973).
9. M. Dräger, B. Mathiasch, *Angew. Chem., Int. Ed. Engl.*, 20, 1029 (1981).
10. J. C. Green, M. L. H. Green, G. E. Morris, *J. Chem. Soc., Chem. Commun.*, 212 (1974).
11. E. A. V. Ebsworth, J. M. Edward, D. W. H. Rankin, *J. Chem. Soc., Dalton Trans.*, 1673 (1976).
12. E. R. Zabolatny, H. Gesser, *J. Am. Chem. Soc.*, 81, 6091 (1959).
13. M. Baudler, *Angew. Chem., Int. Ed. Engl.*, 21, 492 (1982).
14. A. W. Frank, *Chem. Rev.*, 61, 389 (1961).
15. J. E. Drake, C. Riddle, *Q. Rev. Chem. Soc.*, 263 (1970).
16. W. L. Jolly, A. D. Norman, *Prep. Inorg. React.* 4, 1 (1968).

1.5.4.2.3. Involving Exchange–Cleavage.

Phosphorus–hydrogen bonds form in intermolecular exchange. Trimethylstannane or Me_3SiH with PF_5 reacts:

$$Me_3MH + PF_5 \rightarrow Me_3SnF + HPF_4 \qquad (a)^{1,2}$$

where M = Si, Sn;

$$2\ Me_3SnH + PF_5 \rightarrow 2\ Me_3SnF + H_2PF_4 \qquad (b)^1$$

Reactions at 25°C between Me_3SiPMe_2 or $Me_3SiAsMe_2$ and $(CF_3)_2PH$ yield[3] Me_2PH:

$$Me_3SiEMe_2 + (CF_3)_2PH \xrightarrow{Me_4Si} Me_3SiE(CF_3)_2 + Me_2PH \qquad (c)$$

where E = P,As.

(A.D. NORMAN)

1. P. M. Treichel, R. A. Goodrich, S. B. Pierce, *J. Am. Chem. Soc.*, 89, 2017 (1967).
2. A. H. Cowley, R. W. Braun, *Inorg. Chem.*, 12, 491 (1973).
3. J. E. Byrne, C. R. Russ, *J. Inorg. Nucl. Chem.*, 36, 35 (1974).

1.5.4.2.4. Involving Redistribution–Disproportionation.

Disproportionation of P—P bond-containing phosphines can produce new P—H bonds[1,2]. Gaseous P_2H_4 in a hot–cold reactor[3] or neat liq P_2H_4 at[4] 30–40°C reacts to form PH_3, P_3H_5:

$$2\ P_2H_4 \rightarrow P_3H_5 + PH_3 \qquad (a)$$

42   1.5.4. by Reaction of Hydrides
            1.5.4.2. with Compounds of Phosphorus
            1.5.4.2.4. Involving Redistribution–Disproportionation.

and the higher mol wt phosphorus subhydrides P_nH_{n+2}, P_nH_n, P_nH_{n-2} etc.:

$$P_2H_4 \xrightarrow{\Delta} PH_3 + \text{ higher hydrides} \qquad (b)$$

Diphosphine and n-BuLi react in THF to form[4] PH_3:

$$9\ P_2H_4 + 3\ \text{n-BuLi} \xrightarrow[-20°C]{} Li_3P_7 + 11\ PH_3 + 3\ BuH \qquad (c)$$

Phenylphosphine and $(PhP)_5$ in solution exist in equilibrium with[5] $(PhPH)_2$:

$$5\ (PhPH)_2 \rightleftharpoons (PhP)_5 + 5\ PhPH_2 \qquad (d)$$

Mono- and bis(borane)diphosphines at 174°C and 130°C, respectively, undergo thermolysis to form Me_2PHBH_3:

$$H_3BP(Me)_2P(Me)_2BH_3 \rightarrow \tfrac{1}{x}\ (BH_2PMe_2)_x + Me_2PHBH_3 \qquad (e)$$

and trimeric or tetrameric $(BH_2PMe_2)_x$ polymers[9]. The triphosphine $CF_3P(PH_2)_2$ disproportionates stepwise, yielding di- and monophosphine products[7]:

$$CF_3P(PH_2)_2 \xrightarrow{25°C} \tfrac{1}{x}\ (PH)_x + CF_3PHPH_2 \qquad (f)$$

$$CF_3PHPH_2 \xrightarrow{25°C} \tfrac{1}{x}\ (PH)_x + CF_3PH_2 \qquad (g)$$

β-Hydrogen elimination from the product of Me_3CLi reaction with $[(Me_3Si)_2P]_2PCl$ yields a triphosphine:

$$[(Me_3Si)_2P]_2PCl + Me_3CLi \rightarrow [(Me_3Si)_2P]_2PH + LiCl + Me_2C{=}CH_2 \qquad (h)[8]$$

The cyclotriphosphine $(t\text{-}BuP)_2PH$ disproportionates to mono-, di- and tetraphosphines[4]. Thermolysis of SiH_3PH_2 at 400°C yields PH_3 in addition to SiH_4, H_2 and solid polymeric subhydrides[2]. Under controlled conditions at 210°C, PH_3, in addition to Si_2H_6, SiH_3PH_2 and Si_3PH_9, forms in the decomposition of $Si_2H_5PH_2$. Pyrolysis of SiH_3PH_2 in the presence of Me_2SiD_2 yields $Me_2Si_2H_2D_2$ and $Me_2SiDP(H)D$:

$$SiH_3PH_2 \xrightarrow{\Delta,\ Me_2SiD_2} Me_2Si_2H_2D_2 + Me_2SiDP(H)D \qquad (i)$$

in a 2.6:1 ratio, the latter product containing a P—D bond formed through P—H insertion in a Si—D bond of[9] Me_2SiD_2. In the presence of LiOEt methylsilylphosphine redistributes spontaneously[10]:

$$2\ MeSiH_2PH_2 \rightarrow (MeSiH_2)PH + PH_3 \qquad (j)$$

Lithiated silylphosphine[10] equilibrates with $LiPH_2$ and $(SiH_3)_2PLi$:

$$2\ SiH_3PHLi \rightleftharpoons (SiH_3)_2PLi + LiPH_2 \qquad (k)$$

Above $-30°C$ lithiocyclopentaphosphine disproportionates:

$$2\ LiH_4P_5 \rightarrow Li_2HP_7 + P_2H_4 + PH_3 \qquad (l)[10]$$

At 25°C in the presence of H_2O catalyst, germylphosphine and $(GeH_3)_3P$ yield PH_3 and $(GeH_2)_x$:

$$GeH_3PH_2 \rightarrow \tfrac{1}{x}\ (GeH_2)_x + PH_3 \qquad (m)$$

$$(GeH_3)_3P \rightarrow \frac{3}{x}(GeH_2)_x + PH_3 \qquad (n)$$

The $(GeH_2)_x$, being unstable, decomposes[1,2] to $(GeH)_x$, GeH_4, traces of higher germanes and H_2. Silylphosphine[1,2] or $Si_2H_5PH_2$[2] react below $-78°C$ when coordinated to BF_3. Phosphine is produced quantitatively.

Alkyl(phosphino)germanes with H_3PO_3 catalyst present eliminate PH_3 upon redistribution as more highly condensed germylphosphines are formed[11].

$$6\ R_2Ge(PH_2)_2 \rightarrow (R_2Ge)_6P_4 + 8\ PH_3 \qquad (o)$$

where R = Me, Et.

(A.D. NORMAN)

1. J. E. Drake, C. Riddle, *Q. Rev. Chem. Soc.*, 263 (1970).
2. W. L. Jolly, A. D. Norman, *Prep. Inorg. React.*, 4, 1 (1968).
3. T. P. Fehlner, *J. Am. Chem. Soc.*, 88, 2613 (1966).
4. M. Baudler, *Angew. Chem., Int. Ed. Engl.*, 21, 492 (1982). An excellent review of the preparation and properties of the higher phosphorus hydrides.
5. J. P. Albrand, D. Gagnaire, *J. Am. Chem. Soc.*, 94, 8630 (1972).
6. A. B. Burg, *J. Am. Chem. Soc.*, 83, 2226 (1961).
7. R. Demuth, J. Grobe, *Z. Naturforsch., Teil B*, 28, 219 (1973).
8. G. Fritz, J. Härer, *Z. Anorg. Allg. Chem.*, 481, 185 (1981).
9. L. E. Elliot, P. Estacio, M. A. Ring, *Inorg. Chem.*, 12, 2193 (1973).
10. G. Fritz, H. Schäeffer, W. Hölderich, *Z. Anorg. Allg. Chem.*, 407, 266 (1974).
11. A. R. Dahl, A. D. Norman, H. Shenav, R. Schaeffer, *J. Am. Chem. Soc.*, 97, 6364 (1975).

1.5.4.3. with Compounds of Arsenic.

Several hydride reactions lead to As—H bonds; however, these are not preferred routes[1]. Ethylstibine and $MeAsCl_2$ react at 25°C to form[2] $MeAsHCl$:

$$EtSbH_2 + 2\ MeAsCl_2 \rightarrow EtSbCl_2 + 2\ MeAsHCl \qquad (a)$$

Benzene, in the presence of $AlCl_3$ catalyst and $i\text{-}C_5H_{12}$, reacts with $AsCl_3$ to form phenylarsines[3]:

$$PhH + AsCl_3 \xrightarrow{\ AlCl_3\ catal.\ } PhAsH_2,\ Ph_2AsH,\ Ph_3As \qquad (b)$$

Reaction stoichiometry is not established. Thermolysis at 400°C of an Na arsenite–Na formate mixture yields[4] AsH_3:

$$Na_3AsO_3 + 3\ NaCHO_2 \rightarrow 3\ Na_2CO_3 + AsH_3 \qquad (c)$$

Dimethylphosphine cleavage of the As—As bond in $(CF_3)_4As_2$:

$$(CF_3)_4As_2 + Me_2PH \rightarrow (CF_3)_2AsPMe_2 + (CF_3)_2AsH \qquad (d)^5$$

or Me_3SnH cleavage of As—As or As—P bonds in the metal-coordinated analogues[6]

$$Me_2AsAs(CF_3)_2 \cdot M(CO)_5 + Me_3SnH \rightarrow Me_3SnAs(CF_3)_2 + (CO)_5M \cdot Me_2AsH \qquad (e)$$

$$(CF_3)_2PAsMe_2 \cdot M(CO)_5 + Me_3SnH \rightarrow Me_3SnP(CF_3)_2 + (CO)_5M \cdot Me_2AsH \qquad (f)$$

where M = Cr, Mo, produce metal-coordinated Me_2AsH.

Arsine redistribution reactions yield compounds containing new As—H bonds. Germylarsine in the presence of B_2H_6 or H_2O or SiH_3AsH_2 with B_2H_6 produce AsH_3 quantitatively[7]:

$$3 \ H_3MAsH_2 \rightarrow (H_3M)_3As + 2 \ AsH_3 \tag{g}$$

where M = Si, Ge. However, H_2O cleavage of Ge—As bonds is not fast, as is the case with Si—As bonds[7]. Similarly, unsymmetrically substituted silyl- and germyl-arsines redistribute readily[8]:

$$3 \ Me_3MAsHR \rightarrow (Me_3M)_2AsR + RAsH_2 \tag{h}$$

where M = Si, Ge; M = Me, Ph.

Hydride transfer in the decomposition of $CF_3As(PH_2)_2$ yields[9] CF_3AsH_2:

$$CF_3As(PH_2)_2 \rightarrow \tfrac{2}{n} \ (PH)_n + CF_3AsH_2 \tag{i}$$

β-Hydrogen transfer during thermolysis[10] of $Me_3CAs(SiMe_3)_2$ results in $(Me_3Si)_2AsH$:

$$Me_3CAs(SiMe_3)_2 \rightarrow Me_2CCH_2 + (Me_3Si)_2AsH \tag{j}$$

The decomposition of AsH_3, $AsH_3–SiH_4$ or $AsH_3–GeH_4$ mixtures in ozonizer discharge reactions yields products that are the result of AsH, SiH_2 or GeH_2 insertion into As—H bonds. Major As—H bond-containing products are[7,11] As_2H_4, SiH_3AsH_2 and GeH_3AsH_2.

(A.D. NORMAN)

1. G. O. Doak, L. D. Freedman, *Organometallic Compounds of Arsenic, Antimony, and Bismuth*, Wiley-Interscience, New York, 1970.
2. P. Chaudhury, M. F. El-Shazley, C. Spring, A. Rheingold, *Inorg. Chem.*, 18, 543 (1979).
3. L. Schmerling, U.S. Pat. 2,842,579 (1958); *Chem. Abstr. 55*, 497 (1961).
4. D. T. Hurd, *Chemistry of the Hydrides*, John Wiley and Sons, New York, 1952.
5. R. G. Cavell, R. C. Dobbie, *J. Chem. Soc., A*, 1406 (1968).
6. J. Grobe, D. LeVan, *Z. Naturforsch., Teil B, 36*, 666 (1981).
7. J. E. Drake, C. Riddle, *Q. Rev. Chem. Soc.*, 263 (1970).
8. J. W. Drake, J. E. Anderson, *J. Inorg. Nucl. Chem.*, 35, 1032 (1973).
9. R. Demuth, J. Grobe, *Z. Naturforsch., Teil B, 28*, 219 (1973).
10. G. Becker, G. Gutenkunst, H. J. Wessely, *Z. Anorg. Allg. Chem.*, 462, 113 (1980).
11. W. L. Jolly, A. D. Norman, *Prep. Inorg. React.*, 4, 1 (1968).

1.5.4.4. with Compounds of Antimony.

Examples of Sb—H bond formation by reactions of Sb compounds with either covalent or ionic binary hydrides are rare[1]. The alkynlstibine, $n\text{-}Bu_2SbC\equiv CH$, reacts with Ph_3SnH, where instead of hydrostannylation of the alkyne bond, tin hydride cleavage of the Sb—C bond:

$$n\text{-}Bu_2SbH + C_2H_2 \tag{a}$$

$$n\text{-}Bu_2SbC\equiv CH + Ph_3SnH$$

$$n\text{-}Bu_2SbCH\equiv CHSnBu_2 \tag{b}$$

leads to $n\text{-}Bu_2SbH$ and acetylene[2]. The generality of this reaction for R_2SbH synthesis is not tested. The germane, $n\text{-}Bu_3GeH$, reacts with $n\text{-}Bu_2SbC\equiv CH$ to form $n\text{-}Bu_3GeC\equiv CH$ and $n\text{-}Bu_2SbCHCHSbBu_2$, as a result of $n\text{-}Bu_2SbH$ intermediate formation[2]:

$$n\text{-}Bu_2SbC\equiv CH + n\text{-}Bu_3GeH \rightarrow n\text{-}Bu_2SbH + n\text{-}Bu_3GeC\equiv CH \qquad (c)$$

$$n\text{-}Bu_2SbH + n\text{-}Bu_2SbC\equiv CH \rightarrow (n\text{-}Bu_2SbCH)_2 \qquad (d)$$

(A.D. NORMAN)

1. G. O. Doak, L. D. Freedman, *Organometallic Compounds of Arsenic, Antimony, and Bismuth*, Wiley-Interscience, New York, 1970.
2. A. Tyschach, W. Fisher, *Z. Chem.*, 7, 196 (1967).

1.5.4.5. with Compounds of Bismuth.

Reactions of bismuth compounds with hydrides to form Bi—M-containing compounds are not known[1,2].

(A.D. NORMAN)

1. L. D. Freedman, G. O. Doak, *Chem. Rev.*, 82, 15 (1982).
2. G. O. Doak, L. D. Freedman, *Organometallic Compounds of Arsenic, Antimony and Bismuth*, Wiley-Interscience, New York, 1970.

1.5.5. by Reaction of Complex Hydrides

1.5.5.1. with Compounds of Nitrogen.

Complex metal hydroborates, hydroaluminates, hydridoferrates and modified complex hydrides (e.g., $NaBH_4$–$AlCl_3$) are used extensively in syntheses of organic compounds containing N—H bonds[1-5], but few inorganic or organometallic N—H bonds are prepared this way.

Reaction of h^5-CpCo-coordinated nitroso compounds with $LiAlH_4$ in THF followed by treatment with H_2O produces diamine products in 90% yield[6]:

$$h^5\text{-}CpCo(NO)_2C_2R_4 \xrightarrow[\text{2. }H_2O]{\text{1. }LiAlH_4} h^5\text{-}CpCoH_2 + R_2C(NH_2)C(NH_2)R_2 \qquad (a)$$

where R = H, Me, Et. Magnesium-bonded imines, formed from RMgX reaction with nitriles, are reduced by $LiAlH_4$ to primary amines:

$$R\ R'C{=}NMgBr \xrightarrow[\text{2. }H_2O]{\text{1. }LiAlH_4} R\ R'CHNH_2 \qquad (b)[3]$$

Transition-metal-coordinated N_2R complexes react with complex hydrides to form NHNR complexes or free NH_3 or amines in reactions whose stoichiometries are not established:

$$[h^5\text{-}CpRe(CO)_2N_2R]BF_4 \xrightarrow[\text{THF}_{(aq)}]{NaBH_4} h^5\text{-}CpRe(CO)_2N_2 + h^5\text{-}CpRe(CO)_2NHNR \qquad (c)[7]$$

where R = p-MeC_6H_4, o-$CF_3C_6H_4$, p-MeC_6H_4;

$$WBr_2(N_2CMe_2)(PMe_2Ph)_3 \xrightarrow[Et_2O]{LiAlH_4} WH_6(PMe_2Ph)_3,\ NH_3,\ i\text{-}PrNH_2 \qquad (d)[8]$$

$$MoBr(N_2\text{-t-Bu})(dppe)_2 \xrightarrow{NaBH_4} NH_3, \text{t-BuNH}_2 \qquad (e)^9$$

where dppe $= Ph_2PCH_2CH_2PPh_2$. These reactions produce NH_3 and primary amines in high yields.

Complex hydride reductions of N_2, N_2O, $[CN]^-$ and nitriles in reactions containing complex mixtures of metal-ion species produce N—H bonded products[10,11] by reactions that may relate to those of the biological nitrogenases[11]. Complete reaction stoichiometries are not well established. These reactions are not competitive with other methods for NH_3 or amine synthesis. Nitrogen reacts in H_2O with a mixture of $NaBH_4$, S-donor ligands (e.g., $NH_2C_2H_4SH$) and Mo and Fe salts to form NH_3 and N_2H_4 in low yield[10]. In similar systems, nitriles and isonitriles are reduced to NH_3 and amines in low yield[10].

(A.D. NORMAN)

1. A. Hajós, *Comples Hydrides*, Elsevier, Amsterdam, 1979.
2. H. C. Brown, *Boranes in Organic Chemistry*, Cornell Univ. Press, Ithaca, NY, 1972.
3. J. R. Malpass, in *Comprehensive Organic Chemistry*, D. Barton, W. D. Ollis, eds., Vol. 2, I. O. Sutherland, ed., Pergamon Press, New York, 1979, p. 3.
4. L. F. Fieser, M. Fieser, *Advanced Organic Chemistry*, Rheinhold, New York, 1961.
5. H. O. House, *Modern Synthetic Reactions*, 2nd ed., Benjamin, Menlo Park, CA, 1972.
6. P. N. Becker, M. A. White, R. G. Bergman, *J. Am. Chem. Soc., 102*, 5676 (1980).
7. C. F. Barrientos-Penna, F. W. B. Einstein, T. Jones, D. Sutton, *Inorg. Chem., 21*, 2578 (1982).
8. P. C. Bevon, J. Chatt, M. Hidai, G. J. Leigh, *J. Organomet. Chem., 160*, 165 (1978).
9. G. E. Bossard, D. C. Busby, M. Chang, T. A. George, S. D. A. Iske, *J. Am. Chem. Soc., 102*, 1001 (1980).
10. J. Chatt, J. R. Dilworth, R. L. Richards, *Chem. Rev., 78*, 589 (1978).
11. D. Coucouvanis, *Acc. Chem. Res., 14*, 201 (1981).

1.5.5.2. with Compounds of Phosphorus

1.5.5.2.1. Involving Halides.

Lithium tetrahydroaluminate reduction of phosphorus halides is a route to PH_3 and primary and secondary phosphines[1-3]. Phosphorus trichloride at $-115°C$ in Me_2O is reduced to PH_3 in 79% yield:

$$4 \text{ PCl}_3 + 3 \text{ LiAlH}_4 \rightarrow 3 \text{ LiAlCl}_4 + 4 \text{ PH}_3 \qquad (a)$$

Under similar conditions, $LiAlH_4$ reaction with $P(O)Cl_3$ at $-115°C$ yields[4] PH_3. Mono- and dichlorophosphorus(III) compounds are reduced in ether (Me_2O, Et_2O, THF, glymes) to the corresponding secondary and primary phosphines:

$$2 \text{ RPCl}_2 + \text{LiAlH}_4 \rightarrow \text{LiAlCl}_4 + 2 \text{ RPH}_2 \qquad (b)$$

$$4 \text{ R}_2\text{PCl} + \text{LiAlH}_4 \rightarrow \text{LiAlCl}_4 + 4 \text{ R}_2\text{PH} \qquad (c)$$

e.g., $MePH_2$, 55% yield[5]; $PhPH_2$, 75% yield[3] and Ph_2PH, 70% yield[6]. Reaction of Ph_2PCl_3 with $LiAlH_4$ yields[2] Ph_2PH, but no Ph_2PH_3. The fluoroaminophosphine, $(Me_3Si)_2NPF_2$, is reduced by a $LiAlH_4$–secondary amine mixture:

$$(Me_3Si)_2NPF_2 \xrightarrow[\substack{-\text{LiAlF}_x(\text{NPr})_{4-x} \\ -H_2}]{\text{LiAlH}_4-\text{Pr}_2\text{NH}} (Me_3Si)_2NPH_2 \qquad (d)^7$$

1.5.5. by Reaction of Complex Hydrides   47
1.5.5.2. with Compounds of Phosphorus
1.5.5.2.1. Involving Halides.

Phosphonyl dichlorides react with $LiAlH_4$ forming mono- and bis(phosphino)-substituted products. From $PhP(O)Cl_2$ reduction, $PhPH_2$ is obtained[3]:

$$2 \text{ RP(O)Cl}_2 + 2 \text{ LiAlH}_4 \rightarrow \text{LiAlO}_2 + \text{LiAlCl}_4 + 2 \text{ H}_2 + 2 \text{ RPH}_2 \qquad \text{(e)}$$

and from $RR'C[P(O)Cl_2]_2$ compounds, phosphines such as $H_2C(PH_2)_2$, $Me_2C(PH_2)_2$ and $(n\text{-}Bu)_2C(PH_2)_2$ form[8]. Diphenylphosphine is formed[3] in 93% yield by $LiAlH_4$ reduction of $Ph_2P(O)Cl$. Reduction of phosphinic chlorides yields the cyclic secondary phosphines[3,9]:

$$\text{(f)}$$

$$\text{(g)}$$

The substituted Na hydroaluminate, $Na[(MeOCH_2CH_2O)_2AlH_2]$, reacts with Ph_2PCl, forming Ph_2PH along with the undesired alkylation–cleavage products[10] $MePh_2P$ and $Ph_2PCH_2CH_2OH$.

Alkali-metal tetrahydroborate reduction of phosphorus halides is of limited utility[1,3]. Lithium tetrahydroborate reaction with PCl_3 or PCl_3 in Et_2O at $-80°C$ produces PH_3, but yields are low[11]:

$$2 \text{ PCl}_3 + 6 \text{ LiBH}_4 \rightarrow 3 \text{ B}_2\text{H}_6 + 6 \text{ LiCl} + 2 \text{ PH}_3 \qquad \text{(h)}$$

$$2 \text{ PCl}_5 + 10 \text{ LiBH}_4 \rightarrow 5 \text{ B}_2\text{H}_6 + 10 \text{ LiCl} + 2 \text{ PH}_3 + \text{H}_2 \qquad \text{(i)}$$

Reactions of R_2PCl or $RPCl_2$ with $LiBH_4$ yield only the borane-coordinated phosphine products[11].

(A.D. NORMAN)

1. L. Maier, *Prog. Inorg. Chem.*, 5, 27 (1963).
2. E. Wiberg, E. Amberger, *Hydrides of the Elements of Main Groups I-IV*, Elsevier, Amsterdam, 1971.
3. A. Hajós, *Complex Hydrides*, Elsevier, Amsterdam, 1979.
4. E. Wiberg, G. Müller-Schiedmayer, *Z. Anorg. Allg. Chem.*, 308, 352 (1961).
5. L. J. Malone, R. W. Parry, *Inorg. Chem.*, 6, 176 (1967).
6. W. Kuchen, H. Buchwald, *Chem. Ber.*, 91, 2871 (1958).
7. E. Niecke, R. Rüger, *Angew. Chem., Int. Ed. Engl.*, 21, 62 (1982).
8. H. R. Hays, T. J. Logan, *J. Org. Chem.*, 31, 339 (1966).
9. S. D. Venkataramu, G. D. MacDonell, W. R. Purdum, M. El-Deek, K. D. Berlin, *Chem. Rev.*, 77, 121 (1977).
10. M. Gallagher, G. Pollard, *Phosphorus*, 6, 61 (1975).
11. B. D. James, M. G. H. Wallbridge, *Prog. Inorg. Chem.*, 11, 99 (1970).

48   1.5. Formation of Bonds between Hydrogen and N, P, As, Sb, Bi
       1.5.5. by Reaction of Complex Hydrides
       1.5.5.2. with Compounds of Phosphorus

1.5.5.2.2. Involving Oxygen Compounds.

Primary and secondary phosphines are prepared by $LiAlH_4$ reduction of organo-phosphorus acids, or esters[1-3]. Reactions of phosphoryl halides are described in §1.5.5.2.1. Phosphinic acids:

$$2 R_2P(O)(OH) + LiAlH_4 \rightarrow LiAl(OH)_4 + 2 R_2PH \tag{a}$$

phosphinate esters:

$$4 R_2P(O)(OR') + 2 LiAlH_4 \rightarrow LiAl(OH)_4 + LiAl(OR')_4 + 4 R_2PH \tag{b}$$

phosphonous acids:

$$4 RP(O)(OH)_2 + 3 LiAlH_4 \rightarrow 3 LiAl(OH)_4 + 4 RPH_2 \tag{c}$$

phosphonate esters:

$$6 RP(O)(OR')_2 + 4 LiAlH_4 \rightarrow LiAl(OH)_4 + 3 LiAl(OR')_4 + 6 RPH_2 \tag{d}$$

are reduced to phosphines. These reactions often allow syntheses of complex phosphines that cannot be obtained easily by other methods. Phosphines that can be prepared by this method are listed in Table 1. Reactions are carried out under N_2 in ether at or below RT. After reduction, aq acid is added to hydrolyze intermediate alumi-nophosphorus species, e.g., $[Al(PPh_2)_4]^-$ in the $Ph_2P(O)(OEt)$ reduction[1].

Reaction of an alkynylphosphonate with $LiAlH_4$ reduces both the alkyne and the phosphoryl groups[9]:

$$PhC \equiv CP(O)(OEt)_2 \xrightarrow{\text{LiAlH}_4} PhCH = CHPH_2 \tag{e}$$

Phosphoramides react variously with $LiAlH_4$. Phosphoryl reduction and P—N bond cleavage can occur as[1]:

$$Ph(Me)NP(O)MePh \xrightarrow{\text{LiAlH}_4} Ph(Me)NH + Ph(Me)PH \tag{f}$$

TABLE 1. PHOSPHONATE OR PHOSPHINATE REDUCTIONS

Reactant	Product	Yield (%)	Ref.
$MeP(O)(OMe)_2$	$MePH_2$	87	4
$EtP(O)(OEt)_2$	$EtPH_2$	65	1
$PhP(O)(OH)_2$	$PhPH_2$	13	3
$PhP(O)(OEt)_2$	$PhPH_2$	62	3
$PhCH_2P(O)(OEt)_2$	$PhCH_2PH_2$	48	1,3
$PhP(H)CH_2P(O)(OEt)_2$	$PhP(H)CH_2PH_2$	30	5
$o\text{-}NH_2C_6H_4P(O)(OEt)_2$	$o\text{-}NH_2C_6H_4PH_2$	85	6
$(EtO)_2P(O)(CH_2)_2P(O)(OEt)_2$	$H_2P(CH_2)_2PH_2$	57	7
$Ph_2PCH_2CH_2P(O)(OEt)_2$	$Ph_2PCH_2CH_2PH_2$	21	8
$PhP[CH_2CH_2P(O)(OEt)_2]_2$	$PhP(CH_2CH_2PH_2)_2$	55	8
$P[CH_2CH_2P(O)(OEt)_2]_3$	$P(CH_2CH_2PH_2)_3$	28	8
$Ph_2P(O)OH$	Ph_2PH	80	1
$(C_6H_{11})_2P(O)OH$	$(C_6H_{11})_2PH$	50	1
$Me_3SnCH_2P(O)(Ph)(OEt)$	$Me_3SnCH_2P(H)Ph$	48	5

1.5. Formation of Bonds between Hydrogen and N, P, As, Sb, Bi   49
1.5.5. by Reaction of Complex Hydrides
1.5.5.2. with Compounds of Phosphorus

In contrast, an analogous cyclic compound is reduced without P—N bond cleavage[1]:

$$\text{(g)}$$

When heated with a deficit of $LiAlH_4$, solid powdered P_4O_{10} produces[10] small quantities of PH_3:

$$P_4O_{10} + 5\ LiAlH_4 \xrightarrow{148°C} 5\ LiAlO_2 + 4\ H_2 + 4\ PH_3 \qquad \text{(h)}$$

Metal phosphites in H_2O are reduced in low yield[11] to PH_3.

(A.D. NORMAN)

1. A. Hajós; *Complex Hydrides*, Elsevier, Amsterdam, 1979. Excellent review of hydride reductions.
2. E. Wiberg, E. Amberger, *Hydrides of the Elements of Main Groups I–IV*, Elsevier, Amsterdam, 1971.
3. L. Maier, *Prog. Inorg. Chem.*, 5, 27 (1963).
4. K. D. Crosbie, G. M. Sheldrick, *J. Inorg. Nucl. Chem.*, 31, 3684, (1969).
5. H. Weichmann, B. Ochsler, I. Duchek, A. Tzschach, *J Organomet. Chem.*, 182, 465 (1979).
6. K. Issleib, R. Vollmer, *Z. Chem.*, 18, 451 (1978).
7. R. C. Taylor, D. B. Walters, *Inorg. Synth.*, 15, 10 (1973).
8. R. B. King, J. C. Cloyd, P. N. Kapoor, *J. Chem. Soc., Perkin Trans. 1*, 2226 (1973).
9. I. Ionin, G. M. Bogolyubov, A. A. Petrov, *Russ. Chem. Rev. (Engl. Transl.)*, 36, 249 (1967).
10. J. M. Bellama, A. G. MacDiarmid, *Inorg. Chem.*, 7, 2070 (1968).
11. B. D. James, M. G. H. Wallbridge, *Prog. Inorg. Chem.*, 11, 99 (1970).

1.5.5.2.3. Involving Other Derivatives.

Cleavage of diphosphines with $LiAlH_4$ can be a route to secondary phosphines, e.g., diphosphine disulfides react:

$$R_2P(S)P(S)R_2 + LiAlH_4 \xrightarrow{Et_2O} LiAlS_2 + H_2 + 2\ R_2PH \qquad \text{(a)}^{1,2}$$

where R = Me, Et, CH_2Ph, n—Bu, C_6H_{11}, Ph. Yields range from 68 to 80%. Because $Me_2P(S)P(S)Me_2$ is readily available, the reaction is good for preparing large quantities[2] (50–400 mmol) of Me_2PH. Tetraphenyldiphosphine is reduced[3] to Ph_2PH:

$$Ph_2PPPh_2 \xrightarrow[Et_2O]{LiAlH_4} Ph_2PH \qquad \text{(b)}$$

Germanium–P bond cleavage by $LiAlH_4$ yields[4] Ph_2PH:

$$Et_3GePPh_2 \xrightarrow{LiAlH_4} Et_3GeH + Ph_2PH \qquad \text{(c)}$$

or[5] Et_2PH:

$$\text{(d)}^5$$

Lithium tetrahydroborate can substitute for $LiAlH_4$. The $LiBH_4$ reduction occurs with inversion of configuration at Ge, whereas $LiAlH_4$ cleavage occurs with retention of configuration.

Lithium tetrahydroaluminate cleavage of Ph(Me)NP(O)MePh occurs[6] at 25°C in THF:

$$Ph(Me)NP(O)MePh \xrightarrow[THF]{LiAlH_4} MePhNH + PhMeP(O)H \tag{e}$$

In contrast, $LiAlH_4$ reduction of $(Me_3Si)_2NPNSiMe_3$ followed by reaction workup in the presence of H_2O (or D_2O) yields a bis(amino)phosphine product[7].

$$(Me_3Si)_2NPNSiMe_3 \xrightarrow[2.\ H_2O]{1.\ LiAlH_4,\ Et_2O} (Me_3Si)_2NP(H)N(H)SiMe_3 \tag{f}$$

Lithium tetrahydroaluminate or $LiBH_4$ react with the spirophosphonum salt to form the corresponding P—H-phosphorane[8]:

Reduction of phosphonium salts as a route to tertiary phosphines is common, but its use for the synthesis of P—H compounds is unusual.

(A.D. NORMAN)

1. A. Hajós, *Complex Hydrides,* Elsevier, Amsterdam, 1979. Excellent review of complex hydride reductions.
2. G. W. Parshall, *Inorg. Synth., 11,* 157 (1968). An excellent synthesis of PH_3.
3. K. Issleib, A. Tzschach, *Chem. Ber., 92,* 704 (1979).
4. E. W. Abel, S. M. Illingworth, *Organomet. Chem. Rev., A, 5,* 143 (1970).
5. J. Duboc, J. Escudié, C. Couret, J. Cavezzan, J. Satgé, P. Mazerolles, *Tetrahedron, 37,* 1141 (1981).
6. P. D. Henson, S. B. Ochrymiek, R. F. Markham, *J. Org. Chem., 39,* 2296 (1974).
7. A. H. Cowley, R. A. Kemp, *J. Chem. Soc., Chem. Commun.,* 319 (1982).
8. D. Hellwinkel, *Chem. Ber., 102,* 528 (1969).

1.5.5.3. with Compounds of Arsenic

1.5.5.3.1. Involving Halides.

Complex-hydride reductions of chloroarsines can be used to prepare AsH_3 and primary $(RAsH_2)$ and secondary organoarsines (R_2AsH). These syntheses are efficient and convenient for laboratory quantities of arsines[1-3].

Arsenic(III) trichloride reacts[1,2] with $LiAlH_4$ or $LiBH_4$ in ethers at $-80°C$:

$$AsCl_3 + 3\ LiBH_4 \rightarrow AsH_3 + \tfrac{3}{2} B_2H_6 + 3\ LiCl \tag{a}$$

$$4 \text{ AsCl}_3 + 3 \text{ LiAlH}_4 \rightarrow 4 \text{ AsH}_3 + 3 \text{ LiAlCl}_4 \tag{b}$$

Because only 1 equiv of hydride from $LiBH_4$ reacts, the reaction is inefficient, although AsH_3 yields of 93% based on Eq. (a) occur. At higher T, yields are lower. At RT, $AsCl_3$ is reduced to elemental As. Reduction by $LiAlH_4$ proceeds more rapidly. At $-78°C$, yields of 10–15% are obtained.

Reduction by $LiAlH_4$ of $RAsCl_2$ or R_2AsCl (R = alkyl, aryl) occurs in ethers:

$$4 \text{ R}_{3-n}\text{AsCl}_n + n \text{ LiAlH}_4 \rightarrow n \text{ LiAlCl}_4 + 4 \text{ R}_{3-n}\text{AsH}_n \tag{c}$$

yields[2,3] (%) are: CF_3AsH_2, 49; $PhAsH_2$, 54; Me_2AsH, 68; Et_2AsH, 80; $n\text{-}Pr_2AsH$, 57; $n\text{-}Bu_2AsH$, 73; $(C_6H_{11})_2AsH$, 77; Ph_2AsH and $(CF_3)_2AsH$, 16. Phenylarsines (n = 1, 2) also form in low-T $LiBH_4$ reductions of the respective chloroarsines:

$$\text{Ph}_{3-n}\text{AsCl}_n + n \text{ LiBH}_4 \xrightarrow[-60°C]{Et_2O} \frac{n}{2} \text{ B}_2\text{H}_6 + n \text{ LiCl} + \text{Ph}_{3-n}\text{AsH}_n \tag{d[4]}$$

(A.D. NORMAN)

1. W. L. Jolly, A. D. Norman, *Prep. Inorg. React., 4,* 1 (1968).
2. E. Wiberg, E. Amberger, *Hydrides of the Elements of Main Groups I–IV,* Elsevier, Amsterdam, 1971.
3. G. O. Doak, L. D. Freedman, *Organometallic Compounds of Arsenic, Antimony, and Bismuth,* Wiley-Interscience, New York, 1970.
4. E. Wiberg, K. Mödritzer, *Z. Naturforsch., Teil B, 12,* 127 (1957).

1.5.5.3.2. Involving Oxygen Compounds.

Alkali-metal tetrahydroborates reduce oxyarsenic(III) species in H_2O to AsH_3 in high yields[1,2]. Potassium tetrahydroborate in base, when added dropwise to aq acid (H_2SO_4), reacts as:

$$4 \text{ [As(OH)}_4\text{]}^- + 3 \text{ KBH}_4 + 7 \text{ H}^+ \rightarrow 3 \text{ K}^+ + 3 \text{ H}_3\text{BO}_3 + 7 \text{ H}_2\text{O} + 4 \text{ AsH}_3 \tag{a}$$

Arsine yields of 59%, along with traces of As_2H arising from As_2H_4 decomposition, are claimed[3].

Solid As_4O_{10} reacts with $LiAlH_4$ at 148–170°C, yielding[4] AsH_3:

$$\text{As}_4\text{O}_{10} + 5 \text{ LiAlH}_4 \rightarrow 5 \text{ LiAlO}_2 + 4 \text{ H}_2 + 4 \text{ AsH}_3 \tag{b}$$

(A.D. NORMAN)

1. B. D. James, M. G. H. Walbridge, *Prog. Inorg. Chem., 11,* 99 (1970).
2. W. L. Jolly, A. D. Norman, *Prep. Inorg. React., 4,* 1 (1968).
3. W. L. Jolly, J. E. Drake, *Inorg. Synth., 7,* 34 (1966).
4. J. M. Bellama, A. G. MacDiarmid, *Inorg. Chem., 7,* 2070 (1968).

1.5.5.4. with Compounds of Antimony.

Reactions of complex hydrides with Sb compounds are effective for stibine syntheses[1]. Metal tetrahydroborate reductions of Sb compounds in both H_2O and nonaqueous media occur[2]. Dropwise addition of a basic $[Sb(OH)_4]^-$–KBH_4 soln to aqueous acid forms SbH_3 in 95% yield[3]:

$$4 \text{ [Sb(OH)}_4\text{]}^- + 3 \text{ KBH}_4 + 7 \text{ H}^+ \rightarrow 3 \text{ H}_3\text{BO}_3 + 7 \text{ H}_2\text{O} + 3 \text{ K}^+ + 4 \text{ SbH}_3 \tag{a}$$

Product yield is optimized by slow reagent addition. Antimony trichloride in saturated

52   1.5. Formation of Bonds between Hydrogen and N, P, As, Sb, Bi
        1.5.5. by Reaction of Complex Hydrides
        1.5.5.4. with Compounds of Antimony.

aq NaCl is reduced to SbH_3 by aq $NaBH_4$ soln[4]. Yields of SbH_3 up to 70% using a $NaBH_4:Na_3SbCl_6$ ratio greater than 10 are claimed.

Alkali-metal tetrahydroborates in ethers reduce $SbCl_3$:

$$2 \ SbCl_3 \ + \ 6 \ LiBH_4 \ \xrightarrow{-70°C} \ 6 \ LiCl \ + \ 3 \ B_2H_6 \ + \ 2 \ SbH_3 \qquad (b)[1]$$

The synthesis of SbH_3 from MBH_4 (M = Li, Na, K) reduction of $SbCl_5$ in diglyme at 25°C is claimed[5].

Methylstibine and Me_2SbH form from reaction of Me_2SbBr with $NaBH_4$ in diglyme. From the cleaner reduction using $LiBH(OCH_3)_3$:

$$Me_2SbBr \ + \ LiBH(OCH_3)_3 \ \xrightarrow{<-40°C} \ Me_2SbH \ + \ (MeO)_3B \ + \ LiBr \qquad (c)$$

Me_2SbH yields of 35% are obtained[6]. Lithium tetrahydroborate reduction of $PhSbI_2$:

$$2 \ PhSbI_2 \ + \ 4 \ LiBH_4 \ \rightarrow \ 4 \ LiI \ + \ B_2H_6 \ + \ 2 \ PhSbH_2 \qquad (d)$$

and Ph_2SbCl in ether below $-50°C$ produces $PhSbH_2$ and Ph_2SbH, respectively[7,8]:

$$2 \ Ph_2SbCl \ + \ 2 \ LiBH_4 \ \rightarrow \ 2 \ LiCl \ + \ B_2H_6 \ + \ 2 \ Ph_2SbH \qquad (e)$$

Lower yields ($<40\%$) occur when $LiBH_4$ is replaced with $LiAlH_4$. Similarly, $[PhSbCl_5]NH_4$ reacts with $LiBH_4$ or $LiAlH_4$ to form $PhSbH_2$, with no evidence[1] of $PhSbH_4$.

Lithium tetrahydroaluminate reduces[2] $SbCl_3$ in ether to SbH_3:

$$3 \ LiAlH_4 \ + \ 4 \ SbCl_3 \ \xrightarrow{-78°C} \ 3 \ LiAlCl_4 \ + \ 4 \ SbH_3 \qquad (f)$$

An analogous $LiAlH_4$–$SbCl_5$ reaction yields[2] SbH_3, but no SbH_5. Reaction of solid powdered Sb_2O_3 with a deficit of $LiAlH_4$ at 148–170°C yields[9] SbH_3.

$$2 \ Sb_2O_3 \ + \ 3 \ LiAlH_4 \ \rightarrow \ 3 \ LiAlO_2 \ + \ 4 \ SbH_3 \qquad (g)$$

High yields of dialkyl- and monoalkylstibines result from the respective halo-stibines with $LiAlH_4$ in ethers (n-Bu_2O or Et_2O)[1,10,11] below $-20°C$:

$$4 \ R_2SbX \ + \ LiAlH_4 \ \xrightarrow[-20°C]{Et_2O} \ LiAlX_4 \ + \ 4 \ R_2SbH \qquad (h)$$

where X = Cl, Br; R = Me, Et, t-Bu, C_6H_{11};

$$2 \ RSbCl_2 \ + \ LiAlH_4 \ \xrightarrow{n\text{-}Bu_2O} \ LiAlCl_4 \ + \ 2 \ RSbH_2 \qquad (i)$$

where R = Me, Et, Bu.

(A.D. NORMAN)

1. G. O. Doak, L. D. Freedman, *Organometallic Compounds of Arsenic, Antimony, and Bismuth*, Wiley-Interscience, New York, 1970.
2. W. L. Jolly, A. D. Norman, *Prep. Inorg. React.*, 4, 1 (1968).
3. W. L. Jolly, J. E. Drake, *Inorg. Synth.*, 7, 34 (1963). A preferred synthesis for AsH_3.
4. A. D. Zorm, I. A. Frolov, U. S. Zaburdyaev, S. A. Nosyrev, *J. Appl. Chem. USSR (Engl. Transl.)*, 47, 1193 (1974); *Chem Abstr.*, 81, 173,481 (1974).
5. R. G. Gordon, Belg. Pat. 890,356 (1982); *Chem. Abstr.*, 96, 199,887 (1982).
6. A. B. Burg, L. R. Grant, *J. Am. Chem. Soc.*, 81, 1 (1959).
7. E. Wiberg, K. Mödritzer, *Z. Naturforsch., Teil B*, 12, 128 (1957).

8. E. Wiberg, K. Mödritzer, *Z. Naturforsch., Teil B, 12,* 131 (1957).
9. J. M. Bellama, A. G. MacDiarmid, *Inorg. Chem., 7,* 2070 (1968).
10. K. Issleib, B. Hamann, *Z. Anorg. Allg. Chem., 339,* 289 (1965).
11. A. L. Reingold, P. Choudhury, M. F. El-Shazly, *Synth. React. Inorg. Metal-Org. Chem., 8,* 453 (1978).

1.5.5.5. with Compounds of Bismuth.

Bismuthine, $MeBiH_2$ and Me_2BiH can be prepared by $LiAlH_4$ reduction of halobismuthines[1,2]. Yields are low, even at $-110°C$:

$$4\ BiCl_3 + 3\ LiAlH_4 \xrightarrow{Me_2O} 3\ LiAlCl_4 + 4\ BiH_3 \tag{a}$$

$$2\ MeBiCl_2 + LiAlH_4 \xrightarrow{Me_2O} LiAlCl_4 + 2\ MeBiH_2 \tag{b}$$

$$4\ Me_2BiCl + LiAlH_4 \xrightarrow{Me_2O} LiAlCl_4 + 4\ Me_2BH \tag{c}$$

The alkylbismuthines are thermally unstable[3]; e.g., $MeBiH_2$ disproportionates above $-45°C$:

$$3\ MeBiH_2 \rightarrow 2\ Me_3Bi + Bi + 3\ H_2 \tag{d}$$

Compounds containing Bi—H bonds are not isolated[4] from reactions of $LiAlH_4$ with $PhBiBr_2$, Ph_2BiCl or Ph_3BiCl_2.

(A.D. NORMAN)

1. G. O. Doak, L. D. Freedman, *Organometallic Compounds of Arsenic, Antimony, and Bismuth,* Wiley-Interscience, New York, 1970.
2. E. Amberger, *Chem. Ber.,* 1447 (1961).
3. L. D. Freedman, *Chem. Rev., 82,* 15 (1982).
4. E. Wiberg, H. Nöth, *Z. Naturforsch., Teil B, 12,* 132 (1957).

1.5.6. by Industrial Processes

1.5.6.1. Involving Compounds of Nitrogen.

Ammonia is obtained industrially from reaction of N_2 and H_2 gas at high P $(1 \times 10^5$ Pa) and T ($500°C$) over Fe or Fe/Fe_2O_3 catalysts[1-3]:

$$N_2 + 3\ H_2 \rightleftharpoons 2\ NH_3 \tag{a}$$

Ammonia-D_3 can be prepared by replacement of H_2 by D_2; however, it is better prepared by the deuterolysis of a metal nitride:

$$Mg_3N_2 + 6\ D_2O \rightarrow 3\ Mg(OD)_2 + 2\ ND_3 \tag{b[4,5]}$$

Ammonium salts form by direct reaction of amines with mineral acids, e.g.:

$$Et_3N + HCl \rightarrow Et_3NHCl \tag{c[1,5]}$$

Reactions of nitriles or isocyanates with protonic acids yield imines:

$$PhC{\equiv}N + 2\ HX \rightarrow [PhCX{=}NH_2]X \tag{d[6,7]}$$

where X = Cl, Br; amines:

$$RNCO + H_2O \xrightarrow{HCl_{(aq)}} CO_2 + RNH_2 \qquad (e)[8]$$

where R = alkyl, aryl; or amides:

$$RNCO + R'OH \rightarrow RNHCO_2R' \qquad (f)[9,10]$$

Reduction of nitro compounds with metals (Fe, Sn, Zn, Mg) in aq acid is the major industrial synthetic route to primary amines:

$$RNO_2 \xrightarrow{Sn-HCl_{(aq)}} RNH_2 \qquad (g)[7-10]$$

where R = Et, n-Bu, Ph, etc. Similarly, reduction by carbon in hot steam forms $PhNH_2$ from $PhNO_2$ in high yields:

$$PhNO_2 \xrightarrow[200-400°C]{C-H_2O} PhNH_2 \qquad (h)[10]$$

Reductions using H_2 directly, in the presence of homogenous[11] [e.g., $Fe(CO)_5$] or heterogeneous (e.g., Pt)[10,12] catalysts, also yield amines:

$$RNO_2 \xrightarrow{H_2-catal.} RNH_2 \qquad (i)$$

Hydrogen reduction of nitriles:

$$RCN \xrightarrow{H_2-Ni} RCH_2NH_2 \qquad (j)[10,12]$$

where R = alkyl, aryl; or imines:

$$R_2CNR' \xrightarrow{H_2-Pt} R_2CHNHR' \qquad (k)[6,10]$$

where R and R' = alkyl, aryl, in the presence of catalysts is used to produce amines. Similar reduction of pyridine or its derivatives yields piperidines:

$$C_5H_5N \xrightarrow{H_2-Pt} C_5H_{10}NH \qquad (l)[10]$$

Nitrogen compounds containing N—H bonds for specialty chemical, drug or pharmaceutical applications are synthesized using the reactions (especially hydride reductions) discussed in §1.5.4.1 and 1.5.5.1. These small-scale syntheses are highly specialized and usually relatively expensive. Further information can be found in treatises on synthetic organic chemistry.

(A.D. NORMAN)

1. F. A. Cotton, G. Wilkinson, *Advanced Inorganic Chemistry*, 4th ed., Wiley-Interscience, New York, 1980.
2. A. Ozaki, *Acc. Chem. Res.*, *14*, 16 (1981).
3. L. Axelrod, *Catal. Rev. Sci. Eng.*, *23*, 53 (1981).
4. L. K. Krannich, U. Thewalt, W. J. Cook, S. R. Jain, H. H. Sisler, *Inorg. Chem.*, *12*, 2304 (1973).
5. W. L. Jolly, *The Inorganic Chemistry of Nitrogen*, Benjamin, New York, 1964.
6. G. Tennant, in *Comprehensive Organic Chemistry*, D. Barton, W. D. Ollis, eds., Vol. 2, I. O. Sutherland, ed., Pergamon Press, New York, 1979, p. 385.

7. R. G. Coombes, in *Comprehensive Organic Chemistry*, D. Barton, W. D. Ollis, eds., Vol. 2, I.
   O. Sutherland, ed., Pergamon Press, New York, 1979, p. 305.
8. R. J. Lindsay, in *Comprehensive Organic Chemistry*, D. Barton, W. D. Ollis, eds., Vol. 2, I. O.
   Sutherland, ed., Pergamon Press, New York, 1979.
9. L. F. Fieser, M. Fieser, *Advanced Organic Chemistry*, Reinhold, New York, 1961.
10. R. T. Morrison, R. N. Boyd, *Organic Chemistry*, 4th ed., Allyn and Bacon, Boston, 1983.
11. H. Alper, J. T. Edward, *Can. J. Chem.*, 48, 1543 (1970).
12. J. R. Malpass, in *Comprehensive Organic Chemistry*, D. Barton, W. D. Ollis, eds., Vol. 2, I. O.
   Sutherland, ed., Pergamon Press, New York, 1979, p. 3.

1.5.6.2. Involving Compounds of Phosphorus.

Phosphine is obtained by the alkaline hydrolysis of P_4:

$$P_4 + 2\,H_2O + 4\,[OH]^- \rightarrow 2\,[HPO_3]^{2-} + 2\,PH_3 \qquad (a)^{1,2}$$

or the reduction of P_4 by the active metals (e.g., Zn, Mg) in H_2O:

$$P_4 + 12\,e^- + 12\,H_2O \rightarrow 12\,[OH]^- + 4\,PH_3 \qquad (b)^{1,3}$$

Large quantities of PH_3 can be obtained from metal-phosphide hydrolyses:

$$AlP + 3\,H_2O \rightarrow Al(OH)_3 + PH_3 \qquad (c)^{1,2}$$

This method is adapted to large-scale PD_3 synthesis by substitution[4] of D_2O for H_2O.

Primary and secondary phosphines, e.g., $PhPH_2$ and Ph_2PH, are prepared stepwise from the substituted phosphinous chlorides by metal reduction and hydrolysis:

$$2\,Ph_2PCl + 2\,Zn \rightarrow ZnCl_2 + (Ph_2P)_2Zn \qquad (d)^5$$

$$(Ph_2P)_2Zn + 2\,H_2O \rightarrow Zn(OH)_2 + 2\,Ph_2PH \qquad (e)^5$$

Disproportionation of alkylphosphinic acid or alkylphosphinates yields primary phosphines.

$$3\,RPH(O)OH \xrightarrow{150°C} 2\,RP(O)(OH)_2 + RPH_2 \qquad (f)^{6,7}$$

where R = Me, Et, Ph, etc. Reaction of P_4 with $MeNH_2$ at 350°C, over active carbon yields methylphosphines:

$$P_4 + MeNH_2 \xrightarrow[C]{350°C} Me_3P,\ Me_2PH,\ MePH_2 \qquad (g)^8$$

Specialty-chemical level production (kilograms) of phosphines is achieved by hydride reductions of halophosphines, phosphonyl halides and esters and phosphinyl halides and esters, e.g.:

$$n\text{-}Bu_2P(O)Cl \xrightarrow[Me_3N/C_6H_6]{SiHCl_3} n\text{-}Bu_2PH \qquad (h)^9$$

$$n\text{-}Bu_2P(O)OH \xrightarrow{Ph_2SiH_2} n\text{-}Bu_2PH \qquad (i)^{10}$$

$$EtP(OEt)_2 \xrightarrow{LiAlH_4} EtPH_2 \qquad (j)^{11}$$

Phosphorus acids or their salts containing P—H bonds are obtained commercially from hydrolysis of elemental phosphorus or phosphorus halides, under controlled conditions. Variations result in different product mixtures. Phosphorous acid forms in reactions[1,6,11] of PCl_3 with H_2O at 185–195°C:

$$PCl_3 + 3 H_2O \rightarrow 3 HCl + H_3PO_3 \tag{k}$$

At lower T (e.g., 0°C) in carbonate-buffered H_2O, mixtures of acid salts, such as $[H_2P_2O_5]^{2-}$ and $[HP_2O_3]^{2-}$ in addition to $[HPO_3]^{2-}$, $[P_2O_6]^{4-}$, $[PO_4]^{3-}$ and $[P_2O_7]^{4-}$, are formed[11]. Reaction of P_4 with basic H_2O at 80–90°C yields hypophosphites:

$$P_4 + 4 [OH]^- + 4 H_2O \rightarrow 2 H_2 + 4 [H_2PO_2]^- \tag{l}[1,11]$$

$$P_4 + 4 [OH]^- + 2 H_2O \rightarrow 2 PH_3 + 2 [HPO_3]^{2-} \tag{m}[1,11]$$

Salts of hypodiphosphorous acid form in reactions[6,11] of red phosphorus with aq $NaClO_2$:

$$2 P + 2 NaClO_2 + 2 H_2O \rightarrow 2 HCl + Na_2(H_2P_2O_4) \tag{n}$$

and in lesser amounts in complex mixtures from the reaction of P_2I_4 with aq base[12] H_2O.

Alcohols react with PCl_3 to form o-dialkylphosphonates:

$$PCl_3 + 3 ROH \rightarrow RCl + 2 HCl + (RO)_2P(O)H \tag{o}$$

where R = alkyl, or o-monoalkylphosphonates:

$$3 ROH + PCl_3 \rightarrow 2 RCl + HCl + (RO)(HO)P(O)H \tag{p}$$

depending on reaction conditions[11,13]. At high T or upon thermolysis of $(RO)_2P(O)H$, o-monoalkylphosphonates form[11,13].

Alkyl- or arylphosphinic acids form from reaction of H_2O with dialkylphosphinous halides at 100°C[1,7]:

$$RPCl_2 + 2 H_2O \rightarrow 2 HCl + RPH(O)OH \tag{q}$$

where R = Me, Et, i-Pr, t-C_4H_9, CF_3, $PhCH_2$, Ph, etc. Alcoholysis of $RPCl_2$ yields o-alkylphosphinates:

$$PhPCl_2 + 2 EtOH \rightarrow EtCl + HCl + PhPH(O)Et \tag{r}[7,14,15]$$

Reaction of olefins with P_4, O_2 and H_2O yields β-hydroxyphosphinic acids in a complex reaction[7]:

$$2 PhCHCH_2 + P_4 + 4 O_2 + 6 H_2O \rightarrow 2 H_3PO_4 + 2 PhCH(OH)CH_2PH(O)OH \tag{s}$$

Hydrolysis of phosphinous chlorides yields the corresponding phosphine oxides, as:

$$Ph_2PCl + H_2O \rightarrow HCl + Ph_2PH(O) \tag{t}[6,16]$$

(A.D. NORMAN)

1. A. F. Toy, in *Comprehensive Inorganic Chemistry*, A. F. Trotman-Dickenson, ed., Vol. 2, Pergamon Press, Oxford, 1973, p. 389.
2. E. Fluck, *Top. Curr. Chem.*, 35, 1 (1973).
3. M. G. Palmer, Br. Pat. 943,281 (1963); *Chem. Abstr.*, 60, 6524 (1964).
4. R. C. Marriott, J. D. Odom, C. T. Sears Jr., *Inorg. Synth.*, 14, 1 (1973).
5. L. Maier, *Prog. Inorg. Chem.*, 5, 27 (1963).
6. J. Emsley, D. Hall, *The Chemistry of Phosphorus*, Harper and Row, London, 1976.
7. A. W. Frank, *Chem. Rev.*, 61, 389 (1961).
8. K. Hestermann, J. Joedden, G. Heymer, Ger. Pat. 2,721,425 (1979); *Chem. Abstr.*, 90, 152,353 (1979).

9. H. Fritzsche, U. Hasserodt, F. Korte, U.S. Pat. 3,261,871 (1966); *Chem. Abstr.*, 65, 18,619 (1966).
10. H. Fritzsche, U. Hasserodt, F. Korte, *Chem. Ber.*, 98, 1681 (1965).
11. D. S. Payne, in *Topics in Phosphorus Chemistry*, Vol. 4, M. Grayson, E. J. Griffith, eds., Wiley-Interscience, New York, 1967, p. 85.
12. M. Baudler, M. Mengel, *Z. Anorg. Allg. Chem.*, 374, 159 (1970).
13. D. H. Chadwick, R. S. Watt, in *Phosphorus and Its Compounds*, J. R. Van Wazer, ed., Vol. II, Interscience, New York, 1961.
14. K. Weissermel, H. J. Kleiner, M. Finke, U.-H. Felcht, *Angew. Chem., Int. Ed., Engl.*, 20, 223 (1981).
15. M. Finke, H-J. Kleiner, *Justus Leibigs Ann. Chem.*, 741 (1974).
16. H. J. Kleiner, *Justus Leibigs Ann. Chem.*, 751 (1974).

1.5.6.3. Involving Compounds of Arsenic.

Arsine is prepared electrochemically by reduction of As(III) species:

$$As^{3+} + 6\ e^- + 3\ H^+ \xrightarrow{H^+_{(aq)}} AsH_3 \qquad (a)^{1,2}$$

or by $NaBH_4$ reduction of As(III) species in acidic H_2O, e.g.:

$$4\ [As(OH)_4]^- + 3\ NaBH_4 + 7\ H^+ \rightarrow 3\ H_3BO_3 + 7\ H_2O + 3\ Na^+ + 4\ AsH_3 \quad (b)^{1,3}$$

Limited commercial use is made of the Zn reduction of $Me_2AsO(OH)$ to form[4] Me_2AsH:

$$Me_2AsO_2H + 2\ Zn + 4\ HCl \xrightarrow{H_2O} 2\ ZnCl_2 + 2\ H_2O + Me_2AsH \qquad (c)$$

In the presence of $AlCl_3$, benzene reacts with $AsCl_3$, forming a mixture of products[5] that includes $PhAsH_2$ and Ph_2AsH:

$$PhH + AsCl_3 \xrightarrow[35°C]{AlCl_3} Me_2CHCH_2CH_3,\ PhC_5H_{11},\ Ph_2AsH,\ PhAsH_2 \qquad (d)$$

(A.D. NORMAN)

1. W. L. Jolly, A. D. Norman, *Prep. Inorg. React.*, 4, 1 (1968).
2. V. P. Gladyshev, *Tr. Inst. Khim. Nauk., Akad. Nauk. Kaz. SSR*, 35, 74 (1973); *Chem. Abstr.*, 80, 43,447 (1974).
3. J. D. Smith, in *Comprehensive Inorganic Chemistry*, A. F. Trotman-Dickenson, ed., Vol. 2, Pergamon Press, Oxford, 1973, p. 547.
4. R. D. Feltham, W. Silverthorne, *Inorg. Synth.*, 10, 159 (1967).
5. L. Schmerling, U.S. Pat. 2,842,579, (1958); *Chem. Abstr.*, 55, 497e (1961).

1.5.6.4. Involving Compounds of Antimony.

Stibine for use in industrial processes is prepared electrochemically:

$$Sb^{3+} + 6\ e^- + 3\ H^+ \xrightarrow{H^+_{(aq)}} SbH_3 \qquad (a)$$

or by reaction of Sb(III) species with KBH_4[1,2]:

$$4\ [Sb(OH)_4]^- + 3\ NaBH_4 + 7\ H^+ \rightarrow 3\ H_3BO_3 + 7\ H_2O + 3\ Na^+ + 4\ SbH_3 \quad (b)$$

(A.D. NORMAN)

1. W. L. Jolly, A. D. Norman, *Prep. Inorg. React.*, *4*, 1 (1968).
2. I. A. Frolov, V. S. Zaburdyaev, S. I. Kulakov, G. M. Borchenko, USSR Pat. 468,880 (1975); *Chem. Abstr.*, *83*, 134,437 (1975).

1.5.7. The Synthesis of Deuterium Derivatives

1.5.7.1. by Interconversion of Deuterated Compounds

1.5.7.1.1. Involving Nitrogen.

Protonic (deuterio) acid solvolysis of -ide salts is a route to compounds with N–D bonds. Alkali or alkaline-earth nitrides or amides react with D_2O or DCl to form products in high isotopic purity, e.g.:

$$Mg_3N_2 + 3\ D_2O \rightarrow 3\ Mg(OD)_2 + 2\ ND_3 \qquad (a)^{1-3}$$

$$[M][R_2N] + DCl \rightarrow MCl + R_2ND \qquad (b)$$

where R = alkyl, aryl; M = Li, Na, K. Such reactions with DCl can proceed to deuterio ammonium $[NH_4]^+$ ions if xs DCl is present.

Cleavages of metal– or nonmetal–N bonds by D_2O or ROD yield N—D bonds, although these are seldom synthetically important[4]. Reactions of primary or secondary amines can produce mixed hydro–deuterio products:

$$(Me_3Si)_2NH + 2\ DCl \rightarrow 2\ Me_3SiCl + ND_2H \qquad (c)^5$$

Complex metal hydroborate-d_4 and hydroaluminate-d_4 reductions are used in syntheses of deuterioamines. Reactions of nitro compounds:

$$PhNO_2 \xrightarrow[\text{2. } D_2O]{\text{1. } LiAlD_4-THF} PhND_2 \qquad (d)$$

and imines:

$$Et_2C=NPh \xrightarrow[\text{2. } D_2O]{\text{1. } LiAlD_4-THF} Et_2CDNDPh \qquad (e)$$

with $LiAlD_4$ illustrate these. For details, treatises on organic chemistry should be consulted[6-8].

Reaction[9] of N_2 with D_2 gas at high T and P yields ND_3:

$$N_2 + 3\ D_2 \rightleftharpoons 2\ ND_3 \qquad (f)$$

however, the process is useful mainly only for large quantities of ND_3 because the apparatus required is complex.

Catalytic deuterium reduction of nitrogen compounds occurs readily, e.g.:

$$PhNO_2 + 3\ D_2 \xrightarrow{Pt} 2\ D_2O + PhND_2 \qquad (g)^{6,10}$$

$$PhNO + 2\ D_2 \xrightarrow[EtOD]{Pt} D_2O + PhND_2 \qquad (h)^{10}$$

$$PhCN + 2\ D_2 \xrightarrow{RhD(PPh_3)_3} PhCD_2ND_2 \qquad (i)^{10}$$

1.5. Formation of Bonds between Hydrogen and N, P, As, Sb, Bi   59
1.5.7. The Synthesis of Deuterium Derivatives
1.5.7.1. by Interconversion of Deuterated Compounds

Details can be found in organic synthesis texts[6,10].

(A.D. NORMAN)

1. G. Brauer, M. Baudler, *Handbook of Preparative Inorganic Chemistry*, G. Brauer, ed., 2nd ed., Vol. 1, Academic Press, New York, 1963.
2. L. K. Krannich, U. Thewalt, W. J. Cook, S. R. Jain, H. H. Sisler, *Inorg. Chem.*, *12*, 2304 (1973).
3. H. L. Crespi, J. J. Katz, in *Inorganic Isotopic Synthesis*, R. H. Herber, ed., Benjamin, New York, 1962, p. 14.
4. D. C. Bradley, *Adv. Inorg. Chem. Radiochem.*, *15*, 259 (1972).
5. D. A. Armitage, *Organomet. Chem.*, *10*, 86 (1980).
6. A. Hajós, *Complex Hydrides*, Elsevier, Amsterdam, 1979.
7. H. C. Brown, *Boranes in Organic Chemistry*, Cornell Univ. Press, Ithaca, NY, 1972.
8. J. R. Malpass, in *Comprehensive Organic Chemistry*, D. Barton, W. D. Ollis, eds., Vol. 2, I. O. Sutherland, ed., Pergamon Press, New York, 1979, p. 3.
9. W. L. Jolly, *The Inorganic Chemistry of Nitrogen*, Benjamin, New York, 1964.
10. C. A. Beuhler, D. E. Pearson, in *Survey of Organic Synthesis*, Vol. 2, Wiley-Interscience, New York, 1977, p. 391.

1.5.7.1.2. Involving Phosphorus.

Ionic hydride reductions of halophosphorus compounds that are effective for phosphine syntheses are adaptable to deuteriophosphine syntheses. Lithium-deuteride reductions yield phosphorus subhydride, phosphine or alkyl- or arylphosphines[1,2]:

$$PBr_3 + 3 \ LiD \rightarrow 3 \ LiBr + \frac{1}{n} (PD)_n + D_2 \qquad \text{(a)}$$

$$R_{3-n}PCl_n + n \ LiD \rightarrow n \ LiCl + R_{3-n}PD \qquad \text{(b)}$$

where R = alkyl, aryl; n = 1–3. However, covalent or complex hydride reductants are preferred.

Covalent deuterio-hydrides, e.g., silanes or stannanes, are used in deuteriophosphine synthesis[3], e.g.:

$$2 \ R_2P(O)Cl + Ph_2SiD_2 \xrightarrow{200\text{-}275°C} -(-Ph_2SiO-)- + Ph_2SiCl_2 + 2 \ R_2PD \qquad \text{(c)}[2]$$

Complex hydride reductions are adapted to deuteriophosphine syntheses[2,3]. The isotopic purity of the products is high. Reactions of $LiAlD_4$ with chlorophosphines:

$$4 \ PCl_3 + 3 \ LiAlD_4 \rightarrow 3 \ LiAlCl_4 + 4 \ PD_3 \qquad \text{(d)}$$

phosphonyl chlorides:

$$2 \ PhP(O)Cl_2 + 2 \ LiAlD_4 \rightarrow LiAlO_2 + LiAlCl_4 + 2 \ D_2 + 2 \ PhPD_2 \qquad \text{(e)}$$

and phosphinate esters:

$$4 \ Me_2P(O)OEt + 2 \ LiAlD_4 \rightarrow LiAl(OD)_4 + LiAl(OEt)_4 + 4 \ Me_2PD \qquad \text{(f)}$$

illustrate these.

Solvolysis of -ide salts with D_2O yields deuteriophosphines. Reaction of D_2SO_4 or D_2O with AlP [4] or Ca_3P_2[5] produces PD_3:

$$2 \ AlP + 3 \ D_2SO_4 \xrightarrow{-78°C} Al_2(SO_4)_3 + 2 \ PD_3 \qquad \text{(g)}[5]$$

60   1.5. Formation of Bonds between Hydrogen and N, P, As, Sb, Bi
      1.5.7. The Synthesis of Deuterium Derivatives
      1.5.7.1. by Interconversion of Deuterated Compounds

From the $D_2O-Ca_3P_2$ reaction, P_2D_4 is obtained also. When treated with D_2O alkali-metal organophosphides produce deuteriophosphines in high yields ($> 75\%$) and isotopically pure, e.g.:

$$Ph_2PLi + D_2O \xrightarrow{THF} LiOD + Ph_2PD \qquad (h)[6]$$

$$PhPNa_2 + 2 D_2O \rightarrow 2 NaOD + PhPD_2 \qquad (i)[7]$$

Cleavage of Si—P bonds by D_2O produces P—D bonds:

$$2 (R_3Si)_3P + 3 D_2O \rightarrow 3 (R_3Si)_2O + 2 PD_3 \qquad (j)[1,8]$$

where R = H, Me;

$$3 [h^5\text{-}C_5H_5NiP(SiMe_3)_2]_2 + 12 MeOD \rightarrow 12 Me_3SiOMe + 2 (h^5\text{-}C_5H_5NiPD_2)_3 \qquad (k)[9]$$

From primary or secondary silylphosphine deuterolysis, mixed hydro–deuterio products form:

$$(SiH_3)_2PH + D_2O \rightarrow (SiH_3)_2O + PD_2H \qquad (l)[1]$$

Deuterolysis of PI_3 produces[4] PD_3, whereas D_2O and PCl_3 yield D_3PO_3:

$$PCl_3 + 3 D_2O \rightarrow 3 DCl + D_3PO_3 \qquad (m)[4]$$

Reaction of basic D_2O with elemental phosphorus produces $Ba(D_2PO_2)_2$ and PD_3 in high yields[4]:

$$3 Ba(OD)_2 + 6 D_2O + 2 P_4 \rightarrow 3 Ba(D_2PO_2)_2 + 2 PD_3 \qquad (n)$$

(A.D. NORMAN)

1. W. L. Jolly, A. D. Norman, *Prep. Inorg. React.*, 4, 1 (1968).
2. A. Hajós, *Complex Hydrides*, Elsevier, Amsterdam, 1979.
3. S. D. Venkataramu, G. D. McDonnell, W. R. Purdum, M. El-Deek, K. D. Berlin, *Coord. Chem. Rev.*, 77, 121 (1977).
4. H. L. Crespi, J. J. Katz, in *Inorganic Isotopic Syntheses*, R. H. Herber, ed., Benjamin, New York, 1962, p. 14.
5. R. C. Marriott, J. D. Odom, C. T. Sears, Jr., *Inorg. Synth.*, 14, 1 (1973).
6. V. D. Bianco, S. Doronzo, *Inorg. Synth.*, 16, 161 (1976).
7. L. Horner, P. Beck, H. Hofman, *Chem. Ber.*, 92, 2088 (1959).
8. H. Bürger, U. Goetze, *J. Organomet. Chem.*, 12, 451 (1968).
9. H. Schäffer, Z. *Anorg. Allg. Chem.*, 459, 157 (1979).

1.5.7.1.3. Involving Arsenic.

High yields of isotopically pure AsD_3 form upon dropwise addition[1] of D_2O to Na_3As:

$$Na_3As + 3 D_2O \rightarrow 3 NaOD + AsD_3 \qquad (a)$$

Lithium tetrahydroborate-d_4 reduction of chloroarsines yields deuteroarsines, in reactions[1-3] more efficient than those using $LiAlD_4$.

$$R_{3-n}AsCl_n + n LiBD_4 \rightarrow \tfrac{n}{2} B_2D_6 + n LiCl + R_{3-n}AsD_n \qquad (b)$$

Deuterolysis of Si—As or Ge—As bonds produces As—D bonds:

$$2 \text{ MH}_3\text{AsH}_2 + \text{D}_2\text{O} \rightarrow (\text{MH}_3)_2\text{O} + 2 \text{ AsH}_2\text{D} \qquad (c)^{4,5}$$

where M = Si, Ge.

(A.D. NORMAN)

1. J. E. Drake, C. Riddle, *Inorg. Synth.*, *13*, 14 (1972).
2. G. O. Doak, L. D. Freedman, *Organometallic Compounds of Arsenic, Antimony, and Bismuth*, Wiley-Interscience, New York, 1970.
3. E. Wiberg, K. Mödritzer, *Z. Naturforsch.*, *Teil B, 12*, 127 (1957).
4. J. E. Drake, C. Riddle, *J. Chem. Soc., A.*, 2452 (1968).
5. H. Bürger, U. Goetze, *J. Organomet. Chem.*, *12*, 451 (1968).

1.5.7.1.4. Involving Antimony.

High isotopic purity stibines form from LiAlD_4 with halostibines in ether, e.g.:

$$\text{R}_{3-n}\text{SbCl}_n + \tfrac{n}{4} \text{ LiAlD}_4 \rightarrow \tfrac{n}{4} \text{ LiAlCl}_4 + \text{R}_{3-n}\text{SbD}_n \qquad (a)^{1-4}$$

where $n = 1\text{—}3$; R = alkyl, aryl. Stibine-d_3 also forms in the LiBD_4 reduction of SbCl_3 at $-70°C$ in ethers.

$$2 \text{ SbCl}_3 + 6 \text{ LiBD}_4 \xrightarrow{-70°C} 6 \text{ LiCl} + 6 \text{ B}_2\text{D}_6 + 2 \text{ SbD}_3 \qquad (b)^1$$

(A.D. NORMAN)

1. G. O. Doak, L. D. Freedman, *Organometallic Compounds of Arsenic, Antimony, and Bismuth*, Wiley- Interscience, New York, 1970.
2. K. Tamaru, *J. Phys. Chem.*, *59*, 1084 (1955).
3. H. L. Crespi, J. J. Katz, in *Inorganic Isotopic Syntheses*, R. H. Herber, ed., Benjamin, New York, 1962, p. 14.
4. A. L. Reingold, P. Choudhury, M. F. El-Shazly, *Synth. React. Inorg. Metal-Org. Chem.*, *8*, 453 (1978).

1.5.7.1.5. Involving Bismuth.

Deuteriobismuthines form in the LiAlD_4 reduction of the corresponding chlorides; however, bismuthines[1] are thermally unstable above $-45°C$:

$$\text{Me}_{3-n}\text{BiCl}_n + \tfrac{n}{4} \text{ LiAlD}_4 \xrightarrow{\text{Me}_2\text{O}} \tfrac{n}{4} \text{ LiAlCl}_4 + \text{Me}_{3-n}\text{BiD}_n \qquad (a)$$

where n = 1,2.

(A.D. NORMAN)

1. L. D. Freedman, G. O. Doak, *Chem. Rev.*, *82*, 15 (1982).

1.5.7.2. by Isotopic Enrichment Using Chemical Reactions

1.5.7.2.1. of Nitrogen Compounds.

Exchange of amine hydrogen atoms with deuterated hydroxylic solvents occurs rapidly, especially upon acid catalysis:

$$\text{R}_{3-n}\text{NH}_n \underset{}{\overset{\text{D}^+-\text{D}_2\text{O}}{\rightleftharpoons}} \text{R}_{3-n}\text{ND}_n \qquad (a)^{1-3}$$

where R = alkyl, aryl; n = 1–3. Under similar conditions organic amides and metal amines undergo exchange with D_2O also.

$$[Co(NH_3)_6]^{3+} \underset{[OD]^- -D_2O}{\rightleftharpoons} [Co(ND_3)_6]^{3+} \qquad (b)^2$$

(A.D. NORMAN)

1. W. L. Jolly, *The Inorganic Chemistry of Nitrogen,* Benjamin, New York, 1964.
2. F. A. Cotton, G. Wilkinson, *Advanced Inorganic Chemistry,* 4th ed., Wiley-Interscience, New York, 1980.
3. A. I. Shatenshtein, *Isotopic Exchange and the Replacement of Hydrogen in Organic Compounds,* Consultants Bureau, New York, 1962.

1.5.7.2.2. of Phosphorous Compounds.

Hydrogen exchange between phosphines and deuterated hydroxylic solvents occurs slowly, the rate depending on the acidity of the phosphine. In the presence of acid or base catalysts, exchange is faster and the reactions can be used for deuteriophosphine synthesis[1], e.g.:

$$PH_3 \underset{D_2O-D^+}{\rightleftharpoons} PD_3 \qquad (a)$$

Phosphonium ions exchange with acidic deuterium sources to form P—D bond-containing products. Alkylphosphoniums react with D_2O as:

$$[MePH_3]^+ \underset{D_2O}{\rightleftharpoons} [MePD_3]^+ \qquad (b)^{1,2}$$

Acidic phosphido metal complexes exchange similarly, in instances where the complex is resistant to solvolysis:

$$\{[h^5\text{-}Cp(CO)_2Fe]_2PH_2\}Cl \underset{D_2O}{\rightleftharpoons} \{[h^5\text{-}Cp(CO)_2Fe]_2PD_2\}Cl \qquad (c)^3$$

Phosphinites and phosphinates exchange with deuterated media:

$$(RO)_2P(O)H \underset{D_2O}{\rightleftharpoons} (RO)_2P(O)D \qquad (d)^2$$

$$(RO)P(O)H_2 \underset{D_2O}{\rightleftharpoons} (RO)P(O)D_2 \qquad (e)^2$$

where R = alkyl, aryl. The P—H bonds in P—H phosphazenes exchange rapidly in MeOD to form deuterio products:

$$(Ph_2PN)_2[MeP(H)N] \underset{MeOD}{\rightleftharpoons} (Ph_2PN)_2[MeP(D)N] \qquad (f)^4$$

(A.D. NORMAN)

1. A. I. Shatenshtein, *Isotopic Exchange and the Replacement of Hydrogen in Organic Compounds,* Consultants Bureau, New York, 1962.
2. J. Emsley, D. Hall, *The Chemistry of Phosphorus,* Harper and Row, New York, 1976.
3. H. Schaeffer, *Angew. Chem., Int. Ed. Engl., 20,* 608 (1981).
4. A. Schmidpeter, K. Blank, H. Eiletz, H. Smetana, C. Weinand, *Synth. React. Inorg. Metal-Org. Chem., 7,* 1 (1977).

1.5.7.2.3. of Arsenic Compounds.

Hydrogen exchange between AsH_3 and D_2O occurs, at rates enhanced by either acid or base catalysts:

$$R_{3-n}AsH_n \xrightleftharpoons{\quad D^+-D_2O \quad} R_{3-n}AsD_n \qquad (a)^{1,2}$$

where R = alkyl, aryl; $n = 1$–3.

(A.D. NORMAN)

1. A. I. Shatenshtein, *Isotopic Exchange and the Replacement of Hydrogen in Organic Compounds,* Consultants Bureau, New York, 1962.
2. G. O. Doak, L. D. Freedman, *Organometallic Compounds of Arsenic, Antimony, and Bismuth,* Wiley-Interscience, New York, 1970.

1.6. Formation of Bonds between Hydrogen and Elements of Group IVB (C, Si, Ge, Sn, Pb)

1.6.1. Introduction

This chapter presents reactions by which bonds between hydrogen and C, Si, Ge, Sn or Pb are formed. The chemistry encompassed is large, because C—H bond formation is important in organic, organometallic, inorganic and biochemistry. Bond formations of interest in inorganic and organometallic chemistry receive greatest attention. Reactions in the realm of classical organic or biochemistry are treated more briefly.

Reactions are classed according to reagent or reaction type. Arbitrary characterization of systems must be made to allow their classification; thus X—H reagents are classed as protonic reagents if they participate in reactions as protonic acids, as is the case for molecules in which the electronegativity of X is appreciably greater than that of H. Otherwise, they are classed as simple hydrides if they contain hydrogen and one main element (binary) or hydrogen and two main elements (ternary, e.g., SiH_3PH_2) and complex hydrides if they contain hydride (H^-) moieties coordinated to a central element to form a complex anionic species, e.g., $LiBH_4$.

(A.D. NORMAN)

1.6.2. from the Elements

1.6.2.1. Giving Hydrides of Carbon

1.6.2.1.1. from Elemental Carbon.

Carbon does not react with H_2 at an appreciable rate under ordinary conditions[1]. At high T and in the presence of catalysts reaction occurs as:

$$C_{(graph)} + 2 H_2 \rightarrow 2 CH_4 \tag{a}$$

where $\Delta G°(25°C) = -50.7$ kJ mol^{-1}. Powdered $C_{(graph)}$ impregnated[2] with Ni, or C in the presence[1] of powdered Ni or Fe–Cr sponge[3], reacts at $500-1000°C$ to form CH_4 along with small quantities of other hydrocarbons, e.g., C_2H_2. At higher T, C_2H_2 becomes the dominant product:

$$2 C_{(graph)} + H_2 \xrightarrow{> 1000°C} C_2H_2 \tag{b}$$

Atomic hydrogen, produced from H_2 in an electrical or microwave discharge, reacts with a carbon surface to form mainly C_2H_2 and CH_4 [e.g., Eq. (c)] along with small quantities of higher mol wt materials[1,4]:

$$2 C_{(graph)} + 2 H \rightarrow C_2H_2 \tag{c}$$

1.6.2. from the Elements
1.6.2.1. Giving Hydrides of Carbon
1.6.2.1.1. from Elemental Carbon.

65

Reaction of H with ICN yields[5] HCN:

$$H + ICN \rightarrow I + HCN \qquad (d)$$

Carbon atoms, produced by thermal vaporization or laser irradiation of $C_{(graph)}$ or in a carbon arc, react with H-containing substrates (e.g., alkanes, silanes, boranes, phosphines) to form C—H bonds[1,6]. The carbon atoms, mixtures of (^3P), (^1D) and (^1S) electronic-state species, react in the gas or condensed phase in a low-T matrix. Condensation of carbon atoms onto a cooled surface ($-196°C$) along with substrate results in reactions of $C(^1S)$ atoms, before decay to the less reactive $C(^3P)$ species occurs. Condensation of C-arc produced carbon atoms with an alkane (e.g., iso-C_4H_{10}) at $-196°C$ results in reaction of both singlet (^1S) and triplet (^3P) species to form singlet and triplet carbenes:

$$\overset{..}{\underset{..}{C}} + Me_2CHCH_3 \rightarrow Me_2CHCH_2\overset{..}{C}H \qquad (e)$$

$$\overset{.}{:}\overset{.}{C}\cdot + Me_2CHCH_3 \rightarrow Me_2CHCH_2\overset{.}{C}H \qquad (f)$$

Subsequent abstraction of hydrogen atoms from the substrate by the triplet carbene or insertion into a C—H bond by the singlet species can occur:

$$Me_2CHCH_2\overset{..}{C}H \rightarrow Me_2C(CH_2)_2 \qquad (g)$$

Laser-evaporated carbon atoms react[7] with H_2O:

$$C + H_2O \rightarrow CO, H_2, C_2H_2 \qquad (h)$$

or[8] B_5H_9:

$$C + B_5H_9 \rightarrow H_2, B_5C_2H_7 \qquad (i)$$

at low T to form primarily C_2H_2 and $B_5C_2H_7$, respectively. Phosphine passing through a carbon arc yields HCP:

$$C + PH_3 \rightarrow H_2, C_2H_2, C_2H_4, HCP \qquad (j)^9$$

Thermal $C(^3P)$ and hydrogen atoms in the gas phase, from C_3O_2 and H_2 in a microwave discharge, yield CH which subsequently reacts[1] to CH_4 and C_2H_2. Carbon atoms from the $^{14}N(n,p)^{14}C$ reaction with NH_3 yield[1] CH_4 and CH_3NH_2.

Small carbon molecules, C_2, C_3 and C_4, react with H-containing substrates, e.g.:

$$C_3 + 2\ ROH \xrightarrow{ROH} (RO)_2CHCCH \qquad (k)^{1,5}$$

$$C_4 + x\ H \xrightarrow{CH_3CHO} EtCCH, MeCHCCH_2, C_4H_6, C_4H_4 \qquad (l)^{1,6}$$

(A.D. NORMAN)

1. A. K. Holliday, G. Hughes, S. M. Walker, in *Comprehensive Inorganic Chemistry*, Vol. 6, A. F. Trotman-Dickenson, ed., Pergamon Press, Oxford, 1973, p. 1173.
2. S. D. Robertson, N. Mulder, R. Prins, *Carbon*, 13, 348 (1975).
3. M. A. Qayzon, D. A. Reeve, *Carbon*, 14, 199 (1976).
4. R. K. Gould, *J. Chem. Phys.*, 63, 1825 (1975).
5. G. P. Horgan, M. R. Dunn, C. G. Freeman, M. J. McEwan, L. F. Phillips, *J. Phys. Chem.*, 76, 1392 (1972).
6. P. S. Skell, J. S. Havel, M. J. McGlinchey, *Acc. Chem. Res.*, 6, 97 (1973).
7. P. H. Kim, K. Taki, S. Namba, *Bull. Chem. Soc. Jpn.*, 2953 (1975).

66   1.6. Formation of Bonds between Hydrogen and C, Si, Ge, Sn, Pb
       1.6.2. from the Elements
       1.6.2.1. Giving Hydrides of Carbon

8. S. R. Prince, R. Schaeffer, *J. Chem. Soc., Chem. Commun.*, 451 (1968).
9. T. E. Gier, *J. Am. Chem. Soc., 83,* 1769 (1961).

1.6.2.1.2. from Elemental Hydrogen.

Main-group element or transition metal–carbon σ bonds can be cleaved by H_2. Group IVA metal alkyls[1], e.g., Zr, react:

$$(h^5\text{-}Cp)_2Zr(H)CH_2C_6H_{11} + H_2 \xrightarrow{25°C} (h^5\text{-}Cp)_2ZrH_2 + CH_3C_6H_{11} \qquad (a)$$

In the presence of Pd, H_2 reacts with $[PhCH_2Cr(H_2O)_5]^{2+}$ or $[CH_2=CHCH_2Co(CN)_5]^{3-}$ forming $PhCH_3$ and $CH_3CH=CH_2$, respectively.[2] The tetraironcarbonylcarbide cluster reacts with H_2:

$$[Fe_4C(CO)_{12}]^- + H_2 \xrightarrow{0°C} [HFe_4(h^2\text{-}CH)(CO)_{12}]^- \qquad (b)^3$$

Cleavage of the hydrocarbon moieties from Pt or Os complexes occurs at 20°C, e.g.:

$$Os(CO)_4(H)CH_3 + H_2 \rightarrow Os(CO)_4H_2 + CH_4 \qquad (c)^4$$

Triphenylarsine and Ph_3Bi with H_2 in the presence of Ni yield[5] C_6H_6, e.g.:

$$2\ (C_6H_5)_3As + 3\ H_2 \xrightarrow{Ni} 2\ As + 6\ C_6H_6 \qquad (d)$$

Phenylmercury(II) acetate with H_2 in the presence of Rh yields[6] acetic acid and C_6H_6:

$$C_6H_5HgOCOMe + H_2 \rightarrow Hg + C_6H_6 + MeCO_2H \qquad (e)$$

Cleavage of π-bonded alkenes and alkynes from a metal complex can occur:

$$[(Ph_4C_4)PdCl_2]_2 \xrightarrow{H_2} PhCHC(Ph)C(Ph)CHPh \qquad (f)^7$$

The formal cleavage of C—C bonds by H_2 in hydrogenolysis or cracking reactions, usually in the presence of metal catalysts[8], e.g.:

$$H_2 + C_2H_6 \xrightarrow{Ni-Cu} 2\ CH_4 \qquad (g)$$

is common (see treatises on organic chemistry)[9].

Homogeneous or heterogeneous catalyzed H_2 reduction of unsaturated organic compounds arene, alkyne, alkene, ketone, oxime, nitrile and imine occurs readily[1,4,6,8-12], e.g.:

$$C_6H_6 + 3\ H_2 \xrightarrow{(h^3\text{-}C_3H_5)Co[P(OMe)_3]_3} C_6H_{12} \qquad (h)^{13}$$

$$MeC\equiv CMe + H_2 \xrightarrow{Pd-C} cis\text{-}MeCH=CHMe \qquad (i)^{11}$$

$$Me_2CO + H_2 \xrightarrow{RhCl_2(bipy)Cl\cdot 2\ H_2O} Me_2CHOH \qquad (j)^{1,12}$$

$$PhCN + 2\ H_2 \xrightarrow{RhH(PPh_3)_3} PhCH_2NH_2 \qquad (k)^{9,12}$$

Hydrogenations using heterogeneous catalysts usually require above ambient T and H_2 pressures $\geq 10^3$ Pa. Homogeneous catalysts are often more selective and involve lower T (e.g., 25°C) and H_2 (ca. 10^2 Pa). Highly useful selectivity in hydrogenation is shown in reactions of H_2 with prochiral substrates in the presence of chiral catalysts, e.g., optically active dopa, 3-(3,4-dihydroxyphenyl)-L-alanine, can be obtained in high optical purity (enantiomeric xs $> 90\%$):

where Rh* $= [Rh\{(S),(S)-[Ph_2P(Me)C_2H_2(Me)PPh_2]\}(1,5-C_8H_{12})]Cl.$

Reaction of H_2 with Zr-carbonyl complexes yields CH_3O-substituted products:

$$(h^5\text{-}Cp)_2Zr(CO)_2 + 2 H_2 \xrightarrow{110°C} CO + (h^5\text{-}Cp)_2ZrH (OCH_3) \qquad (m)[15]$$

Under similar conditions the CO groups of the Ti complex, $(h^5\text{-}C_5H_5)_2Ti(CO)_2$, are reduced to CH_4[15].

Reactions of H_2 with CO in the presence of catalysts produce hydrocarbon products ranging from CH_3OH and CH_4 to fuel oils[16-20]:

$$n\ CO + 2n\ H_2 \rightarrow (CH_2)_n + n\ H_2O \qquad (n)$$

Products depend critically on reaction conditions. In the presence of Pd–La$_2$O$_3$ or Zn–Cr$_2$O$_3$, CO and H_2 form mainly CH_3OH:

$$CO + 2 H_2 \rightarrow CH_3OH \qquad (o)[17]$$

Over Ni at 500–700°C, CH_4 is the main product[18]. Other transition-metal heterogeneous catalysts produce higher yields of $C>1$ products. Using the homogeneous catalyst, $Rh(CO)_2(MeCOCH_2CO_2)$, $(CH_2OH)_2$ forms in 70% yield[19]:

$$2 CO + 3 H_2 \rightarrow HOCH_2CH_2OH \qquad (p)$$

Hydroformylation, the addition of CO and H_2 to an olefin[1,4,20], occurs in the presence of homogeneous, $[RhCl_3(PPh_3)_3, HCo(CO)_4]$, heterogeneous (Ni, Co, Fe, Rh) or supported catalysts $[Rh(CO)_2Cl$ on $SiO_2][8,20]$:

$$RCH{=}CH_2 \xrightarrow{H_2-CO} \begin{array}{l} RCH_2CH_2CHO,\ RCH_2CH_2CH_2OH \qquad (q)\\[1em] RCH(CH_3)CHO,\ RCH(CH_3)CH_2OH \qquad (r) \end{array}$$

Aldehydes form prior to alcohols, and both branched and linear isomers can arise; however, depending on the olefin substrate and under controlled reaction conditions, conditions highly specific for one compound often exist. From 1-hexene with an equimol CO and H_2 mixture at 5 to 9×10^3 Pa in EtOH–C$_6$H$_6$ at 110°C, in the presence of $RhCl_3(PPh_3)_3$, n-C$_7$H$_{15}$OH forms nearly quantitatively[1]:

$$n\text{-}C_4H_9CH{=}CH_2 + 2 H_2 + CO \rightarrow n\text{-}C_7H_{15}OH \qquad (s)$$

Reaction of prochiral olefins with CO and H_2 in the presence of chiral catalysts results in asymmetric hydroformylation[21].

Carbon dioxide reacts with H_2 over a CuO–ZnO catalyst to form CH_3OH:

$$CO_2 + 3 H_2 \rightarrow H_2O + CH_3OH \qquad (t)[22]$$

and over Ni to form CH_4:

$$CO_2 + 4 H_2 \rightarrow 2 H_2O + CH_4 \qquad (u)[22]$$

Hydrogen and $[C_2H_5]^+$ react in the gas phase to $[C_2H_7]^+$:

$$[C_2H_5]^+_{(g)} + H_2 \rightarrow [C_2H_7]^+_{(g)} \qquad (v)^{23}$$

(A.D. NORMAN)

1. P. J. Brother, *Prog. Inorg. Chem.*, 28, 1 (1981).
2. G. W. Parshall, *Adv. Organomet. Chem.*, 7, 157 (1968).
3. M. Tachikawa, E. L. Muetterties, *Prog. Inorg. Chem.*, 28, 203 (1981).
4. F. A. Cotton, G. Wilkinson, *Advanced Inorganic Chemistry*, 4th ed., Wiley-Interscience, New York, 1980.
5. G. O. Doak, L. D. Freedman, *Organometallic Compounds of Arsenic, Antimony, and Bismuth*, Wiley-Interscience, New York, 1970.
6. F. H. Jardine, *Prog. Inorg. Chem.*, 28, 63 (1981).
7. P. M. Maitlis, *Adv. Organomet. Chem.*, 4, 95 (1966).
8. S. C. Davis, K. J. Klabunde, *Chem. Rev.*, 82, 153 (1982).
9. L. F. Fieser, M. Fieser, *Advanced Organic Chemistry*, Rheinhold, New York, 1961.
10. P. N. Rylander, *Organic Synthesis with Noble Metal Catalysts*, Academic Press, New York, 1973.
11. D. A. MacNeil, N. K. Roberts, B. Bosnich, *J. Am. Chem. Soc.*, 103, 2273 (1981).
12. C. A. Buehler, D. E. Pearson, *Survey of Organic Syntheses*, Vol. 2, Wiley-Interscience, New York, 1977, p. 391.
13. E. L. Muetterties, J. R. Bleeke, *Acc. Chem. Res.*, 12, 324 (1979).
14. M. D. Fryzuk, B. Bosnich, *J. Am. Chem. Soc.*, 99, 6262 (1977).
15. P. T. Wolczanski, J. E. Bercaw, *Acc. Chem. Res.*, 13, 121 (1980).
16. W. A. Hermann, *Angew. Chem., Int. Ed. Engl.*, 21, 117 (1982).
17. E. Ramarson, R. Kieffer, A. Kiennemann, *J. Chem. Soc., Chem. Commun.*, 645 (1982).
18. G. A. Sommerjai, *Catal. Rev.-Sci. Eng.*, 23, 189 (1981).
19. C. Master, *Adv. Organomet. Chem.*, 17, 61 (1979).
20. R. L. Pruett, *Adv. Organomet. Chem.*, 17, 1 (1979).
21. C. U. Pittman, Jr., Y. Kawahata, L. I. Flowers, *J. Chem. Soc., Chem. Commun.*, 473 (1982).
22. J. Haggin, *Chem. Eng. News*, 60, 13 (1982).
23. B. Bethell, in *Comprehensive Organic Chemistry*, D. Barton, W. D. Ollis, eds., Vol. 1, J. F. Stoddart, ed., Pergamon Press, Oxford, 1979, p. 411.

1.6.2.2. Giving Hydrides of Silicon.

Silicon at 400°C reacts with anhyd HCl yielding[1,2] $SiHCl_3$ and $SiCl_4$:

$$Si \xrightarrow{\text{HCl}} SiCl_4, SiHCl_3 \qquad (a)$$

Similarly, alkylsilanes, e.g., $MeSiCl_2H$, are produced from the reaction of MeCl and H_2 with a hot Cu–Si mixture[2].

Ground-state $Si(^3P)$ in the gas phase with SiH_4 yields Si_2H_6, along with lesser amounts of Si_3H_8 through a combination of H abstraction and Si—H bond insertion[3]:

$$Si_{(g)} + SiH_4 \rightarrow Si_2H_6, Si_3H_8 \qquad (b)$$

Silicon atoms in a matrix react above -196°C in the presence of Si—H bond-containing substrates (e.g., $MeSiH_3$, Me_2SiH_2, Me_3SiH, Si_2H_6) to form complex mixtures of products[4]. From the Si atom–Me_2SiH_2 reaction, a product mixture including CH_4, $Me_2Si_2H_4$, $Me_3Si_2H_3$ and $Me_2SiHSiH_2SiHMe_2$ arises.

Hydrogen reacts with SiO_2 at 800°C in the presence of Al and $AlCl_3$ forming SiH_4 in high yield[5]. Optimum yields (ca. 80%) are attained with H_2 at 9×10^4 Pa.

$$3 \, SiO_2 + 4 \, Al + 2 \, AlCl_3 + 6 \, H_2 \rightarrow \frac{6}{n} \, (AlOCl)_n + 3 \, SiH_4 \qquad (c)$$

Hydrogen passed over Si at 1100–1200°C transports Si as a result of formation[2] of unstable SiH. Hydrogen reduces $SiCl_4$ to $SiHCl_3$ in a d.c. arc:

$$SiCl_4 + H_2 \xrightarrow{arc} HCl + SiHCl_3 \qquad (d)^2$$

or to SiH_4 in an LiCl–KCl eutectic containing LiH. With the latter, reduction may involve LiH as the active species (see §1.6.4.2.1)[2].

Hydrogen or H_2–CO cleavage of Si—Si or Si–metal bonds can occur:

$$Ph_3SiSiPh_3 + H_2 \xrightarrow{Cr} 2\ Ph_3SiH \qquad (e)^6$$

$$[HRu_3(CO)_{10}(SiEt_3)_2]^- + CO + H_2 \rightarrow [HRu_3(CO)_{11}]^- + 2\ Et_3SiH \qquad (f)^7$$

Atomic H reacts with SiH_4 in the gas phase, forming mixtures of silanes in low yield:

$$H + SiH_4 \rightarrow H_2,\ Si_2H_6,\ Si_3H_8,\ Si_4H_{10} \qquad (g)^8$$

Silane is produced in low yields from reactions of at H with a Si surface[2]:

$$4\ H + Si \rightarrow SiH_4 \qquad (h)$$

(A.D. NORMAN)

1. E. A. V. Ebsworth, *Volatile Silicon Compounds,* Pergamon Press, New York, 1963.
2. E. Wiberg, E. Amberger, *Hydrides of the Elements of Main Groups I–IV,* Elsevier, Amsterdam, 1971.
3. P. P. Gaspar, K. Y. Choo, E. Y. Y. Lam, A. P. Wolf, *J. Chem. Soc., Chem. Commun.,* 1012 (1971).
4. P. S. Skell, P. W. Owen, *J. Am. Chem. Soc.,* 94, 5434 (1972).
5. W. L. Jolly, A. D. Norman, *Prep. Inorg. React.,* 4, 1 (1968).
6. H. Gilman, W. H. Atwell, F. K. Cartledge, *Adv. Organomet. Chem.,* 4, 1 (1966).
7. G. Suss-Fink, *Angew. Chem., Int. Ed. Engl.,* 21, 73 (1982).
8. E. R. Austin, F. R. Lampe, *J. Phys. Chem.,* 80, 2811 (1976).

1.6.2.3. Giving Hydrides of Germanium.

Germanium at 480–500°C reacts with anhyd HCl forming a mixture[1] of $GeCl_3H$ and $GeCl_4$:

$$Ge \xrightarrow{HCl} GeCl_4\ (70\%),\ GeCl_3H\ (30\%) \qquad (a)$$

Tribromogermane is produced similarly in the reaction of HBr with Ge at 450°C in the presence of Cu powder. Trimethylsilane reacts with Ge atoms:

$$Ge_{(g)} + 4\ Me_3SiH \rightarrow Me_6Si_2 + H_2 + (Me_3Si)_2GeH_2 \qquad (b)^2$$

Hydrogen passed over liq Ge at 1000–1100°C transports Ge, probably through GeH_2 formation[3]. Hydrogen at 10^2 to 10^4 Pa and 20–100°C cleaves the metal–Ge bonds in cis-Pt and -Pd complexes as:

$$cis\text{-}M(P_2)(GePh_3)_2 + H_2 \rightarrow cis\text{-}M(P_2)(H)GePh_3 + Ph_3GeH \qquad (c)^{1,4}$$

where $P_2 = 2\ Et_3P,\ Ph_2PCH_2CH_2PPh_2$; M = Pt, Pd.

(A.D. NORMAN)

1. F. A. Glockling, *The Chemistry of Germanium,* Academic Press, London, 1969.
2. R. T. Conlin, S. H. Lockhard, P. P. Gaspar, *J. Chem. Soc., Chem. Commun.,* 825 (1975).

3. E. Wiberg, E. Amberger, *Hydrides of the Elements of Main Groups I–IV*, Elsevier, Amsterdam, 1971.
4. D. M. Roundhill, *Adv. Organomet. Chem.*, *13*, 273 (1975).

1.6.2.4. Giving Hydrides of Tin.

Reactions of Sn, $SnCl_2$ or $SnCl_4$ with at H produce SnH_4:

$$Sn + 4 H \rightarrow SnH_4 \tag{a}$$

$$SnCl_2 + 6 H \rightarrow 2 HCl + SnH_4 \tag{b}$$

although these are not preferred preparative methods[1]. Hydrogen cleaves[2] Sn—Sn:

$$R_3SnSnR_3 + H_2 \rightarrow 2 R_3SnH \tag{c}$$

where R = Me, Et, Ph; or Sn—Pt bonds, forming stannanes[3]:

$$Pt(SnMe_3)_2(PPh_3)_2 + H_2 \rightarrow PtH(SnMe_3)(PPh_3)_2 + Me_3SnH \tag{d}$$

(A.D. NORMAN)

1. W. L. Jolly, A. D. Norman, *Prep. Inorg. React.*, *4*, 1 (1968).
2. A. K. Sawyer, in *Organotin Compounds*, Vol. 3, A. K. Sawyer, ed., Marcel-Dekker, New York, 1971, p. 823.
3. D. M. Roundhill, *Adv. Organomet. Chem.*, *13*, 273 (1975).

1.6.2.5. Giving Hydrides of Lead.

A highly reactive Pb film, formed by vapor deposition of Pb, reacts[1] with atomic H to form a Pb subhydride that decomposes to Pb and H_2 above 160°C:

$$Pb + 0.19 H \rightarrow PbH_{0.19} \tag{a}$$

(A.D. NORMAN)

1. B. R. Wells, M. W. Roberts, *Proc. Chem. Soc.*, 173 (1964).

1.6.3. by Group IVB Anionic Derivatives

1.6.3.1. Giving Hydrides of Carbon

1.6.3.1.1. from Protonic Species in Water.

Carbides that in the solid contain discrete carbon anions yield hydrocarbons upon hydrolysis. Methanides of Be and Al hydrolyze forming[1-3] CH_4:

$$Be_2C + 4 H_2O \rightarrow 2 Be(OH)_2 + CH_4 \tag{a}$$

$$Al_4C_3 + 12 H_2O \rightarrow 4 Al(OH)_3 + 3 CH_4 \tag{b}$$

The latter reaction is a preferred synthesis for laboratory quantities of CH_4. Hydrolysis of alkali, alkaline-earth and other electropositive metal (Cu, Ag, Au, Zn, Cd, Hg, Al) acetylides produces C_2H_2 in high yield, e.g.:

$$M_2C_2 + 2 H_2O \rightarrow 2 MOH + C_2H_2 \tag{c}^{1,2}$$

where M = Li, Na, K, Rb, Cs, Cu, Ag, Au.

Hydrolysis of Mg_2C_3 yields $CH_3C\equiv CH$:

$$Mg_2C_3 + 4 H_2O \rightarrow 2 Mg(OH)_2 + CH_3C\equiv CH \qquad (d)$$

Lanthanide and actinide carbides with H_2O yield complex hydrocarbons mixtures. From ThC_2 in neutral H_2O, H_2 and C_1–C_6 hydrocarbons form[1] in % yields of: H_2, 35; CH_4, 4; C_2H_6, 29; C_4H_{10}, 3; C_2H_4, 5; C_4H_8, 6; C_2H_2, 8 and C_4H_4, 4.

Cyanide ion reacts[2,4] with H_2O forming the weak acid HCN ($pK_a = 4\times10^{-10}$):

$$[CN]^- + H_2O \rightarrow [OH]^- + HCN \qquad (e)$$

Protonation of resonance-stabilized carbanions occurs readily:

$$[R_3C]^- + H_2O \rightarrow [OH]^- + R_3CH \qquad (f)[2]$$

where R = CN, $C(CN)_2$, NO_2.

Metalated carbon compounds containing ionic, σ-covalent or partial σ-covalent bonds are hydrolyzed[6] in reactions useful mainly for preparing D-containing compounds. Such reactions are highly preferred thermodynamically, because relatively strong C—H and metal–oxygen bonds are formed. Alkali-metal derivatives[7] react, e.g.:

$$C_{10}H_7Na + H_2O \rightarrow NaOH + C_{10}H_8 \qquad (g)[5]$$

$$RLi + H_2O \rightarrow LiOH + RH \qquad (h)[5]$$

where R = Me, Et, i-Pr, n-Bu, t-Bu;

$$Ph_3SiOC(Li)Ph_2 + H_2O \rightarrow LiOH + Ph_3SiOCHPh_2 \qquad (i)[8]$$

Other organometallic compounds, e.g., organo-Mg, Al, and Nd compounds, are hydrolyzed:

$$n\text{-}BuMgBr + H_2O \rightarrow Mg(OH)Br + n\text{-}BuH \qquad (j)[5]$$

$$Me_3Al + 3 H_2O \rightarrow Al(OH)_3 + 3 MeH \qquad (k)[2]$$

$$Nd(C_5H_5\text{-}h^5)_3 + H_2O \rightarrow 3 Nd(OH)_3 + 3 C_5H_6 \qquad (l)[9]$$

The fluorocarbon group VB (P, As, Sb, Bi) compounds are stable in H_2O, but are hydrolyzed in aqueous base:

$$(CF_3)_3E + 3 NaOH \xrightarrow{[OH]^-_{(aq)}} Na_3EO_3 + 3 CF_3H \qquad (m)[10,11]$$

where E = P, As, Sb, Bi. Silicon, Ge and Sn compounds behave similarly. Aryl groups are removed in aqueous base; alkyl groups are not cleaved:

$$2 Et_3SnPh + H_2O \xrightarrow{[OH]^-_{(aq)}} (Et_3Sn)_2O + 2 PhH \qquad (n)[12]$$

Active metal reduction of organic compounds in H_2O, e.g., alkyl halides with Zn[5,13], occurs:

$$2 EtBr + 2 Zn + 2 H^+ \xrightarrow{H_2O} ZnBr_2 + Zn^{2+} + 2 EtH \qquad (o)[5]$$

Similar reduction of $[(Me_4C_4)NiCl_2]_2$ produces $Me_4C_4H_4$ in 90% yield[14].

Electrolytic reduction of CO_2 at an Hg cathode in H_2O yields the $[HCO_2]^-$ ion:

$$H_2O + CO_2 + 2 e^- \rightarrow [OH]^- + [HCO_2]^- \qquad (p)[15]$$

Numerous hydrolyses of unsaturated organic[5,16] or organometallic intermediate species[17] result in formation of C—H bonds (see treatises on organic synthetic chemistry).

(A.D. NORMAN)

1. A. K. Holliday, G. Hughes, S. M. Walker, in *Comprehensive Inorganic Chemistry*, Vol. 1, A. F. Trotman-Dickenson, ed., Pergamon Press, Oxford, 1973, p. 1173.
2. F. A. Cotton, G. Wilkinson, *Advanced Inorganic Chemistry*, 4th ed., Wiley-Interscience, New York, 1980.
3. D. T. Hurd, *Chemistry of the Hydrides*, John Wiley and Sons, New York, 1952.
4. W. L. Jolly, *The Inorganic Chemistry of Nitrogen*, Benjamin, New York, 1964.
5. L. F. Fieser, M. Fieser, *Advanced Organic Chemistry*, Rheinhold, New York, 1961. General coverage of organic chemistry.
6. D. J. Cram, *Fundamentals of Carbanion Chemistry*, Academic Press, New York, 1965.
7. D. Magnus, G. Roy, *Organometallics, 1*, 553 (1982).
8. F. Glockling, *The Chemistry of Germanium*, Academic Press, London, 1969.
9. T. J. Marks, *Prog. Inorg. Chem., 24*, 51 (1978).
10. G. O. Doak, L. D. Freedman, *Organometallic Compounds of Arsenic, Antimony, and Bismuth*, Wiley-Interscience, New York, 1970.
11. J. Emsley, D. Hall, *The Chemistry of Phosphorus*, Harper and Row, London, 1976.
12. A. G. Davies, P. J. Smith, *Adv. Inorg. Chem. Radiochem., 23*, 1 (1980).
13. R. D. Chambers, S. R. James, in *Comprehensive Organic Chemistry*, D. Barton, W. D. Ollis, eds., Vol. 1, J. F. Stoddart, ed., Pergamon Press, Oxford, 1979, p. 493.
14. G. Burkhart, H. Hoberg, *Angew. Chem., Int. Ed. Engl., 21*, 76 (1982).
15. J. Ryu, T. N. Anderson, H. Eyring, *J. Phys. Chem., 76*, 3278 (1972).
16. G. H. Whitham, in *Comprehensive Organic Chemistry*, D. Barton, W. D. Ollis, eds., Vol. 1, J. F. Stoddart, ed., Pergamon Press, Oxford, 1979, p. 121.
17. A. E. Jukes, *Adv. Organomet. Chem., 12*, 215 (1975).

1.6.3.1.2. from Protonic Species in Liquid Ammonia.

Reactions of electropositive metal alkyls with NH_3 in liq NH_3 yield the hydrocarbon and metal amide[1,2], in reactions more useful for the amide than for the hydrocarbon, e.g.:

$$\text{n-BuLi} + NH_3 \xrightarrow{NH_{3(l)}} \text{Li}[NH_2] + \text{n-BuH} \tag{a}$$

$$Zn(C_2H_5)_2 + 2\,NH_3 \xrightarrow{NH_{3(l)}} Zn(NH_2)_2 + 2\,C_2H_6 \tag{b}$$

Reductions of unsaturated organic molecules by alkali metals in liq NH_3 are used in organic syntheses[3,4], e.g.:

$$\text{PhC(Et)CO} \xrightarrow{Na-NH_{3(l)}-EtOH} \text{Ph(Et)CHOH} \tag{c}^{3,4}$$

$$\text{PhCH}{=}\text{CH}_2 \xrightarrow{Li-NH_{3(l)}} \text{PhCH}_2\text{CH}_3 \tag{d}^{1,3}$$

$$\text{MeC}{\equiv}\text{CMe} \xrightarrow{Na-NH_3} \text{trans-MeC(H)}{=}\text{C(Me)H} \tag{e}^{3,4}$$

(A.D. NORMAN)

1. W. L. Jolly, *The Inorganic Chemistry of Nitrogen*, Benjamin, New York, 1964.
2. F. A. Cotton, G. Wilkinson, *Advanced Inorganic Chemistry*, 4th ed., Wiley-Interscience, New York, 1980.
3. C. A. Buehler, D. E. Pearson, *Survey of Organic Syntheses*, Vol. 2, Wiley-Interscience, New York, 1977.
4. H. O. House, *Modern Synthetic Reactions*, Benjamin, Menlo Park, CA, 1972.

1.6.3.1.3. from Protonic Species in Other Solvents.

Methanides react with protonic acids (HX) to form[1,2] CH_4, e.g.:

$$Be_2C + 4\,HX \rightarrow 2\,BeX_2 + CH_4 \tag{a}$$

where X = Cl, Br, H_2PO_4, HSO_4, although CH_4 synthesis is better accomplished using H_2O (see §1.6.3.1.1). Acetylides and higher carbides[1] react similarly; however, subsequent reactions with the unsaturated products can occur. Carbide species formed in the gas phase, e.g., C^-, C_2^- and C_4^-, react with acidic substrates to form neutral molecules, which can further react to final products[3]:

$$C^- \xrightarrow{(CH_3)_2SiH_2} CH \qquad (b)$$

Protonic acids react with cyanide ion[2]:

$$[CN]^- + HCl \rightarrow Cl^- + HCN \qquad (c)$$

resonance- stabilized carbanions[2,4]:

$$[(NC)_3C]^- + HCl \rightarrow Cl^- + (NC)_3CH \qquad (d)$$

and an iron carbide cluster as:

$$[HFe_5(CO)_{14}C]^- + 3\ HCl \rightarrow 2\ CO + Cl^- + 2\ FeCl_2 + H_2 + HFe_4(CO)_{12}CH \quad (e)$$

Protonation of hydrocarbons in the gas phase:

$$CH_{4(g)} + H^+_{(g)} \rightarrow [CH_5]^+_{(g)} \qquad (f)$$

or in solution by strong acids results in carbocation formation[4,6,7]:

$$C_6H_6 + H^+ \xrightarrow{HF-SbF_6} [C_6H_7]^+ \qquad (g)[8]$$

Organometallic compounds having varying degrees of carbanionic character are cleaved by protonic acids, e.g.:

$$C_5H_5Li + HCl \rightarrow LiCl + C_5H_6 \qquad (h)[4,8-10]$$

$$(C_6F_5Cu)_4 + HCl \rightarrow (C_6F_5)_3Cu_4Cl + C_6F_5H \qquad (i)[11]$$

$$n\text{-}C_4H_9MgBr + HCl \rightarrow MgClBr + n\text{-}C_4H_{10} \qquad (j)[4,8,12]$$

$$C_6F_5HgCH_3 + HCl \rightarrow CH_3HgCl + C_6F_5H \qquad (k)[2,13]$$

$$Me_2Cd + 2\ MeOH \rightarrow Cd(OMe)_2 + 2\ CH_4 \qquad (l)[14]$$

$$2\ CH_3Zr(Cl)(C_5H_5\text{-}h^5)_2 + H_2O \rightarrow [(h^5\text{-}C_5H_5)_2ZrCl]_2O + 2\ CH_4 \qquad (m)[15]$$

$$Gd(C_5H_5\text{-}h^5)_3 + HCl \rightarrow Gd(C_5H_5\text{-}h^5)_2Cl + C_5H_6 \qquad (n)[16]$$

$$cis\text{-}Pt(PhC{\equiv}CPh)(PPh_3)_2 + 2\ HCl \rightarrow PtCl_2(PPh_3)_2 + trans\text{-}PhCH{=}CHPh \quad (o)[17]$$

Reactions of compounds containing more than one R group proceed stepwise, allowing intermediate products to be obtained if a deficiency of HX is used [e.g., Eq. (i)]. Rates of bond scission vary also, usually the ease of cleavage is M—aryl > M—alkyl.

Cleavage of nonmetal element–C bonds with protonic acids can occur:

$$(CH_2{=}CH)_3B + 3\ HCl \rightarrow BCl_3 + 3\ CH_2{=}CH_2 \qquad (p)[18]$$

Such reactions can be important in syntheses (e.g., hydroboration) of hydrocarbons[19].

Organometallic complexes containing σ-bonded unsaturated moieties undergo protonation at the multiple bond. Reactions of HCl with σ-allylic complexes proceed as:

$$h^5\text{-}C_5H_5Fe(CO)_2(CH_2CH{=}CH_2) + HCl$$
$$\rightarrow [h^5\text{-}C_5H_5Fe(CO)_2(h^2\text{-}CH_2{=}CHCH_3)]Cl \qquad (q)[2]$$

74   1.6.3. by Group IVB Anionic Derivatives
1.6.3.1. Giving Hydrides of Carbon
1.6.3.1.3. from Protonic Species in Other Solvents.

or with σ-alkynic complexes proceed as:

$$h^5\text{-}C_5H_5Fe(PPh_3)_2C\equiv CH + HCl \rightarrow [h^5\text{-}C_5H_5Fe(PPh_3)_2C=CH_2]Cl \qquad (r)^2$$

Strong-acid protonation of a σ-allylic Co_3 cluster yields the σ-alkyl cationic product:

$$Co_3(CO)_9CC(CH_3)=CH_2 + HPF_6 \xrightarrow{\text{(EtCO)}_2O} [Co_3(CO)_9CC(CH_3)_2]PF_6 \qquad (s)^{20}$$

Protonation of bridging CH_2 groups in diruthenium complexes[2] proceeds:

$$\text{trans-}[Ru_2(CO)_3(h^5\text{-}C_5H_5)_2(h^3\text{-}C_3H_4)] + HBF_4$$
$$\rightarrow [Ru_2(CO)_2(h^5\text{-}C_5H_5)_2C(\mu\text{-}CO)(CH_3)CH_2]BF_4 \qquad (t)^{21}$$

Formylate metal complexes can be converted to carbenes:

$$2\,[Os_3(CO)_{11}CHO]^- + 2\,H^+ \rightarrow H_2O + Os_3(CO)_{12} + Os_3(CO)_{11}CH_2 \qquad (u)^{22}$$

The $[(CO)_4Fe(HCO)]^-$ion reacts to form H_2CO, in a reaction of undetermined stoichiometry[22]. Reaction of a π-complexed h^2-butadiene-Fe complex with HCl occurs:

$$(h^2\text{-}C_4H_6)Fe(CO)_4 + HCl \rightarrow [(h^2\text{-}C_4H_7)Fe(CO)_4]Cl \qquad (v)^2$$

However, h^4-dienes react to π-allyl (h^3) complexes:

$$h^4\text{-}C_4H_6Fe(CO)_3 + HCl \rightarrow h^3\text{-}CH_3C_3H_4FeCl(CO)_3 \qquad (w)^2$$

Cyclooctatetraene $\cdot$ $Mo(CO)_3$ reacts with HCl to form a homotropylium ion complex:

$$C_8H_8Mo(CO)_3 + H^+ \rightarrow [(CO)_3MoC_8H_9]^+ \qquad (x)^2$$

Trifluoroacetic acid protonates an h^5-Me_5C_5 group of $(h^5\text{-}Me_5C_5)_2Ni$:

$$(h^5\text{-}Me_5C_5)_2Ni + CF_3CO_2H \rightarrow [CF_3CO_2]^- + [h^5\text{-}Me_5C_5NiMe_5C_5H]^+ \qquad (y)^{23}$$

Protonation of nonmetal–C multiple bonds, in phosphaethynes or phospha- or arsaethenes:

$$Ph_2C=ER + HCl \rightarrow Ph_2CHE(R)Cl \qquad (z)^{24,25}$$

where E = P, R = Cl; E = As, R = 2,4,6-$Me_3C_6H_2$, yields C—H bonds. Phosphorus and arsenic ylids react similarly.

$$Ph_3PCH_2 + HCl \rightarrow [Ph_3PCH_3]Cl \qquad (aa)^{5,26}$$

Species containing Si=C bonds, formed as reaction intermediates, are added to by HX molecules:

$$Me_2Si=CH_2 + HX \rightarrow Me_2Si(X)CH_3 \qquad (ab)^{27}$$

where X = OH, t-BuO, Cl. Electrochemical cathodic reductions, e.g., of organohalides, carboxylic acids, alcohols or arenes proceed readily[28]:

$$C_6H_5CH_2CH=CH_2 \xrightarrow[\text{(Me}_2\text{N)}_3\text{PO}-\text{EtOH}]{e^-,\ \text{cathode}} C_6H_7CH_2CH=CH_2 \qquad (ac)$$

Active metal reductions, e.g., that of an amide, occur[29]:

$$PhC(O)NH_2 \xrightarrow[\text{POCl}_3]{\text{Zn}-\text{EtOH}} PhCH_2NH_2 \qquad (ad)$$

(A.D. NORMAN)

1. A. K. Holliday, G. Hughes, S. M. Walker, in *Comprehensive Inorganic Chemistry*, Vol. 1, A. F. Trotman-Dickenson, ed., Pergamon Press, Oxford, p. 1173.
2. F. A. Cotton, G. Wilkinson, *Advanced Inorganic Chemistry*, 4th ed, Wiley-Interscience, New York, 1980.
3. J. D. Rayzant, K. Tanaka, C. D. Betowski, D. K. Bohme, *J. Am. Chem. Soc., 98*, 894 (1976).
4. B. Bethell, in *Comprehensive Organic Chemistry*, D. Barton, W. D. Ollis, eds., Vol. 1, J. F. Stoddart, ed., Pergamon Press, Oxford, 1979, p. 411.
5. M. Tachikawa, E. L. Muetterties, *J. Am. Chem. Soc., 102*, 4542 (1980).
6. H. Schwartz, *Angew. Chem., Int. Ed. Engl., 20*, 991 (1981).
7. D. Farcasiu, *Acc. Chem. Res., 15*, 46 (1982).
8. L. F. Fieser, M. Fieser, *Advanced Organic Chemistry*, Rheinhold, New York, 1964. Basic, thorough coverage of organic reactions.
9. D. J. Cram, *Fundamentals of Carbanion Chemistry*, Academic Press, New York, 1965.
10. B. J. Wakefield, in *Comprehensive Organic Chemistry*, D. Barton, W. D. Ollis, eds., Vol. 3, D. N. Jones, ed., Pergamon Press, Oxford, 1979, p. 943.
11. A. E. Jukes, *Adv. Organomet. Chem., 12*, 215 (1974). Reviews organic chemistry of Cu.
12. B. J. Wakefield, in *Comprehensive Organic Chemistry*, D. Barton, W. D. Ollis, eds., Vol. 3, D. N. Jones, ed., Pergamon Press, Oxford, 1979, p. 969.
13. P. M. Treichel, F. G. A. Stone, *Adv. Organomet. Chem., 1*, 145 (1964).
14. P. R. Jones, P. J. Desio, *Chem. Rev., 78*, 491 (1978).
15. G. W. Parshall, J. J. Mrowca, *Adv. Organomet. Chem., 7*, 157 (1968).
16. T. J. Marks, *Prog. Inorg. Chem., 24*, 51 (1976).
17. D. M. Roundhill, *Adv. Organomet. Chem., 13*, 273 (1975).
18. D. Seyferth, *Prog. Inorg. Chem., 3*, 129 (1962).
19. A. Hajós, *Complex Hydrides*, Elsevier, Amsterdam, 1979.
20. D. Seyferth, *Adv. Organomet. Chem., 14*, 97 (1976).
21. A. F. Dyke, S. A. R. Knox, P. J. Naish, *J. Organomet. Chem., 199*, C47 (1980).
22. J. A. Gladysz, *Adv. Organomet. Chem., 20*, 1 (1982).
23. U. Kölle, F. Khouzami, H. Lueken, *Chem. Ber., 115*, 1178 (1982).
24. R. Appel, G. Maier, H. P. Reisenauer, A. Westerhaus, *Angew. Chem., Int. Ed. Engl., 20*, 197 (1981).
25. T. C. Klebach, H. von Dongen, F. Bickelhaupt, *Angew. Chem., Int. Ed. Engl., 18*, 395 (1979).
26. H. Schmidbaur, G. Blaschke, H. P. Scherm, *Chem. Ber., 112*, 3311 (1979).
27. L. E. Gusel'nikov, N. S. Nametkin, *Chem. Rev., 78*, 529 (1979).
28. A. J. Bard, L. R. Faulkner, *Electrochemical Methods*, Wiley-Interscience, New York, 1980.
29. L. G. Wade Jr., *Compendium of Organic Synthetic Methods*, Vol. 4, Wiley-Interscience, New York, 1980.

1.6.3.2. Giving Hydrides of Silicon

1.6.3.2.1. from Protonic Species in Water.

Protonation of Mg silicides or Mg-Si alloys by aq HCl, H_2SO_4 or H_3PO_4 yields the silanes[1] Si_nH_{2n+2}. Product distribution and yields depend on the method of alloy preparation and reaction conditions. Lower silanes (i.e., Si_1-Si_6) are formed primarily, although silanes up to $Si_{15}H_{32}$ can be obtained[2]. Typically SiH_4 (20%):

$$Mg_2Si + 4 H^+ \xrightarrow{H_2O} 2 Mg^{2+} + SiH_4 \tag{a}$$

along with Si_2H_6, Si_3H_8 and small quantities of higher hydrides form[1-3]. Hydrolysis of $CaSi_2$ by dil HCl in aq i-C_3H_7OH yields polymeric hydridosiloxane[1]:

$$3 CaSi_2 + 6 HCl + 3 H_2O \rightarrow 3 CaCl_2 + 3 H_2 + \frac{1}{x} (Si_6H_6O_3)_x \tag{b}$$

Disilane forms[1] upon hydrolysis of Li_6Si_2[1]:

$$Li_6Si_2 + 6 H_2O \rightarrow 6 LiOH + Si_2H_6 \tag{c}$$

Ternary Ge-Si hydrides form from aq HF hydrolysis of Mg-Ge-Si alloys. From an Si-rich alloy, nominally $Mg_{20}Si_9Ge$, are obtained ternary hydrides in % yields, based on Si conversion, of Si_2GeH_8, 3; n-Si_3GeH_{10}, 1.5; n-Si_4GeH_{12}, 0.3 and n-Si_5GeH_{14}, 0.07.

Hydrolysis of metal alkyl- or arylsilanes produces the corresponding silanes in reactions useful for preparation of deuterated silanes:

$$R_3SiLi + H_2O \rightarrow LiOH + R_3SiH \qquad (d)[1]$$

Polysilanes form in hydrolyses of dialkali-metal polysilicides:

$$M(R_2Si)_nM + 2 H^+ \xrightarrow{H_2O} 2 M^+ + H(R_2Si)_nH \qquad (e)[1,4]$$

where M = Li, Na; R = Me, Et, Ph; n = 2–6. Silylaluminate-ion hydrolysis cleaves Si—Al bonds and forms Ph_3SiH:

$$Li[H_3AlSiPh_3] + 4 H_2O \rightarrow LiAl(OH)_4 + 3 H_2 + Ph_3SiH \qquad (f)[4]$$

Divalent silicon species react with H_2O, forming silanes. Dilute aq HF (10%) reacts with SiO or SiO–GeO mixtures producing silanes (SiH_4, Si_2H_6, Si_3H_8, Si_4H_{10}, Si_5H_{12}, Si_6H_{14}) or ternary Si–Ge hydrides ($SiGeH_6$, Si_2GeH_8, etc.), respectively[1]. Dimethylsilylene, obtained by photolysis of $(Me_2Si)_6$, upon insertion into the OH bonds of H_2O yields siloxane:

$$2 Me_2Si + H_2O \rightarrow Me_2Si(H)OSi(H)Me_2 \qquad (g)[5]$$

1. E. Wiberg, E. Amberger, *Hydrides of the Elements of Main Groups I-IV,* Elsevier, Amsterdam, 1971.
2. F. Fehér, D. Schinkitz, J. Schaaf, *Z. Anorg. Allg. Chem., 383,* 303 (1971).
3. G. Brauer, ed., *Handbook of Preparative Inorganic Chemistry,* 2nd ed., Vol. 1, Academic Press, New York, 1963.
4. H. Gillman, F. W. G. Fearon, R. L. Harrell, *J. Organomet. Chem., 5,* 592 (1966).
5. P. G. Harrison, *Coord. Chem. Rev., 40,* 179 (1982).

1.6.3.2.2. from Protonic Species in Liquid Ammonia.

Ammonium halide (NH_4Cl or NH_4Br) protonation of Mg-Si alloys in liq NH_3 yields the silanes Si_nH_{2n+2}, although SiH_4 is the main product (yields up to 90%)[1,2].

$$Mg_2Si + 4 NH_4Br \xrightarrow{NH_3(l)} 2 MgBr_2 + 4 NH_3 + SiH_4 \qquad (a)$$

Disilyl- and trisilylamine disproportionate in $NH_{3(l)}$ to oligomeric and polymeric silazanes and SiH_4, e.g.:

$$2 (SiH_3)_3N \rightarrow [(SiH_3)_2N]_2SiH_2 + SiH_4 \qquad (b)[3]$$

(A.D. NORMAN)

1. E. Wiberg, E. Amberger, *Hydrides of the Elements of Main Groups I-IV,* Elsevier, Amsterdam, 1971.
2. G. Brauer, ed., *Handbook of Preparative Inorganic Chemistry,* 2nd ed., Vol. 1, Academic Press, New York, 1963.
3. A. D. Norman, W. L. Jolly, *Inorg. Chem., 18,* 1594 (1979).

1.6.3.2.3. from Protonic Species in Other Solvents.

Reaction of CaSi with $MeCO_2H$ or HCl in EtOH yields an amorphous Si subhydride of composition[1] $SiH_{0.7-0.9}$:

$$CaSi \xrightarrow{\text{HCl–EtOH}} SiH_{0.7-0.9} \tag{a}$$

Calcium disilicide reacts with HCl in EtOH to form two-dimensional polysilanes as:

$$CaSi_2 + 2\ H^+ \rightarrow Ca^{2+} + \tfrac{2}{x}\ (SiH)_x \tag{b}[1]$$

The ternary alloy $Ca(Si_{0.5}Ge_{0.5})$ with acid yields $SiGeH_6$ and $SiGe_2H_8$, along with large quantities[1] of SiH_4.

Protonation of alkali-metal silyl compounds produces the parent silanes in reactions that are useful for Si—D bond formation[1]:

$$R_3SiM + HX \rightarrow MX + R_3SiH \tag{c}$$

where R = H, alkyl, aryl; X = Cl, Br; M = Li, Na, K. Hydrogen-chloride cleavage of Si—Si or Si—metal bonds yields silanes:

$$(Me_2Si)_6 + HCl \rightarrow H(Me_2Si)_6Cl \tag{d}[1,3]$$

$$Ph_3SiMgBr + HCl \rightarrow MgClBr + Ph_3SiH \tag{e}[2]$$

$$Ni(bipy)(SiCl_3)_2 + 2\ HCl \rightarrow Ni(bipy)Cl_2 + 2\ SiCl_3H \tag{f}[3]$$

Divalent silicon species react readily with protonic acids, e.g., Me_2Si from $(Me_2Si)_6$ photolysis reacts with alcohols and amines:

$$Me_2Si + HX \rightarrow Me_2Si(X)H \tag{g}[4]$$

where X = OMe, OEt, NEt_2, n-BuO. Silicon difluoride reacts similarly with HBr above $-196°C$, forming a disilane:

$$2\ SiF_2 + HBr \rightarrow BrF_2SiSiF_2H \tag{h}[5]$$

(A.D. NORMAN)

1. E. Wiberg, E. Amberger, *Hydrides of the Elements of Main Groups I–IV*, Elsevier, Amsterdam, 1971.
2. Y. Kiso, K. Tamao, M. Kumada, *J. Organomet. Chem.*, 76, 95 (1974).
3. P. G. Harrison, *Coord. Chem. Rev.*, 40, 179 (1982).
4. K. P. Steele, W. P. Weber, *J. Am. Chem. Soc.*, 102, 6095 (1980).
5. K. G. Sharp, J. F. Bald, *Inorg. Chem.*, 14, 2553 (1975).

1.6.3.3. Giving Hydrides of Germanium

1.6.3.3.1. from Protonic Species in Water.

Protonation of Mg or Ca germanides or Mg-Ge alloys (Mg_2Ge) with acidic H_2O yields germanes of formula[1,2] Ge_nH_{2n+2}. Germane forms from Mg_2Ge:

$$Mg_2Ge + 4\ H^+ \xrightarrow{H_2O} 2\ Mg^{2+} + GeH_4 \tag{a}$$

In practice a mixture of hydrides is obtained. Typically from the $HCl_{(aq)}$ hydrolysis of Mg_2Ge arise[1], in % yield, GeH_4, 15; Ge_2H_6, 5; Ge_3H_8, 5; Ge_4H_{10}, 1.5; Ge_5H_{12}, 0.8 and higher germanes (0.9%).

Ternary hydrides GeH_3PH_2 and GeH_3AsH_2 can be prepared from hydrolysis of $CaGe–Ca_3P_2$ or $CaGe–Ca_3As_2$, respectively[2]. Similarly, Mg-Ge-Si alloys react with 10% HF in H_2O, forming mixtures of silanes, germanes, and ternary Si–Ge hydrides[2,3]. From an alloy of nominal composition Mg_4SiGe is obtained the mixture, in % yield, $SiGeH_6$, 0.6; $SiGe_2H_8$, 0.4; n-$SiGe_3H_{10}$, 0.03; n-$Si_2Ge_2H_{10}$, 0.5 and n-$Si_3Ge_2H_{12}$.

Hydrolysis of alkali-metal alkyl- or arylgermanides yields the corresponding Ge hydrides in reactions most useful for Ge—D bond formation[4]:

$$R_3GeM + H_2O \xrightarrow{H_2O} MOH + R_3GeH \qquad (b)$$

where M = Li, Na, K; R = Ph, $PhCH_2$, Et, etc. Treatment of sodium germanide (NaGe) with aqueous acid yields a polymeric Ge subhydride[1]:

$$(NaGe)_x + x\,H^+ \xrightarrow{H^+_{(aq)}} x\,Na^+ + (GeH)_x \qquad (c)$$

Germylmagnesium halides are hydrolyzed easily to the parent germanes:

$$R_3GeMgCl + H_2O \rightarrow Mg(OH)Cl + R_3GeH \qquad (d)^4$$

where R = Et, i-Pr; C_6H_{11}, Ph, etc.

Active metal reduction (Zn, Mg or Zn amalgam) of trialkyl- and triarylgermanium halides or $(Me_2GeS)_2$ in aq acid yields the respective germanes, e.g.:

$$R_3GeBr + Zn + H^+ \xrightarrow{H^+_{(aq)}} Zn^{2+} + Br^- + R_3GeH \qquad (e)^4$$

where R = Me, Et, Ph. Sodium amalgam, Zn, or Mg reductions of Ge(IV) species in aq sulfuric acid produce GeH_4 in low yields[2].

Electrolytic reduction of aq alkaline $[GeO_3]^{2-}$ solns or Ph_2GeCl_2 produces $GeH_4{}^2$ or $Ph_2GeH_2{}^5$, respectively, in low yields.

Hydrolysis of SiO–GeO mixtures produces silanes, germanes, and ternary Si–Ge hydrides[2]. Germanium(II) hydroxide with aq HBr yields[4] $GeBr_3H$:

$$Ge(OH)_2 + 3\,HBr \xrightarrow{H_2O} 2\,H_2O + GeBr_3H \qquad (f)$$

Acid hydrolysis of $KGeH_3BH_3$ or $KGeH_3BMe_3$ produces GeH_4 in high yield[6]:

$$K[GeH_3BMe_3] + H_2O \rightarrow K[HOBMe_3] + GeH_4 \qquad (g)$$

Cleavage of the Ge—C bond in GeH_3CO_2H by acetic acid, followed by decarbonylation, yields CO, GeH_4 and a Ge subhydride[7].

(A.D. NORMAN)

1. E. Wiberg, E. Amberger, *Hydrides of the Elements of Main Groups I–IV*, Elsevier, Amsterdam, 1971.
2. W. L. Jolly, A. D. Norman, *Prep. Inorg. React., 4*, 1 (1968).
3. P. L. Timms, C. C. Simpson, C. S. G. Phillips, *J. Chem. Soc.*, 1467 (1964).
4. M. Lesbre, D. Mazerolles, J. Satgé, *The Organic Compounds of Germanium*, Wiley-Interscience, New York, 1971.
5. R. E. Dessey, W. Kitching, T. Chivers, *J. Am. Chem. Soc., 88*, 453 (1966).
6. E. R. DeStaruco, C. Riddle, W. L. Jolly, *J. Inorg. Nucl. Chem., 35*, 297 (1973).
7. P. G. Harrison, *Coord. Chem. Rev., 30*, 137 (1979).

1.6.3.3.2. from Protonic Species in Liquid Ammonia.

Protonation of alkyl- or aryl-substituted germanides in $NH_{3(l)}$ by NH_4Br produces germanes in high yields:

$$Ph_3GeLi + NH_4Br \xrightarrow{NH_{3(l)}} LiBr + Ph_3GeH \qquad (a)^1$$

Trialkylgermanides react similarly; however, being more basic they react directly with NH_3:

$$Et_3GeM + NH_3 \xrightarrow{NH_{3(l)}} MNH_2 + Et_3GeH \qquad (b)^1$$

Ammonium bromide protonation of Mg_2Ge in $NH_{3(l)}$ yields GeH_4 (60–70% yield), with only small quantities of higher hydrides[2]:

$$Mg_2Ge + 4 NH_4Br \xrightarrow{NH_{3(l)}} 2 MgBr_2 + 4 NH_3 + GeH_4 \qquad (c)$$

Reactions of GeH_4 derivatives or Ge_2H_6 in $NH_{3(l)}$ result in GeH_4 and sometimes Ge subhydrides:

$$NaGeH_3 + NH_4Br \xrightarrow{NH_{3(l)}} NaBr + NH_3 + GeH_4 \qquad (d)^3$$

$$3 NaGeH_3 + 3 PhBr \xrightarrow{NH_{3(l)}} 3 NaBr + 3 PhH + \tfrac{2}{x} (GeH)_x + GeH_4 \qquad (e)^3$$

$$Ge_2H_6 \xrightarrow{NH_{3(l)}} GeH_4 + \tfrac{1}{x} (GeH_2)_x \qquad (f)^4$$

Sodium germanide (NaGe) protonation by NH_4Br yields[3] $(GeH)_x$:

$$\tfrac{1}{x} (NaGe)_x + NH_4Br \xrightarrow{NH_{3(l)}} NH_3 + NaBr + \tfrac{1}{x} (GeH)_x \qquad (g)$$

Electrolysis of $NaGeH_3$ in $NH_{3(l)}$ using a Pt anode and an Hg cathode produces small quantities[5] of GeH_4.

(A.D. NORMAN)

1. M. Lesbre, P. Mazerolles, J. Satgé, *The Organic Compounds of Germanium,* Wiley-Interscience, New York, 1971.
2. G. Brauer, ed., *Handbook of Preparative Inorganic Chemistry,* 2nd ed., Vol. 1, Academic Press, New York, 1963.
3. E. Wiberg, E. Amberger, *Hydrides of the Elements of Main Groups I–IV,* Elsevier, Amsterdam, 1971.
4. R. Dreyfuss, W. L. Jolly, *Inorg. Chem., 7,* 2645 (1968).
5. W. L. Jolly, A. D. Norman, *Prep. Inorg. React., 4,* 1, (1968).

1.6.3.3.3. from Protonic Acids in Other Solvents.

Protonation of germanides by hydrogen halides, water or alcohols in nonprotonic solvents yields the parent germanes, e.g.[1,2]:

$$R_3GeM + HX \rightarrow MX + R_3GeH \qquad (a)$$

where R = H, alkyl, aryl; X = Cl, Br; M = Li, Na, K;

$$(Ph_3Ge)_3GeLi + H_2O \xrightarrow{Et_2O} LiOH + (Ph_3Ge)_3GeH \qquad (b)^2$$

Hydrogen halides add to Ge(II) hydrides, halides, oxides or sulfides:

$$GeO + 3 HCl \xrightarrow{150-175°C} H_2O + GeCl_3H \qquad (c)^3$$

$$(GeH_2)_x \xrightarrow[Et_2O]{HCl} GeCl_2, H_2, Ge_nH_{2n+2} \qquad (d)^{1,3}$$

$$GeCl_2 + HCl \xrightarrow{Et_2O} GeCl_3H \qquad (e)^3$$

Protonic acid cleavage of metal–Ge bonds when the electronegativity of Ge exceeds that of the metal results in germane formation, e.g.:

$$(Et_3Ge)_2Cd + MeCO_2H \rightarrow Cd + MeCO_2GeEt_3 + Et_3GeH \qquad (f)[2]$$

Reaction of $Ph_3PAuGePh_3$ with HCl yields Ph_3GeH:

$$Ph_3PAuGePh_3 + HCl \rightarrow Ph_3PAuCl + Ph_3GeH \qquad (g)[3]$$

$$(h^5\text{-}C_5H_5)_2TiCl(GePh_3) + HCl \rightarrow (h^5\text{-}C_5H_5)_2TiCl_2 + Ph_3GeH \qquad (h)[2]$$

In an unsymmetrically substituted germylplatinum compound one equiv of HCl selectively cleaves the Me_3Ge group[2]:

$$(Et_3P)_2Pt(GePh_3)GeMe_3 + HCl \rightarrow (Et_3P)_2Pt(GePh_3)Cl + Me_3GeH \qquad (i)$$

Hydrogen-chloride cleavage of Ge—Ge bonds in polymethylated germanes yields a mixture of methylgermanes[2], including $Me_2Ge(Cl)H$:

$$Me_3Ge(GeMe_2)_nGeMe_3 \xrightarrow[250°C]{HCl} Me_3GeCl, Me_2GeCl_2, Me_2Ge(Cl)H \qquad (j)$$

(A.D. NORMAN)

1. E. Wiberg, E. Amberger, *Hydrides of the Elements of Main Groups I–IV*, Elsevier, Amsterdam, 1971.
2. M. Lesbre, P. Mazerolles, J. Satgé, *The Organic Compounds of Germanium*, Wiley-Interscience, New York, 1971.
3. T. Birchall, I. Drummond, *Inorg. Chem., 11*, 250 (1972).

1.6.3.4. Gring Hydrides of Tin

1.6.3.4.1. from Protonic Species in Water.

Stannane forms in low yields from the aq HCl or H_2SO_4 hydrolysis of Mg-Sn alloy[1]:

$$Mg_2Sn + 4 H^+ \xrightarrow{H^+_{(aq)}} 2 Mg^{2+} + SnH_4 \qquad (a)$$

Metal triorganostannides (R = Et, n-Bu, Ph, etc.) are hydrolyzed to the parent stannanes. Alkali-metal derivatives react:

$$R_3SnM + H^+ \xrightarrow{H^+_{(aq)}} M^+ + R_3SnH \qquad (b)[2]$$

where R = n-Bu, Ph; M = Li, Na. Hydrolysis of Mg derivatives, e.g., produces triorganostannanes in high yield:

$$(Ph_3Sn)_2Mg + 2 H_2O \rightarrow Mg(OH)_2 + 2 Ph_3SnH \qquad (c)[2,3]$$

Triethylstannane results from hydrolysis of $Et_3Sn(Me_2N)_2B$:

$$Et_3Sn(Me_2N)_2B + 3 H_2O \rightarrow 2 Me_2NH + B(OH)_3 + Et_3SnH \qquad (d)[2]$$

Aluminum amalgam reduction of triorganochlorostannanes forms the hydrides in < 60% yields[2]:

$$3 R_3SnCl + 2 Al + 3 H_2O \xrightarrow{H_2O} Al(OH)_3 + Al^{3+} + 3 Cl^- + 3 R_3SnH \qquad (e)$$

Dichloro- and trichlorostannanes are reduced less efficiently. Stannane is obtained[4] from reactions of Mg powder with Sn in H_2SO_4. Electrolytic reduction of Sn(IV) sulfate solutions or of Ph_2SnCl_2 yields SnH_4:

$$Sn^{4+} + 8\ e^- + 4\ H^+ \xrightarrow{\ H_2SO_{4(aq)}\ } SnH_4 \qquad (f)^{1,4}$$

or Ph_2SnH_2, respectively:

$$Ph_2SnCl_2 + 2\ H_2O + 4\ e^- \rightarrow 2\ [OH]^- + 2\ Cl^- + Ph_2SnH_2 \qquad (g)^5$$

(A.D. NORMAN)

1. E. Wiberg, E. Amberger, *Hydrides of the Elements of Main Groups I–IV*, Elsevier, Amsterdam, 1971.
2. E. J. Kupchik, in *Organotin Compounds*, Vol. 1, A. K. Sawyer, ed., Marcel Dekker, New York, 1971, p. 7.
3. A. G. Davies, P. J. Smith, *Adv. Inorg. Chem. Radiochem.*, **23**, 1 (1980).
4. W. L. Jolly, A. D. Norman, *Prep. Inorg. React.*, **4**, 1 (1968).
5. R. E. Dessy, W. Kitching, T. Chivers, *J. Am. Chem. Soc.*, **88**, 453 (1966).

1.6.3.4.2. from Protonic Species in Liquid Ammonia.

Ammonium bromide protonation of Mg_2Sn in liq NH_3 yields SnH_4, although inefficiently[1]:

$$Mg_2Sn + 4\ NH_4Br \xrightarrow{\ NH_{3(l)}\ } 2\ Mg^{2+} + 4\ NH_3 + 4\ Br^- + SnH_4 \qquad (a)$$

Calcium stannide with NH_4I in $NH_{3(l)}$ form a solid polystannane[1]:

$$CaSn + 2\ NH_4I \xrightarrow{\ NH_{3(l)}\ } CaI_2 + 2\ NH_3 + \frac{1}{x}\ (SnH_2)_x \qquad (b)$$

Reaction of Na triorganostannides with NH_4Cl or NH_4Br in $NH_{3(l)}$ yields the corresponding stannanes:

$$R_3SnNa + NH_4Br \xrightarrow{\ NH_{3(l)}\ } NH_3 + NaBr + R_3SnH \qquad (c)^2$$

where R = Me, Et, Ph.

(A.D. NORMAN)

1. E. Wiberg, E. Amberger, *Hydrides of the Elements of Main Groups I–IV*, Elsevier, Amsterdam, 1971.
2. E. J. Kupchik, in *Organotin Compounds*, Vol. 1, A. K. Sawyer, ed., Marcel Dekker, New York, 1971, p. 7.

1.6.3.4.3. from Protonic Species in Other Solvents.

Protonation of organostannides in aprotic solvents yields stannanes[1]. Protonation of R_2Sn species with HCl or $[NH_4][HF_2]$ occurs:

$$R_2Sn + HCl \rightarrow R_2SnHCl \qquad (a)^2$$

where R = Me, Et, Ph. Diphenylstannane in MeOH reacts losing H_2 to form an acyclic product[1]:

$$6\ Ph_2SnH_2 \xrightarrow{\ MeOH\ } 5\ H_2 + H(Ph_2Sn)_6H \qquad (b)$$

(A.D. NORMAN)

1. E. J. Kupchik, in *Organotin Compounds*, Vol. 1, A. K. Sawyer, ed., Marcel-Dekker, New York, 1971, p. 7.
2. A. G. Davies, P. J. Smith, in *Comprehensive Organometallic Chemistry*, Vol. 2, G. Wilkinson, F. G. A. Stone, E. W. Abel, eds., Pergamon Press, Oxford, 1982, p. 519.
3. W. P. Newmann, K. König, *Angew. Chem., Int. Ed. Engl.*, *1*, 212 (1962).

1.6.3.5. Giving Hydrides of Lead.

Traces of PbH_4 form in reactions of Mg-Pb alloy with aq acid[1]:

$$Mg_2Pb + 4 H^+ \xrightarrow{H_2O} 2 Mg^{2+} + PbH_4 \qquad (a)$$

or in the electrochemical reduction of lead anodes[2]:

$$Pb + 4 H^+ + 4 e^- \xrightarrow{H_2SO_{4(aq)}} PbH_4 \qquad (b)$$

Aluminum-foil reduction of $MPb(OH)_3$ (M = alkali metals) in H_2O yields a solid Pb subhydride, Pb_2H_2, which decomposes in vacuum[2]. Trialkyl- and triarylplumbides with NH_4Br in $NH_{3(l)}$ form plumbanes in low yield[3]:

$$R_3PbLi + NH_4Br \xrightarrow{NH_{3(l)}} LiBr + NH_3 + R_3PbH \qquad (c)$$

where R = Et, Ph, C_6H_{11}.

1. F. E. Saalfeld, H. J. Svec, *Inorg. Chem.*, *2*, 46 (1963).
2. E. Wiberg, E. Amberger, *Hydrides of the Elements of Main Groups I–IV*, Elsevier, Amsterdam, 1971.

1.6.4. by Hydride Reduction

1.6.4.1. Giving Hydrides of Carbon

1.6.4.1.1. from Halides.

Binary ionic or covalent hydride reduction of organic halides is important in organic syntheses[1,2]. Reactions of metal hydrides with alkyl halides in ethers occur:

$$PhCH_2Cl + MH \rightarrow MCl + PhCH_3 \qquad (a)$$

where M = Li, Na, K. Sodium hydride with aryl iodides forms the parent arenes[2]:

$$PhI + NaH \rightarrow NaI + PhH \qquad (b)$$

Diborane reacts only slowly with most alkyl halides[3]; however, reduction of $PhCH_2Br$ occurs readily[4] in $PhNO_2$:

$$6 PhCH_2Br + B_2H_6 \xrightarrow[PhNO_2]{15°C} 2 BBr_3 + 6 PhCH_3 \qquad (c)$$

Organosilanes, at elevated T (ca. 600°C) or with catalysts, (Pd on C), react:

$$Ph_3SiH + CH_2{=}CHCl \rightarrow Ph_3SiCl + CH_2{=}CH_2 \qquad (d)[2]$$

Organogermane and organostannane reductions occur under milder conditions:

$$Et_2GeH_2 + 2 EtI \rightarrow R_2GeI_2 + 2 EtH \qquad (e)[5]$$

$$R_3CCl + \text{n-Bu}_3SnH \rightarrow \text{n-Bu}_3SnCl + R_3CH \tag{f}[2]$$

Stannanes supported on SiO_2 or Al_2O_3 reduce alkyl halides cleanly to alkanes in 40% yields[6].

(A.D. NORMAN)

1. L. F. Fieser, M. Fieser, *Advanced Organic Chemistry*, Rheinhold, New York, 1961.
2. A. Hajós, *Complex Hydrides*, Elsevier, Amsterdam, 1979.
3. C. H. Long, *Adv. Inorg. Chem. Radiochem.*, 16, 201 (1974).
4. S. Matsimura, N. Takura, *Tetrahedron Lett.*, 363 (1969).
5. M. Lesbre, P. Mazerolles, J. Satgé, *The Organic Chemistry of Germanium*, Wiley-Interscience, New York, 1971.
6. H. Schuman, B. Pachaly, *Angew. Chem., Int. Ed. Engl.*, 20, 1043 (1981).

1.6.4.1.2. from Oxygen Compounds.

Carbon monoxide reacts with CaH_2 at high T ($> 600°C$) to form CH_4 and small quantities of CH_2O:

$$2\ CaH_2 + 2\ CO \rightarrow 2\ CaO + C + CH_4 \tag{a}[1]$$

$$CaH_2 + CO \rightarrow Ca + CH_2O \tag{b}[1]$$

Reaction with KH at 240–270°C yields formate ion[1]:

$$KH + CO \rightarrow KHCO \tag{c}$$

Zirconium and Ta hydride complexes react with CO forming formyl:

$$(h^5\text{-Cp})_2ZrHCl + CO \rightarrow (h^5\text{-Cp})_2Zr(Cl)CHO \tag{d}[2]$$

and μ-CH complexes:

$$[(h^5\text{-Me}_4EtC_5)TaCl_2]_2H_2 + CO \rightarrow (h^5\text{-Me}_4EtC_5)_2Ta_2Cl_4(\mu\text{-h}^2\text{-CH})(\mu\text{-h}^2\text{-H})(\mu\text{-h}^2\text{-O}) \tag{e}[3]$$

Tungsten-coordinated CO reacts with $(h^5\text{-Me}_5C_5)_2ZrH_2$:

$$(h^5\text{-Cp})_2WCO + (h^5\text{-Me}_5C_5)_2ZrH_2 \rightarrow (h^5\text{-Cp})_2WC(H)OZr(H)(C_5Me_5\text{-h}^5)_2 \tag{f}[2]$$

Reduction of formyl or acyl metal complexes with BH_3:

$$h^5\text{-CpRe(NO)(CO)CHO} \xrightarrow{\text{BH}_3\text{-THF}} h^5\text{-CpRe(NO)(CO)CH}_3 \tag{g}[2,4]$$

t-Bu_2AlH:

$$h^5\text{-CpRe(NO)(CO)CHO} \xrightarrow{\text{t-Bu}_2AlH} h^5\text{-CpRe(NO)(CO)CH}_2OH \tag{h}[3]$$

or Et_3SiH:

$$(CO)_9Co_3CCOR \xrightarrow{\text{Et}_3SiH\text{-CF}_3CO_2H} (CO)_9Co_3CCH_2R \tag{i}[5]$$

where R = Me, Et, i-Pr, Ph, proceeds easily.

Alkali-metal hydrides (M = Li, Na, K) and CO_2 yield metal formates in reactions that normally require elevated T:

$$MH + CO_2 \rightarrow MO_2CH \tag{j}[1]$$

84

1.6.4. by Hydride Reduction
1.6.4.1. Giving Hydrides of Carbon
1.6.4.1.2. from Oxygen Compounds.

Calcium hydride and CO_2 or $NaHCO_3$ produce CH_4:

$$4\ CaH_2\ +\ 2\ CO_2\ \rightarrow\ 4\ CaO\ +\ C\ +\ 2\ H_2\ +\ CH_4 \qquad (k)[1]$$

or $NaO_2CH[1,2]$:

$$CaH_2\ +\ 2\ NaHCO_3\ \rightarrow\ Ca(OH)_2\ +\ 2\ NaO_2CH \qquad (l)[2]$$

Group VIA metal-carbonyl hydrides react with CO_2 or COS, forming metalloformate:

$$[M(CO)_5H]^-\ +\ CO_2\ \rightarrow\ [M(CO)_5OC(O)H]^- \qquad (m)$$

where M = Cr, Mo,W, or thioformate complexes[6].

Binary hydride (boranes, alanes, silanes, stannanes) reductions of oxygen-containing compounds are used in organic synthesis[7-9], e.g., with ketones :

$$PhCOMe\ \xrightarrow{\ Et_3SnH\ }\ Ph(Me)CHOH \qquad (n)$$

aldehydes:

$$EtCHO\ \xrightarrow{\ AlH_3-THF\ }\ EtCH_2OH \qquad (o)$$

alcohols:

$$PhCH_2OH\ \xrightarrow{\ Et_3SiH-CF_3CO_2H\ }\ PhCH_3 \qquad (p)$$

carboxylic acids:

$$MeCO_2H\ \xrightarrow{\ BH_3-THF\ }\ MeCH_2OH \qquad (q)$$

and esters:

$$EtCO_2Me\ \xrightarrow{\ BH_3-THF\ }\ EtCH_2OH \qquad (r)$$

The final products shown are obtained after hydrolysis of intermediate addition species, e.g.:

$$6\ Me_2CO\ +\ B_2H_6\ \rightarrow\ 2\ (Me_2CHO)_3B \qquad (s)$$

$$(Me_2CHO)_3B\ +\ 3\ H_2O\ \xrightarrow{\ H^+_{(aq)}\ }\ B(OH)_3+\ 3\ Me_2CHOH \qquad (t)$$

Variation of substituent groups and reaction conditions allows considerable reaction selectivity, as in:

$$PhCOMe\ \xrightarrow[\ [Rh(1,5-C_8H_{12})Cl]_2-PPh_3\ ]{\ Ph_2SiH_2\ }\ Ph(Me)CHOH \qquad (u)$$

where Ph_2SiH_2 in the presence of a chiral catalyst reduces PhCOMe to chiral product in 57% optical yield[10,11].

Thermolysis of alkoxymetal compounds often yields alkanes, e.g.:

$$Ph_2AlOC_6H_{11}\ \rightarrow\ \tfrac{1}{x}\ (PhAlO)_x\ +\ C_6H_{10}\ +\ PhH \qquad (v)[12,13]$$

Metal carboxylates decarboxylate upon heating:

$$2 \ M(HCO_2)_2 \xrightarrow{200°C} MO + CO_2 + H_2CO \qquad (w)^{14}$$

(A.D. NORMAN)

1. E. Wiberg, E. Amberger, *Hydrides of the Elements of Main Groups I–IV*, Elsevier, Amsterdam, 1971.
2. J. A. Gladysz, *Adv. Organomet. Chem.*, 20, 1, (1982).
3. M. R. Churchill, H. J. Wasserman, *J. Chem. Soc., Chem. Commun.*, 274 (1981).
4. C. Master, *Adv. Organomet. Chem.*, 17, 61 (1979).
5. D. Seyferth, *Adv. Organomet. Chem.*, 14, 97 (1976).
6. D. J. Darensbourg, A. Rokicki, *J. Am. Chem. Soc.*, 104, 349 (1982).
7. L. F. Fieser, M. Fieser, *Advanced Organic Chemistry*, Rheinhold, New York, 1961. Comprehensive treatment of organic syntheses and reactions.
8. H. C. Brown, *Boranes in Organic Chemistry*, Cornell Univ. Press, Ithaca, NY, 1972. Basic, readable review of borane reductions.
9. A. Hajós, *Complex Hydrides*, Elsevier, Amsterdam, 1979. Excellent summary of hydrides in organic reductions.
10. H. Brunner, G. Riepl, *Angew. Chem., Int. Ed. Engl.*, 21, 377 (1982).
11. D. M. Roundhill, *Adv. Organomet. Chem.*, 13, 273 (1975).
12. F. A. Cotton, G. Wilkinson, *Advanced Inorganic Chemistry*, 4th ed., Wiley-Interscience, New York, 1980.
13. E. C. Ashby, G. F. Willard, A. B. Goel, *J. Org. Chem.*, 44, 1221 (1979).
14. M. N. Ray, N. D. Sinnarka, *J. Inorg. Nucl. Chem.*, 35, 1373 (1973).

1.6.4.1.3. from Nitrogen Compounds.

Boranes, alanes, silanes, germanes and stannanes, and their substituted derivatives, reduce C—N multiple bonds[1-3], e.g., with nitriles:

$$3 \ PhC{\equiv}N + BH_3 \xrightarrow{THF} (PhCH{=}N)_3B \qquad (a)$$

cyanates:

$$PhNCO + Ph_3SnH \rightarrow Ph_3SnN(Ph)CHO \qquad (b)$$

thiocyanates:

$$EtNCS + Et_3SiH \rightarrow Et_3SiN(Et)CHS \qquad (c)$$

imines:

$$PhCH{=}NPh + n\text{-}Bu_3SnH \rightarrow PhCH_2N(Ph)Sn(n\text{-}Bu)_3 \qquad (d)$$

and carbodimides:

$$C_6H_{11}N{=}C{=}NC_6H_{11} + Et_3SnH \rightarrow C_6H_{11}N{=}CHN(SnEt_3)C_6H_{11} \qquad (e)$$

Reaction of NaH with the zwitterionic ammonium carboranes results in cleavage of the NMe$_3$ moiety and formation of the anionic carborane product:

$$B_{10}H_{12}CNMe_3 + NaH \xrightarrow{THF} Me_3N + Na[B_{10}H_{12}CH] \qquad (f)^4$$

(A.D. NORMAN)

1. H. C. Brown, *Boranes in Organic Chemistry*, Cornell Univ. Press, Ithaca, NY, 1972.
2. A. Hajós, *Complex Hydrides*, Elsevier, Amsterdam, 1979.
3. C. A. Beuhler, D. E. Pearson, *Organic Syntheses*, Vol. 2, Wiley-Interscience, New York, 1977.
4. D. E. Hyatt, F. R. Scholer, L. J. Todd, J. L. Warner, *Inorg. Chem.*, 6, 2229 (1967).

1.6.4.1.4. from Compounds with Carbon–Carbon Multiple Bonds.

Group IIIB (B, Al), IVB (Si, Ge, Sn, Pb), VB (P, As, Sb) and VIB (S, Se) hydrides add to $C=C$ or $C\equiv C$ bonds under thermal, photolytic, radical-initiated and catalyzed (homogeneous or heterogeneous) conditions. Such reactions are routes to alkyl- or alkenyl-element compounds and to intermediates subsequent to hydrolyzed or reduced products. Addition places the hydrogen on the carbon with least hydrogen, except under highly polar-reagent addition (e.g., base-catalyzed) conditions, where hydrogen adds to the carbon with most hydrogens.

Group IIIB hydrides[1,2] react thermally, e.g.:

$$6\ MeCH=CH_2\ +\ B_2H_6\ \xrightarrow{THF}\ 2\ (MeCH_2CH_2)_3B \tag{a}$$

$$PhCH=CH_2\ +\ i\text{-}Bu_2AlH\ \rightarrow\ i\text{-}Bu_2AlCH_2CH_2Ph \tag{b}$$

Reactions can be highly selective. Excess i-Bu$_2$AlH reacts with n-C$_7$H$_{15}$CH=CH-CH=CH$_2$ almost quantitatively first at the more highly substituted $C=C$ bond. Only 4% reaction with the terminal $C=C$ bond occurs[3]:

$$n\text{-}C_7H_{15}CH=CHCH=CH_2\ \xrightarrow{i\text{-}Bu_2AlH}\ n\text{-}C_7H_{15}CH(i\text{-}Bu_2Al)CH_2CH=CH_2\ (96\%)$$
$$+\ n\text{-}C_7H_{15}CH(i\text{-}Bu_2Al)CH_2CH_2CH_2Al(Bu\text{-}i)_2\ (4\%) \tag{c}$$

Boranes (B > 1) react with olefins and alkynes, e.g.:

$$B_5H_9\ +\ C_2H_4\ \rightarrow\ B_5H_8C_2H_5 \tag{d}[4]$$

Group IVB hydrides react with alkynes and alkenes under thermal, photolytic, radical-initiated and catalyzed (homogeneous or heterogeneous) conditions[2,4], e.g.:

$$Me_3SiH\ +\ CH_2=CHCH=CH_2\ \xrightarrow{PdCl_2(NCPh)_2}\ Me_3SiCH_2CH_2CH=CH_2\ \tag{e}[2,5,6]$$

$$Me_2SiH_2\ +\ 2\ CF_2=CF_2\ \xrightarrow{h\nu}\ Me_2Si(C_2F_2H)_2 \tag{f}[4,7]$$

$$n\text{-}Bu_3GeH\ +\ CH_2=CHCO_2Me\ \rightarrow\ n\text{-}Bu_3GeCH_2CH_2CO_2Me \tag{g}[4,8]$$

$$n\text{-}Bu_3GeH\ +\ HC\equiv CPh\ \xrightarrow{H_2PtCl_6}\ n\text{-}Bu_3GeCH=CHPh \tag{h}[2,4,8]$$

$$Et_3SnH\ +\ PhCH=CH_2\ \xrightarrow[C_6H_6]{AIBN}\ Et_3SnCH_2CH_2Ph \tag{i}[4,9]$$

where AIBN = 2,2'-[Me$_2$(CN)C]$_2$N$_2$;

$$n\text{-}Bu_2SnH_2\ +\ MeC\equiv CC\equiv CMe\ \rightarrow\ \begin{matrix} CH=C(Me) \\ | \qquad\qquad\quad \\ CH=C(Me) \end{matrix} Sn(Bu\text{-}n)_2 \tag{j}[9,10]$$

$$Me_3PbH\ +\ EtCH=CH_2\ \rightarrow\ Me_3PbCH_2CH_2Et \tag{k}[11]$$

By variation of substituents, reactivity of the hydride toward $C=C$ or $C\equiv C$ bonds can be modified. Asymmetric reduction of prochiral olefins[2], in the presence of chiral catalysts can be achieved. Reaction of styrene:

$$PhC(Me)=CH_2\ +\ MeSiCl_2H\ \xrightarrow{NiCl_2(L^\bullet)_2}\ PhC(Me)HCH_2SiCl_2Me \tag{l}[12]$$

1.6.4. by Hydride Reduction
1.6.4.1. Giving Hydrides of Carbon
1.6.4.1.4. from Compounds with Carbon–Carbon Multiple Bonds.

87

where $L^* = $ (R)-PhCH$_2$(Me)(Ph)P, after hydrolysis to products, yields PhC(Me)HCH$_3$ in optical purity up to 18%.

Group VB and VIB hydrides react with alkenes and alkynes, thermally or under radical, photolytic or base catalysis, in reactions used primarily for substituted element (P, As, Sb) compound synthesis, e.g.:

$$2 \ Ph_2PCH=CH_2 \ + \ H_2P(CH_2)_2CN \xrightarrow{AIBN} NC(CH_2)_2P[(CH_2)_2PPh_2]_2 \qquad (m)^{13}$$

$$Me_2AsH \ + \ CF_3C\equiv CCF_3 \xrightarrow{h\nu} Me_2AsC(CF_3)=C(CF_3)H \qquad (n)^{14}$$

$$(EtO)_2P(S)SH \ + \ Et_3SiCH=CH_2 \ \rightarrow \ (EtO)_2P(S)SCH_2CH_2SiEt_3 \qquad (o)^{15}$$

Metal-coordinated phosphines react similarly, as in the metal-templated cyclization[16] of $(CO)_3Mo(PH_2CH_2CH=CH_2)_3$ in the presence of 2.2'-azobis(isobutyronitrile)(AIBN):

$$fac\text{-}(CH_2=CHCH_2PH_2)_3Mo(CO)_3 \xrightarrow{AIBN} fac\text{-}(CO)_3Mo[PH(CH_2)_3]_3 \qquad (p)$$

Numerous organic synthetic reactions[17,18] occur, e.g.:

$$ROH \ + \ R'CH=CH_2 \xrightarrow{H^+} R'CH_2CH_2OR \qquad (q)^{17,18}$$

$$R_2NH \ + \ CH_2=CHCH=CH_2 \xrightarrow{Pd(OAc)_2} CH_2CHCH(NR_2)CH_3 \qquad (r)^{17,18}$$

$$CH_2=CH_2 \xrightarrow{Ni(PCl_3)_4-AlBr_3-BuLi} CH_2CHCH_2CH_3, \ trans\text{-}CH_3CHCHCH_3,$$
$$cis\text{-}CH_3CHCHCH_3 \qquad (s)^{19}$$

Transition-metal hydrides react with olefins to form σ-bonded derivatives:

$$h^5\text{-}CpFe(CO)_2H \ + \ C_2F_4 \ \rightarrow \ h^5\text{-}CpFe(CO)_2C_2F_4H \qquad (t)^{7,20}$$

$$PtH(Cl)(PEt_3)_2 \ + \ C_2H_4 \ \rightarrow \ P(C_2H_5)Cl(PEt_3)_2 \qquad (u)^6$$

$$RhH(CO)(PPh_3)_3 \ + \ MeCHCH_2 \ \rightarrow \ PPh_3 \ + \ Rh(MeCH_2CH_2)(CO)(PPh_3)_2 \quad (v)^{20}$$

Both addition of hydrogen to carbon with the most hydrogens in unsymmetrical olefins and its reverse can occur[20]. Intramolecular hydride transfer, e.g., in HMo(dppe)$_2$-(N$_2$)$_2$C$_3$H$_5$-h^3, produces the h^2-propene:

$$HMo(dppe)_2(N_2)_2C_3H_5\text{-}h^3 \ \rightarrow \ Mo(dppe)_2(N_2)_2CH_3C_2H_3\text{-}h^2 \qquad (w)^{20}$$

where dppe is (C$_6$H$_5$)$_2$P(CH$_2$)$_2$P(C$_6$H$_5$)$_2$. Cleavage of the reduced product occurs in the reaction of HMn(CO)$_5$ with PhC(Me)=CH$_2$:

$$2 \ HMn(CO)_5 \ + \ PhC(Me)=CH_2 \ \rightarrow \ 2 \ Mn_2(CO)_{10} \ + \ PhCHMe_2 \qquad (x)^{21}$$

Reaction of a diene with [HNiL$_4$]$^+$ yields the h^3-product:

$$CH_3CH=CHCH=CH_2 \ + \ [HNiL_4]^+ \ \rightarrow \ [h^3\text{-}CH_3CHCHCHCH_3)NiL_4]^+ \quad (y)^{20}$$

Hydride-ion reactions with the ligands of some cationic organometallic complexes results in C—H bond formation, e.g.:

$$[(CO)_3Mn(h^6\text{-}C_6H_6)]^+ \ + \ H^- \ \rightarrow \ (CO)_3MnC_6H_7\text{-}h^5 \qquad (z)^{20}$$

$$[Ru_2(CO)_3(h^5\text{-}Cp)_2\mu\text{-}h^3\text{-}C_3H_5]^+ + CsH \rightarrow Cs^+$$
$$+ [Ru_2(CO)_2(h^5\text{-}Cp)_2\mu\text{-}CO(\mu\text{-}CMe_2)] \qquad (aa)^{22}$$

(A.D. NORMAN)

1. H. C. Brown, *Boranes in Organic Chemistry*, Cornell Univ. Press, Ithaca, NY, 1972.
2. A. Hajós, *Complex Hydrides*, Elsevier, Amsterdam, 1979.
3. M. Montury, J. Gore, *Tetrahedron Lett.*, *21*, 51 (1980).
4. E. Wiberg, E. Amberger, *Hydrides of the Elements of Main Groups I–IV*, Elsevier, Amsterdam, 1971.
5. J. Tsuji, *Adv. Organomet. Chem.*, *17*, 141 (1979).
6. D. M. Roundhill, *Adv. Organomet. Chem.*, *13*, 273 (1975).
7. P. M. Treichel, F. G. A. Stone, *Adv. Organomet. Chem.*, *1*, 145 (1964).
8. F. Glockling, *The Chemistry of Germanium*, Academic Press, London, 1969.
9. A. G. Davies, P. J. Smith, *Adv. Inorg. Chem. Radiochem.*, *23*, 1 (1980).
10. A. J. Ashe III, T. P. Diephouse, *J. Organomet. Chem.*, *202*, C95 (1980).
11. W. P. Neumann, K. Kühlein, *Adv. Organomet. Chem.*, *7*, 242 (1968).
12. K. Yamamoto, Y. Kiso, R. Ito, K. Tamao, M. Kumada, *J. Organomet. Chem.*, *210*, 9 (1981).
13. R. Uriaite, T. J. Mazanec, K. D. Tau, D. W. Meek, *Inorg. Chem.*, *19*, 79 (1980).
14. W. R. Cullen, *Adv. Organomet. Chem.*, *4*, 145 (1966).
15. D. Seyferth, *Prog. Inorg. Chem.*, *3*, 129 (1962).
16. B. N. Diel, R. C. Haltiwanger, A. D. Norman, *J. Am. Chem. Soc.*, *104*, 4700 (1982).
17. L. F. Fieser, M. Fieser, *Advanced Organic Chemistry*, Rheinhold, New York, 1961.
18. G. H. Whitham, in *Comprehensive Organic Chemistry*, D. Barton, W. D. Ollis, eds., Vol. 1, J. F. Stoddart, ed., Pergamon Press, Oxford, 1979, p. 121.
19. G. P. Chiusoli, G. Salerno, *Adv. Organomet. Chem.*, *17*, 195 (1979).
20. F. A. Cotton, G. Wilkinson, *Advanced Inorganic Chemistry*, 4th ed., Wiley-Interscience, New York, 1980.
21. R. L. Sweany, J. Halpern, *J. Am. Chem. Soc.*, *99*, 8335 (1977).
22. D. D. Davies, A. F. Dyke, A. Endesfelder, S. A. R. Knox, P. J. Naish, A. G. Oipen, D. Plass, G. E. Taylor, *J. Organomet. Chem.*, *198*, C43 (1980).

1.6.4.1.5. from Other Derivatives.

Hydride cleavage of carbon–element σ-bonds frequently results in C—H bond formation[1], although often the other cleavage product is of primary interest, e.g.:

$$3 Ph_3SiH + Et_3Sb \rightarrow (Ph_3Si)_3Sb + 3 EtH \qquad (a)^2$$

$$HMn(CO)_5 + MeAuPPh_3 \rightarrow (CO)_5MnAuPPh_3 + MeH \qquad (b)^3$$

Cleavage of organic moieties from Cd, Zn, Hg or Mg in their organometallic compounds can be important in organic synthesis, e.g.:

$$2 R_3SiH + R'_2Cd \rightarrow (R_3Si)_2Cd + 2 R'H \qquad (c)^4$$

$$R_3GeH + R'MgCl \rightarrow R_3GeMgCl + R'H \qquad (d)^5$$

where R and R' = alkyl or aryl.

Reductive elimination (thermal or photochemical) can form C—H bonds, although it is not a preferred synthesis of C—H bond-containing products[1-5,8,9], e.g.:

$$2 (CF_3)_2AsH \xrightarrow{\Delta} C_2F_6 + 2 As + 2 CF_3H \qquad (e)^2$$

$$2 Mn(CO)_5H + Ph_3As \rightarrow Mn_2(CO)_9(AsPh_3) + H_2CO \qquad (f)^7$$

$$2 Os(CO)_4H(Me) \rightarrow Os_2(CO)_8(H)Me + MeH \qquad (g)^8$$

1.6.4. by Hydride Reduction
1.6.4.1. Giving Hydrides of Carbon
1.6.4.1.5. from Other Derivatives.

89

$$Yb(C_5H_5\text{-}h^5)_3NH_3 \rightarrow Yb(C_5H_5\text{-}h^5)_2NH_2 + C_5H_6 \qquad (h)^9$$

$$2\ Me_3SiOSiMe_3 \xrightarrow{h\nu(185\ nm)} (Me_3SiOMeSiCH_2)_2 + 2\ MeH \qquad (i)^{1,10}$$

Intramolecular rearrangements or isomerizations of organic compounds primarily important in organic syntheses[1,11] produce new C—H bonds, e.g.:

$$(j)^{12}$$

$$CH_2{=}CHCH_2CH_3 \xrightarrow{Ni} cis,trans\text{-}CH_3CH{=}CHCH_3 \qquad (k)^{11}$$

Similarly, rearrangements of organometallic compounds lead to new C—H bond formation:

$$BrMgCH_2CH(Et)CH{=}CH_2 \rightarrow EtCH(MgBr)CH_2CH{=}CH_2 \qquad (l)^{13}$$

$$[PhC(SiMe_3)_2CH(Ph)]^- \rightarrow [PhCH(SiMe_3)C(SiMe_3)Ph]^- \qquad (m)^{14}$$

Carbon-centered free radicals abstract H atoms from substrate molecules to form C—H bonds[2,15]:

$$[R_3C]^\cdot \xrightarrow{H\text{-substrate}} R_3H \qquad (n)$$

Triplet electronic-state carbenes react similarly. Singlet carbenes insert into substrate-H bonds, e.g.:

$$R_2C{:} + R'OH \rightarrow R_2CHOR' \qquad (o)^{16}$$

$$R_2C{:} + R'H \rightarrow R_2CHR' \qquad (p)^{16}$$

where R and R' = alkyl.

(A.D.NORMAN)

1. F. A. Cotton, G. Wilkinson, *Advanced Inorganic Chemistry*, 4th ed., Wiley-Interscience, New York, 1980.
2. G. O. Doak, L. D. Freedman, *Organometallic Compounds of Arsenic, Antimony, and Bismuth*, Wiley-Interscience, New York, 1970.
3. M. H. Chisholm, I. P. Rothwell, *Prog. Inorg. Chem.*, 29, 1 (1982).
4. L. Räsch, H. Müller, *Angew. Chem., Int. Ed. Engl.*, 15, 670 (1976).
5. M. Lesbre, P. Mazerolles, J. Satgé, *The Organic Compounds of Germanium*, Wiley-Interscience, New York, 1971.
6. E. C. Ashby, G. F. Willard, A. B. Goel, *J. Org. Chem.*, 44, 1221 (1979).
7. H. Berke, M. Stumpp, *J. Organomet. Chem.*, 192, 385 (1980).
8. J. R. Norton, *Acc. Chem. Res.*, 12, 139 (1979).
9. T. J. Marks, *Prog. Inorg. Chem.*, 24, 78 (1978).
10. L. E. Gusel'nikov, N. S. Nametkin, *Chem. Rev.*, 79, 529 (1979).
11. L. F. Fieser, M. Fieser, *Advanced Organic Chemistry*, Rheinhold, New York, 1961.
12. L. T. Scott, *Acc. Chem. Res.*, 15, 52 (1982).
13. E. A. Hill, *Adv. Organomet. Chem.*, 16, 131 (1977).
14. R. West, *Adv. Organomet. Chem.*, 16, 1 (1977).
15. J. T. Sharp, *Comprehensive Organic Chemistry*, D. Barton, W. D. Ollis, eds., Vol. 1, J. F. Stoddart, ed., Pergamon Press, Oxford, 1979, p. 455.
16. W. Kirmse, *Carbene Chemistry*, Academic Press, New York, 1964.

1.6.4.2. Giving Hydrides of Silicon

1.6.4.2.1. from Halides.

Alkali-metal hydrides, boranes or alanes reduce halosilanes in high yield[1-3]:

$$R_{4-n}SiCl_n + n \ LiH \rightarrow n \ LiCl + R_{4-n}SiH_n \tag{a}$$

$$R_{4-n}SiCl_n + n \ Et_2AlH \rightarrow n \ Et_2AlCl + R_{4-n}SiH_n \tag{b}$$

where R = alkyl, aryl; n = 1–4;

$$3 \ SiCl_4 + 2 \ B_2H_6 \rightarrow 4 \ BCl_3 + 3 \ SiH_4 \tag{c}$$

Representative reactions and products are listed in Table 1. Lithium and Na hydride reductions are performed in ether, although with $SiCl_4$ eutectic salts can be used[2]. In large-scale syntheses NaH is prepared in situ from reaction of Na with H_2. In the Al–$AlCl_3$–H_2 or $AlCl_3$–NaCl–H_2 reductant systems, intermediate alanes AlH_nCl_{3-n} are the active reductant.

Halosilane redistributions–disporportionations produce new Si—H bonds. Fluorosilane reactions occur at 25°C, e.g.[4]:

$$SiF_3H \rightarrow SiH_xF_{4-x} \tag{d}$$

At higher T or in the presence of $AlCl_3$ catalyst, other halosilanes react similarly[3,4]. Reactions of halosilanes with organo-Mg:

$$Ph_3SiCl + C_6H_{11}MgCl \rightarrow C_6H_{10} + MgCl_2 + Ph_3SiH \tag{e}^{[3,5]}$$

or alkyl-Li reagents:

$$\text{i-Pr}_2ClSiSiCl(\text{Pr-i})_2 + 2 \ \text{i-PrLi} \rightarrow 2 \ LiCl + 2 \ C_3H_6 + \text{i-Pr}_2Si(H)SiH(\text{Pr-i})_2 \quad (f)^{[6]}$$

through reductive elimination produce silanes.

TABLE 1. HYDRIDE REDUCTION OF HALOSILANES

Halosilane	Reductant, solvent	Product	Yield (%)	Ref.
$SiCl_4$	LiH, Et_2O	SiH_4	85	1,2
$SiCl_4$	Al–NaH–H_2 $AlCl_3$–NaCl	SiH_4	64	2
$SiCl_4$	NaH–Et_3B, THF	SiH_4	94	2
$Si(OEt)_2F_2$	NaH, C_8H_{18}	SiH_4	> 80	2,3
Si_2Cl_6	AlH_3, Et_2O	Si_2H_6	90	2,3
$CH_2CHCH_2SiCl_3$	LiH, $(C_5H_{11})_2O$	$CH_2CHCH_2SiH_3$	85	3
$MeCHClSiCl_3$	LiH, dioxane	$MeCHClSiH_3$	a	2,3
Me_2SiCl_2	Et_2AlH	Me_2SiH_2	100	2,3
Et_2SiF_2	NaH^b	Et_2SiH_2	88	2,3
Et_3SiF	Et_2AlH	Et_3SiH	100	2,3
Et_3SiF	$Na–H_2^b$	Et_3SiH	90	2,3
Ph_3SiCl	NaH, C_8H_{18}	Ph_3SiH	81	2,3

a Not available.
b No solvent.

Insertion of halosilenes into substrate–H bonds yields silanes:

$$SiF_2 + PH_3 \xrightarrow{> -196°C} SiHF_2PH_2, Si_2F_5H \qquad (g)^7$$

$$SiF_2 + Et_2SiH_2 \rightarrow Et_2Si(H)SiF_2H \qquad (h)^8$$

Thermolysis of a halodisilane at 600–700°C produces insertion products containing SiH bonds:

$$Me_3SiSiClMe_2 \rightarrow Me_2Si(CH_2)_2SiMe_2, MeSi(H)(CH_2)_2Si(H)Me \qquad (i)^9$$

(A.D. NORMAN)

1. E. Wiberg, E. Amberger, *Hydrides of the Elements of Main Groups I–IV*, Elsevier, Amsterdam, 1971. An excellent, comprehensive review of silane syntheses.
2. W. L. Jolly, A. D. Norman, *Prep. Inorg. React.*, 4, 1 (1968).
3. D. A. Armitage, in *Comprehensive Organometallic Chemistry*, Vol. 2, G. Wilkinson, F. G. A. Stone, E. W. Abel, eds., Pergamon Press, Oxford, 1982, p. 1.
4. K. Moedritzer, *Adv. Organomet. Chem.*, 6, 171 (1968).
5. E. A. V. Ebsworth, *Volatile Silicon Compounds*, Pergamon Press, New York, 1963.
6. M. Weidenbruch, W. Peter, *J. Organomet. Chem.*, 84, 151 (1975).
7. G. R. Lanford, D. C. Moody, J. D. Odom, *Inorg. Chem.*, 14, 134 (1975).
8. R. L. Jenkins, A. J. Vanderwielen, S. P. Ruis, S. R. Gird, M. A. Ring, *Inorg. Chem.*, 12, 2968 (1973).
9. P. G. Harrison, *Coord. Chem. Rev.*, 40, 179 (1982).

1.6.4.2.2. from Oxygen Compounds.

Alkali-metal hydrides, boranes or alanes react with alkoxysilanes or silanols to form Si—H bonds in reactions that often are preferred syntheses, depending on the availability of the silane reactant[1-4]. Lithium hydride or alanate reductions used are:

$$R_{4-n}Si(OR')_n + n\ LiH \rightarrow n\ LiOR' + R_{4-n}SiH_n \qquad (a)$$

$$R_{4-n}Si(OR')_n + n\ R_2''AlH \rightarrow R_2''AlOR' + R_{4-n}SiH_n \qquad (b)$$

Typical reactions and products are listed in Table 1. Reductions involving H_2 in NaCl–AlCl$_3$ solvents or H_2 in the presence of Na involve $AlCl_{3-n}H_n$ or NaH, respectively, as reductant species.

TABLE 1. HYDRIDE REDUCTION OF SILICON–OXYGEN COMPOUNDS

Reactant silane	Reductant, solvent	Product	Yield (%)	Ref.
NaSiO$_3$	AlCl$_3$–NaCl–H$_2$, AlCl$_3$–NaCl	SiH$_4$	20	1,2
Si(OEt)$_4$	AlCl$_3$–NaCl–H$_2$, AlCl$_3$–NaCl	SiH$_4$	16	1
(SiH$_3$)$_2$O	B$_2$H$_6$[a]	SiH$_4$	95	2,3
SiCl$_2$(OEt)$_2$	NaH, octane	SiH$_4$	80	2,4
Et$_2$Si(OEt)$_2$	LiH, (C$_5$H$_{11}$)$_2$O	Et$_2$SiH$_2$	45	2
Me$_2$Si(OEt)$_2$	Et$_2$AlH[a]	Me$_2$SiH$_2$[b]	100	2
(Me$_3$Si$_2$)$_2$O	(i-C$_4$H$_9$)$_2$AlH[a]	Me$_3$SiH	81	2,4

[a] No solvent.

[b] Some Et$_2$SiH$_2$ forms also.

Hydridosiloxanes redistribution in the presence of catalysts yields new Si—H products[4]:

$$3 \; EtSiH(OEt)_2 \xrightarrow[100°C]{NaOEt} 2 \; EtSi(OEt)_3 + EtSiH_3 \qquad (c)$$

(A.D. NORMAN)

1. W. L. Jolly, A. D. Norman, *Prep. Inorg. React.*, 4, 1 (1968).
2. E. Wiberg, E. Amberger, *Hydrides or the Elements of Main Groups I–IV*, Elsevier, Amsterdam, 1971.
3. L. H. Long, *Adv. Inorg. Chem. Radiochem.*, 16, 201 (1974).
4. D. A. Armitage, in *Comprehensive Organometallic Chemistry*, Vol. 2, G. Wilkinson, F. G. A. Stone, E. W. Abel, eds., Pergamon Press, Oxford, 1982, p. 1.

1.6.4.2.3. from Other Derivatives.

Silicon–hydrogen bonds form in reactions[1] of alkali-metal hydrides with Si_2H_6:

$$KH + Si_2H_6 \rightarrow KSiH_3 + SiH_4 \qquad (a)$$

or[2] Ph_3SiCH_2Ph:

$$NaH + Ph_3SiCH_2Ph \rightarrow Ph_4Si, \; PhMe, \; Ph_3SiH, \; Ph_2SiH_2 \qquad (b)$$

Cleavage of Si—Bi bonds by Et_3GeH produces a silane product[3]:

$$(Et_3Si)_3Bi + 3 \; Et_3GeH \xrightarrow{180°C} (Et_3Ge)_3Bi + 3 \; Et_3SiH \qquad (c)$$

Thermolysis or photolysis of silanes produces new Si—H bonds; the reactions involve silylenes:

$$(Me_2Si)_6 + 6 \; HCl \xrightarrow[(254 \; nm)]{h\nu, \; 2537 \; Å} 6 \; Me_2SiHCl \qquad (d)^{4,7}$$

$$(Me_2Si)(Me_2C)_2 + Et_3SiH \xrightarrow{108°C} Me_2CCMe_2, \; Et_3SiSiMe_2H \qquad (e)^{5,7}$$

$$Si_2H_6 + MeSiH_3 \xrightarrow{300°C} SiH_4 + MeSi_2H_5 \qquad (f)^{6,7}$$

$$(Me_2Si)_6 + 6 \; C_6H_{10} \xrightarrow[(254 \; nm)]{h\nu, \; 2537 \; Å} 6 \; C_6H_9(Me)_2SiH \qquad (g)^8$$

Through Si—Si double-bond-containing intermediates, new Si—H bonds can form:

$$(Me_2SiSiMe_2) \xrightarrow{360°C} Me_2Si(CH_2)_2SiH_2, \; MeHSi(CH_2)_2SiMeHC_6H_4(C_4H_4), \; Me_3SiH \qquad (h)^9$$

Other less well-defined thermolysis or photolysis reactions also produce silanes:

$$Ph_3SiSiPh_2Me + MeOH \xrightarrow[(254 \; nm)]{h\nu, \; 2537 \; Å} Ph_2Si(OMe)Me + Ph_3SiH \qquad (i)^{7,10}$$

$$Me_3SiSiH_2SiMe_3 \xrightarrow{750°C} Me_3SiH, \; MeSiH(CH_2)_2SiH_2 \qquad (j)^{7,11}$$

$$Me_3SiH \xrightarrow{\Delta} MeSiH_2CH_2SiH_2Me, \; Me_2SiHCH_2SiH_2Me, \; Me_3SiCH_2SiH_3,$$
$$Me_3SiCH_2SiH_2Me, \; (Me_2SiCH_2)_3 \qquad (k)^{12}$$

The Hg-sensitized photolysis of Si_2H_6 yields SiH_4 and higher silanes:

1.6.4. by Hydride Reduction
1.6.4.2. Giving Hydrides of Silicon
1.6.4.2.3. from Other Derivatives.

93

$$Si_2H_6 \xrightarrow{h\nu, Hg^*} H_2, SiH_4, Si_3H_8, Si_4H_{10} \qquad (l)^{7,13}$$

Pyrolysis of silane or silane–germane mixtures by rapid passage through a 350–370°C hot zone yields mixtures of binary and ternary silanes. Typical pyrolysis mixtures and major hydrides (in parentheses) obtained are[14]: SiH_4 (Si_2H_6, Si_3H_8), Si_2H_6 (SiH_4, Si_3H_8, Si_4H_{10}), SiH_4–GeH_4 (Si_2H_6, $SiGeH_6$), Ge_2H_6–Si_2H_6 (Si_2GeH_8) and Ge_2H_6–Si_3H_8(i- and n-Si_3GeH_{10}).

Passage of SiH_4, SiH_4–GeH_4, SiH_4–PH_3, or SiH_4–AsH_3 through a 15 kV silent electric discharge results in products, some of which involve new Si—H bond formation. Pyrolysis mixtures and major products (in parentheses) from such reactions are[15]: SiH_4 (Si_2H_6, Si_3H_8), SiH_4–GeH_4 (Si_2H_6, $SiGeH_6$, Ge_2H_6), SiH_4–PH_3 [Si_2H_6, P_2H_4, $(SiH_3)_2PH$], Si_2H_6–PH_3 (SiH_4, Si_3H_8, $Si_2H_5PH_2$) and SiH_4–AsH_3 (Si_2H_6, SiH_3AsH_2).
Redistribution:

$$2\ SiH_3X \rightarrow SiH_4 + SiH_2X_2 \qquad (m)^{1,7}$$

where X = halogen, alkoxy, NR_2, or disproportionation yields new Si—H bonds[1,7]:

$$3\ Me_3SiPMe_2BH_3 \xrightarrow{150°C} (Me_2PBH_2)_3 + 3\ Me_3SiH \qquad (n)^{16}$$

Such reactions often proceed rapidly in the liquid but only slowly in the gas phase. These reactions often can be either acid catalyzed:

$$2\ (SiH_3)_3N \xrightarrow{B_5H_9} [(SiH_3)_2N]_2SiH_2 + SiH_4 \qquad (o)^{17}$$

or base catalyzed:

$$(SiH_3)_3N \xrightarrow{NH_3} \frac{1}{x}\ [SiH_3NSiH_2]_x + SiH_4 \qquad (p)^1$$

Thermal elimination of silyl and hydride moieties from silyl–metal complexes yields silanes:

$$h^5\text{-}MeC_5H_4(CO)_2HMnSiClPh_2 \xrightarrow{100°C} h^5\text{-}MeC_5H_4Mn(CO)_3 + Ph_2SiClH \qquad (q)^{7,18}$$

$$2\ RhHCl(SiPh_3)(PPh_3)_2 \xrightarrow{100°C} [RhCl(PPh_3)_2]_2 + 2\ Ph_3SiH \qquad (r)^7$$

Intramolecular rearrangement produces a Pt—H—Si bridge compound in low yield[19]:

$$\{Pt(\mu\text{-}H)(SiMe_2Ph)[P(C_6H_{11})_3]\}_2 \xrightarrow{\Delta} \{PtH(\mu\text{-}SiMe_2)[P(C_6H_{11})_3]\}_2 \qquad (s)$$

Ion–molecule reactions in the gas phase yield new Si—H bonds in reactions that are not useful synthetically:

$$[MeSi]^+ + MeSiH_3 \rightarrow H_2, [MeSi_2CH_2]^+, [MeSiH]^+, [Me_2Si_2H]^+, \text{etc.} \qquad (t)^{20}$$

$$SiH_4 + [CH_5]^+ \rightarrow CH_4 + [SiH_5]^+ \qquad (u)^{21}$$

(A.D. NORMAN)

1. B. J. Ayelett, *Adv. Inorg. Chem. Radiochem.*, *11*, 249 (1968).
2. K. Ruhlmann, *Z. Chem.*, 6, 421 (1966).
3. F. Glockling, *The Chemistry of Germanium*, Academic Press, London, 1969.

4. P. G. Harrison, *Coord. Chem. Rev., 40*, 179 (1982).
5. D. Seyferth, D. C. Annarelli, *J. Am. Chem. Soc., 97*, 7162 (1975).
6. M. D. Sefcik, M. A. Ring, *J. Am. Chem. Soc., 95*, 5168 (1973).
7. D. Armitage, in *Comprehensive Organometallic Chemistry*, Vol. 2, G. Wilkinson, F. G. A. Stone, E. W. Abel, eds., Pergamon Press, Oxford, 1982, p. 1.
8. M. Ishikawa, M. Ishiguro, M. Kumada, *J. Organomet. Chem., 49*, C71 (1973).
9. W. D. Wulff, W. F. Goure, T. J. Barton, *J. Am. Chem. Soc., 100*, 6236 (1978).
10. P. Boudjouk, J. R. Roberts, C. M. Golino, L. H. Sommer, *J. Am. Chem. Soc., 94*, 7926 (1972).
11. S. H. M. Ho, J. D. Halten, III, S. Konieczny, E. C-L. Ma, P. P. Gaspar, *J. Am. Chem. Soc., 104*, 1424 (1982).
12. G. Ritz, J. Maas, A. Hornung, *Z. Anorg. Allg. Chem., 460*, 115 (1980).
13. T. L. Pollack, H. S. Sandhur, A. Jodhan, O. P. Strausz, *J. Am. Chem. Soc., 95*, 1017 (1973).
14. E. Wiberg, E. Amberger, *Hydrides of the Elements of Main Groups I–IV*, Elsevier, Amsterdam, 1971.
15. W. L. Jolly, A. D. Norman, *Prep. Inorg. React., 4*, 1 (1968).
16. E. W. Abel, S. M. Illingworth, *Organomet. Chem. Rev., A., 5*, 143 (1970).
17. W. M. Scantlin, A. D. Norman, *J. Chem. Soc., Chem. Commun.*, 1246 (1971).
18. U. Schubert, B. Wörle, P. Jandik, *Angew. Chem., Int. Ed. Engl., 20*, 695 (1981).
19. M. Auburn, M. Ciriano, J. A. K. Howard, M. Murray, N. J. Pugh, J. L. Spencer, F. G. A. Stone, P. Woodward, *J. Chem. Soc., Dalton Trans.*, 659 (1980).
20. T. M. Mayer, F. W. Lampe, *J. Phys. Chem., 78*, 2422 (1972).
21. M. D. Sefcik, J. M. S. Henis, P. P. Gaspar, *J. Chem. Phys., 61*, 4329 (1974).

1.6.4.3. Giving Hydrides of Germanium

1.6.4.3.1. from Halides.

Alkali-metal hydrides in ether reduce[1] $GeCl_4$:

$$GeCl_4 + 4\ LiH \rightarrow 4\ LiCl + GeH_4 \tag{a}$$

or alkyl- and aryl-substituted halogermanes:

$$R_{4-n}GeX_n + n\ MH \rightarrow n\ MX + R_{4-n}GeH_n \tag{b}$$

where R = alkyl, aryl; X = Cl, Br, I; M = Li, Na, K, to germanes[2,3]. From these reactions germanes such as $R_{4-n}GeH_n$ (n = 1–3; R = Me, Et, i-Pr), $EtMe_2GeH$ and Ph_3GeH are readily obtained. Yields are better using complex hydride reducing agents (see §1.6.5.3.1). In some cases, e.g., NaH reactions, reduction is catalyzed[3] by the presence of electron-pair acceptor acids such as R_2AlCl.

Substituted alane reduction of $GeCl_4$ produces[4] GeH_4:

$$GeCl_4 + 4\ i\text{-}Bu_2AlH \rightarrow 4\ i\text{-}Bu_2AlCl + GeH_4 \tag{c}$$

Organosilane reactions with Ge tetrahalides provide near quantitative yields of trihalogermanes:

$$R_2SiH_2 + GeCl_4 + 2\ Et_2O \rightarrow R_2SiClH + (Et_2O)_2 \cdot GeHCl_3 \tag{d}$$[3]

where R = Me, Et, Ph; or:

$$(Me_2SiH)_2O + 2\ GeCl_4 + 2\ Et_2O \xrightarrow{Et_2O} (Me_2SiCl)_2O + 2\ Et_2O \cdot GeCl_3H \tag{e}$$[5]

Halogermane redistribution–disproportionation forms Ge—H bonds. Fluorogermanes react at 25°C:

$$2\ RGeH_2F \rightarrow RGeH_3 + RGeF_2H \tag{f}$$[3]

where R = Me, Et, n-Bu. Chlorogermane reactions occurs similarly and are catalyzed by $AlCl_3$:

$$R_2GeH_2 + R_2GeCl_2 \xrightarrow{AlCl_3} 2 R_2GeHCl \qquad (g)^2$$

where R = Me, Et, Ph. However, with $AlCl_3$ present, Ge—C bond cleavage in arylgermanes also can occur. Triethylgermane reacts with $GeCl_4$ in Et_2O to form $GeCl_3H$ in 80% yield[3]:

$$Et_3GeH + GeCl_4 + 2 Et_2O \xrightarrow{Et_2O} Et_3GeCl + (Et_2O)_2 \cdot GeCl_3H \qquad (h)$$

Insertion of GeF_2 into the M—H bond of trialkylsilanes, -germanes or -stannanes yields marginally stable fluorogermanes:

$$R_3MH + GeF_2 \xrightarrow{THF} R_3MGeF_2H \qquad (i)^6$$

where M = Si, Ge, Sn; R = Et, n-Bu.

(A.D. NORMAN)

1. E. Wiberg, E. Amberger, *Hydrides of the Elements of Main Groups I–IV*, Elsevier, Amsterdam, 1971.
2. M. Lesbre, D. Mazerolles, J. Satgé, *The Organic Compounds of Germanium*, Wiley-Interscience, New York, 1971.
3. P. Rivière, M. Rivière-Baudet, J. Satgé, *Comprehensive Organometallic Chemistry*, Vol. 2, G. Wilkinson, F. G. A. Stone, E. W. Abel, eds., Pergamon Press, Oxford, 1982, p. 399.
4. G. G. Devyatykh, A. D. Zorin, I. A. Frolov, R. P. Rostinova, *Russ. J. Inorg. Chem. (Engl. Transl.)*, 1396 (1971).
5. V. F. Mironov, T. K. Gar, *J. Gen. Chem. USSR (Engl. Transl.)*, 45, 94 (1975).
6. P. Rivière, J. Satgé, A. Boy, *J. Organomet. Chem.*, 96, 25 (1975).

1.6.4.3.2. from Oxygen Compounds.

Germanium dioxide reacts with $(i\text{-}Bu)_2AlH$ in the absence of solvent to form[1] GeH_4:

$$GeO_2 + 4 \text{ i-Bu}_2AlH \rightarrow 2 [\text{i-Bu}_2Al]_2O + GeH_4 \qquad (a)$$

(A.D. NORMAN)

1. G. G. Devyatykh, A. D. Zorin, I. A. Frolov, R. P. Rostunova, *Russ. J. Inorg. Chem. (Engl. Transl.)*, 16, 1396 (1971).

1.6.4.3.3. from Other Derivatives.

Germanium–hydrogen bonds are formed by trialkylstannane cleavage of Ge—Sb, Ge—Bi, Ge—Cd or Ge—Tl bonds[1,2]:

$$3 Me_3SnH + (Et_3Ge)_3M \rightarrow (Me_3Sn)_3M + 3 Et_3GeH \qquad (a)$$

where M = Sb, Bi;

$$2 Et_3SnH + (Et_3Ge)_2Cd \rightarrow Cd + Et_6Sn_2 + 2 Et_3GeH \qquad (b)$$

$$6 Et_3SnH + 2 (Et_3Ge)_3Tl \rightarrow 2 Tl + 3 Et_6Sn_2 + 6 Et_3GeH \qquad (c)$$

Silane or germane cleavage of Ge—C bonds, in H_2PtCl_6-catalyzed reactions e.g.:

$$Et_2Ge(CH_2)_3 + R_2R'MH \rightarrow Et_2Ge(H)CH_2CH_2CH_2MR_2R' \qquad (d)^2$$

where $M = Ge$; $R = Me$, $R' = Et$: $M = Si$; $R_2R' = Et_3$, Ph_2Me, Me_2Cl, results in germabutane ring opening to the acyclic germane products.

Thermolysis of germanes yields products containing new Ge—H bonds. Germane decomposes to Ge, H_2 and Ge subhydride when heated at 280°C. However, if heated at low P in a gas-circulating system, small quantities of Ge_2H_6 and Ge_3H_8 are formed[3]:

$$2\ GeH_4 \rightarrow H_2 + Ge_2H_6 \qquad\qquad (e)$$

$$GeH_4 + Ge_2H_6 \rightarrow H_2 + Ge_3H_8 \qquad\qquad (f)[3]$$

Digermane at 195–220°C undergoes thermolysis:

$$Ge_2H_6 \rightarrow 0.82\ GeH_{0.3} + 0.49\ H_2 + 1.18\ GeH_4 \qquad\qquad (g)[3]$$

If heated in the presence of trapping reagents, e.g., Me_3GeH:

$$Ge_2H_6 + Me_3GeH \rightarrow GeH_4 + Me_3GeGeH_3 \qquad\qquad (h)$$

or Me_3SiH:

$$Ge_2H_6 + Me_3SiH \rightarrow GeH_4 + Me_3SiGeH_3 \qquad\qquad (i)$$

products result from GeH_2 insertion into Ge—H or Si—H bonds, respectively[4]. Similarly, Me_5Ge_2H undergoes thermolysis to form a trigermane[5]:

$$2\ Me_5Ge_2H \rightarrow Me_3GeH + Me_3GeGe(Me)_2GeMe_2H \qquad\qquad (j)$$

Pyrolysis of germane–silane mixtures by their rapid passage through a 350–370°C hot zone yields mixtures of binary and ternary hydrides[3,6]. Typical pyrolysis mixtures and the major ternary hydride products obtained are[3]: SiH_4–GeH_4, $SiGeH_6$; Ge_2H_6–Si_2H_6, Si_2GeH_8; Si_3H_8–Ge_2H_6, i- and n-Si_3GeH_{10} and n-Si_4H_{10}–Ge_3H_8, n-Si_4GeH_{12}. Such reactions, likely involving GeH_2 intermediates, proceed:

$$GeH_4 + SiH_4 \rightarrow H_2 + SiH_3GeH_3 \qquad\qquad (k)$$

$$Ge_3H_8 + (SiH_3)_2SiHSiH_3 \rightarrow Ge_2H_6 + (SiH_3)_2SiHSiH_2GeH_3 \qquad\qquad (l)$$

The Hg-sensitized photolysis of GeH_4 yields[7] some Ge_2H_6, and the X-irradiation of Ge_2H_6–C_2H_4 mixtures yields[8] $EtGe_2H_5$, $EtGe_3H_7$, etc., in processes that may involve formation of new Ge–H bonds.

Phenylgermane at 200°C and Ph_3GeH at 300°C disproportionate:

$$2\ PhGeH_3 \rightarrow Ph_2GeH_2 \rightarrow GeH_4 \qquad\qquad (m)[2]$$

$$2\ Ph_3GeH \rightarrow Ph_4Ge + GeH_4 \qquad\qquad (n)[2]$$

Reactions are instantaneous in the presence of $AlCl_3$.

h^1-Cyclopentadienylgermane disproportionates in the presence of Et_2NH to a mixture of Ge_2H_6, Ge_3H_8, C_5H_6 and solid Ge hydrides[9].

Passage of GeH_4, GeH_4–SiH_4, GeH_4–PH_3 or GeH_4–AsH_3 through a 15-kV silent electric discharge gives products, some of which are likely the result of new Ge—H bond formation[3], e.g., GeH_4 (Ge_2H_6, Ge_3H_8), GeH_4–SiH_4 (Ge_2H_6, Si_2H_6 and SiH_3GeH_3), GeH_4–PH_3 (Ge_2H_6, P_2H_4 and GeH_3PH_2) and GeH_4–AsH_3 (Ge_2H_6 and GeH_3AsH_2). Lesser products form also. From the GeH_4 and GeH_4–SiH_4 reactions, germanes to Ge_5H_{12} and silylgermanes to $SiGe_4H_{12}$ are formed[6].

Disproportionation of GeH_3-substituted compounds, e.g., $(GeH_3)_3P$, $(GeH_3)_2O$ and GeH_3CO_2H, yields GeH_4 along with higher mol wt products or uncharacterized polymers:

$$2 \; (GeH_3)_3P \xrightarrow{25°C} [(GeH_3)_2P]_2GeH_2 \; + \; GeH_4 \qquad (o)^{10}$$

$$(GeH_3)_2O \rightarrow \tfrac{1}{x} (GeH_2O)_x \; + \; GeH_4 \qquad (p)^{11}$$

$$GeH_3CO_2H \rightarrow CO \; + \; polymer \; + \; GeH_4 \qquad (q)^{12}$$

Similar equilibrium disproportionations of $MeGe(Ph_2)_2H$ and $Me_2Ge(PH_2)H$ yields mixtures of phosphinogermanes[13]:

$$2 \; Me_2Ge(PH_2)H \rightarrow Me_2Ge(PH_2)_2 \; + \; MeGeH_2 \qquad (r)$$

$$MeGe(PH_2)_2H \rightarrow MeGe(PH_2)_3, \; MeGe(PH_2)H_2, \; MeGeH_3 \qquad (s)$$

Thermal elimination of germyl and hydride moieties from germyl metal hydrides yields germanes:

$$(dppe)Pt(Cl)_2H(GePh_3) \rightarrow (dppe)PtCl_2 \; + \; Ph_3GeH \qquad (t)^1$$

where $dppe = Ph_2PCH_2CH_2PPh_2$.

(A.D. NORMAN)

1. F. Glockling, *The Chemistry of Germanium,* Academic Press, London, 1969.
2. M. Lesbre, P. Mazerolles, J. Satgé, *The Organic Chemistry of Germanium,* Wiley-Interscience, New York, 1971.
3. W. L. Jolly, A. D. Norman, *Prep. Inorg. React., 4,* 1 (1968).
4. M. D. Sefcik, M. A. Ring, *J. Organomet. Chem., 59,* 167 (1973).
5. P. G. Harrison, *Coord. Chem. Rev., 40,* 179 (1982).
6. E. Wiberg, E. Amberger, *Hydrides of the Elements of Main Group I–IV,* Elsevier, Amsterdam, 1971.
7. R. Varma, K. R. Ramaprosod, A. J. Signorelli, B. K. Solray, *J. Inorg. Nucl. Chem., 37,* 563 (1975).
8. J. K. Khandelwal, J. W. Purson, *Inorg. Nucl. Chem. Lett., 9,* 393 (1973).
9. P. C. Angus, S. R. Stobart, *J. Chem. Soc., Chem. Commun.* 127 (1973).
10. J. E. Drake, C. Riddle, *Q. Rev. Chem. Soc.,* 263 (1970).
11. D. W. H. Rankin, *J. Chem. Soc., Chem. Commun.,* 194 (1969).
12. P. M. Kuznesof, W. L. Jolly, *Inorg. Chem., 7,* 2574 (1968).
13. A. R. Dahl, C. A. Heil, A. D. Norman, *Inorg. Chem., 14,* 1095 (1975).

1.6.4.4. Giving Hydrides of Tin

1.6.4.4.1. from Halides.

Reductions of chlorostannanes by substituted alanes constitute excellent syntheses of stannanes[1,2], e.g.:

$$R_{4-x}SnCl_x + x \; Et_2AlH \xrightarrow{Et_2O} x \; Et_2AlCl + R_{4-x}SnH_x \qquad (a)$$

where $R = Me, Et, n-Bu, Ph; x = 1–3$. Representative hydrides formed and % yields are[2]: $EtSnH_3$, 97; $n-BuSnH_3$, 62; $PhSnH_3$, 72; Et_2SnH_2, 84; $i-Bu_2SnH_2$, 72 and Et_3SnH, 89.

Diorganostannannes are prepared by exchange reactions, e.g., $n-Bu_3SnH$ with organodihalostannanes:

$$2 \; n-Bu_3SnH + R_2SnCl_2 \rightarrow 2 \; n-Bu_3SnCl + R_2SnH_2 \qquad (b)^{1,3}$$

where $R = Me, Et, Ph$. Exchange is useful for the synthesis of halohydrides because the equilibria involved often favor the mixed-substituent products[4]:

$$R_2SnH_2 + R_2SnX_2 \rightleftharpoons 2 \; R_2SnH(X) \tag{c}$$

where R = Et, n-Bu, Ph; X = F, Cl, Br.

(A.D. NORMAN)

1. E. J. Kupchik, in *Organotin Compounds*, Vol. 1, A. K. Sawyer, ed., Marcel Dekker, New York, 1971, p. 7.
2. E. Wiberg, E. Amberger, *Hydrides of the Elements of Main Groups I–IV*, Elsevier, Amsterdam, 1971.
3. H.-J. Albert, W. P. Neumann, *Synthesis*, 942, (1980).
4. K. Moedritzer, *Adv. Organomet. Chem.*, 6, 171 (1968).

1.6.4.4.2. from Oxygen Compounds.

Substituted alanes or B_2H_6 react with alkoxystannanes forming stannanes[1,2] in high yields:

$$Sn(OEt)_4 + 4 \; Et_2AlH \xrightarrow{\text{xylene}} 4 \; Et_2AlOEt + SnH_4 \tag{a}[2]$$

$$2 \; R_2Sn(OMe)_2 + B_2H_6 \xrightarrow{C_5H_{12}} 2 \; BH(OMe)_2 + 2 \; R_2SnH_2 \tag{b}[2]$$

where R = Me, Et, n-C_3H_7, Ph. Typical products and % yields are[1,2]: SnH_4, 32; Et_2SnH_2, 91; n-Bu_3SnH, 100 and Ph_3SnH, 76.

Reduction of organostannyl oxides or alkoxides is effected with hydridosiloxanes, such as $(MeSiHO)_n$ or $(Ph_2SiH)_2O$:

$$(n\text{-}Bu_2Sn)_2O + 2 \; (MeSiHO)_x \rightarrow [(MeSi)_2O_3]_x + 2 \; n\text{-}Bu_3SnH \tag{c}[1,3]$$

$$(Ph_2SiH)_2O + (n\text{-}Bu_3Sn)_2O \rightarrow \frac{2}{x} (Ph_2SiO_2)_x + 2 \; n\text{-}Bu_3SnH \tag{d}[1,3]$$

Exchange can result in new Sn—H bond formation in synthetically useful reactions, e.g.:

$$n\text{-}Bu_2SnO + 2 \; n\text{-}Bu_2Sn(Cl)H \overset{100°C}{\rightleftharpoons} (n\text{-}Bu_2SnCl)_2O + n\text{-}Bu_2SnH_2 \tag{e}[4]$$

Trialkylformylstannanes upon thermolysis eliminate CO_2:

$$R_3SnCO_2H \xrightarrow{160-180°C} CO_2 + R_3SnH \tag{f}[2]$$

where R = i-Pr, n-Bu.

(A.D. NORMAN)

1. E. J. Kupchik, in *Organotin Compounds*, Vol. 1, A. K. Sawyer, ed., Marcel Dekker, New York, 1971, p. 7.
2. E. Wiberg, E. Amberger, *Hydrides of Elements of the Main Groups I–IV*, Elsevier, Amsterdam, 1971.
3. W. P. Neumann, K. Kühlein, *Adv. Organomet. Chem.*, 7, 242 (1968).
4. A. K. Sawyer, J. E. Brown, S. L. Fredrickson, G. A. Scott, *Synth. React. Inorg. Metal-Org. Chem.*, 6, 281 (1976).

1.6.4.4.3. from Other Derivatives.

N-Diethylaminostannanes are converted[1] to stannanes in yields up to 99% by n-Bu_2AlH or B_2H_6.

$$R_{4-x}Sn(NEt_2)_x + x \; n\text{-}Bu_2AlH \rightarrow x \; n\text{-}Bu_2AlNEt_2 + R_{4-x}SnH_x \tag{a}$$

$$2 R_{4-x} Sn(NEt_2)_x + x/2\ B_2H_6 \rightarrow x\ H_2BNEt_2 + 2\ R_{4-x}SnH_x \qquad (b)$$

where R = Me, Et, n-Bu, Ph; x = 1–3.

Dialkyl- or diarylstannanes react with formamidostannanes to form distannanes:

$$R_3SnN(Ph)CHO + R_2'SnH_2 \rightarrow PhNHCHO + R_2SnSnR_2'H \qquad (c)[1]$$

where R = alkyl; R′ = alkyl, aryl.

Exchange occurs between organostannanes and $(R_3Sn)_2Hg$ compounds:

$$(R_3Sn)_2Hg + R_3'SnH \rightarrow R_3SnHgSnR_3' + R_3SnH \qquad (d)[1]$$

where R = Me, Et, n-Pr, n-Bu; R′ = Ph. Reaction of Et_3SnPPh_2 with Ph_3SnH yields Et_3SnH.

$$Et_3SnPPh_2 + Ph_3SnH \rightarrow Ph_3SnPPh_2 + Et_3SnH \qquad (e)[3]$$

Dialkylstannylenes insert into Sn—H bonds forming new Sn—H bond-containing products:

$$n\text{-}Bu_2Sn + Me_2SnH_2 \rightarrow n\text{-}Bu_2Sn(H)SnMe_2(H) \qquad (f)[2]$$

Similarly, $[(Me_3Si)_2CH]_2Sn$ and $h^5\text{-}CpMo(CO)_3H$ react:

$$[(Me_3Si)_2CH]_2Sn + h^5\text{-}CpMo(CO)_3H \rightarrow h^5\text{-}Cp(CO)_3MoSn[CH(SiMe_3)_2]_2H \qquad (g)[4]$$

(A.D. NORMAN)

1. E. J. Kupchik, *Organotin Compounds*, Vol. 1, A. K. Sawyer, ed., Marcel Dekker, New York, 1971, p. 7.
2. A. G. Davies, P. J. Smith, *Adv. Inorg. Chem. Radiochem.*, 23, 1 (1980).
3. E. W. Abel, S. M. Illingworth, *Organomet. Chem. Rev. A.*, 5, 143 (1970).
4. J. D. Cotton, P. J. Davison, D. E. Goldberg, M. F. Lappert, K. M. Thomas, *J. Chem. Soc., Chem. Commun.*, 893 (1974).

1.6.4.5. Giving Hydrides of Lead

1.6.4.5.1. from Halides.

Reactions of trialkylchloroplumbanes with Et_2AlH or $i\text{-}Bu_2AlH$ at $-60°C$ yield plumbanes:

$$R_3PbCl + Et_2AlH \rightarrow Et_2AlH + R_3PbH \qquad (a)$$

where R = Me, Et, i-Pr, n-Bu; however, the products are difficult to separate from the reaction mixtures[1]. Triethylplumbane forms in an equilibrium:

$$Et_3PbCl + n\text{-}Bu_3SnH \rightleftharpoons n\text{-}Bu_3SnCl + Et_3PbH \qquad (b)[1]$$

(A.D. NORMAN)

1. W. P. Neumann, K. Kühlein, *Adv. Organomet. Chem.*, 7, 241 (1968).

1.6.4.5.2. from Oxygen Compounds.

Organotin hydrides react with $n\text{-}Bu_3PbX$ and $(n\text{-}Bu_3Pb)_2O$ to form products in 10–20% yield[1,2]:

$$n\text{-}Bu_3PbX + R_3SnH \rightarrow R_3SnX + n\text{-}Bu_3PbH \qquad (a)$$

where X = OAc, R = Et, Ph:

$$(\text{n-Bu}_3\text{Pb})_2\text{O} \xrightarrow{\text{R}_3\text{SnH}} 2 \text{ n-Bu}_3\text{PbH} \qquad (b)^2$$

Trialkyllead methoxides react quantitatively with diborane in ether or pentane to form boronates which in methanol above $-78°C$ convert cleanly to the Pb hydrides[3]:

$$\text{R}_3\text{Pb}(\text{BH}_4) + 3 \text{ MeOH} \rightarrow 3 \text{ H}_2 + \text{B}(\text{OMe})_3 + \text{R}_3\text{PbH} \qquad (c)$$

where R = Me, Et, n-Pr, n-Bu.

(A.D. NORMAN)

1. W. P. Neumann, K. Kühlein, *Adv. Organomet. Chem.*, 7, 241 (1968).
2. H. Shapiro, F. W. Frey, *Organic Compounds of Lead*, Interscience, New York, 1968.
3. E. Amberger, R. Hönigschmidt-Grossich, *Chem. Ber.*, 99, 1673 (1966).

1.6.4.5.3. from Other Derivatives.

Reactions of alkyl- or vinyl-substituted plumbanes with Ph_3SnH result in Pb—C bond cleavage and formation of Pb—H bonds[1]:

$$\text{Et}_3\text{PbR} + \text{Ph}_3\text{SnH} \rightarrow \text{Ph}_3\text{SnR} + \text{Et}_3\text{PbH} \qquad (a)$$

where R = alkyl, vinyl. Diborane reacts similarly with $\text{R}_4\text{Pb}(\text{R} = \text{vinyl})$ at $80°C$[2]:

$$\text{Pb}(\text{C}_2\text{H}_3)_4 + \text{B}_2\text{H}_6 \rightarrow \text{C}_2\text{H}_3\text{B}_2\text{H}_5 + (\text{C}_2\text{H}_3)_3\text{PbH} \qquad (b)$$

Self-association of Me_3PbH in liq NH_3 results in $[\text{Me}_3\text{PbH}_2]^+$ formation[2]:

$$2 \text{ Me}_3\text{PbH} \rightleftharpoons [\text{Me}_3\text{Pb}]^- + [\text{Me}_3\text{PbH}_2]^+ \qquad (c)$$

(A.D. NORMAN)

1. W. P. Neumann, K. Kühlein, *Adv. Organomet. Chem.*, 7, 241 (1968).
2. R. Duffy, J. Feeney, A. K. Holiday, *J. Chem. Soc.*, 1144 (1962).

1.6.5. by Complex Hydride Reduction

1.6.5.1. Giving Hydrides of Carbon

1.6.5.1.1. from Halides.

Complex metal hydroborates and hydroaluminates, Cu hydrides, and modified complex hydrides reduce C–halogen (X = Cl, Br, I) to C—H bonds[1-3], e.g.:

$$\text{EtBr} \xrightarrow{\text{LiAlH}_4-\text{Et}_2\text{O}} \text{EtH} \qquad (a)$$

$$\text{Me}_2\text{CHBr} \xrightarrow{\text{NaBH}_4-\text{C}_4\text{H}_8\text{O}_2} \text{Me}_2\text{CH}_2 \qquad (b)$$

$$\text{Me}(\text{CH}_2)_6\text{CH}_2\text{Br} \xrightarrow{(\text{KCuH})_n} \text{Me}(\text{CH}_2)_6\text{CH}_3 \qquad (c)$$

$$\text{m-NO}_2\text{C}_6\text{H}_4\text{COCl} \xrightarrow{\text{LiHAl}(\text{t-BuO})_3} \text{m-NO}_2\text{C}_6\text{H}_4\text{CHO} \qquad (d)$$

Relative reactivities of organic halides toward complex hydrides, conditions for optimum yields and limitations on the reactions are detailed in treatises on organic chemistry[2-4].

(A.D. NORMAN)

1. E. Wiberg, E. Amberger, *Hydrides of the Elements of Main Groups I–IV*, Elsevier, Amsterdam, 1971.
2. A. Hajós, *Complex Hydrides*, Elsevier, Amsterdam, 1979.
3. H. C. Brown, *Boranes in Organic Chemistry*, Cornell Univ. Press, Ithaca, NY, 1972.
4. C. A. Buehler, D. E. Pearson, *Survey of Organic Synthesis*, Vol. 2, Wiley-Interscience, New York, 1977.

1.6.5.1.2. from Oxygen Compounds.

Lithium tetrahydroborate and CO_2 react at 25°C and 125°C[1,2], respectively:

$$LiBH_4 + 2 CO_2 \xrightarrow[25°C]{Et_2O} LiBO(OCH_3)O_2CH \tag{a}$$

$$5 LiBH_4 + 8 CO_2 \xrightarrow{125°C} 3 LiBO_2 + 2 LiB(OCH_3)_3O_2CH \tag{b}$$

In contrast, $NaBH_4$ and CO_2 at 25°C yield only the formatohydroborate ion[2]:

$$NaBH_4 + 3 CO_2 \xrightarrow[25°C]{Me_2O} Na[BH(O_2CH)_3] \tag{c}$$

Reduction of CO_2 with 3, 2, or 1 equiv of $LiAlH_4$ occurs:

$$4 CO_2 + 3 LiAlH_4 \rightarrow 2 LiAlO_2 + LiAl(OCH_3)_4 \tag{d}[3]$$

$$2 CO_2 + LiAlH_4 \rightarrow LiAl(OCH_2O)_2 \tag{e}[3]$$

$$4 CO_2 + LiAlH_4 \rightarrow LiAl(O_2CH)_4 \tag{f}[3]$$

Carbon monoxide in MeOH reacts with $LiAlH_4$:

$$2 CO + 2 LiAlH_4 \rightarrow 2 [H_2AlCH_2O]Li \tag{g}[3]$$

Borane carbonyls, e.g., $B_{10}H_8(CO)_2$, are reduced[1] by $LiAlH_4$:

$$B_{10}H_8(CO)_2 \xrightarrow{LiAlH_4} [B_{10}H_8(CH_3)_2]Li_2 \tag{h}$$

Metal-coordinated carbonyl moieties are reduced by hydroborates or hydroaluminates to formyl:

$$[Re(CO)_5PPh_3]^+ \xrightarrow{LiEt_3BH} [Ph_3P(CO)_3Re(CHO)_2]^- \tag{i}[4,5]$$

alcohol:

$$[h^5\text{-}CpFe(CO)_3]^+ \xrightarrow[2. ROH]{1. NaBH_3CN} h^5\text{-}CpFe(CO)_2CH_2OH \tag{j}[6]$$

or alkyl derivatives in high yields[4,5]:

$$[h^5\text{-}CpRe(CO)_2(NO)]^+ \xrightarrow{NaBH_4\text{–}THF} h^5\text{-}CpRe(CO)(NO)CH_3 \tag{k}[4,5]$$

Reduction of a diiron complex by $LiAlH_4$ yields hydrocarbons:

$$[h_5\text{-}Cp(CO)Fe]_2(\mu\text{-}CO)\mu\text{-}MeCH \xrightarrow{\text{LiAlH}_4} CH_4, C_2H_4, C_2H_6, C_3H_6, C_3H_8, n\text{-}C_4H_{10} \quad (1)^7$$

Intramolecular reductive elimination from $[[(RO)_3P]_4Co(CH_3)H]^+$ yields CH_4:

$$[cis\text{-}[(RO)_3P]_4Co(CH_3)H]^+ \rightarrow [[(RO)_3P]_4Co]^+ + CH_4 \quad (m)^4$$

Organic oxygen compounds, e.g., aldehydes:

$$PhCHO \xrightarrow{\text{LiBEt}_3\text{H}} PhCH_2OH \quad (n)$$

ketones:

$$Et_2CO \xrightarrow{\text{NaBH}_4-\text{EtOH}} Et_2CHOH \quad (o)$$

acid chlorides:

$$MeCOCl \xrightarrow{\text{LiAlH}_4} MeCH_2OH \quad (p)$$

carboxylic acids:

$$PhCO_2H \xrightarrow{\text{LiAlH}_4} PhCH_2OH \quad (q)$$

and alcohols:

$$p\text{-}MeC_6H_4OH \xrightarrow{\text{LiAlH}_4} p\text{-}MeC_6H_5 \quad (r)$$

are reduced by complex hydrides[1-3,8,9]. Some reactions occur directly to final reduced products:

$$(EtO)_3CH + Al(BH_4)_3 \rightarrow (EtO)_2AlBH_4 + B_2H_6 + EtOCH_3 \quad (s)^2$$

However, most are two step, reduction followed by hydrolysis.

(A.D. NORMAN)

1. E. Wiberg, E. Amberger, *Hydrides of the Elements of Main Groups I–IV*, Elsevier, Amsterdam, 1971.
2. B. D. James, M. G. H. Walbridge, *Prog. Inorg. Chem.*, *11*, 99 (1970).
3. A. Hajós, *Complex Hydrides*, Elsevier, Amsterdam, 1979.
4. F. A. Cotton, G. Wilkinson, *Advanced Inorganic Chemistry*, 4th ed., Wiley-Interscience, New York, 1980.
5. J. A. Gladysz, *Adv. Organomet. Chem.*, *20*, 1 (1982).
6. T. Bodnar, E. Coman, K. Menarad, A. Cutler, *Inorg. Chem.*, *21*, 1275 (1982).
7. S. C. Kao, P. P. Y. Liu, R. Pettit, *Organometallics*, *1*, 911 (1982).
8. H. C. Brown, *Boranes in Organic Chemistry*, Cornell Univ. Press, Ithaca, NY, 1972.
9. C. A. Buehler, D. E. Pearson, *Survey of Organic Syntheses*, Vol. 2, Wiley-Interscience, New York, 1977.

1.6.5.1.3. from Nitrogen Compounds.

Complex metal hydroborates and hydroaluminates, hydridoferrates and modified complex hydrides (e.g., $NaBH_4-AlCl_3$) reduce organic N-containing compounds[1-4] (see also §1.5.5.1), e.g., amides:

$$4 \, RCONR_2' + LiAlH_4 \rightarrow [[RCH(NR_2')O]_4Al]Li \tag{a}$$

and nitriles:

$$4 \, RC{\equiv}N + LiAlH_4 \rightarrow [(RCH{=}N)_4Al]Li \tag{b}$$

Hydrolysis yields final reduction products[1-3]. Reaction mixtures containing Mo salts and sulfur ligands, in aqueous base, with $NaBH_4$ as the reducing agent, when treated with nitriles, isonitriles or $[CN]^-$ react to produce hydrocarbons (CH_4, C_2H_2, etc.) in reactions of undetermined stoichiometry[5].

(A.D. NORMAN)

1. A. Hajós, Complex Hydrides, Elsevier, Amsterdam, 1979.
2. H. C. Brown, Boranes in Organic Chemistry, Cornell Univ. Press, Ithaca, NY, 1972.
3. J. R. Malpass, in Comprehensive Organic Chemistry, D. Barton, W. D. Ollis, eds., Vol. 2, I. O. Sutherland, ed., Pergamon Press, Oxford, 1979, p. 3.
4. F. A. Cotton, G. Wilkinson, Advanced Inorganic Chemistry, 4th ed., Wiley-Interscience, New York, 1980.
5. J. Chatt, J. R. Dilworth, R. L. Richards, Chem. Rev., 78, 589 (1978).

1.6.5.1.4. from Compounds with Carbon–Carbon Multiple Bonds.

Alkenes and alkynes react with complex hydrides only under special conditions[1]. Reaction products are often intermediates prior to solvolysis to the final hydrocarbons desired in organic syntheses[1-5], e.g.:

$$LiAlH_4 + ClC{\equiv}CPh \xrightarrow[-30°C]{THF} [trans\text{-}PhCH{=}C(Cl)AlH_3]Li \tag{a}$$

$$2 \, [trans\text{-}PhCH{=}C(Cl)AlH_3]Li + 8 \, MeOH \rightarrow 3 \, H_2 + \\ 2 \, LiAl(OMe)_4 + 2 \, trans\text{-}PhCH{=}CHCl \tag{b}$$

Such reactions, especially those of MBH_4 (M = Li, Na), occur most easily in the presence of acids (e.g., $AlCl_3$ or $BF_3 \cdot OEt_2$) as catalysts:

$$9 \, RCH{=}CH_2 + 3 \, NaBH_4 + AlCl_3 \rightarrow AlH_3 + 3 \, NaCl + 3 \, (RCH_2CH_2)_3B \tag{c}[3,5]$$

Aluminum tetrahydroborate reacts with alkenes or alkynes:

$$Al(BH_4)_3 + 12 \, C_2H_4 \xrightarrow{140°C} Al(C_2H_5)_3 + 3 \, B(C_2H_5)_3 \tag{d}[3]$$

Metal tetrahydroborates and tetrahydroaluminates can react with coordinated unsaturated organic molecules without ligand displacement:

$$[h^5\text{-}CpFe(CO)_2CH_2CHCH_3\text{-}h^2)]^+ \xrightarrow{NaBH_4\text{-}THF} h^5\text{-}CpFe(CO)CH(CH_3)_2 \tag{e}[7]$$

$$[h^6\text{-}Me_6C_6Re(CO)_3]^+ \xrightarrow{LiAlH_4} h^5\text{-}Me_6C_6HRe(CO)_3 \tag{f}[8]$$

In other cases, reduction accompanied by cleavage of the organic moiety from the metal occurs.

$$Ph_4C_4NiBr_2 \xrightarrow{LiAlH_4\text{-}Et_2O} Ph_4C_4H_2 \tag{g}[7]$$

(A.D. NORMAN)

1. A. Hajós, *Complex Hydrides*, Elsevier, Amsterdam, 1979.
2. C. A. Buehler, D. E. Pearson, *Survey of Organic Syntheses*, Vol. 2, Wiley-Interscience, New York, 1977.
3. B. D. James, M. G. H. Wallbridge, *Prog. Inorg. Chem.*, *11*, 99 (1970).
4. E. Wiberg, E. Amberger, *Hydrides of the Elements of the Main Groups I–IV*, Elsevier, Amsterdam, 1971.
5. H. C. Brown, *Boranes in Organic Synthesis*, Cornell Univ. Press, Ithaca, NY, 1972.
6. E. C. Ashby, S. A. Nading, *J. Org. Chem.*, *45*, 1035 (1980).
7. F. A. Cotton, G. Wilkinson, *Advanced Inorganic Chemistry*, 4th ed., Wiley-Interscience, New York, 1980.
8. H. W. Quin, J. H. Tsai, *Adv. Inorg. Chem. Radiochem.*, *12*, 217 (1969).

1.6.5.2. Giving Hydrides of Silicon

1.6.5.2.1. from Halides.

Silicon halides react with complex hydrides to form silanes. Complex hydrides used are $LiAlH_4$ and to a lesser extent MBH_4 (M = Li, Na, K)[1,2]. Reductions are carried out in aprotic ether solvents, e.g., Et_2O, glymes, THF or n-Bu_2O. These reactions are adaptable to Si—D bond synthesis (see §1.6.7.1.2). Typical syntheses of monosilanes are shown in Table 1. Chlorosilanes are preferred; the fluorides react less readily. Bromo- and iodosilanes react easily but offer no advantage over the more available chlorides[1].

Reduction of halosilanes by $LiAlH_4$ occurs:

$$R_{4-n}SiX_n + \frac{n}{4} LiAlH_4 \xrightarrow{Et_2O} \frac{n}{4} LiAlX_4 + R_{4-n}SiH_n \tag{a}$$

where R = H, alkyl, aryl; X = F, Cl, Br; n = 1–4. Yields are high (> 50%), and consequently this method is the preferred synthesis for substituted silanes.

Reduction of Si—Cl bonds in polysilanes is without extensive Si—Si bond cleavage. Reduction of Si_nCl_{2n+2} (n = 2–5) halides occurs, although Si_5Cl_{12} reduction even at $-100°C$ is accompanied by some bond cleavage[1,6]. High yields of cyclosilanes (n = 5)[7]:

$$Si_nCl_{2n+2} + \frac{2n + 2}{4} LiAlH_4 \rightarrow \frac{2n + 2}{4} LiAlCl_4 + Si_nH_{2n+2} \tag{b}$$

TABLE 1. COMPLEX HYDRIDE REDUCTIONS OF HALOSILANES

Reactant	Complex hydride, solvent	Product	Yield (%)	Ref.
$SiCl_4$	$LiAlH_4$, Et_2O	SiH_4	99	1–3
$SiHCl_3$	$NaAlH_4$, diglyme	SiH_4	100	1
Si_2Cl_6	$LiAlH_4$, n-Bu_2O	Si_2H_6	80	1,3,4
$MeSiCl_3$	$LiAlH_4$, dioxane	$MeSiH_3$	90	1,2
$EtSiCl_3$	$LiAlH_4$, dioxane	$EtSiH_3$	90	1,2
$PhSiCl_3$	$LiAlH_4$, Et_2O	$PhSiH_3$	70	1,2,5
Me_2SiCl_2	$LiAlH_4$, dioxane	Me_2SiH_2	90	1,2
Ph_2SiCl_2	$LiAlH_4$, THF	Ph_2SiH_2	76	1,2
Et_3SiCl	$LiAlH_4$, dioxane	Et_3SiH	90	1,2
$(p\text{-}ClC_6H_4)Ph_2SiCl$	$LiAlH_4$, Et_2O	$(p\text{-}ClC_6H_4)Ph_2SiH$	83	1,5

and oligomeric silanes $(n = 1-5)^8$ are obtained in $LiAlH_4$ reductions:

$$4 \ (SiBr_2)_5 + 10 \ LiAlH_4 \xrightarrow{Et_2O} 10 \ LiAlBr_4 + 4 \ (SiH_2)_5 \qquad (c)$$

$$(Ph_2Si)_4 \xrightarrow{LiAlH_4} (Ph_2Si)_5, \ H(SiPh_2)_nH \qquad (d)$$

Tetrahydroborates also reduce halosilanes to the parent silanes. Lithium tetrahydroborate reduces $SiCl_4$ to SiH_4 in high yield; in addition, it is more selective than $LiAlH_4$. Lithium tetrahydroborate reacts with alkoxychlorosilanes to reduce Si—Cl bonds but without reduction of Si—O bonds, providing a good route to alkoxysilanes[9]:

$$2 \ (EtO)_2SiHCl + 2 \ LiBH_4 \rightarrow 2 \ LiCl + B_2H_6 + 2 \ (EtO)_2SiH_2 \qquad (e)$$

Alkylchlorosilanes react with $Al(BH_4)_3$ forming alkysilanes, e.g.:

$$6 \ Me_3SiCl + 2 \ Al(BH_4)_3 \rightarrow B_2H_6 + Al_2Cl_6 + 6 \ Me_3SiH \qquad (f)$$

However, reaction with $SiCl_4$ does not easily produce[9] SiH_4.

(A.D. NORMAN)

1. E. Wiberg, E. Amberger *Hydrides of the Elements of Main Groups I–IV*, Elsevier, Amsterdam, 1971.
2. D. A. Armitage, *Comprehensive Organometallic Chemistry*, Vol. 2, G. Wilkinson, F. G. A. Stone, E. W. Abel, eds., Pergamon Press, Oxford, 1982, p. 1.
3. A. D. Norman, J. R. Webster, W. L. Jolly, *Inorg. Synth.*, *11*, 170 (1968).
4. M. Kumada, K. Tamao, *Adv. Organomet. Chem.*, *6*, 19 (1968).
5. L. G. L. Ward, *Inorg. Synth.*, *11*, 159 (1968).
6. F. Höfler, R. Jannach, *Inorg. Nucl. Chem. Lett.*, *9*, 723 (1973).
7. E. Hengge, G. Bauer, *Angew. Chem., Int. Ed. Engl.*, *12*, 316 (1973).
8. H. Gilman, W. H. Atwell, F. K. Cartledge, *Adv. Organomet. Chem.*, *4*, 1 (1966).
9. B. D. James, M. G. H. Wallbridge, *Prog. Inorg. Chem.*, *11*, 99 (1970).

1.6.5.2.2. from Oxygen Compounds.

Tetrahydroaluminate reduction of siloxanes or alkoxysilanes is an effective route to SiH_4 or alkyl or arylsilanes in ethers (Table 1)[1,2]:

$$(EtO)_4Si + NaAlH_4 \xrightarrow{diglyme} NaAl(OEt)_4 + SiH_4 \qquad (a)$$

TABLE 1. COMPLEX HYDRIDE REDUCTION OF SILOXANES

Reactant	Complex hydride, solvent	Product	Yield (%)	Ref.
$(EtO)_4Si$	$NaAlH_4$, diglyme	SiH_4	100	1,2
$(SiCl_3)_2O$	$LiAlH_4$, Et_2O	SiH_4	100	1
$SiHCl(OMe)_2$	$LiBH_4$[a]	$(MeO)_2SiH_2$	63	1,3
$EtOSiHCl_2$	$LiBH_4$[a]	$EtOSiH_3$	40	1,3
$Me_2Si(OEt)_2$	$LiAlH_4$, Et_2O	Me_2SiH_2	100	1,3
$(Ph_2SiH)_2O$	$LiAlH_4$, Et_2O	Ph_2SiH_2	59	1
$i\text{-}Pr_3SiOEt$	$LiAlH_4$, $n\text{-}Bu_2O$	$i\text{-}Pr_3SiH$	[b]	1

[a] No solvent.
[b] Not reported.

$$2 \ Ph_2Si(OEt)_2 \ + \ LiAlH_4 \ \xrightarrow{\text{n-Bu}_2O} \ LiAl(OEt)_4 \ + \ 2 \ Ph_2SiH_2 \tag{b}$$

In contrast, $LiBH_4$ does not reduce Si—O bonds, making it useful for selective reduction of alkoxychlorosilanes.

Silica and silica gel react with $LiAlH_4$ in the solid to form SiH_4 in low yields[1]:

$$SiO_2 \ + \ LiAlH_4 \ \rightarrow \ LiAlO_2 \ + \ SiH_4 \tag{c}$$

although Na_2SiO_3 with KBH_4 or $NaBH_4$ in H_2O do not react[1].

Reaction of $Al(BH_4)_3$ with $(EtO)_4Si$ yields SiH_4 nearly quantitatively:

$$(EtO)_4Si \ + \ 2 \ Al(BH_4)_3 \ \rightarrow \ 2 \ (EtO)_2AlBH_4 \ + \ 2 \ B_2H_6 \ + \ SiH_4 \tag{d}[3]$$

(A.D. NORMAN)

1. E. Wiberg, E. Amberger, *Hydrides of the Elements of Main Groups I–IV*, Elsevier, Amsterdam, 1971.
2. D. A. Armitage, *Comprehensive Organometallic Chemistry*, Vol. 2, G. Wilkinson, F. G. A. Stone, E. W. Abel, eds., Pergamon Press, Oxford, 1982, p. 1.
3. B. D. James, M. G. H. Wallbridge, *Prog. Inorg. Chem.*, 11, 99 (1970).
4. J. M. Bellama, A. G. MacDiarmid, *Inorg. Chem.*, 7, 2070 (1968).

1.6.5.2.3. from Other Derivatives.

Cleavage of Si—N bonds with $LiBH_4$ results in silane formation, although the reactions are not often synthetically useful[1,2]:

$$Me_2NSiMe_2Cl \ + \ LiBH_4 \ \rightarrow \ LiCl \ + \ Me_2NBH_2 \ + \ Me_2SiH_2 \tag{a}$$

From $LiAlH_4$ reduction of dialkylaminosilanes, 90% yields of SiH_4 are attained[3]. Cleavage of Si—Si bonds by $LiAlH_4$ can yield silanes:

$$Me_3SiSiPh_3 \ + \ LiAlH_4 \ \rightarrow \ [H_3AlSiPh_3]Li \ + \ Me_3SiH \tag{b}[4]$$

$$(Ph_2Si)_4 \ \xrightarrow[\text{Et}_2O]{\text{LiAlH}_4} \ (Ph_2Si)_5, \ H(Ph_2Si)_nH \tag{c}[4]$$

(A.D. NORMAN)

1. E. Wiberg, E. Amberger, *Hydrides of the Elements of Main Groups I–IV*, Elsevier, Amsterdam, 1971.
2. B. D. James, M. G. H. Wallbridge, *Prog. Inorg. Chem.*, 11, 99 (1970).
3. W. L. Jolly, A. D. Norman, *Prep. Inorg. React.*, 4, 1 (1968).
4. H. Gilman, W. H. Atwell, F. K. Cartledge, *Adv. Organomet. Chem.*, 4, 1 (1966).

1.6.5.3. Giving Hydrides of Germanium

1.6.5.3.1. from Halides.

Halogermane reductions by complex hydrides are efficient, preferred methods for germane synthesis[1-5]. The complex hydrides used are MBH_4 (M = Li, Na, K) and $LiAlH_4$. Reduction of Ge—X bonds by this method can be used for any molecule that otherwise is unsusceptible to complex hydride reduction or reaction. Lithium tetrahydroaluminate reduction of chiral halogermanes and alkoxygermanes results in inversion and retention of configuration, respectively. The $LiBH_4$ and $LiAlH_4$ reactions require aprotic solvents, such as Et_2O, THF, n-Bu_2O or glyme ethers. Sodium and K

1.6.5. by Complex Hydride Reduction
1.6.5.3. Giving Hydrides of Germanium
1.6.5.3.1. from Halides.

107

tetrahydroborate reductions can be carried out in H_2O. The hydroborate or hydroaluminate reductions are adaptable to Ge—D synthesis (see §1.6.7.1.3).

Reduction of halogermanes by $LiAlH_4$ occurs (Table 1):

$$R_{4-n}GeX_n + \tfrac{n}{4} LiAlH_4 \xrightarrow{Et_2O} \tfrac{n}{4} LiAlX_4 + R_{4-n}GeH_n \tag{a}$$

where n = 1–4; R = H, alkyl, aryl; X = Cl, Br, I. Yields are high. Reduction of ring compounds, e.g., diiodogermacyclopentane, occurs without Ge—C bond cleavage[3].

Unsubstituted halogermanes, e.g., GeH_3GeH_2I:

$$4\ GeH_3GeH_2I + LiAlH_4 \xrightarrow{-78°C} LiAlI_4 + 4\ Ge_2H_6 \tag{b}[4]$$

and $GeCl_3SiCl_3$:

$$2\ SiCl_3GeCl_3 + 3\ LiAlH_4 \xrightarrow{-80°C} 3\ LiAlCl_4 + 2\ SiH_3GeH_3 \tag{c}[4]$$

are reduced at low T. Reactions of $LiAlH_4$ with $GeCl_3H$ in Et_2O yields $(GeH_2)_x$:

$$4\ HGeCl_3 + 3\ LiAlH_4 \xrightarrow{Et_2O} 3\ LiAlCl_4 + 4\ H_2 + \tfrac{4}{x} (GeH_2)_x \tag{d}[4]$$

Halogermane reduction by $Li[(t\text{-}BuO)_3AlH]$ or $Li[As(Me_2)_2AlH]$ yields germanes, the former reagent forming $Me_{4-n}GeH_n$ (n = 1–4) in yields up to 70% from the respective chlorides[3]:

$$Me_{4-n}GeCl_n + n\ Li[(t\text{-}BuO)_3AlH] \xrightarrow{dioxane} n\ Li[(t\text{-}BuO)_3AlCl] + Me_{4-n}GeH_n \tag{e}$$

Reduction of a halogermane by its addition to a basic aq $NaBH_4$ or KBH_4 or to MBH_4 (M = Li, Na, K) in ether, followed by hydrolysis in neutral or acid H_2O, germanes in high yields of (Table 1). Germylboronate intermediate species are hydrolyzed in the last step[1-3].

TABLE 1. COMPLEX HYDRIDE REDUCTIONS OF HALOGERMANES

Reactant	Complex hydride, solvent	Product	Yield (%)	Ref.
$GeCl_4$	$NaBH_4$, THF[a]	GeH_4	40	2,4
$MeGeBr_3$	$NaBH_4$, $HBr_{(aq)}$	$MeGeH_3$	99	2,3
$EtGeCl_3$	$LiAlH_4$, Et_2O	$EtGeH_3$	80	3,4
$n\text{-}C_5H_{11}GeBr_3$	$LiAlH_4$, $n\text{-}Bu_2O$	$n\text{-}C_5H_{11}GeH_3$	90–100	3,4
Me_2GeCl_2	$LiAlH_4$, $n\text{-}Bu_2O$	Me_2GeH_2	95	6
Et_2GeCl_2	$LiAlH_4$, Et_2O	Et_2GeH_2	90–100	3,4
Ph_2GeBr_2	$LiAlH_4$, Et_2O	Ph_2GeH_2	67	3,4
Me_3GeBr	$NaBH_4$, H_2O	Me_3GeH	95	2,3
Et_3GeCl	$LiAlH_4$, Et_2O	Et_3GeH	90–100	3,4
Ph_3GeBr	$LiAlH_4$, Et_2O	Ph_3GeH	60	3,4
$(i\text{-}Pr_2GeCl)_2$	$LiAlH_4$, Et_2O	$(i\text{-}Pr_2GeH)_2$	50	7

[a] Water added after reaction to hydrolyze the germylboronate.

$$R_{4-n}GeX_n + \frac{n}{4} NaBH_4 \xrightarrow[2. H_2O]{1. Et_2O} \frac{n}{4} NaX + \frac{n}{4} BX_3 + R_{4-n}GeH_n \qquad (f)$$

An Fe–Ge dichloride complex is reduced with $NaBH_4$:

$$[h^5\text{-}Cp(CO)_2Fe]_2GeCl_2 \xrightarrow{NaBH_4\text{–}THF} [h^5\text{-}Cp(CO)_2Fe]_2GeH_2 \qquad (g)^4$$

(A.D. NORMAN)

1. W. L. Jolly, A. D. Norman, *Prep. Inorg. React., 4,* 1 (1968).
2. E. Wiberg, E. Amberger, *Hydrides of the Elements of Main Groups I–IV,* Elsevier, Amsterdam, 1971.
3. P. Rivière, M. Rivière-Baudet, J. Satgé, in *Comprehensive Organometallic Chemistry,* Vol. 2, G. Wilkinson, F. G. A. Stone, E. W. Abel, eds., Pergamon Press, Oxford, 1982, p. 399.
4. F. Glockling, *The Chemistry of Germanium,* Academic Press, London, 1969.
5. A. E. Finholt, A. C. Bond, K. E. Wilzbach, H. I. Schlesinger, *J. Am. Chem. Soc., 69,* 2692 (1947).
6. J. E. Drake, B. M. Glavincevski, R. T. Hemmings, H. E. Henderson, *Inorg. Synth., 18,* 154 (1978).
7. J. C. Mendelsohn, F. Metras, J. C. Labournère, J. Valade, *J. Organomet. Chem., 12,* 327 (1968).

1.6.5.3.2. from Oxygen Compounds.

Potassium or Na tetrahydroborate reduction of aqueous germanate, accomplished by adding a basic Na_2GeO_3–KBH_4 soln dropwise to $MeCO_2H$ or H_2SO_4, produces GeH_4 in up to 73% yield[1]:

$$[HGeO_3]^- + [BH_4]^- + 2\,H^+ \xrightarrow{H_2O} H_3BO_3 + GeH_4 \qquad (a)$$

Small quantities of Ge_2H_6:

$$8\,[HGeO_3]^- + 7\,[BH_4]^- + 15\,H^+ \xrightarrow{H_2O} 7\,H_3BO_3 + 3\,H_2O + 4\,Ge_2H_6 \qquad (b)$$

Ge_3H_8 and $(GeH_2)_x$ form also. Germanium dioxide, when heated with a deficit of powdered $LiAlH_4$ at 148–170°C, produces GeH_4, Ge_2H_6 and Ge_3H_8 in 4, 3 and 1% yield, respectively[2].

Germyl ether[3,4]:

$$2\,(Ph_3Ge)_2O + LiAlH_4 \rightarrow LiAlO_2 + 4\,Ph_3GeH \qquad (c)$$

germylhydroxide[3,4]:

$$4\,Ph_3GeOH + LiAlH_4 \rightarrow LiAl(OH)_4 + 4\,Ph_3GeH \qquad (d)$$

or alkoxygermane[3,4] reduction by $LiAlH_4$ yields germanes nearly quantitatively:

$$4\,(+)\text{-}Me(Ph)(1\text{-}C_{10}H_7)GeOC_{10}H_{12} + LiAlH_4 \rightarrow LiAl(OC_{10}H_{12})_4 +$$
$$4\,(+)\text{-}Me(Ph)(1\text{-}C_{10}H_7)GeH \qquad (e)$$

Chiral alkoxides are reduced with retention of configuration; reduction of chiral halogermanes results in inversion of configuration[4].

(A.D. NORMAN)

1. W. L. Jolly, J. E. Drake, *Inorg. Synth., 7,* 34 (1966).
2. J. M. Bellama, A. G. MacDiarmid, *Inorg. Chem., 7,* 2070 (1968).

3. E. Wiberg, E. Amberger, *Hydrides of the Elements of the Main Groups I–IV*, Elsevier, Amsterdam, 1971.
4. M. Lesbre, P. Mazerolles, J. Satgé, *The Organic Chemistry of Germanium*, Wiley-Interscience, New York, 1971.

1.6.5.3.3. from Other Derivatives.

Lithium tetrahydroaluminate reduction of $(Me_2GeS)_x$ occurs:

$$\frac{2}{x} (Me_2GeS)_x + LiAlH_4 \rightarrow LiAlS_2 + 2 Me_2GeH_2 \qquad (a)^1$$

Cleavage of Ge—P bonds in acyclic[2] and cyclic[3] germyl phosphines yields Ge—H-containing products, e.g.:

$$Et_3GePPh_2 \xrightarrow{\quad LiAlH_4 \quad} Ph_2PH + Et_3GeH \qquad (b)$$

Dialkylgermacyclobutanes are cleaved[4] slowly by $LiAlH_4$ in refluxing Et_2O:

$$R_2Ge(CH_2)_3 \xrightarrow[Et_2O]{LiAlH_4} R_2Ge(H)CH_2CH_2CH_3 \qquad (c)$$

Reaction of $(n-Pr_3P)_2Pt(GePh_3)_2$ with $LiAlH_4$ yields Ph_3GeH and uncharacterized Pt-containing products[4].

(A.D. NORMAN)

1. E. W. Abel, D. A. Armitage, *Adv. Organomet. Chem.*, 5, 2 (1967).
2. E. W. Abel, S. M. Illingworth, *Organomet. Chem. Rev., A*, 5, 143 (1970).
3. J. Duboc, J. Escudié, C. Couret, J. Cavezzan, J. Satgé, P. Mazerolles, *Tetrahedron, 37*, 1141 (1981).
4. M. Lesbre, P. Mazerolles, J. Satgé, *The Organic Compounds of Germanium*, Wiley-Interscience, New York, 1971.

1.6.5.4. Giving Hydrides of Tin

1.6.5.4.1. from Halides.

Reactions of halostannanes with complex hydrides provide routes to stannanes[1,2] (Table 1). Complex hydrides most used are MBH_4 (M = Li, Na, K) and $LiAlH_4$. The $LiBH_4$ and $LiAlH_4$ reductions are carried out in aprotic ether solvents (THF, Et_2O, monoglyme). Sodium and K hydroborate reactions may be carried out in H_2O. Most complex hydride reductions are adaptable to Sn—D bond synthesis (see also §1.6.7.1.4). Halostannanes are reduced by $LiAlH_4$:

$$R_{4-n}SnX_n + \tfrac{n}{4} LiAlH_4 \xrightarrow{\quad Et_2O \quad} \tfrac{n}{4} LiAlX_4 + R_{4-n}SnH_n \qquad (a)$$

where R = H, alkyl, aryl; X = Cl, Br; n = 1–4. Product yields are high[1], although in some cases H_2O addition upon completion of a reaction increases the yield. Stannane forms by $Li[(t-BuO)_3AlH]$ reduction of $SnCl_4$ at $-80°C$:

$$SnCl_4 + 4 Li[(t-BuO)_3AlH] \rightarrow 4 LiCl + 4 Al(t-BiO)_3 + SnH_4 \qquad (b)^1$$

Aqueous Na or K tetrahydroborates reduce Sn(II) or Sn(IV) species to SnH_4 along with traces[1,5] of Sn_2H_6. From $NaBH_4$ and $SnCl_2$ in H_2O {Sn(II) present as

TABLE 1. COMPLEX HYDRIDE REDUCTIONS OF HALOSTANNANES

Reactant	Complex hydride, solvent	Product	Yield (%)	Ref.
$SnCl_4$	$LiAlH_4$, Et_2O	SnH_4	30	1,3
$MeSnCl_3$	$LiAlH_4$, Et_2O	$MeSnH_3$	5	1,2
$n\text{-}BuSnCl_3$	$NaBH_4$, monoglyme	$n\text{-}BuSnH_3$	16	1,2
Me_2SnCl_2	$LiAlH_4$, $C_4H_8O_2$	Me_2SnH_2	72	1,2
Et_2Sncl_2	$LiAlH_4$, Et_2O	Et_2SnH_2	90	1,2
Ph_2SnCl_2	$LiAlH_4$, Et_2O	Ph_2SnH_2	72	1,2
Me_3SnBr	$LiAlH_4$, $C_4H_8O_2$	Me_3SnH	40	1,2
$n\text{-}Pr_3SnCl$	$LiAlH_4$, Et_2O	$n\text{-}Pr_3SnH$	75	1,2
$n\text{-}Bu_3SnCl$	$NaBH_4$, monoglyme	$n\text{-}Bu_3SnH$	62	4
Ph_3SnCl	$NaBH_4$, monoglyme	Ph_3SnH	82	4
$(n\text{-}Bu_2SnCl)_2$	$LiAlH_4$, Et_2O	$(n\text{-}Bu_2SnH)_2$	76	1

$[HSnO_2]^-$}, using xs $NaBH_4$, SnH_4 forms in 84% yield:

$$4 [HSnO_2]^- + 3 [BH_4]^- + 7 H^+ + H_2O \rightarrow 3 B(OH)_3 + 4 SnH_4 \qquad (c)^1$$

(A.D. NORMAN)

1. E. Wiberg, E. Amberger, *Hydrides of Elements of the Main Groups I–IV*, Elsevier, Amsterdam, 1971.
2. E. J. Kupchik, in *Organotin Compounds*, Vol. 1, A. K. Sawyer, ed., Marcel Dekker, New York, 1971, p. 7.
3. A. D. Norman, J. R. Webster, W. L. Jolly, *Inorg. Synth.*, *11*, 170 (1968).
4. E. R. Birnbaum, P. H. Javora, *Inorg. Synth.*, *12*, 45 (1970).
5. A. D. Zorin, I. A. Frolov, T. V. Morozova, *J. Gen. Chem. USSR*, (*Engl. Transl.*), *42*, 890 (1972).

1.6.5.4.2. from Oxygen Compounds.

Sodium tetrahydroborate reduction of stannites in H_2O produces SnH_4:

$$4 [HSnO_2]^- + 3 [BH_4]^- + 7 H^+ + H_2O \rightarrow 3 H_3BO_3 + 4 SnH_4 \qquad (a)^{1,2}$$

and Sn_2H_6:

$$8 [HSnO_2]^- + 5 [BH_4]^- + 13 H^+ \rightarrow 5 H_3BO_3 + H_2O + 4 Sn_2H_6 \qquad (b)^{1,2}$$

Phenylstannane and $MeSnH_3$ form readily from $NaBH_4$ reduction in H_2O of $PhSnCl_3$ or $K[MeSnO_2]$, respectively[3].

Lithium tetrahydroaluminate reduction of acetoxystannanes e.g.:

$$(MeCO_2)_3SnGe(CO_2Me)_3 + 6 LiAlH_4 \rightarrow 6 Li[MeCO_2]$$
$$+ 6 AlH_3 + H_3GeSnH_3 \qquad (c)$$

yields Sn-H bonds effectively[1]. Alkoxystannanes are reduced to the corresponding organostannanes[4]:

$$R_{4-x}Sn(OR')_x + \tfrac{x}{4} LiAlH_4 \rightarrow \tfrac{x}{4} LiAl(OR')_4 + R_{4-x}SnH_x \qquad (d)$$

where x = 1–3; R = aryl, alkyl; R' = alkyl.

(A.D. NORMAN)

1. E. Wiberg, E. Amberger, *Hydrides of the Elements of Main Groups I–IV*, Elsevier, Amsterdam, 1971.
2. W. L. Jolly, J. E. Drake, *Inorg. Synth. 7*, 34 (1966).
3. E. J. Kupchik, in *Organotin Compounds*, Vol. 1, A. K. Sawyer, ed., Marcel Dekker, New York, 1971, p. 7.
4. A. G. Davies, P. J. Smith, *Adv. Inorg. Chem. Radiochem., 23*, 1 (1980).

1.6.5.4.3. from Other Derivatives.

The Sn—C bond in stannylmethylphosphonate is cleaved by $LiAlH_4$:

$$R_3SnCH_2P(O)Ph(OEt) \xrightarrow{LiAlH_4} MeP(Ph)H, R_3SnH \qquad (a)^1$$

where R = Me, Et, Ph.

(A.D. NORMAN)

1. H. Weichmann, B. Ochsler, I. Duchek, A. Tyschach, *J. Organomet. Chem., 182*, 465 (1979).

1.6.5.5. Giving Hydrides of Lead.

Plumbane, because of its low thermal stability, is not isolated from reactions of complex hydride with Pb halides[1]. Alkylplumbanes (R_3PbH, R_2PbH_2) form in reactions of dialkyl- or trialkylhaloplumbanes with $LiAlH_4$ in ethers (Me_2O or monoglyme) at -60 to $-110°C$ in up to 90% yield, e.g.:

$$2 R_2PbCl_2 + LiAlH_4 \rightarrow LiAlCl_4 + 2 R_2PbH_2 \qquad (a)^2$$

where R = Me, Et, i-Pr, n-Bu, C_6H_{11}.

Potassium tetrahydroborate reduces[4] trialkylhaloplumbanes in liq NH_3:

$$R_3PbCl + KBH_4 + NH_3 \xrightarrow{NH_3(l)} BH_3NH_3 + KCl + R_3PbH \qquad (b)$$

where R = Me, Et, i-Pr, n-Bu. Complex hydride reactions with alkyllead halides are preferred syntheses of alkylplumbanes.

(A.D. NORMAN)

1. E. Wiberg, E. Amberger, *Hydrides of the Elements of the Main Groups I–IV*, Elsevier, Amsterdam, 1971.
2. H. Shapiro, F. W. Frey, *Organic Compounds of Lead*, Wiley-Interscience, New York, 1968.
3. W. P. Neumann, K. Kühlein, *Adv. Organomet. Chem., 7*, 241 (1968).
4. R. Duffy, J. Feeny, A. K. Holliday, *J. Chem. Soc.*, 1144 (1962).

1.6.6. by Industrial Processes

1.6.6.1. Involving Compounds of Carbon.

Reactions of H_2 or H_2 sources (e.g., Zn–HCl) with alkenes, alkynes, arenes, ketones, nitriles, carboxylic acids and esters are used industrially for C—H bond formation[1-8]. Heterogeneous reaction catalysts (e.g., Ni, Pt, Pd, Fe, Ni–Cu) are used, e.g.:

$$CH_2=CHCH_2CH_2CH_3 \xrightarrow{Pt, H_2} C_5H_{12} \qquad (a)$$

$$MeC \equiv CMe \xrightarrow{\text{Pt, } H_2} cis\text{-}MeCH = CH(Me) \tag{b}$$

$$MeC_6H_5 \xrightarrow{\text{Pt-SiO}_2, \, H_2} MeC_6H_{11} \tag{c}$$

$$C_4H_4O \xrightarrow{\text{Ni, } H_2} C_4H_8O \tag{d}$$

$$RCO_2R' \xrightarrow{\text{CuO/CuCrO}_4} RCH_2OH + R'OH \tag{e}$$

Reactions proceed at $> 25°C$ ($25–300°C$) at $> 10^3$ Nm^{-2} of H_2.

Examples of hydrogenation using homogeneous catalysts in moderate-scale industrial syntheses exist. Such catalysts are more selective and involve milder reaction conditions (i.e., $25°C$ and ca. 10^2 Pa H_2) than their heterogeneous counterparts[4,6]. Asymmetric hydrogenation of a prochiral amino acid using chiral Rh catalysts yields L-dopa in high optical purity[9,10]:

$$(f)$$

where $Rh^* = [Rh\{S,S\text{-}[Ph_2P(Me)C_3H_4(Me)PPh_2]\} 1,5\text{-}C_8H_{12}]Cl$. Reactions in which H_2 formally cleaves C—C bonds occur in hydrogenolysis processes[2,4]. Heterogeneous catalysts are used, e.g.:

$$CH_3CH_2CH_2CH_3 \xrightarrow[H_2]{\text{Ni–Cu}} CH_4, \, C_2H_6, \, C_3H_8 \tag{g}$$

These reactions are discussed in treatises on organic chemistry. Hydrogenolysis of metal– or nonmetal–carbon bonds are used for small-scale specialty chemical syntheses:

$$PhHgOCOR + H_2 \xrightarrow{\text{Rh}} Hg + PhH + RCO_2H \tag{h)[1]}$$

Hydrogen reacts with CO in the presence of catalysts to produce hydrocarbon products ranging from CH_3OH and CH_4 to fuel oils[6,11-16]:

$$n \, CO + 2n \, H_2 \rightarrow (CH_2)_n + n \, H_2O \tag{i}$$

Products obtained depend on reaction conditions. Using $Pd–La_2O_3$ or $Zn–Cr_2O_3$ as catalysts, CH_3OH is the main product[11,13]:

$$CO + 2 \, H_2 \rightarrow CH_3OH \tag{j}$$

Over Ni at $500–700°C$, principally CH_4 forms[11]. Other transition-metal catalysts result in higher yields of $C_{>1}$ products. Ethylene glycol is produced effectively from CO and H_2 using homogeneous Rh catalysts:

$$2 \, CO + 3 \, H_2 \xrightarrow{\text{Rh}} HOCH_2CH_2OH \tag{k)[13,14]}$$

Hydroformylation, the addition of CO and H_2 to an alkene[6,14,16-18], occurs with homogeneous[18] [e.g., $RhCl(PPh_3)_3$, $HCo(CO)_4$], heterogeneous[18] (e.g., Ni, Co, Fe, Rh) or supported catalysts[3,14] [e.g., $Rh(CO)_2Cl$ on SiO_2] to form alcohols and/or aldehydes:

$$RCH=CH_2 \quad \xrightarrow{H_2 + CO} \quad \begin{cases} RCH_2CH_2CHO, \ RCH_2CH_2CH_2OH & \text{(l)} \\ \\ RCH(CH_3)CHO, \ RCH(CH_3)CH_2OH & \text{(m)} \end{cases}$$

Aldehydes form prior to alcohols. Both branched and linear isomers form. Depending on the alkene substrate, reaction conditions, and catalyst selected, conditions selective for specific compound synthesis can be found[15,17,18].

Carbide hydrolysis is used to produce CH_4 and/or C_2H_2 in limited, special situations. Aluminum methanide hydrolyzes to form CH_4:

$$Al_4C_3 + 12 \ H_2O \rightarrow 4 \ Al(OH)_3 + 3 \ CH_4 \qquad \text{(n)}[6,19]$$

Electropositive metal (Cu, Zn, Al) acetylide hydrolysis produces C_2H_2, e.g.:

$$ZnC_2 + 2 \ H_2O \rightarrow Zn(OH)_2 + C_2H_2 \qquad \text{(o)}$$

Protonation of aq $[CN]^-$ is a route[6,20] to HCN:

$$[CN]^- + H_2O \rightarrow [OH]^- + HCN \qquad \text{(p)}$$

Addition of protonic reagents to $C=C$ and/or $C\equiv C$ bonds is a route to $C-H$ bond-containing products[2,7,21-23]. Acetaldehyde is produced by hydrolysis of C_2H_4 in the presence of catalysts:

$$C_2H_4 \quad \xrightarrow[H_2O]{|PdCl_4|^{2-}-CuCl_2} \quad CH_3CHO \qquad \text{(q)}[8]$$

Hydrolysis or alcoholysis occurs with mineral acid catalysts:

$$CH_3CH=CH_2 + H_2O \xrightarrow{H+} CH_3CH(OH)CH_3 \qquad \text{(r)}[2,7,21]$$

$$MeC\equiv CMe + H_2O \xrightarrow{H+} MeCH_2C(O)Me \qquad \text{(s)}[2,7,22]$$

Reaction of alkenes with conc H_2SO_4 is a major route to sulfonates[7,21]:

$$MeCH=CH_2 + H_2SO_4 \rightarrow MeCH_2CH_2OSO_3H \qquad \text{(t)}$$

Addition of HCN to alkenes, in the presence of heterogeneous metal catalysts, produces substituted nitriles[23].

$$HCN + C_2H_4 \xrightarrow{Ni} CH_3CH_2CN \qquad \text{(u)}$$

Hydrogen halide addition to alkenes and alkynes yields organohalides:

$$Me_3CCH=CH_2 + HCl \rightarrow Me_3CCH(Cl)CH_3 \qquad \text{(v)}[7]$$

Halide addition to the most substituted carbon occurs [Eq. (v)], except in cases of radical reaction promotion, in which case addition of halide to the least substituted carbon is observed:

$$MeCH=CH_2 + HBr \xrightarrow{peroxide} MeCH_2CH_2Br \qquad \text{(w)}[7,21]$$

Addition of nonmetal– or metal–H bonds to $C=C$ or $C\equiv C$ bonds result in $C-H$ bond formation and production of organoelement products. Boranes:

$$B_2H_6 + 6\ C_2H_4 \rightarrow 2\ (C_2H_5)_3B \qquad (x)^{24,25}$$

alanes:

$$(C_2H_5)_2AlH + MeCH{=}CH_2 \rightarrow (C_2H_5)_2AlCH_2CH_2CH_3 \qquad (y)^{24}$$

and silanes add directly to form organo derivatives, e.g.:

$$(CH_3)_3SiH + CH_2{=}CHCH_3 \rightarrow (CH_3)_3SiCH_2CH_2CH_3 \qquad (z)^{8,23,24}$$

Reactions of silanes are often best catalyzed by hydrosilation catalysts[8] such as Pt or Pd. Similarly, C—H bonds are formed in the addition of phosphines or amines to C=C or C≡C bonds. Reactions are catalyzed (acid, base, or metal) or free radical promoted, e.g.:

$$(RO)_2P(O)H + CH_2{=}CHR' \xrightarrow{\text{AIBN}} (RO)_2P(O)CH_2CH_2R'$$

$$\text{AIBN} = 2,2'\text{-azobis(isobutyronitrile)} \qquad (aa)^{26}$$

$$R_2NH + C_2H_2 \xrightarrow{\text{(CuCl)}_2} CH_2{=}CHNR_2 \qquad (ab)^{22,27}$$

Carbon–hydrogen bonds add to alkenes or alkynes, e.g., iso-butane adds to iso-butene in the presence of conc H_2SO_4:

$$Me_3CH + Me_2C{=}CH_2 \xrightarrow{H_2SO_4} Me_2CHCH_2CMe_3 \qquad (ac)^{2,21}$$

Such reactions are important in fuel-upgrading processes.

Hydride, either binary or complex, reductions of organic functional groups are expensive routes to C—H bonds, consequently such reactions are used only for specialty chemical and expensive product (pharmacueticals, etc.) synthesis. Reactions as described in §1.6.4 and 1.6.5 are used. Commonly used reactions involve NaH, KH, boranes, alanes, silanes and occasionally stannanes. Detailed discussions of hydride reductions can be found in standard treatises on organic chemistry[2,4,21,24,25,27]. Molecular rearrangement:

(ad)

and/or isomerizations are important in petroleum refining and industrial chemical processes[2,3,7,21]:

$$CH_2{=}CHCH_2CH_3 \xrightarrow{\text{Ni}} \text{cis,trans-}CH_3CH{=}CHCH_3 \qquad (ae)$$

These, along with the molecule degradation and reformation processes of thermal and catalytic cracking[2,3,7,16], e.g.:

$$(CH_2)_n \xrightarrow{\text{SiO}_2\text{-Al}_2O} C_2H_4,\ C_3H_6,\ C_4H_6,\ C_4H_8,\ H_2,\ \text{etc.} \qquad (af)$$

involve formation of new C—H bonds, albeit often in nonspecific reaction processes.

(A.D. NORMAN)

1. F. H. Jardine, *Prog. Inorg. Chem.*, *28*, 63 (1981).
2. L. F. Fieser, M. Fieser, *Advanced Organic Chemistry*, Reinhold, New York, 1961.
3. S. C. Davis, K. J. Klabunde, *Chem. Rev.*, *82*, 153 (1982).
4. T. Clark, M. A. McKervey, in *Comprehensive Organic Chemistry*, D. Barton, W. D. Ollis, eds., Vol. 1, J. F. Stoddart, ed., Pergamon Press, Oxford, 1979, p. 37.
5. P. N. Rylander, *Organic Synthesis with Noble-Metal Catalysts*, Academic Press, New York, 1973.
6. F. A. Cotton, G. Wilkinson, *Advanced Inorganic Chemistry, 4th ed.*, Wiley-Interscience, New York, 1980.
7. G. H. Whitham, in *Comprehensive Organic Chemistry*, D. Barton, W. D. Ollis, eds., Vol. 1, J. F. Stoddart, ed., Pergamon Press, Oxford, 1979, p. 121.
8. P. M. Henry, *Adv. Organomet. Chem.*, *13*, 363 (1975).
9. D. A. MacNeil, N. K. Roberts, B. Bosnich, *J. Am. Chem. Soc.*, *103*, 2273 (1981).
10. B. R. James, *Adv. Organomet. Chem.*, *17*, 319 (1979).
11. G. A. Sommerjai, *Cat. Rev.-Sci. Eng.*, *23* 189 (1981).
12. W. A. Hermann, *Angew. Chem., Int. Ed. Engl.*, *21*, 117 (1982).
13. C. Master, *Adv. Organomet. Chem.*, *17*, 61 (1979).
14. R. L. Pruett, *Adv. Organomet. Chem.*, *17*, 1 (1979).
15. M. E. Dry, J. C. Haagendoorn, *Catal. Rev.-Sci. Eng.*, *23*, 265 (1981).
16. J. D. Downer, K. I. Beynan, in *Rodds' Chemistry of Carbon Compounds*, S. Coffey, ed., Vol. 1, 2nd ed., Elsevier, New York, 1964, p. 357.
17. I. Tkatchenko, in *Comprehensive Organometallic Chemistry*, Vol. 8, G. Wilkinson, F. G. A. Stone, E. W. Abel, eds., Pergamon Press, Oxford, 1982, p. 101.
18. M. Orchin, *Acc. Chem. Res.*, *14*, 259 (1981).
19. A. K. Holliday, G. Hughes, S. M. Walker, in *Comprehensive Inorganic Chemistry*, Vol. 1, A. F. Trotman-Dickenson, ed., Pergamon Press, Oxford, 1973, p. 1173.
20. W. L. Jolly, *The Inorganic Chemistry of Nitrogen*, Benjamin, New York, 1964.
21. R. T. Morrison, R. N. Boyd, *Organic Chemistry*, 4th ed., Allyn and Bacon, Boston, 1983.
22. J. Tsuji, *Adv. Organomet. Chem.*, *17*, 141 (1975).
23. D. M. Roundhill, *Adv. Organomet. Chem.*, *13*, 273 (1975).
24. A. Hajós, *Complex Hydrides*, Elsevier, Amsterdam, The Netherlands, 1979.
25. H. C. Brown, *Boranes in Organic Chemistry*, Cornell University Press, Ithaca, NY, 1972.
26. C. Walling, M. S. Pearson, in *Topics in Phosphorus Chemistry*, Vol. 3, M. Grayson, E. J. Griffith, eds., Wiley-Interscience, New York, 1966, p. 1.
27. J. R. Malpass, in *Comprehensive Organic Chemistry*, D. Barton, W. D. Ollis, eds., Vol. 2, I. O. Sutherland, eds., Pergamon Press, Oxford, 1979, p. 3.

1.6.6.2. Involving Compounds of Silicon.

Silane is prepared industrially by the reaction of Mg_2Si with strong protonic acids[1,2] such as H_3PO_4 or H_2SO_4:

$$Mg_2Si + 4 H^+ \rightarrow 2 Mg^{2+} + SiH_4 \tag{a}$$

by the reaction of H_2 with $SiCl_4$ and Al in an $AlCl_3$ melt[1]:

$$3 SiCl_4 + 4 Al + 6 H_2 \xrightarrow{AlCl_3} 4 AlCl_3 + 3 SiH_4 \tag{b}$$

or by the reaction of H_2 with $SiCl_4$ and LiH or NaH in an LiCl–KCl eutectic[1,3,4]:

$$SiCl_4 + 4 LiH \xrightarrow[\text{LiCl-KCl}]{H_2} 4 LiCl + SiH_4 \tag{c}$$

The latter two methods are preferred because they are adapted to the continuous production of SiH_4.

Trichlorosilane is prepared from reaction of Si with anhyd HCl at 400°C:

$$Si \xrightarrow{HCl} SiCl_4, SiHCl_3 \qquad (d)^{1,5}$$

Along with alkylsilanes, including $MeSiCl_2H$, it is produced from the reaction of MeCl with a hot Cu–Si mixture[5].

Organosilanes for commercial use are prepared by either LiH or $LiAlH_4$ reduction of the corresponding chlorosilanes, e.g., $PhSiH_3$:

$$PhSiCl_3 + 3 \ LiH \rightarrow 3 \ LiCl + PhSiH_3 \qquad (e)$$

or $(C_6H_{13})_3SiH$:

$$4 \ (C_6H_{13})_3SiCl + LiAlH_4 \rightarrow LiAlCl_4 + 4 \ (C_6H_{13})_3SiH \qquad (f)$$

Mono-, di- and triorganosilanes produced by these methods include $RSiH_3$ (R = C_5H_{11}, Ph, C_6H_{13}), R_2SiH_2 (R = Et, Ph) and R_3SiH (R = Et, C_6H_{13}, EtO, and n-Bu)[1,5]. Calcium hydride reduction of Me_3SiCl in the presence of $AlCl_3$ yields Me_3SiH:

$$2 \ Me_3SiCl + CaH_2 \xrightarrow{AlCl_3} CaCl_2 + 2 \ Me_3SiH \qquad (g)^3$$

(A.D. NORMAN)

1. E. Wiberg, E. Amberger, *Hydrides of the Elements of Main Groups I–IV*, Elsevier, Amsterdam, 1971.
2. F. Feher, D. Schinkitz, J. Schaaf, *Z. Anorg. Allg. Chem., 383*, 303 (1971).
3. E. G. Rochow, in *Comprehensive Inorganic Chemistry*, Vol. 2, A. F. Trotman-Dickenson, ed., Pergamon Press, Oxford, 1973, p. 1323.
4. A. M. Pavlov, G. N. Bodyagin, I. L. Agafonov, *Tr. Khim. Khim. Techol.*, 175 (1967); *Chem. Abstr., 78*, 83,822 (1969).
5. D. A. Armitage, in *Comprehensive Organometallic Chemistry*, Vol. 2, G. Wilkinson, F. G. A. Stone, E. W. Abel, eds., Pergamon Press, Oxford, 1982, p. 1.

1.6.6.3. Involving Compounds of Germanium.

Germane is prepared in large quantities by reaction of strong protonic acids[1] with Mg_2Ge:

$$Mg_2Ge + 4 \ H^+ \rightarrow 4 \ Mg^{2+} + GeH_4 \qquad (a)$$

Lithium tetrahydroaluminate reduction of organochlorogermanes is used for preparation of limited (kilogram) quantities of selected organogermanes, e.g., Me_3GeH and Ph_3GeH:

$$4 \ RGeCl_3 + 3 \ LiAlH_4 \rightarrow 3 \ LiAlCl_4 + 4 \ RGeH_3 \qquad (b)^{2,3}$$

where R = Me, Ph.

(A.D. NORMAN)

1. E. Wiberg, E. Amberger, *Hydrides of the Elements of Main Groups I–IV*, Elsevier, Amsterdam, 1971.
2. P. Rivière, M. Rivière-Baudet, J. Satgé, in *Comprehensive Organometallic Chemistry*, Vol. 2, G. Wilkinson, F. G. A. Stone, E. W. Abel, eds., Pergamon Press, Oxford, 1982, p. 399.
3. M. Lesbre, D. Mazerolles, J. Satgé, *The Organic Compounds of Germanium*, Wiley-Interscience, New York, 1971.

1.6.6.4. Involving Compounds of Tin.

Specialty chemical quantities of organostannanes are prepared by $LiAlH_4$ reduction of the corresponding chlorostannanes[1]:

$$4 R_3SnCl + LiAlH_4 \rightarrow LiAlCl_4 + 4 R_3SnH \tag{a}$$

where R = Et, n-Bu, Ph. Poly(methylhydrido)siloxane reduction of $(n-Bu_3Sn)_2O$ forms $n-Bu_3SnH$:

$$(n-Bu_3Sn)_2O + \tfrac{2}{n} (MeSiHO)_n \rightarrow \tfrac{2}{n} (MeSiO_2) + 2 \ n-Bu_3SnH \tag{b}[1]$$

(A.D. NORMAN)

1. A. G. Davies, in *Comprehensive Organometallic Chemistry*, Vol. 2, G. Wilkinson, F. G. A. Stone, E. W. Abel, eds., Pergamon Press, Oxford, 1982, p. 519.

1.6.7. The Synthesis of Deuterium Derivatives

1.6.7.1. by Interconversion of Deuterated Compounds

1.6.7.1.1. Involving Carbon Compounds.

Descriptions of C—D bond formation are found in treatises on synthetic organic chemistry[1-5], e.g., in the complex deuteride reductions:

$$EtBr \xrightarrow{\text{LiAlD}_4-\text{Et}_2\text{O}} EtD \tag{a}[1]$$

$$Me_2CHBr \xrightarrow{\text{NaBD}_4-\text{C}_4\text{H}_8\text{O}} Me_2CDH \tag{b}[1,3]$$

$$PhCHO \xrightarrow{\text{LiBEt}_3\text{D}} PhCHDOD \tag{c}[1,3]$$

$$PhCO_2H \xrightarrow{\text{LiAlD}_4} PhCD_2OD \tag{d}[1]$$

$$PhCONEt_2 \xrightarrow{\text{LiAlD}_4} PhCD_2ND_2 \tag{e}[1]$$

$$EtC\equiv N \xrightarrow{\text{LiAlD}_4} RCD=ND \tag{f}[1]$$

$$PhCH=CH_2 \xrightarrow{\text{NaBD}_4-\text{AlCl}_3} PhCHDCH_2D \tag{g}[1,2]$$

Some reactions proceed directly to products [e.g., Eqs. (a), (b)]; however, often the final product arises after D_2O hydrolysis of an intermediate borane, boronate or aluminate species.

Binary ionic (e.g., NaD) or covalent (e.g., Me_2SiD_2) hydrides are reductants for C—D bond formation[1,6], e.g.:

$$PhCH_2Cl + LiD \rightarrow LiCl + PhCH_2D \tag{h}[1]$$

$$CH_2=CHCl + Ph_3SiD \rightarrow Ph_3SiCl + CH_2=CHD \tag{i}[1,6]$$

$$PhCOMe \xrightarrow{Et_3SnD} Ph(Me)CDOD \qquad (j)^1$$

$$EtCHO \xrightarrow{AlD_3-THF} EtCHDOD \qquad (k)^1$$

$$EtCO_2Me \xrightarrow{BD_3-THF} EtCD_2OD \qquad (l)^{1,3}$$

$$6 \ MeCH{=}CH_2 + B_2D_6 \xrightarrow{THF} (MeCHDCH_2)_3B \qquad (m)^{1,3}$$

Hydride reductions can be selective and produce products in high isotopic yields.

Carbides, carbanionic or metalated carbon compounds react with protonic acids, e.g., D_2O or DCl, forming deuteriocarbon compounds:

$$Al_4C_3 + 12 \ D_2O \rightarrow Al(OD)_3 + 3 \ CD_4 \qquad (n)^7$$

$$EtLi + D_2O \rightarrow LiOD + EtD \qquad (o)^9$$

$$C_{10}H_7Na + DCl \rightarrow NaCl + C_{10}H_7D \qquad (p)^9$$

$$Me_3Al + 3 \ D_2O \rightarrow Al(OD)_3 + 3 \ MeD \qquad (q)^{10}$$

$$2 \ Et_3SnPh + D_2O \xrightarrow{|OD|^- -D_2O} (Et_3Sn)_2O + 2 \ PhD \qquad (r)^{11}$$

$$n\text{-BuMgBr} + DCl \rightarrow MgClBr + n\text{-BuD} \qquad (s)^{8,9,12}$$

Reaction of complex metal species can be used also:

$$(h^5\text{-Cp})_2TiMe_2 + 2 \ DCl \rightarrow (h^5\text{-Cp})_2TiCl_2 + 2 \ MeD \qquad (t)^{12}$$

Deuterium halides (DCl, DBr) add to alkenes or alkynes:

$$MeCH{=}CH_2 + DCl \rightarrow MeCH(Cl)CH_2D \qquad (u)^4$$

$$PhC{\equiv}CH + 2 \ DBr \rightarrow PhC(Br)_2CHD_2 \qquad (v)^4$$

(A.D. NORMAN)

1. A. Hajós, *Complex Hydrides,* Elsevier, Amsterdam, 1979.
2. C. A. Buehler, D. E. Pearson, *Survey of Organic Syntheses,* Vol. 2, Wiley-Interscience, New York, 1977.
3. H. C. Brown, *Boranes in Organic Chemistry,* Cornell Univ. Press, Ithaca, NY, 1972.
4. A. I. Shatenshtein, *Isotopic Exchange and the Replacement of Hydrogen in Organic Compounds,* Consultants Bureau, New York, 1962.
5. A. Murray, III, D. L. Williams, *Organic Syntheses with Isotopes,* Part II, Wiley-Interscience, New York, 1958.
6. E. Wiberg, E. Amberger, *Hydrides of the Elements of Main Groups I–IV,* Elsevier, Amsterdam, 1971.
7. A. K. Holliday, G. Hughes, S. M. Walker, in *The Chemistry of Carbon,* Pergamon Press, Oxford, 1973, Ch. 13, p. 1173.
8. D. J. Cram, *Fundamentals of Carbanion Chemistry,* Academic Press, New York, 1965.
9. D. Bethell, in *Comprehensive Organic Chemistry,* D. Barton, W. D. Ollis, eds., Vol. 1, J. F. Stoddard, ed., Pergamon Press, Oxford, 1979, p. 411.
10. F. A. Cotton, G. Wilkinson, *Advanced Inorganic Chemistry,* 4th ed., Wiley-Interscience, New York, 1980.
11. A. G. Davies, P. J. Smith, *Adv. Inorg. Chem. Radiochem. 23,* 1 (1980).
12. E. Klei, J. H. Teichen, *J. Organomet. Chem., 188,* 97 (1980).

1.6. Formation of Bonds between Hydrogen and C, Si, Ge, Sn, Pb   119
1.6.7. The Synthesis of Deuterium Derivatives
1.6.7.1. by Interconversion of Deuterated Compounds

1.6.7.1.2. Involving Silanes.

Lithium tetrahydroaluminate-d_4 or $LiBD_4$ reductions of halosilanes or oxysilanes are preferred routes to alkyl, aryl or unsubstituted deuteriosilanes[1-4]:

$$R_{4-n}SiX_n + \tfrac{n}{4} LiAlD_4 \rightarrow \tfrac{n}{4} LiAlX_4 + R_{4-n}SiD_n \tag{a}$$

where R = H, alkyl, aryl; n = 1–4;

$$2 (SiCl_3)_2O + 4 LiAlD_4 \rightarrow 3 LiAlCl_4 + LiAlO_2 + 4 SiD_4 \tag{b}$$

$$4 \text{ i-Pr}_3SiOEt + LiAlD_4 \rightarrow LiAl(OEt)_4 + 4 \text{ i-Pr}_3SiD \tag{c}$$

Similarly, higher deuterio silanes form:

$$2 Si_2Cl_6 + 3 LiAlD_4 \rightarrow 3 LiAlCl_4 + 2 Si_2D_6 \tag{d}$$

Reaction of $LiAlD_4$ with halohydrides allows synthesis of specifically labeled mixed H–D compounds:

$$4 SiH_3Cl + LiAlD_4 \rightarrow LiAlCl_4 + 4 SiH_3D \tag{e}[5]$$

Alkali-metal deuteride reduction of halosilanes produces deuteriosilanes readily, but yields are less than from $LiAlD_4$ halosilane reactions:

$$Ph_3SiCl + LiD \rightarrow LiCl + Ph_3SiD \tag{f}[4]$$

Alkali-metal silyls react with D_2O:

$$R_3SiNa + D_2O \rightarrow NaOD + R_3SiD \tag{g}[3,4]$$

Silicides with D_2O or $DCl-D_2O$ yield SiD_4, along with lesser quantities of higher deuteriosilanes, i.e., Si_2D_6, Si_3D_8, and n-Si_4D_{10}:

$$Mg_2Si + 4 D_2O \rightarrow 2 Mg(OD)_2 + SiD_4 \tag{h}[1,6]$$

(A.D. NORMAN)

1. E. Wiberg, E. Amberger, *Hydrides of the Elements of Main Groups I–IV*, Elsevier, Amsterdam, 1971.
2. A. D. Norman, J. R. Webster, W. L. Jolly, *Inorg. Synth.*, *11*, 170 (1968).
3. E. A. V. Ebsworth, *Volatile Silicon Compounds*, Pergamon Press, Oxford, 1963.
4. D. A. Armitage, in *Comprehensive Organometallic Chemistry*, Vol. 2, G. Wilkinson, F. G. A. Stone, E. W. Abel, eds., Pergamon Press, Oxford, 1982, p. 1.
5. H. J. Meal, M. R. Wilson, *J. Chem. Phys.*, *24*, 385 (1956).
6. B. Bok, J. Bruhn, J. Rastings-Anderson, *Acta Chem. Scand.*, *8*, 367 (1954).

1.6.7.1.3. Involving Germanes.

Lithium tetrahydroaluminate-d_4 or $LiBD_4$ reduction of halogermanes and oxygermanes are preferred syntheses of deuteriogermanes, e.g.:

$$R_{4-n}GeCl_n + \tfrac{n}{4} LiAlD_4 \rightarrow \tfrac{n}{4} LiAlCl_4 + R_{4-n}GeD_n \tag{a}[1,2]$$

where n = 1–4; R = alkyl, aryl;

$$4 (Ph_3Ge)_2O + LiAlD_4 \rightarrow LiAl(OD)_4 + 4 Ph_3GeD \tag{b}[1,2]$$

Reaction of $LiAlD_4$ with $GeHCl_3$ yields the mixed hydro–deuterio product in high isotopic purity and specificity:

120   1.6. Formation of Bonds between Hydrogen and C, Si, Ge, Sn, Pb
       1.6.7. The Synthesis of Deuterium Derivatives
       1.6.7.1. by Interconversion of Deuterated Compounds

$$3 \text{ LiAlD}_4 + 4 \text{ GeHCl}_3 \rightarrow 3 \text{ LiAlCl}_4 + 4 \text{ GeHD}_3 \qquad \text{(c)}[3]$$

Aklali-metal deuteride reduction of halogermanes yields deuteriogermanes but not in reactions preferred over those using complex hydrides:

$$R_{4-n}\text{GeX}_n + n \text{ LiD} \rightarrow n \text{ LiX} + R_{4-n}\text{GeD}_n \qquad \text{(d)}[2,3]$$

Reaction of germanide salts with D_2O or DCl yields deuteriogermanes in high isotopic purity:

$$R_3\text{GeM} + \text{DCl} \rightarrow \text{MCl} + R_3\text{GeD} \qquad \text{(e)}[2,3]$$

where R = alkyl, aryl; M = Li, Na, K.

Germane-d_4 and lesser amounts of higher germanes form upon DCl–D_2O treatment of Mg_2Ge:

$$\text{Mg}_2\text{Ge} \xrightarrow{\text{D}_2\text{O–DCl}} \text{GeD}_4, \text{ Ge}_2\text{D}_6, \text{ Ge}_3\text{D}_8 \qquad \text{(f)}[1,2,4,5]$$

(A.D. NORMAN)

1. A. D. Norman, J. R. Webster, W. L. Jolly, *Inorg. Synth. 11*, 170 (1968).
2. E. Wiberg, E. Amberger, *Hydrides of the Elements of Main Groups I–IV*, Elsevier, Amsterdam, 1971.
3. M. Lesbre, D. Mazerolles, J. Satgé, *The Organic Compounds of Germanium*, Wiley-Interscience, New York, 1971.
4. L. P. Lindeman, M. K. Wilson, *Z. Phys. Chem. (Leipzig), 929* (1956).
5. A. H. Zeltman, G. C. Fitzgibbon, *J. Am. Chem. Soc., 76*, 2021 (1954).

1.6.7.1.4. Involving Stannanes.

Lithium tetrahydroaluminate-d_4 and LiBD$_4$ reductions of halostannanes constitute preferred syntheses of deuteriostannanes, e.g.:

$$\text{SnCl}_4 + \text{LiAlD}_4 \xrightarrow{\text{Et}_2\text{O}} \text{LiAlCl}_4 + \text{SnD}_4 \qquad \text{(a)}[1,2]$$

$$R_{4-n}\text{SnCl}_n + \tfrac{n}{4} \text{LiAlD}_4 \xrightarrow{\text{Et}_2\text{O}} \tfrac{n}{4} \text{LiAlCl}_4 + R_{4-n}\text{SnH}_n \qquad \text{(b)}$$

where R = alkyl, aryl. Diethylalane-d_1 reacts similarly with halostannanes forming deuterated products:

$$n \text{ Et}_2\text{AlD} + R_{4-n}\text{SnCl}_n \rightarrow \text{Et}_2\text{AlCl} + R_{4-n}\text{SnD}_n \qquad \text{(c)}[3]$$

where n = 1–3; R = alkyl, aryl.

Solvolysis of stannides with deuterated hydroxylic reagents yields deuteriostannanes. Stannane-d_4 and alkyl- and arylstannanes form:

$$\text{Mg}_2\text{Sn} + 4 \text{ D}_2\text{O} \rightarrow 2 \text{ Mg(OD)}_2 + \text{SnD}_4 \qquad \text{(d)}[2]$$

$$R_3\text{SnM} + \text{D}^+ \xrightarrow{\text{D}^+ - \text{D}_2\text{O}} \text{M}^+ + R_3\text{SnD} \qquad \text{(e)}[4]$$

where R = aryl, alkyl; M = Li, Na.

Tri-n-butylstannane, upon treatment with EtMgCl followed by D_2O, forms[4] n-Bu$_3$SnD.

(A.D. NORMAN)

1. A. D. Norman, J. R. Webster, W. L. Jolly, *Inorg. Synth. 11,* 170 (1968).
2. E. Wiberg, E. Amberger, *Hydrides of Elements of the Main Groups I–IV,* Elsevier, Amsterdam, 1971.
3. A. G. Davies, P. J. Smith, in *Comprehensive Organometallic Chemistry,* Vol. 2, G. Wilkinson, F. G. A. Stone, E. W. Abel, eds., Pergamon Press, Oxford, 1982, p. 519.
4. E. J. Kupchik, in *Organotin Compounds,* Vol. 1, A. K. Sawyer, ed., Marcel Dekker, New York, 1971, p. 7.

1.6.7.1.5. Involving Plumbanes.

Reactions of $LiAlD_4$ with di- or trialkylhaloplumbanes in ether (Me_2O or monoglyme) at $-60°C$ to $-110°C$ can produce alkyldeuterioplumbanes:

$$2 \ R_2PbCl_2 + LiAlD_4 \xrightarrow{Et_2O} LiAlCl_4 + 2 \ R_2PbD_2 \qquad \text{(a)}^{1,2}$$

(A.D. NORMAN)

1. E. Wiberg, E. Amberger, *Hydrides of the Elements of the Main Group I–IV,* Elsevier, Amsterdam, 1971.
2. H. Shapiro, F. W. Frey, *Organic Compounds of Lead,* Wiley-Interscience, New York, 1968.

1.6.7.2. by Isotopic Enrichment Using Chemical Reactions

1.6.7.2.1. of Carbon Compounds.

Direct exchange of C—H bonds with D_2 can yield C—D bonds. Alkane C—H exchange occurs on activated metal (e.g., Ni or Pt) surfaces[1]:

$$C_3H_8 \underset{\text{}}{\overset{D_2-\text{Ni film}}{\rightleftharpoons}} C_3D_8 \qquad \text{(a)}$$

Aromatic C—H bond exchange occurs similarly but more slowly[1]. Benzene–D_2 exchange occurs in the presence of metal hydrides such as $(h^5\text{-}C_5H_5)_3TaH_3$, $(h^5\text{-}C_5H_5)_2NbH_3$, or $h^3\text{-}C_3H_5CoH[P(OMe)_3]_3$[1–3].

$$C_6H_6 \underset{\text{}}{\overset{D_2-\text{catal}}{\rightleftharpoons}} C_6D_6 \qquad \text{(b)}$$

Hydrocarbons undergo metal or non-metal halide-catalyzed C—H bond exchange with strong deuterio acids. Reactions are accelerated by metal-halide catalysts such as $AlCl_3$, $AlBr_3$, $FeBr_3$ or $TiBr_4$:

$$C_6H_6 \underset{\text{}}{\overset{DBr-AlBr_3}{\rightleftharpoons}} C_6D_6 \qquad \text{(c)}^{3-6}$$

In strong acid media, i.e., DF or DBr, aromatic C—H bond exchange occurs:

$$C_6H_6 \underset{\text{}}{\overset{DF_{(l)}}{\rightleftharpoons}} C_6D_6 \qquad \text{(d)}^3$$

Deuteriosulfuric acid exchanges slowly with C_2H_4 or saturated hydrocarbons:

$$C_2H_4 \underset{\text{}}{\overset{D_2SO_4-D_2O}{\rightleftharpoons}} C_2D_4 \qquad \text{(e)}^3$$

In the presence of transition-metal hydride catalysts, olefin exhange also can occur[4]:

$$CH_2CHCH_2CH_3 \underset{\{DNi[P(OEt)_3]_4\}^+}{\overset{D_2SO_4-MeOD}{\rightleftharpoons}} CD_2CDCD_2CD_3, \ CD_3CDCDCD_3 \qquad \text{(f)}^1$$

Hydrocarbon C—H bond exchange, in reactions involving them as protonic acids, occurs readily. Ease of reaction depends on the acidity of the hydrocarbon. Exchange between aliphatic hydrocarbons and ND_3 in the presence of $[ND_2]^-$ catalyst occurs:

$$C_3H_8 \underset{ND_3-[ND_2]^-}{\overset{}{\rightleftharpoons}} C_3D_8 \qquad (g)^{5,6}$$

Exchange of the aliphatic C—H bonds in $C_6H_5CH_3$ occurs ca. 250 times faster than that of the aromatic C—H bonds[5,7]:

$$C_6H_5CH_3 \underset{ND_3-[ND_2]^-}{\overset{}{\rightleftharpoons}} C_6H_5CD_3 \qquad (h)$$

$$C_6H_5CD_3 \underset{ND_3-[ND]_2^-}{\overset{}{\rightleftharpoons}} C_6D_5CD_3 \qquad (i)$$

Exchange of more acidic hydrocarbons, e.g., C_2H_2 or C_5H_6 occurs readily in D_2O with an $[OD]^-$ catalyst[5]:

$$C_2H_2 \underset{D_2O-[OD]^-}{\overset{}{\rightleftharpoons}} C_2D_2 \qquad (j)$$

however, most alkanes undergo sufficient activation for rapid exchange only in the presence of metal catalysts (e.g., Pt, Ni) and at elevated $T^{1,8}$:

$$C_4H_{10} \underset{D_2O-CH_3CO_2D}{\overset{}{\rightleftharpoons}} C_4D_{10} \qquad (k)$$

The coordination of aromatic rings to transition metals, such as in $h^6\text{-}C_6H_6Cr(CO)_3$, increases C—H bond acidity enough to allow C—H bond exchange with EtOD:

$$h^6\text{-}C_6H_6Cr(CO)_3 \underset{EtOD-[EtO]^-}{\overset{}{\rightleftharpoons}} h^6\text{-}C_6D_6Cr(CO)_3 \qquad (l)^9$$

(A.D. NORMAN)

1. S. C. Davis, K. J. Klabunde, *Chem. Rev., 82,* 153 (1982).
2. B. R. James, *Adv. Organomet. Chem., 17,* 319 (1979).
3. F. A. Cotton, G. Wilkinson, *Advanced Inorganic Chemistry,* 4th ed., Wiley-Interscience, New York, 1980.
4. D. M. Roundhill, *Adv. Organomet. Chem., 13,* 273 (1975).
5. A. I. Shatenstein, *Isotopic Exchange and the Replacement of Hydrogen in Organic Compounds,* Consultants Bureau, New York, 1962.
6. L. F. Fieser, M. Fieser, *Advanced Organic Chemistry,* Reinhold, New York, 1961.
7. W. L. Jolly, *The Inorganic Chemistry of Nitrogen,* Benjamin, New York, 1964.
8. D. E. Webster, *Adv. Organomet. Chem., 15,* 147 (1977).
9. W. E. Silverthorne, *Adv. Organomet. Chem., 13,* 48 (1975).

1.6.7.2.2. of Silanes.

Silane exchanges with DCl in the presence of $AlCl_3$:

$$SiH_4 \underset{DCl-AlCl_3}{\overset{}{\rightleftharpoons}} SiD_4 \qquad (a)^1$$

Competing reactions forming chlorosilanes occur also. Alkysilanes exchange similarly, although C—Si bond cleavage also occurs.

Silane reactions with D_2 under thermal, photolytic or silent electric discharge conditions result in H–D exchange. These reactions are not usually synthetically viable[1,2]:

$$SiH_4 + D_2 \rightleftharpoons HD + SiH_3D \qquad \text{(b)}$$

Intermolecular exchange between silanes in the presence of an H_2PtCl_6 catalyst occurs at 100°C:

$$MeCl_2SiH \underset{}{\overset{H_2PtCl_6-SiCl_3D}{\rightleftharpoons}} MeCl_2SiD \qquad \text{(c)}^{2,3}$$

(A.D. NORMAN)

1. E. A. V. Ebsworth, *Volatile Silicon Compounds*, Pergamon Press, Oxford, 1963.
2. E. Wiberg, E. Amberger, *Hydrides of the Elements of Main Groups I–IV*, Elsevier, Amsterdam, 1971.
3. D. M. Roundhill, *Adv. Organomet. Chem.*, *13*, 273 (1975).

1.6.7.2.3. of Germanes.

Germanes are often sufficiently acidic[1-3] that their exchange with deuterio solvents can be an effective route to Ge—D bonds:

$$GeH_4 \underset{}{\overset{D_2O-[OD]^-}{\rightleftharpoons}} GeD_4 \qquad \text{(a)}$$

These reactions seldom offer advantages over hydride reductions, except if GeD bond formation in a complex molecule, such as a germyl-metal species, is involved:

$$(Ph_3P)_3NiGePh_2H \underset{}{\overset{D_2O}{\rightleftharpoons}} (Ph_3P)_3NiGePh_2D \qquad \text{(b)}^3$$

(A.D. NORMAN)

1. T. Birchall, W. L. Jolly, *Inorg. Chem.*, *5*, 2177 (1966).
2. M. Lestre, D. Mazerolles, J. Satgé. *The Organic Compounds of Germanium*, Wiley-Interscience New York, 1971.
3. D. M. Roundhill, *Adv. Organomet. Chem.*, *13*, 273 (1975).

1.7. Formation of Bonds between Hydrogen and Elements of Group IIIB (B, Al, Ga, In, Tl)

1.7.1. Introduction

The chemistry described in this section is dominated by that of boron and to a lesser extent of Al. The hydride chemistry of Ga and, especially, of In and Tl is developed only to a limited extent.

There are few useful reactions in which new B—H bonds are formed. Although the formation of boranes from the protolysis of borides or the reduction of boron compounds with H_2, either in electrical discharges or in the presence of active metals, have historical importance, these methods have no importance or utility today. Indeed, the preparation of boranes is so dominated by the single common starting material, the tetrahydroborate ion, that the only important reactions in which B—H bonds are formed are those in which hydride ion either reduces species with B—O or B–halogen bonds to form boranes or adds to trifunctional boron compounds to form hydroborates.

There are other reactions in which, although in the strictest interpretation new B—H bonds are not formed, B—H—B bridge bonds are formed from B—H terminal bonds. The formation of B—H—B bridge bonds from B—H bonds is considered B—H bond formation for the purpose of this treatment.

The arrangement of this section patterns that of others in these volumes; however, for the group IIIB elements this is not as useful because the important chemical reactions in which B—H bonds are formed from other B—H bonds are not highlighted. Some of the important reactions in which there is no net increase in the number of B—H bonds are found in §1.7.5.1.

Except for the tetrahydroaluminates and -gallates the chemistry of element–H bond formation for the congeners of B is rare. Most electron-pair donor base adducts of AlH_3 and GaH_3 are derived from $LiAlH_4$ and $LiGaH_4$, respectively. The hydride chemistry of In and Tl is sparse. The species $LiInH_4$ and $LiTlH_4$ exist; however, evidence for the existence of the normal hydrides InH_3 and TlH_3, is not convincing.

Reviews describing H–group III element bond formation are listed in refs. 1–14.

(L. BARTON)

1. A. Stock, *The Hydrides of Boron and Silicon*, Cornell Univ. Press, Ithaca, NY, 1933.
2. H. I. Schlesinger, A. B. Burg, *Chem. Rev., 32*, 1 (1942).
3. R. M. Adams, in *Borax to Boranes*, D. L. Martin, ed., Advances in Chemistry Series No. 32, American Chemical Society, Washington, DC, 1961, p. 66.
4. R. M. Adams, ed., *Boron, Metalloboron Compounds and Boranes*, Interscience, New York, 1964.
5. E. C. Ashby, *Adv. Inorg. Chem. Radiochem., 8*, 283 (1966).
6. L. A. Sheka, I. S. Chaus, T. T. Mityureva, *The Chemistry of Gallium*, Elsevier, Amsterdam, 1966, Ch. 2.

7. R. T. Holtzman, R. L. Hughes, I. C. Smith, E. W. Lawless, *Production of the Boranes and Related Research*, Academic Press, New York, 1967.
8. N. N. Greenwood, in *New Pathways in Inorganic Chemistry*, E. A. V. Ebsworth, A. G. Maddock, A. G. Sharpe, eds., Cambridge Univ. Press, Cambridge, 1968, Ch. 3.
9. R. W. Parry, M. K. Walter, *Prep. Inorg. React.*, 5, 45 (1968).
10. B. D. James, M. G. H. Wallbridge, *Prog. Inorg. Chem.*, 11, 99 (1970).
11. H. D. Johnson II, S. G. Shore, *Top. Curr. Chem.* 15, 88 (1970).
12. S. G. Shore, in *Boron Hydride Chemistry*, E. L. Muetterties, ed., Academic Press, New York, 1975, Ch. 3.
13. A. B. Burg, *Chem. Tech.*, Jan., 1977, p. 50.
14. S. G. Shore, in *Rings, Cages, Clusters and Polymers of the Main Group Elements*, A. H. Cowley, ed., ACS Symposium Series No. 232, American Chemistry Society, Washington, DC 1983, p. 1.

1.7.2. from the Elements.

The direct reaction between elemental boron and H_2 gas has limited utility[1]; e.g., Mg_2B_3, which contains elemental boron, does not react with H_2 at high T. However the reaction between boron and H_2 at 840°C forms[2] only traces of B_2H_6, and H_2 reacts little with boron powder[3]. Thermodynamic calculations based on free energy minimization for the chemical-vapor deposition of boron from BX_3–H_2 mixtures (X = Cl, Br) at 1000–1900 K and 0.101 MPa indicate low borane (BH_3) conc at equilibrium, but traces of HBX_2 are predicted in these T ranges[4].

An electrochemical process in which H_2 gas is fed into an electrolyzed metal–halide mixture with a boron anode[5] yields B_2H_6. The reaction proceeds via the intermediacy of metal hydride.

A cyclic process using as raw materials H_2, coke, MgO and BCl_3 is available[6]. The reaction proceeds through the intermediacy of MgH_2 formed at the cathode and boron halide formed at the anode. These react together in the melt with the stoichiometry described in §1.7.4.1.1:

$$C + MgO \rightarrow Mg + CO \tag{a}$$

$$Mg + H_2 \rightarrow MgH_2 \tag{b}$$

$$3\ MgH_2 + 2\ BCl_3 \rightarrow 3\ MgCl_2 + B_2H_6 \tag{c}$$

$$Heat + MgCl_2 + H_2O \rightarrow MgO + HCl \tag{d}$$

The B—H bond is formed from the reaction between H_2 and boron monoxide[7] (B_2O_2) or in systems in which the latter is prepared in situ[8–10]. When H_2 is passed over a mixture[8] of boron and boric oxide at 1200°C:

$$\frac{3}{2}\ H_{2(g)} + B_{(s)} + B_2O_{3(s)} \longrightarrow$$

$$\tag{e}$$

The product boroxine ($B_3H_3O_3$) disproportionates when condensed at 77 K and subsequently warmed to RT. Diborane and boric oxide are formed:

$$B_3O_3H_{3(s)} \rightarrow B_2O_{3(s)} + \tfrac{1}{2} B_2H_{6(g)} \tag{f}$$

Reaction (f) also proceeds with first-order kinetics in the gas phase; the half-life of the process[11] is < 90 min.

Polymeric BO, prepared from the reaction between B_2O_3 and boron or carbon at 1150°C, B_2O_3 and B_4C at 1250°C, or MO_2 (M = Ti, Zr) and B_4C at 1200°C, reacts with H_2 at 1400°C to give, ultimately, diborane in ~10% yields[9-11]. Similarly, when H_2 is passed over mixtures that form $B_2O_{2(g)}$ in situ, B—H bonds are formed. The isolated B_2H_6 is formed via the intermediacy of $B_3O_3H_{3(g)}$. When carbon is in the system, e.g., reaction mixtures C + B_2O_3 or TiO_2 + B_4C, the ultimate product is borane carbonyl[9,10]. The stoichiometry for the latter process is:

$$4\ B_4C + 5\ TiO_2 + 3\ H_2 \xrightarrow{1150°C} 5\ TiB_2 + 2\ H_3B{\cdot}CO + 2\ B_2O_3 + 2\ CO \tag{g}$$

Boroxine is also formed when H_2O vapor is passed over elemental boron at 1150°C[8]:

$$3\ H_2O_{(g)} + 3\ B_{(s)} \rightarrow B_3O_3H_{3(g)} + \tfrac{3}{2} H_{2(g)} \tag{h}$$

Diborane may be prepared[12] in yields of 40–50% by treatment of B_2O_3 with H_2 gas above 150°C at 75.8 MPa in the presence of Al and $AlCl_3$. The hydrogenation proceeds through a chloroalane intermediate.

The hydrogenation of $(HSBS)_3$ to B_2H_6, with finely divided Ni as a catalyst, occurs in inert solvents[13]. The $(HSBS)_3$ is prepared from H_2S and BBr_3 and the scheme is:

$$3\ BBr_3 + 6\ H_2S \rightarrow 9\ HBr + (HSBS)_3 \tag{i}$$

$$2\ (HSBS)_3 + 18\ H_2 \xrightarrow{Ni} 3\ B_2H_6 + 12\ H_2S \tag{j}$$

The first moderate-yield process for the formation of B—H bonds from the reaction of H_2 with boron halides utilizes passage of mixtures of H_2 and BCl_3 through an electrical discharge between Cu electrodes[14]. When H_2: BCl_3 is 10 at 2.67×10^3 Pa, yields of $\leq$ 60% B_2H_6, based on BCl_3 consumed in 2 h, are available. This approach[15] is more convenient when BBr_3 is used instead of BCl_3. The lower volatility of the bromo species makes the process simpler and separation of products more convenient. This discharge process proceeds in steps[16]:

$$BCl_3 \rightarrow BCl_2 + Cl \tag{k}$$

$$BCl_2 + H_2 \rightarrow HBCl_2 + H \tag{l}$$

$$5\ HBCl_2 \rightarrow B_2H_5Cl + 3\ BCl_3 \tag{m}$$

$$6\ B_2H_5Cl \rightarrow 5\ B_2H_6 + 2\ BCl_3 \tag{n}$$

The major initially formed product is B_2H_5Cl, which disproportionates to BCl_3 and B_2H_6 on fractionation. The γ-irradiation-induced reaction between H_2 and BCl_3 does not form boranes[17].

The thermal reaction between BCl_3 and H_2 at 300–450°C in the presence of granular, 20-mesh Al affords B_2H_6 in $\leq$60% yields[18]. The 6:1 H_2–BCl_3 mixture is passed through a heated reactor column. The similar reduction of BBr_3 with H_2 produces only traces of B_2H_6. An analogous process[19] forms B_2H_6 in 8% yields when H_2 and BCl_3 are passed over Al_2Cu alloy at 450°C.

Hydrogenation of borates or boroxines at 81.1 MPa of H_2 in the presence of Al and $NaAlCl_4$ forms diborane[20,21]:

$$Al + \frac{9}{2} H_2 + 3 C_2H_4 + BF_3 \rightarrow AlF_3 + 3 C_2H_6 + \frac{1}{2} B_2H_6 \qquad (o)$$

$$Al + \frac{9}{2} H_2 + 3 C_2H_4 + \frac{1}{2} B_2O_3 + \frac{3}{2} H_2SO_4 \rightarrow$$
$$\frac{1}{2} Al_2(SO_4)_3 + 3 C_2H_6 + \frac{1}{2} B_2H_6 + \frac{3}{2} H_2O \qquad (p)$$

$$Al + \frac{3}{2} H_2 + \frac{1}{2} B_2O_3 + \frac{3}{2} H_2SO_4 \rightarrow Al_2(SO_4)_3 + \frac{1}{2} B_2H_6 + \frac{3}{2} H_2O \qquad (q)$$

Boric acid may be similarly reduced to B_2H_6 under basic conditions on a large scale. Hydrogen in the presence of Al reduces borates to boranes quantitatively. These boranes are trapped as the amine borane, e.g., phenylborate is dissolved in triethylamine and to the solution is added activated Al powder and small amounts of $AlCl_3$ catalyst. The mixture is agitated at 180°C for 1 h at 14.2 MPa of H_2. Quantitative yields of the $(CH_3)_3NBH_3$ species are available from this process[22] with a lower yield of $C_6N_5N(CH_3)_2BH_3$:

$$B(OC_6H_5)_3 + Al + \frac{3}{2} H_2 \xrightarrow{(C_2H_5)_3N} BH_3N(C_2H_5)_3 + Al(OC_6H_5)_3 \qquad (r)$$

A similar process converts complex alkylborates to the tetrahydroborate ion:

$$3 NaB(OCH_3)_4 + 4 Al + 6 H_2 \xrightarrow{diglyme} 3 NaBH_4 + 4 Al(OCH_3)_3 \qquad (s)$$

The B—C bond may be reduced in the presence of H_2 at high P and T to afford boranes, e.g., alkylboranes[23] and arylboranes[24] are converted to organodiboranes and, in the presence of amines, to amine boranes[23], and also in processes[25] in which solid borane polymer is formed from liquid organoboranes and H_2 at 200°C. Partial hydrogenation of trialkylboranes to dialkylboranes is effected[26] by treatment of the borane with H_2 gas at 30.4 MPa and 120–200°C. Amine complexes of haloboranes are reduced by H_2 gas to R_3NBH_3 species and ultimately to diborane[27]. Amine complexes of triorganoboranes are also converted to amineborane by reaction with H_2 gas[28]. Finally, tetrahalodiboranes(4) are converted to the dihaloboranes by reaction with H_2 below 0°C. The H_2BX species disproportionate[29] to yield ultimately B_2H_6 and BX_3.

This hydrogenolysis of compounds containing B—B bonds also may be extended to B_4H_{10}. One mol of B_4H_{10} reacts with 1 mol of H_2 to afford B_2H_6, a reaction in which two new B—H bonds are formed[30]:

$$B_4H_{10} + H_2 \rightarrow 2 B_2H_6 \qquad (t)$$

Boron–H bonds may be formed from the reaction between alkali metals, H_2 and either BF_3 or trimethylborate. Low yields of B_2H_6 and $[BH_4]^-$ are available from the halide; however, reaction with $B(OCH_3)_3$ at 230°C affords the $[BH(OCH_3)_3]^-$ ion[31]:

$$2 Na + H_2 + 2 B(OCH_3)_3 \rightarrow 2 NaBH(OCH_3)_3 \qquad (u)$$

At higher T:

$$4 Na + 2 H_2 + B(OCH_3)_3 \rightarrow NaBH_4 + 3 NaOCH_3 \qquad (v)$$

However, yields of B—H bond-containing products are low.

The Al—H bond forms by passage of an electrical discharge through $Al(CH_3)_3$ gas in the presence of xs H_2 at elevated T as the mixed methylaluminum hydrides[32], $Al_2H_n(CH_3)_{6-n}$. The unsubstituted alane is formed as the $(AlH_3)_n$ polymer or as a base adduct from the same system[33]; gaseous AlH_3 and Al_2H_6 are obtained[34,35] by the passage of H_2 gas over an Al droplet at 1170–1250°C at low P.

Treatment of 1,4-diazabicyclo[2.2.2]octane in tetrahydrofuran (THF) with activated Al powder at 70°C under 34 MPa of H_2 forms[36] the Al—H bond as the amine alane:

$$Al + \frac{3}{2}\,H_2 + \;\text{[diazabicyclooctane]} \longrightarrow \;\text{[AlH}_3\text{ adduct]} \qquad (w)$$

The best route to Al—H bonds from the elements is the formation of tetrahydroaluminates, because these reagents are convenient to handle and may be used to prepare other alanes. The reaction between an alkali- or alkaline-earth metal and activated Al at 110–140°C in THF at 15.2–35.5 MPa H_2 affords the tetrahydroaluminate salt in quantitative yield[37-40]:

$$M + Al + 2\,H_2 \rightarrow MAlH_4\ (M = Ca, Na) \qquad (x)$$

Under similar conditions metal hydrides react with Al and H_2 to afford the Na^+, Li^+, K^+ and Cs^+ salts:

$$MH + Al + \frac{2}{3}\,H_2 \rightarrow MAlH_4 \qquad (y)$$

The most suitable solvents are THF for $NaAlH_4$ and $LiAlH_4$, diglyme for $KAlH_4$ and toluene for $CsAlH_4$.

Treatment of Na, Al and $Al(C_2H_5)_3$ with H_2 at ~33.4 MPa at 165°C in toluene for 10 h affords the hexahydroaluminate salt Na_3AlH_6 in 98% yield[41].

Gallane, GaH_3, is claimed[42] from passage of an electrical discharge through mixtures of $Ga(CH_3)_3$ and H_2; however, these results are questioned[43]. Exposure of H_2 and Ga to ~ 2300°C shows UV spectra[44] assigned to the GaH molecule, and passage of H_2 and Ga vapor at 926–1176°C into a time-of-flight mass spectrometer results[35] in weak spectra attributable to $[GaH_3]^+$. Similar evidence is available for the formation of InH_3 from its elements.

(L. BARTON)

1. A. Stock, *Hydrides of Boron and Silicon*, Cornell Univ. Press, Ithaca, NY, 1933, p. 42.
2. A. E. Newkirk, D. T. Hurd, *J. Am. Chem. Soc.*, 77, 241 (1955).
3. G. V. Tsagreishvili, I. A. Bairamashvili, K. A. Oganezov, M. L. Tabutsidze, O. A. Tsagareishvili, *J. Less-Common Met.*, 82, 131 (1981).
4. R. Naslain, J. Thebault, P. Hagemuller, C. Bernard, *J. Less-Common Met.*, 67, 85 (1979).
5. E. Enk, J. Nickl, Ger. Pat. 1,092,890 (1960); *Chem Abstr.*, 55, 2687 (1961).
6. L. G. Dean, C. W. McCutcheon, A. C. Doumas, U.S. Pat. 3,024,091 (1962); *Chem. Abstr.*, 57, 8198 (1962).
7. C. C. Clark, F. A. Kanda, J. A. King, U.S. Pat. 3,021,197 (1962); *Chem. Abstr.*, 56, 13,810 (1962).
8. W. P. Sholette, R. F. Porter, *J. Phys. Chem.*, 67, 177 (1963).
9. L. Barton, D. Nicholls, *Proc. Chem. Soc.*, 242 (1964).

10. L. Barton, D. Nicholls, *J. Inorg. Nucl. Chem.*, 28, 1367 (1966).
11. L. Barton, *J. Inorg. Nucl. Chem.*, 30, 1683 (1968).
12. T. A. Ford, G. H. Kalb, A. L. McClelland, E. L. Muetterties, *Inorg. Chem.*, 3, 1032 (1964).
13. C. D. Barr, D. G. Hummel, U.S. Pat. 2,965,456 (1960); *Chem. Abstr.*, 55, 22,737 (1961).
14. H. I. Schlesinger, A. B. Burg, *J. Am. Chem. Soc.*, 53, 4321 (1931).
15. A. Stock, H. Martini, W. Sutterlin, *Chem. Ber.*, 67B, 396 (1934).
16. H. W. Myers, R. F. Putnam, *Inorg. Chem.*, 3, 655 (1963).
17. A. J. Levy, J. Williamson, L. W. Steiger, *J. Inorg. Nucl. Chem.*, 17, 26 (1961).
18. D. T. Hurd, *J. Am. Chem. Soc.*, 71, 20 (1949).
19. V. I. Mikheeva, T. N. Dymova, *Zh. Neorg. Khim.*, 2, 2530 (1957).
20. R. M. Adams, in *Boron, Metalloboron Compounds and Boranes*, R. M. Adams, ed., Interscience, New York, 1964, p. 562.
21. R. Köster, K. Ziegler, *Angew. Chem.*, 69, 94 (1957).
22. E. C. Ashby, W. E. Foster, *J. Am. Chem. Soc.*, 84, 3407 (1962).
23. R. Köster, *Angew. Chem.*, 69, 94 (1957).
24. R. Köster, G. Bruno, P. Binger, *Justus Liebigs Ann. Chem.*, 644, 1 (1961).
25. R. Köster, *Angew. Chem.*, 70, 743 (1958).
26. Studiengesellschaft Kohle GmbH., Br. Pat. 854,919 (1960); *Chem. Abstr.*, 55, 15,350 (1961).
27. R. M. Adams, in *Boron, Metalloboron Compounds and Boranes*, R. M. Adams, ed., Interscience, New York, 1964, p. 566.
28. R. Köster, K. Ziegler, *Angew. Chem.*, 69, 94 (1957).
29. G. Urry, T. Wartik, H. I. Schlesinger, *J. Am. Chem. Soc.*, 74, 5809 (1952).
30. H. I. Schlesinger, Univ. Chicago, Navy Contract N173-s-9280, Final Report (1945–46); see ref. 20, p. 560.
31. H. I. Schlesinger, H. C. Brown, A. E. Finholt, *J. Am. Chem. Soc.*, 75, 205 (1953).
32. E. Wiberg, O. Stecher, *Angew. Chem.*, 52, 372 (1939).
33. O. Stecher, E. Wiberg, *Chem. Ber.*, 75B, 2003 (1942).
34. P. Briesacher, B. Siegel, *J. Am. Chem. Soc.*, 86, 5053 (1964).
35. P. Briesacher, B. Siegel, *J. Am. Chem. Soc.*, 87, 4257 (1965).
36. E. C. Ashby, *J. Am. Chem. Soc.*, 68, 1882 (1964).
37. H. Clasen, *Angew. Chem.*, 73, 322 (1961).
38. E. C. Ashby, *Chem. Ind. (London)*, 208 (1962).
39. E. C. Ashby, G. J. Brendel, H. E. Redman, *Inorg. Chem.*, 2, 499 (1963).
40. T. N. Dymova, N. G. Eliseeva, S. I. Bakum, Yu. M. Dergachev, *Dokl. Akad. Nauk SSSR*, 215, 1369 (1974).
41. E. C. Ashby, P. Kobetz, *Inorg. Chem.*, 5, 1615 (1966).
42. E. Wiberg, T. Johannsen, *Die Chemie*, 55, 38 (1942); *Chem. Abstr.*, 37, 3363 (1943).
43. D. F. Shriver, R. W. Parry, N. N. Greenwood, A. Storr, M. G. H. Wallbridge, *Inorg. Chem.*, 2, 867 (1963).
44. W. R. S. Garton, *Proc. Phys. Soc.*, A64, 509 (1951).

1.7.3. from Group IIIB Derivatives, Excluding Reactions of Hydrides and Complex Hydrides

1.7.3.1. Involving Borides.

Acidification[1] of Mg boride yields B_4H_{10}. Reaction between Mg_3B_2, obtained from the high-T reduction of B_2O_3 with Mg, and dil acid forms a borane mixture that is predominately B_4H_{10}. The best yields are obtained when the powered boride is added to 10% aq HCl or phosphoric acid. The Mg boride Mg_3B_2 is a solid soln[2] of MgB_2 in Mg. Acidolysis[3] of the Mg_3B_2 with aq acids indicates that H_3PO_4 gives the highest yield of B_4H_{10}. Hydroxyboranes of composition $Mg_3B_2(OH)_6$ [4] and $[BH_2(OH)_2]^-$ [5] are formed as intermediates. The hydrolysis of MgB_2 prepared from amorphous B and Mg powder[6] affords a borane mixture[1,3] that contains B_6H_{10} in only 5–10% yield based on boron.

Borides of Mn, Cr, Fe, Ni and Co also hydrolyze in aq HCl to afford borane mixtures composed primarily of B_2H_6 and B_4H_{10}; however, the yields are low. The highest yield of borane available from these borides is 2% when Cr_2B is used[7].

The hydrolysis of MgB_2 in strong base forms the tetrahydroborate anion[8]. Digestion of MgB_2 in 3M KOH or 4M $[(CH_3)_4N][OH]$ for 8–12 h followed by evaporation affords KBH_4 or $(CH_3)_4NBH_4$. Conversion of boron to $[BH_4]^-$ occurs to 10–15%.

(L. BARTON)

1. A. Stock, C. Massenez, *Chem. Ber., 45,* 3529 (1912).
2. R. Thompson, *Prog. Boron Chem., 2,* 176 (1970).
3. V. I. Mikeeva, V. Yu. Markina, *Zh. Neorg. Khim., 1,* 619 (1956).
4. R. C. Ray, P. C. Sinha, *J. Chem. Soc.,* 1694 (1935).
5. P. Duhart, *Ann. Chim. (Paris), 7,* 339 (1962).
6. P. L. Timms, C. S. G. Phillips, *Inorg. Chem., 3,* 297 (1964).
7. L. Ya. Markovskii, E. T. Bezruk, *J. Appl. Chem. USSR (Engl. Transl.), 35,* 491 (1962); *Chem. Abstr., 57,* 1853 (1962).
8. A. J. King, F. A. Kanda, V. A. Russell, W. J. Katz, *J. Am. Chem. Soc., 78,* 4176 (1956).

1.7.3.2. Involving the Reduction of Group IIIB Derivatives with Covalent Hydrides.

The reaction between SiH_4 and BCl_3, which should yield diborane(6) according to bond energy calculations[1], occurs at 150°C to afford dichloroborane:

$$SiH_4 + BCl_3 \rightarrow HBCl_2 + SiCl_4 \tag{a}$$

which subsequently disproportionates[2] to B_2H_6:

$$6 \, HBCl_2 \rightarrow B_2H_6 + 4 \, BCl_3 \tag{b}$$

The same reaction proceeds at RT when catalyzed by methyl radicals to afford B_2H_6 in 67% yield[1]. Similar conversions of B–halide bonds to B—H bonds are available from reactions of trichloroborane with disilane[3] or alkylsilanes[4].

Dimethylstibine and 1-bromodiborane react[5] rapidly at −78°C:

$$(CH_3)_2SbH + B_2H_5Br \rightarrow (CH_3)_2SbBr + B_2H_6 \tag{c}$$

this reaction being favorable theromodynamically. At 400°C, haloboranes may be reduced[6] to boranes by formaldehyde over activated Cu:

$$6 \, HCHO + BBr_3 \xrightarrow[400°C]{Cu} B_2H_6 + 6 \, CO + HBr \tag{d}$$

Borate esters are reduced by AlH_3; e.g., isopropylborate gives $Al(BH_4)_3$, whereas phenylborate gives B_2H_6:

$$4 \, AlH_3 + 3 \, B[OCH(CH_3)_2]_2 \rightarrow Al(BH_4)_3 + 3 \, Al[OCH(CH_3)_2]_2 \tag{e}$$

$$2 \, AlH_3 + 2 \, B(OC_6H_5)_3 \rightarrow B_2H_6 + 2 \, Al(OC_6H_5)_3 \tag{f}$$

Reactions (e) and (f) are reversible. If the isopropylborate reaction is carried out in the presence of amine, the BH_3 amine adduct is formed:

$$AlH_3 + B[OCH(CH_3)_2]_3 + N(C_2H_5)_3 \rightarrow H_3B \cdot N(C_2H_5)_3 + Al[OCH(CH_3)_2]_3 \tag{g}$$

1.7. Formation of Bonds between Hydrogen and B, Al, Ga, In, Tl   131
1.7.3. from Group IIIB Derivatives
1.7.3.2. Involving the Reduction of Group IIIB Derivatives.

Alkylaluminum hydrides[7] also reduce borate esters to BH_3; they function as mixtures of aluminum hydrides and aluminum alkyls, yielding both boranes and alkylborane:

$$3 \ (i\text{-}C_3H_7)_2AlH \ + \ 3 \ (OC_3H_7)_3B \ + \ N(C_2H_5)_3 \ \rightarrow$$

$$2 \ B(C_3H_7\text{-}i)_3 + \ H_3B\cdot N(C_2H_5)_3 \ + \ 3 \ Al(OC_3H_7)_3 \qquad (h)$$

The reaction between AlH_3 and BCl_3 affords B_2H_6: however, because B_2H_6 reacts[8,9] with further AlH_3 to give $Al(BH_4)_3$, the appropriate stoichiometry is:

$$4 \ AlH_3 \ + \ 3 \ BCl_3 \ \xrightarrow{\ Et_2O\ } \ Al(BH_4)_3 \ + \ 3 \ AlCl_3 \qquad (i)$$

Reaction between $Al_2(CH_3)_6$ and xs B_2H_6 also produces $Al(BH_4)_3$.

$$Al_2(CH_3)_6 \ + \ 4 \ B_2H_6 \rightarrow 2 \ Al(BH_4)_3 \ + \ 2 \ B(CH_3)_3 \qquad (j)$$

The species contains three $Al{\overset{H}{\underset{H}{<}}}B$ bridge bonds, which define a trigonal prism about the Al atom[11]. The analogous reaction between B_2H_6 and $Ga(CH_3)_3$ produces only the predicted decomposition products of $Ga(BH_4)_3$; however, at[12] $-45°C$,

$(CH_3)_2GaBH_4$ containing two $Ga{\overset{H}{<}}B$ bridge bonds[13] is formed:

$$2 \ (CH_3)_3Ga \ + \ 3 \ B_2H_6 \xrightarrow{-45°C} 2 \ (CH_3)_2GaBH_4 \ + \ 2 \ CH_3B_2H_5 \qquad (k)$$

Treatment[14] with $(CH_3)_3SiH$ in benzene or cyclohexane at $+20°C$ reduces Ga_2Cl_6 to the $HGaCl_2$:

$$(CH_3)_3SiH \ + \ \tfrac{1}{2} \ Ga_2Cl_6 \ \xrightarrow[-20°C]{C_6H_6} \ (CH_3)_3SiCl \ + \ HGaCl_2 \qquad (l)$$

The analogous reaction also occurs with Ga_2Br_6.

The reaction between $In(CH_3)_3$ and B_2H_6 in tetrahydrofuran (THF) at $-40°C$ affords the unstable $In(BH_4)_3\cdot 3$ THF:

$$(CH_3)_3In \ + \ 2 \ B_2H_6 \ \xrightarrow[-40°C]{THF} \ In(BH_4)_3\cdot 3 \ THF \ + \ B(CH_3)_3 \qquad (m)$$

The species decomposes at $-10°C$ to form B_2H_6, In and H_2. Because the structure is unknown, In—H bond formation is not confirmed[15].

Polymeric $(TlH)_x$ is prepared[16] as a brown powder from $TlOC_2H_5$ and B_2H_6 in ether at $-20°C$.

(L. BARTON)

1. R. Schaeffer, L. Ross, *J. Am. Chem. Soc., 81,* 3486 (1959).
2. L. J. Edwards, R. K. Pearson, U.S. Pat. 3,007,768 (1957); cited in *Boron, Metalloborane Compounds and Boranes,* R. M. Adams, ed., Interscience, New York, 1964, p. 567.
3. C. H. Van Dyke, A. G. MacDiarmid, *J. Inorg. Nucl. Chem., 25,* 1503 (1963).
4. H. Jenkner, Ger. Pat. 1,095,797 (1960); for source see ref. 2.
5. A. B. Burg, L. B. Grant, *J. Am. Chem. Soc., 81,* 1 (1959).
6. O. Glemser, Ger. Pat. 949,943 (1956); *Chem. Abstr., 51,* 14,785 (1957).
7. J. Kollonitsch, *Nature (London), 189,* 1005 (1961).
8. A. E. Finholt, A. C. Bond, H. I. Schelsinger, *J. Am. Chem. Soc., 69,* 1199 (1947).

9. O. Stecher, E. Wiberg, *Chem. Ber.*, *75*, 2003 (1942).
10. H. I. Schlesinger, R. T. Sanderson, A. B. Burg, *J. Am. Chem. Soc.*, *62*, 3421 (1940).
11. A. Almenningen, G. Gunderson, A. Haaland, *Acta Chem. Scand.*, *22*, 328 (1968).
12. H. I. Schelsinger, H. C. Brown, G. W. Schaeffer, *J. Am. Chem Soc.*, *65*, 1786 (1943).
13. M. T. Barlow, A. J. Downs, P. D. P. Thomas, D. W. H. Rankin, *J. Chem. Soc., Dalton Trans.*, 1793 (1979).
14. H. Schmidbaur, W. Findeiss, E. Gast, *Angew. Chem., Int. Ed. Engl.*, *4*, 152 (1965).
15. E. Wiberg, H. Nöth, *Z. Naturforsch., Teil B*, *6*, 59 (1957).
16. E. Wiberg, O. Dittmann, H. Nöth, M. Schmidt, *Z. Naturforsch., Teil B 12*, 62 (1957).

1.7.3.3. Involving Other Reactions.

Protonation of B—B bonds affords $B\diagdown^{H}\diagdown B$ bridge bonds, and this is seen in the reaction of borane anions with protons[1]. The reaction between B_6H_{10} and HBr and between $(CO)_3FeB_5H_9$ and HBr both afford a B—H—μ bridge bond at a site where B—B (or Fe—B) bond existed. The hydride B_6H_{10} affords $[B_6H_{11}]^+$, and, when $(CO)_3FeB_5H_9$ is protonated,[3] the unique Fe—B bond is protonated to form $[(CO)_3FeB_5H_{10}]^+$.

Pyrolysis of $[(CH_3)_2N]_4B_2$ at 300°C affords bis(dimethylamino)borane[4]:

$$[(CH_3)_2N]_4B_2 \xrightarrow[48 \text{ h}]{300°C} [(CH_3)_2N]_2BH \qquad (a)$$

As described in §1.7.4.1.4, ionic hydrides react with organoboranes to afford trialkylhydroborates. In some cases this reaction does not occur; however, treatment with t-C_4H_9Li results in the formation of the desired compounds[5]:

$$ \qquad (b)$$

This reaction is general[6,7], with simple routes to trialkylhydroborates available. The method is effective for boranes from $B(C_2H_5)_3$ to such complex ones as phenyl-9-borabicyclo[3.3.1]nonane[8]. Reaction occurs on a slight xs of (t-C_4H_9)Li in pentane is added dropwise to borane in THF at $-78°C$ with vigorous stirring. The lithium trialkylhydroborate is formed quantitatively:

$$BR_3 + t\text{-}C_4H_9Li \xrightarrow[-78°C]{THF} LiBR_3H + i\text{-}C_4H_8 \qquad (c)$$

Use of CH_3Li produces only the so-called -ate complex.

Trialkylalanes decompose to form olefin and dialkylalane[9]:

$$Al(C_2H_4R)_3 \rightarrow HAl(C_2H_4R)_2 + H_2C_2HR \qquad (d)$$

This is a convenient laboratory synthesis of dialkylaluminum hydrides from trialkylalanes[10], especially those with branched-chain alkyl groups. Diisobutylaluminum hydride is prepared by heating triisobutylalane under N_2 for 12 h at 160–180°C, followed by fractional distillation. Yields are quantitative; about 60% of the trialkylalane is converted to the hydride, and the rest is recovered unreacted.

(L. BARTON)

1. R. J. Remmel, H. D. Johnson, I. S. Jaworiwsky, S. G. Shore, *J. Am. Chem. Soc.*, *97*, 5395 (1975).
2. H. D. Johnson, V. T. Brice, G. L. Brubaker, S. G. Shore, *J. Am. Chem. Soc.*, *94*, 6711 (1972).
3. J. D. Ragaini, R. L. Smith, T. Schmitkons, M. Mangion, S. G. Shore, unpublished work.
4. R. J. Brotherton, L. L. Petterson, U.S. Pat. 3,006,730 (1962); *Chem. Abstr.*, *57*, 7105 (1962).
5. E. J. Corey, S. M. Albonico, U. Koelliker, T. K. Schaaf, R. K. Varma, *J. Am. Chem. Soc.*, *93*, 1491 (1971).
6. E. J. Corey, R. K. Varma, *J. Am. Chem. Soc.*, *93*, 7319 (1971).
7. E. J. Corey, K. B. Becker, R. K. Varma, *J. Am. Chem. Soc.*, *94*, 8616 (1972).
8. H. C. Brown, G. W. Kramer, J. L. Hubbard, S. Krishnamurthy, *J. Organomet. Chem.*, *188*, 1 (1980).
9. K. Ziegler, H. G. Gellert, H. Lehmkuhl, W. Pfohl, K. Zosel, *Justus Liebigs Ann. Chem.*, *629*, 1 (1960).
10. J. J. Eisch, W. S. Kaska, *J. Am. Chem. Soc.*, *88*, 2213 (1966).

1.7.4. by Hydride Ion Reduction

1.7.4.1. of Compounds of Boron

1.7.4.1.1. Involving Halides.

High yields of B—H compounds derive from reactions of metal hydrides with boron halides[1].

In the gas phase, LiH and BF_3 gas react only slightly at 180°C to give compounds containing B—H bonds. The reaction in Et_2O proceeds exothermically to afford B_2H_6 in two different stoichiometries[2,3] at 250°C:

$$6 \text{ Li} + 8 \text{ BF}_3 \xrightarrow{\text{Et}_2\text{O}} B_2H_6 + 6 \text{ LiBF}_4 \tag{a}$$

In the presence of traces of $LiBH(OCH_3)_3$ or $B(OCH_3)_3$ as catalysts:

$$6 \text{ LiH} + 2 \text{ BF}_3 \xrightarrow[\text{Et}_2\text{O}]{\text{catal}} B_2H_6 + 6 \text{ LiF} \tag{b}$$

The former reaction affords yields of 40% of diborane(6), whereas in the latter the yields are quantitative. The LiH + BCl_3 reaction proceeds as Eq. (b) without the catalyst.

These reaction mixtures may be used to form the tetrahydroborate ion; e.g., LiH and BF_3 react in nonethereal organic liquids to afford[4] $LiBH_4$; $LiBH_4$ is available in 90% yield[5] in the presence of $B(OCH_3)_3$ as catalyst via the intermediacy of B_2H_6, and similar yields arise from the reaction[6] in an autoclave at 120°C.

In these reactions the effectiveness of the hydrides varies[7] in the sequence LiH > NaH > KH; NaH reduces[8] B halides to diborane(6). Reaction[9] of xs BF_3 with NaH in glyme below 20°C affords B_2H_6 in 97% yields, but BCl_3 is not so effective[10]. However, BF_3 and NaH react at −70°C to form $Na[HBF_3]$, which decomposes at 200°C. Also at 200°C in the presence of NaH, $Na[H_2BF_2]$ is obtained from $Na[HBF_3]$. This method should be a convenient route to $NaBH_4$ but is not. Calcium hydride[11] reacts with $BF_3 \cdot O(C_2H_5)_2$ in $O(C_2H_5)_2$ at 120°C to afford B_2H_6.

134   1.7. Formation of Bonds between Hydrogen and B, Al, Ga, In, Tl
       1.7.4. by Hydride Ion Reduction
       1.7.4.1. of Compounds of Boron

The maximum yield[12,13] of $LiBH_4$ from LiH and BF_3 in $(C_2H_5)_2O$ between -5 and $34°C$ are obtained at $3-10°C$, and at $10-25°C$ the major product is B_2H_6.

(L. BARTON)

1. H. I. Schlesinger, H. C. Brown, B. Abraham, A. C. Bond, N. Davidson, A. E. Finholt, J. R. Gilbreath, H. Hoekstra, L. Horvitz, E. K. Hyde, J. J. Katz, J. Knight, R. A. Lad, D. L. Mayfield, L. Rapp, D. M. Ritter, A. M. Schwartz, I. Sheft, L. D. Tuck, A. O. Walker, *J. Am. Chem. Soc.*, 75, 186 (1953).
2. H. I. Schlesinger, H. C. Brown, J. R. Gilbreath, J. J. Katz, *J. Am. Chem. Soc.*, 75, 195 (1953).
3. J. R. Elliot, E. M. Boldebuck, G. F. Roedel, *J. Am. Chem. Soc.*, 74, 5047 (1952).
4. P. F. Winternitz, U.S. Pat. 2,532,217 (1950); *Chem. Abstr.*, 45, 2162 (1951).
5. E. Wiberg, O. Kleynot, Ger. Pat. 950,062 (1959); *Chem. Abstr.*, 53, 2551 (1959).
6. G. Wittig, P. Hornberger, *Z. Naturforsch., Teil B*, 6, 225 (1951).
7. R. W. Parry, M. K. Walter, *Prep. Inorg. React.*, 5, 68 (1968).
8. H. C. Brown, P. A. Tierney, *J. Am. Chem. Soc.*, 80, 1522 (1954).
9. R. M. Adams, R. K. Pearson, U.S. Pat. 2,968,531 (1961); *Chem. Abstr.*, 55, 13,789 (1961).
10. J. Goubeau, R. Bergman, *Z. Anorg. Allg. Chem.*, 263, 69 (1950).
11. V. I. Mikheeva, E. M. Fedneva, V. I. Alpatova, *Dokl. Akad. Nauk SSSR*, 131, 318 (1959).
12. V. I. Mikheeva, E. M. Fedneva, *Izv. Akad. Nauk SSSR, Otdel Khim. Nauk*, 902 (1956).
13. E. M. Fedneva, *Russ. J. Inorg. Chem. (Engl. Transl.)*, 4, 286 (1959).

1.7.4.1.2. Involving Oxygen Compounds.

The reaction between borate esters and metal hydrides represents the most important development in borane chemistry, and the ultimate syntheses of tetrahydroborate salts have important consequences, especially in organic chemistry.

The presence of $B(OCH_3)_3$ has a solubilizing effect on LiH in ethers,[1] owing to[2,3]:

$$LiH + B(OCH_3)_3 \rightarrow LiBH(OCH_3)_3 \qquad (a)$$

This reaction is general and provides a convenient route to useful reducing agents. Sodium hydride reacts with $B(OCH_3)_3$ on refluxing for 5 h to afford $NaBH(OCH_3)_3$ in quantitative yield. In refluxing tetrahydrofuran (THF) the rate of reaction of NaH with borate esters decreases[4] $CH_3 > C_2H_5 >>> HC(CH_3)_2 > C(CH_3)_3$. The time required for the isopropyl- and t-butylborates is reduced if glyme or diglyme is used as solvent at $130-150°C$. The species formed are stable toward disproportionation and, therefore, are useful reducing reagents. Potassium hydride is more reactive[5] toward borates than either LiH or NaH, e.g., triisopropylborate is nearly inert toward NaH at $20°C$, requiring 83 h at reflux to react with xs hydride, but KH reacts completely in <1 h at $20°C$:

$$KH + B[OCH(CH_3)_2]_3 \xrightarrow[20°C]{THF} KBH[OCH(CH_3)_2] \qquad (b)$$

Sodium trimethoxyhydroborate disproportionates at $250-250°C$ to afford sodium tetrahydroborate:

$$4\ NaBH(OCH_3)_3 \rightarrow NaBH_4 + 3\ NaOCH_3 + 3\ B(OCH_3)_3 \qquad (c)$$

However, this is not a good preparation of $NaBH_4$ because $B(OCH_3)_3$ must be removed continuously from the reaction medium and, furthermore, a competing process forming dimethyoxyborane, $HB(OCH_3)_2$, occurs:

$$NaBH(OCH_3)_3 \rightarrow NaOCH_3 + HB(OCH_3)_2 \qquad (d)$$

1.7. Formation of Bonds between Hydrogen and B, Al, Ga, In, Tl   135
1.7.4. by Hydride Ion Reduction
1.7.4.1. of Compounds of Boron

The most important reaction between hydrides and boron–oxygen compounds is[6] between NaH and $B(OCH_3)_3$ to form $NaBH_4$ at 225–275°C:

$$B(OCH_3)_3 + 4\,NaH \rightarrow NaBH_4 + 3\,NaOCH_3 \qquad (e)$$

Methylborate is added dropwise to NaH under N_2 at 230°C with vigorous stirring to form $NaBH_4$ in 90–96% purity and 94% yield. The $[BH_4]^-$ salt is extracted with liq NH_3, filtered and the NH_3 boiled off. Recrystallization from H_2O or isopropylamine affords product in > 99% yield. Reaction between NaH and $NaBH(OCH_3)_3$ at 250–260°C affords $NaBH_4$ in 78% yield. A reaction analogous to Eq. (e) produces $LiBH_4$ in 70% yield; however, extraction of the crude $LiBH_4$ from the mixture is difficult.

Boron(III) oxide may be reduced to the $[BH_4]^-$ ion by NaH under stringent conditions. Formation of $NaBH_4$ proceeds in 60% yield at 350°C while the reagents are ground together in a glass ball mill for 20–48 h:

$$4\,NaH + 2\,B_2O_3 \rightarrow 3\,NaBO_2 + NaBH_4 \qquad (f)$$

Equation (e) is the most important reaction in which the B–H bond is formed because the product, $NaBH_4$, is the starting material from which all other boranes are formed. The species is also a reagent in organic chemistry. The starting materials for Eq. (e) are readily available, and so this represents a convenient commercial route to B—H bonded compounds.

Other $M(BH_4)_x$ species may be prepared by extension of Eq. (e), e.g., $Ca(BH_4)_2$ is prepared by the reaction between CaH_2 and $B(OCH_3)_3$ in a pressure bomb[7].

(L. BARTON)

1. J. R. Elliot, E. M. Boldebuck, C. F. Roedel, *J. Am. Chem. Soc., 74,* 5047 (1952).
2. H. I. Schlesinger, H. C. Brown, B. Abraham, A. C. Bond, N. Davidson, A. E. Finholt, J. R. Gilbreath, H. Hoekstra, L. Horvitz, E. K. Hyde, J. J. Katz, J. Knight, R. A. Lad, D. L. Mayfield, L. Rapp, D. M. Ritter, A. M. Schwartz, I. Sheft, L. D. Tuck, A. O. Walker, *J. Am. Chem. Soc., 75,* 186 (1953).
3. H. C. Brown, H. I. Schlesinger, I. Sheft, D. M. Ritter, *J. Am. Chem. Soc., 75,* 192 (1953).
4. H. C. Brown, E. J. Mead, C. J. Shoaf, *J. Am. Chem. Soc., 78,* 3616 (1956).
5. C. A. Brown, *J. Am. Chem. Soc., 95,* 4100 (1973).
6. H. I. Schlesinger, H. C. Brown, A. E. Finholt, *J. Am. Chem. Soc., 75,* 205 (1953).
7. H. W. Stone, R. L. Pecsok, E. F. C. Cain, R. Green, B. Griggs, R. Meekev, D. Nail, I. Pearson, M. Ring, *Nucl. Sci. Abstr., 10,* 574 (1956).

1.7.4.1.3. Involving Nitrogen Compounds.

Reduction of B—N to B—H bonds is difficult, and examples of this reaction are rare; e.g., $(CH_3)_2NBCl_2$ is reduced[1] only as far as $[(CH_3)_2NBH_2]_2$ and trichloroborazine is reduced only to the borazine[2], i.e., the B—N bonds in neither example react.

The B—N bond is reduced in amineborane adducts, e.g., $(CH_3)_3NBCl_3$ is converted[3] to $NaBH_4$ by treatment with NaH in ethylene glycol for 2 h at 150°C:

$$4\,NaH + (CH_3)_3NBCl_3 \rightarrow (CH_3)_3N + 3\,NaCl + NaBH_4 \qquad (a)$$

Similarly, H_3BNR_3 (R = CH_3, C_2H_5) is converted[4] to the $[BH_4]^-$ salt by treatment with NaH or CaH_2 at 300°C:

$$NaH + H_3BNR_3 \xrightarrow{\Delta} NaBH_4 + NR_3 \qquad (b)$$

(L. BARTON)

1. R. M. Adams, in *Boron, Metalloboron Compounds and Boranes,* R. M. Adams, ed., Inter-science, New York, 1964, p. 414.
2. R. Schaeffer, M. Steindler, L. F. Hohnstedt, H. S. Smith Jr., L. B. Eddy, H. I. Schlesinger, *J. Am. Chem. Soc., 76,* 3303 (1954).
3. H. H. Bronaugh, U.S. Pat. 2,880,058 (1958); *Chem. Abstr., 53,* 15,503 (1959).
4. R. Köster, *Angew. Chem., 69,* 94 (1957).

1.7.4.1.4. Involving Other Compounds.

Metal hydrides form addition compounds with triorganoboranes[1-4], which are useful reagents[5]; e.g., LiH and NaH react with triorganoboranes to afford the species Li[HBR_3]. However, long reaction times or reflux conditions are required, and with hindered triorganoboranes the reactions are slow[6]. On the other hand, KH reacts quantitatively with trialkyl or triarylboranes in tetrahydrofuran (THF) at 25°C:

$$KH + BR_3 \xrightarrow[25°C]{THF} K[HBR_3] \qquad (a)$$

Therefore, the species where R = C_2H_5, n-C_4H_9, sec-C_4H_9, C_5H_{10}, C_6H_{12}, exo-2-norbornyl, trans-2-methylcyclohexyl, 3-methyl-2-butyl and phenyl are conveniently prepared[7-10]. With the exception of K[$HB(C_6H_5)_3$], the reaction proceeds when THF and then triorganoborane is added to KH under Ar. Stirring for 1 h at 25°C effects 100% conversion to the triorganohydridoborate[7,8]. In the preparation of K[$HB(C_6H_5)_3$] it is necessary to add $B(C_6H_5)_3$ in THF to KH in THF dropwise over 7 h. Additional stirring of the filtrate with [$(C_2H_5)_4N$]Br for 20 h followed by recrystallization affords [$(C_2H_5)_4N$][$HB(C_6H_5)_3$] in 56% yield[10].

The B—P bond may be converted to B—H bonds, e.g.:

$$BPO_4 + 4 NaH \rightarrow NaBH_4 + Na_3PO_4 \qquad (b)$$

when the reaction is allowed to proceed in mineral oil at 288–300°C for 2 h[11].

(L. BARTON)

1. H. C. Brown, H. I. Schlesinger, I. Sheft, D. M. Ritter, *J. Am. Chem. Soc., 75,* 192 (1953).
2. G. Wittig, G. Keichev, A. Ruckert, P. Raff, *Justus Liebigs Ann. Chem., 563,* 110 (1949).
3. J. B. Honeycutt, J. M. Riddle, *J. Am. Chem. Soc., 83,* 369 (1961).
4. P. Binger, G. Benedict, G. W. Rotermund, R. Köster, *Justus Liebigs Ann. Chem., 717,* 21 (1968).
5. H. C. Brown, S. Krishnamurthy, *J. Am. Chem. Soc., 95,* 8486 (1973).
6. H. C. Brown, S. Krishnamurthy, J. L. Hubbard, *J. Am. Chem. Soc., 100,* 3343 (1978).
7. C. A. Brown, *J. Am. Chem. Soc., 95,* 4100 (1973).
8. C. A. Brown, *Inorg. Synth., 17,* 26 (1977).
9. C. A. Brown, S. Krishnamurthy, *J. Organomet. Chem., 156,* 111 (1978).
10. J. M. Burlitch, J. H. Burk, M. E. Leonowicz, R. E. Hughes, *Inorg. Chem., 18,* 1702 (1979).
11. H. J. Bronaugh, U.S. Pat. 2,849,276 (1958); *Chem. Abstr., 51,* 12,171 (1958).

1.7.4.2. of Compounds of Aluminum.

The important reaction for Al—H bond formation is between LiH and $AlCl_3$ to form[1] $LiAlH_4$. The species $LiAlH_4$ is used to form most other Al—H bond-containing species. The reaction between LiH and $AlCl_3$ proceeds when $AlCl_3$ in ether is added to xs finely ground LiH, also in ether, in the presence of traces of $LiAlH_4$. Removal of

1.7. Formation of Bonds between Hydrogen and B, Al, Ga, In, Tl   137
1.7.4. by Hydride Ion Reduction
1.7.4.2. of Compounds of Aluminum.

precipitated LiCl and xs LiH is effected by filtration and $LiAlH_4$ is obtained in $> 95\%$ yields:

$$4 \text{ LiH} + \text{AlCl}_3 \xrightarrow{\text{Et}_2\text{O}} \text{LiAlH}_4 + 3 \text{ LiCl} \tag{a}$$

The $LiAlH_4$ used as promotor is obtained by reaction between LiH and $AlCl_3$ in dioxane or ether. This process affords $LiAlH_4$ in low yields ($\sim 30\%$), and the reaction mixture requires heating to 50°C in dioxane or periodic cooling with liq N_2 when ether is the solvent.

Improvements in this process to eliminate the necessity for finely grinding the LiH or using $LiAlH_4$ as promotor[2] have only limited success. The reaction may be initiated by the addition[3] of I_2, and the substitution[4] of $AlBr_3$ for $AlCl_3$ eliminates the need to grind the LiH; however, this results in product contaminated with $AlBr_3$ because the latter is soluble in diethyl ether. Optimization of the conditions for Eq. (a) involves the intermediates[5] AlH_3 and $AlH_3 \cdot AlCl_3$; optimum conditions are 0–4°C when an induction period is not necessary[6], and a slow rate of addition of $AlCl_3$, which reduces the formation of Cl-containing alanes[7]. An alternative procedure[8] for the preparation of $LiAlH_4$ employs LiH and $AlBr_3$ in the solvent system $C_6H_6-(C_2H_5)_2O$.

Preparation[9] of $NaAlH_4$ in 60% yield results from NaH and $AlBr_3$ in $(CH_3)_2O$. The reaction between NaH and $AlCl_3$ in $(C_2H_5)_2O$ does not proceed well owing to the limited solubility of NaH; however, the presence of $Al(C_2H_5)_3$ as catalyst solubilizes the NaH as $Na[Al(C_2H_5)_3H]$ and facilitates the reaction[10]:

$$\text{NaH} + (\text{C}_2\text{H}_5)_3\text{Al} \xrightarrow{(\text{CH}_3)_2\text{O}} \text{Na}[\text{Al}(\text{C}_2\text{H}_5)_3\text{H}] \tag{b}$$

$$4 \text{ Na}[\text{Al}(\text{C}_2\text{H}_5)_3\text{H}] + \text{AlCl}_3 \rightarrow \text{NaAlH}_4 + 3 \text{ NaCl} + 3 (\text{C}_2\text{H}_5)_3\text{Al} \tag{c}$$

Reaction between KH and $AlCl_3$ produces $K[AlH_4]$ in the solvents C_6H_6- $(C_2H_5)_2O$ when $Al(C_2H_5)_3$ or $(i\text{-}C_4H_9)_2AlH$ is present as catalyst[10].

Dialkylaluminum hydrides, important reagents industrially, are prepared by treatment[11] of dialkylaluminum halides with LiH, e.g., diisobutylaluminum hydride (or deuteride) may be prepared by treatment of $(i\text{-}C_4H_9)_2AlCl$ with LiH in ether[12,13]; LiH or LiD in ether is cooled to 0°C and $(i\text{-}C_4H_9)_2AlCl$ is added dropwise under N_2 so that an xs of LiH exists. The mixture is heated to reflux for 48 h or until a Cl^- test shows negative. After workup $(i\text{-}C_4H_9)_2AlH$ is obtained in 67% yield.

The species $C_2H_5OAlCl_2$ may be converted to the salt $Na[AlH_3OC_2H_5]$ by treatment[14] with NaH in tetrahydrofuran (THF). The $C_2H_5OAlCl_2$ in THF is added dropwise to NaH and the resulting mixture stirred and maintained at 40°C for 20 min by external cooling with solid CO_2. The resulting $Na[AlH_3OC_2H_5]$ is obtained in 68% yield.

(L. BARTON)

1. A. E. Finholt, A. C. Bond, Jr., H. I. Schlesinger, J. Am. Chem. Soc., 69, 1199 (1947).
2. E. C. Ashby, Adv. Inorg. Chem. Radiochem., 8, 283 (1966).
3. E. Wiberg, Z. Naturforsch., Teil B, 6, 393 (1951).
4. E. Wiberg, M. Schmidt, Z. Naturforsch., Teil B, 7, 59 (1952).
5. V. I. Mikheeva, E. M. Fedneva, Z. L. Shnitkova, Zh. Neorg. Khim., 1, 2440 (1956).
6. V. I. Mikheeva, M. S. Selivokhina, V. V. Leonova, Russ. J. Inorg. Chem. (Engl. Transl.), 4, 2436 (1959).

7. V. I. Mikheeva, M. S. Selivokhina, V. V. Leonova, *Russ. J. Inorg. Chem. (Engl. Transl.)*, *4*, 2705 (1959).
8. J. Vit, V. Prochazka, F. Petru, *Khim Prom., (Moscow)*, *10*, 183 (1960); *Chem. Abstr.*, *54*, 20,598 (1960).
9. H. I. Schlesinger, A. E. Finholt, U.S. Pat. 2,576,311 (1951); *Chem. Abstr.*, *46*, (1952).
10. L. I. Zakharkin, V. V. Gavrilenko, *Izv. Akad, Nauk SSSR, Otd. Khim. Nauk*, 2246 (1961).
11. K. Zeigler, H. G. Gellert, H. Martin, K. Nagel, J. Scheider, *Justus Liebigs Ann. Chem.*, *589*, 91 (1954).
12. J. J. Eisch, S. G. Rhee, *J. Am. Chem. Soc.*, *96*, 7276 (1974).
13. J. J. Eisch, *Organometallic Synthesis*, Vol. 2, *Non-Transition Metal Compounds*, Academic Press, New York, 1981, p. 136.
14. G. Hamprecht, M. Schwarzmann, M. Tittel, Ger. Pat. 1,085,515 (1960); *Chem. Abstr.*, *55*, 15,350 (1961).

1.7.4.3. of Compounds of Gallium.

Anhydrous $GaCl_3$ reacts with a fourfold xs of finely ground LiH in ether when the reaction mixture is slowly warmed from $-80°C$ to $25°C$. The white, solid product is isolated in 76% yield after filtration and removal of solvent in vacuo[1]:

$$GaCl_3 + 4 LiH \rightarrow LiGaH_4 + LiCl_3 \qquad (a)$$

The reaction can be carried out[2] at $0°C$, and, if the reaction mixture is warmed to $35°C$ while $N(CH_3)_3$ is added, a 95% yield is obtained[3].

(L. BARTON)

1. A. E. Finholt, A. C. Bond, H. I. Schlesinger, *J. Am. Chem. Soc.*, *69*, 1199 (1947).
2. N. N. Greenwood, A. Storr, M. G. H. Wallbridge, *Inorg. Chem.*, *2*, 1036 (1963).
3. E. Wiberg, M. Schmidt, *Z. Naturforsch., Teil B*, *6*, 171 (1951).

1.7.4.4. of Compounds of Indium.

Treatment of $InCl_3$ or $InBr_3$ with an equimolar quantity of finely divided LiH in ether at $-25°C$ affords $LiInH_4$ in 60–65% or 80% yield, respectively. When LiH reacts with $InCl_3$ in refluxing $(C_2H_5)_2O$, $(InH_3)_x$ and LiCl are formed[1]. When In trihalides react with large granules of LiH, the trihaloindo hydrides are formed[2], e.g., LiH and $InBr_3$ in ether at $0°C$ afford $LiInBr_3H·6 O(C_2H_5)_2$ in 40% yield and LiH and InI_3 afford the corresponding species $LiInI_3H·6 O(C_2H_5)_2$ under the same conditions.

(L. BARTON)

1. E. Wiberg, M. Schmidt, *Z. Naturforsch., Teil B*, *12*, 54 (1957).
2. E. Wiberg, O. Dittmann, H. Nöth, M. Schmidt, *Z. Naturforsch., Teil B*, *12*, 56 (1957).

1.7.4.5. of Compounds of Thallium.

Treatment of $TlCl_3$ with pure, finely divided LiH in $(C_2H_5)_2O$ at $-15°C$ affords $LiTlH_4$. At higher T the $LiTlH_4$ cannot be isolated.

(L. BARTON)

1. E. Wiberg, O. Dittmann, M. Schmidt, *Z. Naturforsch., Teil B*, *12*, 60 (1957).

1.7.5. from Complex Hydrides

1.7.5.1. with Compounds of Boron.

The reaction between BCl_3 and $LiAlH_4$ in ether affords B_2H_6 quantitatively[1]:

$$3 \text{ LiAlH}_4 + 4 \text{ BCl}_3 \xrightarrow{(C_2H_5)_2O} 3 \text{ LiCl} + 3 \text{ AlCl}_3 + 2 \text{ B}_2\text{H}_6 \qquad (a)$$

An xs of BCl_3 is distilled into the $LiAlH_4$ in $(C_2H_5)_2O$ at $-196°C$. The mixture is warmed to $25°C$ and thoroughly mixed, the B_2H_6 distills through a trap maintained at $-112°C$ and the xs BCl_3 forms a complex with the solvent. Because $LiAlH_4$ may be prepared simply from LiH and $AlCl_3$, the process may be considered as a synthesis from LiH and BCl_3 with $AlCl_3$ present as a solubilizing agent for LiH.

The reaction[2,3] between $LiAlH_4$ and BF_3 proceeds in two well-defined steps:

$$3 \text{ LiAlH}_{4(soln)} + 3 \text{ BF}_{3(soln)} \xrightarrow{(C_2H_5)_2O} 3 \text{ LiBH}_{4(soln)} + 3 \text{ AlF}_{3(s)} \qquad (b)$$

$$3 \text{ LiBH}_{4(soln)} + \text{ BF}_{3(soln)} \xrightarrow{(C_2H_5)_2O} 2 \text{ B}_2\text{H}_{6(g)} + 3 \text{ LiF}_{(s)} \qquad (c)$$

Overall:

$$3 \text{ LiAlH}_{4(soln)} + 4 \text{ BF}_3 \xrightarrow{(C_2H_5)_2O} 3 \text{ LiF}_{(s)} + 3 \text{ AlF}_{3(s)} + 2 \text{ B}_2\text{H}_{6(g)} \qquad (d)$$

Other methods involving halide–hydride exchange starting with $[BH_4]^-$ ion are available[4,5]. These are included here because formation of BHB bridge bonds from B—H terminal bonds is interpreted as new B—H bond formation, e.g., BF_3 and $NaBH_4$ react[6] in diglyme to afford B_2H_6 in yields $> 90\%$:

$$3 \text{ NaBH}_4 + \text{ BF}_3 \xrightarrow{diglyme} 2 \text{ B}_2\text{H}_6 + 3 \text{ NaBF}_4 \qquad (e)$$

Hydride abstraction from tetrahydroborate anions is the most convenient route for the preparation of boranes[7,8]. In these processes an unstable intermediate transfers a BH_3 moiety and in so doing forms new BHB bridge bonds. The simplest of these reactions involves the formation of B_2H_6 from $NaBH_4$ and BF_3 at $25°C$. In addition to forming new B—H—B bonds in B_2H_6, 1 mol of $[HBF_3]^-$ is formed per mol of $[BH_4]^-$:

$$2 [\text{BH}_4]^- + 2 \text{ BF}_3 \rightarrow \text{B}_2\text{H}_6 + 2 [\text{HBF}_3]^- \qquad (f)$$

This process also occurs for higher boranes, e.g.:

$$[\text{B}_3\text{H}_8]^- + \text{BCl}_3 \xrightarrow{25°C} \tfrac{1}{2} \text{B}_4\text{H}_{10} + [\text{HBCl}_3]^- + \tfrac{1}{x} (\text{BH}_2)_x \qquad (g)$$

$$[\text{B}_4\text{H}_9]^- + \text{BCl}_3 \rightarrow \tfrac{1}{2} \text{B}_5\text{H}_{11} + [\text{HBCl}_3]^- + \tfrac{1}{2x} (\text{B}_3\text{H}_5)_x \qquad (h)$$

New BHB bridge bonds form from terminal B—H bonds; e.g., $LiBH_4$ and $NaBH_4$ absorb 1 mol equiv of borane[6,9-11]:

140   1.7. Formation of Bonds between Hydrogen and B, Al, Ga, In, Tl
        1.7.5. from Complex Hydrides
        1.7.5.1. with Compounds of Boron.

$$NaBH_4 + BH_3 \xrightarrow[0°C]{diglyme} Na\left[\begin{array}{c} \overset{\displaystyle H}{|} \quad \overset{\displaystyle H}{|} \\ H-B-H-B-H \\ \underset{\displaystyle H}{|} \quad \underset{\displaystyle H}{|} \end{array} \right] \qquad (i)$$

The same product is formed[6] from the reaction of $NaBH_4$ with diborane formed in situ from BF_3 and $NaBH_4$:

$$7 \ NaBH_4 + 4 \ BF_3 \xrightarrow{diglyme} 3 \ NaBF_4 + 4 \ NaB_2H_7 \qquad (j)$$

This reaction[12] does not occur with KBH_4. Similar hexaalkyldiborohydrides are available when LiH is treated with 2 equiv of trialkylborane in tetrahydrofuran (THF), glyme or diglyme[13]:

$$LiH + (CH_3)_3B \xrightarrow[0°C]{THF} Li[(CH_3)_3BH] \qquad (k)$$

$$Li(CH_3)_3BH + (CH_3)_3B \rightarrow Li[(CH_3)_3B-H-B(CH_3)_3] \qquad (l)$$

When such solvents as $(C_2H_5)_2O$ or $(n-C_4H_9)_2O$, which are poor solvating media for the Li^+ ion, are used, the second mol of alkylborane does not add to the $[(CH_3)_3BH]^-$ ion, which must be strongly associated with the Li^+ ion.

Other examples of new $B\overset{H}{\diagup}\diagdown B$ bridge formation involve the condensation of hydroborate anions with neutral boranes to form large boranes or borane anions. Borane will add to the anions $[B_4H_9]^-$, $[B_5H_8]^-$, and $[B_6H_9]^-$ to form the species $[B_5H_{12}]^-$, $[B_6H_{11}]^-$ and $[B_7H_{12}]^-$, respectively[14]:

$$[B_4H_9]^- + \tfrac{1}{2} B_2H_6 \rightarrow [B_5H_{12}]^- \qquad (m)$$

$$[B_5H_8]^- + \tfrac{1}{2} B_2H_6 \rightarrow [B_6H_{11}]^- \qquad (n)$$

$$[B_6H_9]^- + \tfrac{1}{2} B_2H_6 \rightarrow [B_7H_{12}]^- \qquad (o)$$

These reactions proceed at $-78°C$ in ethers by the addition of a BH_3 group to a $B-B$ bond in the anion. The products are fluxional, bridging hydrogens moving into and out of vacant $B-B$ bond sites in solution. New $B-H$ bonds are formed:

The protonation of borane anions affords new $B-H$ bonds; e.g., protonation[14] of $[B_6H_{11}]^-$ formed in Eq. (p) affords B_6H_{12}:

1.7. Formation of Bonds between Hydrogen and B, Al, Ga, In, Tl   141
1.7.5. from Complex Hydrides
1.7.5.1. with Compounds of Boron.

This is not always the case, for when H^+ is added to $[B_5H_{12}]^-$, H_2 is eliminated and B_5H_{11} is formed[14]. Hydride transfer between boranes and hydroborate anions can occur:

$$[B_6H_{11}]^- + B_5H_{11} \rightarrow B_6H_{10} + [B_5H_{12}]^- \tag{r}$$

These polyhedral expansion reactions of boranes have synthetic utility.

The decomposition of $[B_5H_8]^-$ proceeds via the formation[15-17] of $[B_9H_{14}]^-$ in the presence of B_5H_9:

$$[B_5H_8]^- + B_5H_9 \xrightarrow[25°C]{THF} [B_9H_{14}]^- + \tfrac{1}{2} B_2H_6 \tag{s}$$

Although the stoichiometry is not as simple as indicated in Eq. (s), the reaction proceeds[7,8] when 2 mol B_5H_9 are treated with 1 mol NaH in THF at 25°C.

Insertion of boron atoms from larger boranes into the B—B bonds of higher tetrahydroborate anions proceeds with the formation of new B—H bonds and represents a rational approach to the synthesis of higher boranes, e.g., B_8H_{12} combines[18] with $[B_6H_9]^-$ at $-78°C$ to form $[B_{14}H_{21}]^-$. Protonation of the latter affords $B_{14}H_{20}$:

$$K[B_6H_9] + B_8H_{12} \xrightarrow[-78°C]{(C_2H_5)_2O} K[B_{14}H_{21}]$$

$$KCl + H_2 + B_{14}H_{20} \xleftarrow{HCl} \tag{t}$$

Bonds B—O or B—N are not reduced by reaction with complex hydrides, and so selective reductions are possible. Treatment of $(CH_3O)_2BCl$ with $NaBH_4$ in diglyme affords[19] B_2H_6, via the intermediacy of $HB(OCH_3)_2$:

$$3 (CH_3O)_2BCl + 3 NaBH_4 \xrightarrow[25°C]{diglyme} 3 (CH_3O)_2BH + 3 NaCl + \tfrac{3}{2} B_2H_6 \tag{u}$$

$$3 (CH_3O)_2BH \rightarrow 2 B(OCH_3)_3 + \tfrac{1}{2} B_2H_6 \tag{v}$$

Overall:

$$3 (CH_3O)_2BCl + 3 NaBH_4 \rightarrow 3 NaCl + 2 B(OCH_3)_3 + 2 B_2H_6 \tag{w}$$

Similarly, B–trichloroborazine is reduced[19,20] to $H_3B_3N_3H_3$ at 25°C by treatment with $LiBH_4$ in $(n-C_4H_9)_2O$:

$$6 LiBH_4 + 2 Cl_3B_3N_3H_3 \xrightarrow{(n-C_4H_9)_2O} 2 H_3B_3N_3H_3 + 6 LiCl + 3 B_2H_6 \tag{x}$$

142   1.7. Formation of Bonds between Hydrogen and B, Al, Ga, In, Tl
        1.7.5. from Complex Hydrides
        1.7.5.1. with Compounds of Boron.

Moreover, $B_3N_3H_6$ may be reduced to $B_3N_3H_{12}$ by treatment with HCl to form the hydrochloride followed by reaction with $NaBH_4$ in diglyme[21]:

$$2\ B_3N_3H_6 + 6\ HCl \xrightarrow{diglyme} 2\ B_3N_3H_6 \cdot 6\ HCl + 6\ NaBH_4 \xrightarrow{diglyme}$$
$$6\ NaCl + 2\ B_3N_3H_{12} + 3\ B_2H_6 \qquad (y)$$

Boranes may be prepared by reduction of B—O bonds using $LiAlH_4$, e.g., C_6H_5-$B(OC_6H_5)_2$ or $(C_6H_5BO)_3$ reacts with $LiAlH_4$ in $(C_2H_5)_2O$ containing xs C_5H_5N at 70°C to form the air-stable pyridine—borane[22]:

$$C_6H_5B(OC_6H_5)_2 + LiAlH_4 \xrightarrow[-70°C]{pyridine} C_6H_5BH_2 \cdot py \qquad (z)$$

Other pyridine boranes, $RBH_2 \cdot C_5H_5N$, are prepared, where $R = $ p-ClC_6H_4, p-CH_3 C_6H_4, p-$CH_3OC_6H_4$, α-$C_{10}H_7$, 1-C_3H_7, 1-C_4H_9; $R_2BH \cdot NC_5H_5$ also, can be prepared, where $R = C_6H_5$, p-ClC_6H_4, p-$CH_3C_6H_4$ or p-$CH_3OC_6H_4$. Trialkylborates may be reduced to B_2H_6 by treatment with MBH_4 or $MAlH_4$ in ethers or in the absence of solvent[23]:

$$NaAlH_4 + B(OC_3H_7\text{-}i)_3 \rightarrow NaBH_4 + Al(OC_3H_7\text{-}i)_3 \qquad (aa)$$

This reaction may be general for all lower alkylborates; e.g., treatment of $B(OCH_3)_3$ with $LiAlH_4$ affords a mixture not easily identified. Addition of $LiAlH_4$ to $B(OCH_3)_3$ in $(C_2H_5)_2O$ forms $LiBH_4$ and LiB_2H_7 among the products; however, in $(C_2H_5)_3N$ as solvent, the borane adduct is formed[24]:

$$3\ LiAlH_4 + 4\ B(OCH_3)_3 \xrightarrow{(C_2H_5)_3N} 4\ H_3B \cdot N(C_2H_5)_3 + 3\ LiAl(OCH_3)_3 \qquad (ab)$$

When triphenylborate is used, $LiAlH_4$ forms $LiBH_4$ if the borate ester is added to the $LiAlH_4$ solution. If $LiAlH_4$ is added to $B(OC_6H_5)_3$, B_2H_6 is formed:

$$3\ LiAlH_4 + 4\ B(OC_6H_5)_3 \rightarrow 2\ B_2H_6 + 3\ LiAl(OC_6H_5)_4 \qquad (ac)$$

In this reaction only 43% yield of B_2H_6 forms owing to a side reaction in which AlH_3 is formed, so $LiAlH_4$ is more reactive[24] toward B—OR compounds than $NaBH_4$.

Circulation of $B_2H_{6(g)}$ over solid $LiAlH_4$ at 80—90°C affords $Al(BH_4)_3$ in 90% yield.[1] The $Al(BH_4)_3$ is collected at −80°C in a U-trap:

$$LiAlH_4 + 2\ B_2H_6 \xrightarrow{80-90°C} LiBH_4 + Al(BH_4)_3 \qquad (ad)$$

(L. BARTON)

1. A. E. Finholt, A. C. Bond, H. I. Schlesinger, J. Am. Chem. Soc., 69, 1199 (1947).
2. R. C. Lord, E. Nielsen, J. Chem. Phys., 19, 1 (1951).
3. I. Shapiro, H. G. Weiss, M. Schmich, S. Skolnich, G. B. L. Smith, J. Am. Chem. Soc., 74, 901 (1952).
4. R. M. Adams, ed., Boron, Metalloboron Compounds and Boranes, Interscience, New York, 1964, Ch. 7.
5. R. W. Parry, M. K. Walter, Prep. Inorg. React., 5, 46 (1968).
6. H. C. Brown, P. A. Tierney, J. Am. Chem. Soc., 80, 1552 (1958).
7. J. B. Leach, M. A. Toft, F. L. Himpsl, S. G. Shore, J. Am. Chem. Soc., 103, 988 (1981).
8. M. A. Toft, J. B. Leach, F. L. Himpsl, S. G. Shore, Inorg. Chem., 21, 1952 (1982).
9. H. C. Brown, P. F. Stehle, P. A. Tierney, J. Am. Chem. Soc., 79, 2020 (1957).
10. W. G. Evans, C. E. Holloway, K. Sukumarabandhu, D. H. McDaniel, Inorg. Chem., 7, 1746 (1968).
11. R. K. Hertz, H. D. Johnson II, S. G. Shore, Inorg. Synth., 17, 24 (1977).

12. K. R. Pearson, L. L. Lewis, L. J. Edwards, *Reaction of Potassium Borohydride with Boron Trifluoride*, P. M. Maginnity, ed., Project ZIP, Contract No. a(s)-S2-1024-C (1957); *Nucl. Sci. Abstr.*, *12*, 4069 (1958).
13. H. C. Brown, A. Khuri, S. Krishnamurthy, *J. Am. Chem. Soc.*, *99*, 6237 (1977).
14. R. J. Remmel, H. D. Johnson II, I. S. Jaworiwsky, S. G. Shore, *J. Am. Chem. Soc.*, *97*, 5395 (1975).
15. V. T. Brice, H. D. Johnson, II, D. L. Denton, S. G. Shore, *Inorg. Chem.*, *11*, 1135 (1972).
16. C. G. Savory, M. G. H. Wallbridge, *J. Chem. Soc., Dalton Trans.*, 179 (1974).
17. L. Barton, T. Onak, S. G. Shore, *Gmelin Handbuch der Anorganischen Chemie*, Springer-Verlag, Berlin, 1979, Vol. 54, Ch. 4, p. 113.
18. J. C. Huffman, D. C. Moody, R. Schaeffer, *J. Am. Chem. Soc.*, *97*, 1621 (1974).
19 H. Nöth, *Angew. Chem.*, *73*, 371 (1961).
20. R. Schaeffer, M. Steindler, L. Holnstedt, H. S. Smith, L. B. Eddy, H. I. Schlesinger, *J. Am. Chem. Soc.*, *76*, 3303 (1954).
21. G. H. Dahl, R. Schaeffer, *J. Am. Chem. Soc.*, *83*, 3032 (1961).
22. M. F. Hawthorne, *Chem. Ind.* (*London*), 1242 (1957).
23. J. Kollonitsch, *Nature* (*London*), 189, 1005 (1961).
24. E. C. Ashby, *Adv. Inorg. Chem. Radiochem.*, *8*, 283 (1966).

1.7.5.2. with Compounds of Aluminum, Gallium, Indium and Thallium.

Treatment of $AlCl_3$ with $NaBH_4$ at 25°C in ether for 38 h affords[1] $H_3Al \cdot O(C_2H_5)_2$:

$$3\ NaBH_4 + 4\ AlCl_3 \xrightarrow[25°C,\ 38\ h]{(C_2H_5)_2O} 3\ NaCl + 3\ BCl_3 + 4\ H_3Al \cdot O(C_2H_5)_2 \qquad (a)$$

In the absence of solvent at 100–150°C, MBH_4 salts and Al halides afford[2] $Al(BH_4)_3$. The reaction proceeds for X = Cl, Br, and the reactivities of the MBH_4 salts decrease Li > Na > K.

$$MBH_4 + Al_2X_6 \xrightarrow[100-150°C]{no\ solvent} 2\ Al(BH_4)_3 + 3\ NaX \qquad (b)$$

The tetrahydroborates of Ga are well established, although $Ga(BH_4)_3$ is unstable, and only its decomposition products are isolable[3]:

$$GaCl_3 + LiBH_4 \xrightarrow{15°C} Li[Cl_3Ga \cdot BH_4] \qquad (c)$$

$$Li[Cl_3GaBH_4] \xrightarrow{2\ LiBH_4} Ga(BH_4)_3 + 3\ LiCl \qquad (d)$$

Species containing the BH_4 moiety and other atoms covalently bound to Ga are more stable, and their structures contain the $Ga\overset{\displaystyle H}{\underset{\displaystyle H}{<}}\overset{\displaystyle H}{\underset{\displaystyle H}{>}}B$ moiety[4].

Such species are available from reactions at low T, e.g., either powdered Ga_2Cl_6 or $HGaCl_2$ affords $HGa(BH_4)_2$ with xs $LiBH_4$ at −45°C in the absence of solvent[5]:

$$\tfrac{1}{2}\ Ga_2Cl_6 + 3\ LiBH_4 \xrightarrow{-45°C} HGa(BH_4)_2 + 3\ LiCl + \tfrac{1}{2}\ B_2H_6 \qquad (e)$$

$$HGaCl_2 + 2\ LiBH_4 \xrightarrow{-45°C} HGa(BH_4)_2 + 2\ LiCl \qquad (f)$$

The mixed species $(CH_3)_2GaCl$ and $(CH_3)_3N \cdot GaH_2Cl$, prepared from exchange reactions, are converted to the corresponding borohydrides by treatment with $LiBH_4$ at low T, e.g., $(CH_3)_2GaBH_4$ is prepared[6] in the absence of solvent at −15°C and $(CH_3)_3N \cdot GaH_2(BH_4)$ with xs $LiBH_4$ at low T in benzene or ether[7]:

144   1.7. Formation of Bonds between Hydrogen and B, Al, Ga, In, Tl
         1.7.5. from Complex Hydrides
         1.7.5.2. with Compounds of Al, Ga, In and Tl.

$$(CH_3)_2GaCl + LiBH_4 \xrightarrow{-15°C} (CH_3)_2GaBH_4 + LiCl \qquad (g)$$

$$(CH_3)_3N \cdot GaH_2Cl + LiBH_4 \xrightarrow[\text{low T}]{O(C_2H_5)_2} (CH_3)_3N \cdot GaH_2(BH_4) + LiCl \qquad (h)$$

The unstable $Ga(AlH_4)_3$, which contains the $Ga\underset{H}{\overset{H}{<}}Al\underset{H}{\overset{H}{<}}$ moiety, is available from the reaction[8] between $GaCl_3$ and $LiAlH_4$ in ether at 0°C.

$$GaCl_3 + 3 LiAlH_4 \xrightarrow[0°C]{Et_2O} Ga(AlH_4)_3 + 3 LiCl \qquad (i)$$

The $Ga(AlH_4)_3$ decomposes to $H_3Ga \cdot O(C_2H_5)_2$ and $(AlH_3)_x$ in ether above 0°C:

$$Ga(AlH_4)_3 \xrightarrow[> 0°C]{Et_2O} (C_2H_5)_2O \cdot GaH_3 + \frac{1}{x}(AlH_3)_x \qquad (j)$$

Above 35°C, GaH_3 itself decomposes to the elements.

Compounds containing In—H bonds may also be prepared from InX_3 (X = Cl, Br) and complex hydrides, e.g., at −20°C $InCl_3$ reacts[9] with $LiBH_4$ in ether to form $LiInCl_3(BH_4)$:

$$LiBH_4 + InCl_3 \xrightarrow[-20°C]{Et_2O} LiInCl_3(BH_4) \qquad (k)$$

The structure of $LiInCl_3(BH_4)$ is unknown. It is stable at −40°C but at −20°C loses BH_3 to afford $LiInCl_3H$. Similar reactions occur with $InBr_3$. The $LiInX_3H$ species are $LiX \cdot InX_2H$ in ethers, and in the absence of ether, $LiH \cdot InX_3$.

Ethereal $InCl_3$ reacts with $LiAlH_4$ at −70°C to afford[10] $In(AlH_4)_3$:

$$InCl_3 + 3 LiAlH_4 \xrightarrow[-70°C]{Et_2O} In(AlH_4)_3 + LiCl \qquad (l)$$

The $In(AlH_4)_3$ decomposes above −40°C, to form AlH_3, In and H_2 via the intermediacy of InH_3. When $InCl_3$ is treated at 25°C with 1/3 equiv of $LiAlH_4$, $LiInCl_3(AlH_4)$ is obtained, which is stable ≤ 100°C. The ether insolubility of $In(AlH_4)_3$ and $InCl_2$-(AlH_4) suggests that these species do not contain $In\overset{H}{\diagdown}Al$ bonds.

The Tl—H bond may be formed by decomposition of $TlBH_4$. Treatment of $TlOC_2H_5$ with $LiBH_4$ in ether affords[11] $TlBH_4$. This species is ionic and so does not contain Tl—H bonds[12]; however, at ≥ 40°C, $TlBH_4$ decomposes to B_2H_6 and TlH, and $TlCl_3$ is reduced to TiCl on treatment with $LiBH_4$. If the reaction is carried out at −100°C, an unstable intermediate of unknown structure, $TlCl(BH_4)_2$, forms, which decomposes above −95°C.

A species[13,14] of unknown structure, $Tl(GaH_4)_3$, which decomposes above −90°C, may be prepared from $TlCl_3$ and $LiGaH_4$ in ether at −115°C.

(L. BARTON)

1. S. G. Shore, R. W. Parry, *J. Am. Chem. Soc.,* 80, 12 (1958).
2. H. I. Schlesinger, H. C. Brown, E. K. Hyde, *J. Am. Chem. Soc.,* 75, 209 (1953).
3. E. Wiberg, O. Dittman, H. Nöth, M. Schmidt, *Z. Naturforsch., Teil B,* 12, 56 (1957).
4. M. T. Barlow, A. J. Downs, P. D. P. Thomas, D. W. H. Rankin, *J. Chem. Soc., Dalton Trans.,* 1793 (1979).

5. A. J. Downs, P. D. P. Thomas, *J. Chem. Soc., Chem. Commun.*, 825 (1976).
6. A. J. Downs, P. D. P. Thomas, *J. Chem. Soc., Dalton Trans.*, 809 (1978).
7. N. N. Greenwood, A. Storr, *J. Chem. Soc.*, 3420 (1965).
8. E. Wiberg, M. Schmidt, *Z. Naturforsch., Teil B, 6*, 171 (1951).
9. E. Wiberg, M. Schmidt, *Z. Naturforsch., Teil B, 12*, 56 (1957).
10. E. Wiberg, M. Schmidt, *Z. Naturforsch., Teil B, 6*, 172 (1951).
11. E. Wiberg, O. Dittman, H. Nöth, M. Schmidt, *Z. Naturforsch., Teil B, 12*, 62 (1957).
12. T. C. Waddington, *J. Chem. Soc.*, 4783 (1958).
13. E. Wiberg, H. Noth, *Z. Naturforsch., Teil B, 12, 6*, 63 (1957).
14. E. Wiberg, M. Schmidt, *Z. Naturforsch., Teil B*, 355 (1951).

1.7.6. by Industrial Processes.

The important industrial chemicals containing group III element–hydrogen bonds are $NaBH_4$ and $LiAlH_4$ because all industrial products containing such bonds are derived from these two chemicals.

In the U.S., $NaBH_4$ is prepared[1] commercially from NaH and $B(OCH_3)_3$ in high-bp mineral oil at ca. 275°C:

$$4 \ NaH + (CH_3O)_3B \xrightarrow[275°C]{oil} NaBH_4 + 3 \ NaOCH_3 \qquad (a)$$

The NaH is prepared on site in the $NaBH_4$ production plant, and an oil slurry is fed into a mixing tank into which $B(OCH_3)_3$ also flows. The product slurry is added to H_2O, the aq layer is separated from the oil, and the CH_3OH is evaporated. The resulting solution is 12% $NaBH_4$ in aq NaOH, which is itself an important commercial product; $NaBH_4$ is extracted from this solution, dried and recrystallized to a 97+% pure product[2].

One process employs a high-T, dry reaction of borax, sodium metal, H_2 and sand or borosilicate, Na and H_2O:

$$Na_2B_4O_7 + 16 \ Na + 8 \ H_2 + 7 \ SiO_2 \rightarrow 4 \ NaBH_4 + 7 \ Na_2SiO_3 \qquad (b)$$

The $NaBH_4$ is extracted from the mass with aq NH_3; the resulting solution is fed into a drier that drives off the NH_3 and the $NaBH_4$ crystals are further dried and packaged[2].

The major process for the commercial production[3] of $LiAlH_4$ utilizes the procedure in its original discovery[4,5]:

$$4 \ LiH + AlCl_3 \rightarrow LiAlH_4 + 3 \ LiCl \qquad (c)$$

where $AlCl_3$ as an ether slurry is fed into a reactor to which is added a LiH slurry[6]. An induction period ensues because an oxide forms, coating the LiH. This is overcome by adding seeds of $LiAlH_4$ before the $AlCl_3$ is added. Pure product is obtained from this process.

Other commercial processes include the preparation of $NaAlH_4$ from NaH and $AlCl_3$ followed by metathesis with LiCl, and a process utilizing reaction of the elements Li, Al and H_2 at 14.5 MPa and 100°C in the presence of a catalyst[6]. An extension of this method employs the use of 1:1 Al-Li alloys. In this process the alloy is shaken in tetrahydrofuran (THF) in an autoclave at 120–130°C and 10.1 MPa H_2 for 3 days[7]. Pure $LiAlH_4$ is obtained in 80% yield.

(L. BARTON)

1. H. I. Schlesinger, H. C. Brown, U.S. Pat. 2,534,533 (1950); *Chem. Abstr.*, *45*, 4007 (1951).
2. R. C. Wade, in *Speciality Inorganic Chemicals*, Special Publication No. 40, The Royal Society of Chemistry, London, 1981, p. 25.
3. E. C. Ashby, *Adv. Inorg. Chem. Radiochem.*, *8*, 283 (1966).
4. A. E. Finholt, A. C. Bond, H. I. Schlesinger, *J. Am. Chem. Soc.*, *69*, 1199 (1947).
5. H. I. Schlesinger, A. E. Finholt, U.S. Pat. 2,576,311 (1951); *Chem. Abstr.*, *46*, 2761 (1952).
6. M. Grayson, D. Eckroth, eds., *Kirk-Othmer Encyclopedia of Chemical Technology*, 3rd ed., Wiley-Interscience, New York, Vol. 11, 1980, p. 217.
7. H. Hoffman-La Roche, Br. Pat. 888,045 (1962); *Chem. Abstr.*, *56*, 13,801 (1962).

1.7.7. The Synthesis of Deuterium Derivatives

The reagents $NaBD_4$ and $LiAlD_4$ are available commercially, so compounds containing the deuterium–group III element bonds may be prepared by standard methods using these reagents. The $NaBD_4$ is prepared from NaD and $B(OCH_3)_3$, $LiAlD_4$ is prepared from LiD and $AlCl_3$ and $LiGaD_4$ is prepared from LiD and $GaCl_3$.

(L.BARTON)

1.7.7.1. by Isotopic Exchange.

The $[BH_4]^-$ ion undergoes exchange with D_2 at 5.07 MPa and 500°C to afford randomly deuterium-substituted products[1]:

$$[BH_4]^- \rightleftharpoons BH_3 + H^- \tag{a}$$

$$BH_3 + D_2 \rightleftharpoons BH_2D + HD \tag{b}$$

Also, B_2H_6 may be deuterated by treatment with $D_{2(g)}$. Successive treatments for 24 h at 75°C afford[2] 98% B_2D_6. Kinetics[3] between 25 and 75°C reflect a $\frac{3}{2}$ order with respect to B_2H_6 and zero order with respect to D_2. The reaction proceeds between BH_3 and D_2 on the walls of the reaction vessel with an activation energy of 91.3 kJ mol^{-1}. The mechanism is:

$$B_2H_6 \rightleftharpoons 2 BH_3 \tag{c}$$

$$BH_3 + D_2 \rightleftharpoons BH_2D + HD \tag{d}$$

$$BH_2D + B_2H_6 \rightleftharpoons B_2H_5D + BH_3 \tag{e}$$

Exchange also occurs[4] with T_2, DT and TH. Self-exchange between B_2H_6 and B_2D_6 occurs, wherein rapid equilibrium between B_2H_6 and 2 BH_3 is followed by a rate-determining collision between B_2D_6 and a BH_3 molecule[6,7]. Exchange between B_4H_{10} and B_2D_6 at 45°C may proceed by rapid selective deuteration at the 1,3-terminal positions[8], followed by a slower process involving all other positions; however, D is found in the 2,4- and bridge positions[9]. Self-exchange in μ-DB_4H_9 proceeds rapidly at the 1,3- and 2,4-(either axial or equatorial, but not both)-positions and more slowly at the remaining 2,4- and bridge-positions. The numbering system in B_4H_{10} is shown in Fig. 1.

Perdeuterated B_5H_9 may be prepared[5] by reaction between B_5H_9 and D_2 in the presence of Cr_2O_3–Al_2O_3 dehydrogenation catalysts at 25°C. By passage of B_5H_9 through the catalyst chamber in the presence of a 5:1 xs of D_2, B_5D_9 in which 98.5% of the original H has been replaced by D is obtained after five passes.

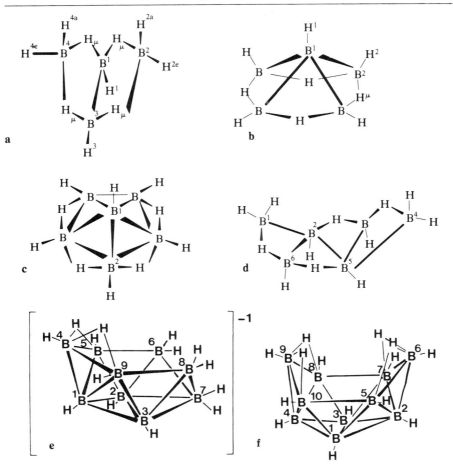

Figure 1. Atomic numbering scheme for borane species: a, B_4H_{10}; b, B_5H_9; c, B_6H_{10}; d, B_6H_{12}; e, $[B_9H_{14}]^-$; f, $B_{10}H_{14}$. Where not indicated, terminal H atoms bear the same numbers as the B atoms to which they are bonded.

Pentaborane(9) and SiD_4 undergo hydrogen transfer at 125°C. The apical position in B_5H_9 is preferentially deuterated, but some traces of D also exchange with basal terminal hydrogens[10]. With B_2D_6 at 80°C, B_5H_9 undergoes rapid exchange of H for D in a process in which only terminal hydrogens are replaced[11,12]. The rate for basal-terminal H–D exchange is the same as for apical-hydrogen exchange[13]. At 45°C exchange of D for H between the bridging and basal terminal positions occurs in $(C_2H_5)_2O$ solns; however, similar exchange between the apical and basal terminal positions does not occur[14], even at 95°C after 15 h. This thermal rearrangement is a high-barrier process and 145–200°C is required for the gas-phase reaction. Both the intra- and intermolecular processes require[15] similarly high T.

Hexaborane(10) may be deuterated selectively in the basal terminal positions by treatment with B_2D_6 in ether at −20°C. This is in contrast to the higher T required[16,17] for B_5H_9. Some H–D exchange occurs in the bridging position in B_6H_{10}, but either higher T or catalysts may be involved[18]. Hexaborane(12) may be deuterated selectively by treatment with B_2D_6 at $-31°C$ to afford 1,1,4,4-tetradeuteriohexaborane(12). Further reaction with liq B_2D_6 at 25°C results[19] in the complete deuteration of B_6H_{12}. Again, atom positions for B_6H_{10} and B_6H_{12} are given in Fig. 1.

Treatment of B_8H_{12} with liq B_2D_6 at $-30°C$ forms n-B_9H_{15} and $B_{10}H_{14}$, which reflect extensive nonspecific deuteration[20]. This reaction proceeds initially by exchange of H for D in B_8H_{12}, and then the $B_8H_nD_{12-n}$ reacts more slowly with B_2D_6 to give n-$B_9H_nD_{15-n}$. The reaction of n-B_9H_{15} with liq B_2D_6 at 27°C yields a nonspecifically deuterated $B_{10}H_{14}$ and a deuterio-n-B_9H_{15} with at least one terminal site not deuterated. This latter reaction does not occur at $-30°C$; however, $B_{10}H_{14}$, formed from the reaction between i-B_9H_{15} and B_2D_6 at $-30°C$, is deuterated extensively.

Exchange between $B_{10}H_{14}$ and B_2D_6 occurs only in processes involving the terminal hydrogens[21].

(L. BARTON)

1. E. E. Mesmer, W. L. Jolly, *J. Am. Chem. Soc.,* 84, 2039 (1962).
2. A. B. Burg, *J. Am. Chem. Soc.,* 74, 1340 (1952).
3. P. C. Maybury, W. S. Koski, *J. Chem. Phys.,* 21, 742 (1953).
4. J. S. Rigden, W. S. Koski, *J. Am. Chem. Soc.,* 83, 3037 (1961).
5. H. J. Hrostowski, G. C. Pimentel, *J. Am. Chem. Soc.,* 76, 998 (1954).
6. W. S. Koski, in *Borax to Boranes,* D. L. Martin, ed., Advances in Chemistry Series No. 32, American Chemical Society, Washington, DC, 1961, p. 78.
7. I. Shapiro, B. Keilen, *J. Am. Chem. Soc.,* 77, 2663 (1955).
8. J. E. Todd, W. S. Koski, *J. Am. Chem. Soc.,* 81, 2319 (1959).
9. R. Schaeffer, L. G. Sneddon, *Inorg. Chem.,* 12, 3098 (1972).
10. M. L. Thompson, R. Schaeffer, *Inorg. Chem.,* 7, 1677 (1968).
11. W. S. Koski, J. J. Kaufman, L. Friedman, A. P. Irsa, *J. Chem. Phys.,* 24, 221 (1956).
12. J. J. Kaufman, W. S. Koski, *J. Chem. Phys.,* 24, 403 (1956).
13. W. S. Koski, J. J. Kaufman, P. C. Lauterbur, *J. Am. Chem. Soc.,* 79, 2382 (1957).
14. J. A. Heppert, D. F. Gaines, *Inorg. Chem.,* 22, 3155 (1983).
15. T. P. Onak, F. J. Gerhart, R. E. Williams, *J. Am. Chem. Soc.,* 85, 1754 (1963).
16. J. C. Carter, N. L. H. Mock, *J. Am. Chem. Soc.,* 91, 5891 (1969).
17. J. D. Odom, R. Schaeffer, *Inorg. Chem.,* 9, 2157 (1970).
18. R. E. Williams, S. G. Gibbins, I. Shapiro, *J. Chem. Phys.,* 30, 353 (1959).
19. A. L. Collins, R. Schaeffer, *Inorg. Chem.,* 9, 2153 (1970).
20. R. Maruca, J. D. Odom, R. Schaeffer, *Inorg. Chem.,* 7, 412 (1968).
21. J. J. Kaufman, W. S. Koski, *J. Am. Chem. Soc.,* 78, 5774 (1956).

1.7.7.2. by Isotopic Enrichment Using Chemical Reactions.

Fully deuterated B_2H_6 may be obtained by treatment[1] of commercially available $NaBD_4$ with hot H_3PO_4 and also by treatment[2,4] of $LiAlD_4$ with $F_3B·O(C_2H_5)_2$.

Tetraborane(10) deuterated in a bridge position is prepared[5] by treatment of $K[B_4H_9]$ with DCl at $-78°C$. When the reaction mixture is warmed over a period of several hours, deuteration at terminal positions is also evident[5,6]. The species μ-DB_4H_9 is also prepared[7] by cleavage of B_5H_{11} with D_2O in a sealed tube at 0°C:

$$B_5H_{11} + 3\ D_2O \xrightarrow[\text{2 min}]{0°C} \mu\text{-}DB_4H_9 + B(OD)_3 + 2\ HD \qquad (a)$$

Treatment of B_4H_8CO with D_2 at 0.2 kPa affords 1,μ-dideuteriotetraborane(10) in yields[8] of $\leq 20\%$:

$$B_4H_8CO + D_2 \rightarrow \mu,1\text{-}D_2B_4H_8 + CO \qquad (b)$$

Pentaborane(9) may be deuterated selectively at the apical position by treatment with DCl in the presence of $AlCl_3$ at RT. 1-Deuteration is complete after 2 h, and 1-DB_5H_8 containing 90% D at the 1-position may be prepared[9]:

$$B_5H_9 + DCl \xrightarrow[25°C]{AlCl_3} 1\text{-}DB_5H_8 + HCl \qquad (c)$$

This reaction does not occur in the absence of $AlCl_3$, nor may 1-$CH_3B_5H_8$ be deuterated by this process[9]. The 1-DB_5H_8 also may be prepared[10] by treatment of B_5H_9 with C_6D_6 in the presence of a catalytic amount of $AlCl_3$ at 25°C for 24 h. The method is quantitative, and pure 1-DB_5H_8 is isolated by trap-to-trap distillation at $-78°$ and $-196°C$, the desired product collecting in the colder trap. Pentaborane(9) may be deuterated selectively[11] in a bridging position by treatment of $Li[B_5H_8]$ with DCl at $-78°C$. Intramolecular H–D exchange[12] takes place in 1-DB_5H_8 catalyzed by 2,6-dimethylpyridine at 25°C. The intermediate in the process is an electron-pair base adduct, and exchange occurs with all H positions in the B_5H_8 moiety.

The 2-DB_5H_8 derivative is prepared by treatment of 2-ClB_5H_8 with $(n\text{-}C_4H_9)_3SnD$ in a sealed tube under N_2. The reactor is allowed to warm from $-78°$ to 25°C over several hours. The product is isolated by vacuum line distillation through $-63°$, $-96°$ and $-196°C$ U-traps. The 2-DB_5H_8 is isolated in the $-96°C$ trap[10].

Hexaborane(10) may be prepared with nonspecific deuteration by shaking B_6H_{12} with a fourfold xs of D_2O in $(C_2H_5)_2O$ for 1 h. The average composition of the product[13] is $B_6H_{7.3}D_{2.7}$.

The B ion in $K[B_9H_{14}]$ undergoes acid-catalyzed exchange of hydrogens at the 4-, 6- and 8-positions[14], e.g., when $K[B_9H_{14}]$ in D_2O is made 10^{-3} M in DCl, deuteration at the 4-, 6- and 8-positions occurs within 20 min (see Fig. 1, §1.7.7.1, for numbering scheme). Proton NMR studies indicate that bridge hydrogens do not exchange under these conditions; however, after a period of several hours at 25°C some collapse of the bridge-region resonances is observed, indicating slow μ-H–D exchange by a secondary process. In solutions 10^{-3} M in NaOD, complete, rapid exchange of bridge hydrogens occurs. When 1 M NaOD is used, both the bridge and the 4-, 6- and 8-terminal positions deuterate. Exchange between bridge and terminal positions in this anion involves only the 4-, 6- and 8-terminal positions; D originally in bridge positions exchanges[14] with terminal H at positions 4, 6 and 8 over a period of 3–4 h.

Decaborane(14) may be prepared with selective deuteration in several different positions. Treatment of $B_{10}H_{14}$ in dioxane with D_2O at 25°C results in initial substitution in the bridge position[15–17]. After ca. 10 min, flash evaporation of the solution affords μ-$D_4B_{10}H_{10}$ containing an average of 3.5 D per molecule. After extended periods, exchange occurs between the bridge and the terminal hydrogen positions; in no case are more than eight D atoms incorporated. Both the 5,7,8,10- and the 6,9-terminal positions participate in this base-catalyzed exchange. The relative rates[18] of base-catalyzed H–D exchange in $B_{10}H_{14}$ are 6,9 > 5,7,8,10 > 1,3 and 2,4. Treatment in ether of the salt $Na[B_{10}H_{13}]$, prepared from NaH and $B_{10}H_{14}$, with DCl affords μ-$DB_{10}H_{13}$ quantitatively[16]; DCl also reacts with $B_{10}H_{14}$ in basic solvents at 25°C to effect substitution of D for H at the bridge and 5,6,7,8,9,10-terminal positions. Thus, dioxane 5.0 M in DCl and 5×10^{-2} M in $B_{10}H_{14}$, left to equilibrate at 25°C for 404 h, affords $B_{10}D_{10}H_4$ in which the bridge and 5,6,7,8,9,10-terminal positions are deuterated completely[20,22]. Shorter reaction times allow the μ-$D_4B_{10}H_{10}$ species to be isolated[19,20]. The species μ-1,2,3,4-$D_8B_{10}H_6$ is prepared[21] by shaking μ-$D_4B_{10}H_{10}$ in a D_2O–CH_3CN (1:1 by vol) solution for 20 min. It is possible to prepare $B_{10}H_{14}$ with deuterium at only the 6-

and 9-positions. The reaction between μ-$D_4B_{10}H_{10}$ and $NaBD_4$ in glyme under N_2 affords$[B_{10}H_{10}D_5]^-$. Treatment with DCl affords $B_{10}H_8D_6$ and H_2, and reaction with H_2O–dioxane affords[23] 6,9-$D_2B_{10}H_{12}$:

$$\mu\text{-}D_4B_{10}H_{10} + [BD_4]^- \rightarrow [B_{10}H_{10}D_5]^- + BH_3 \tag{d}$$

$$[B_{10}H_{10}D_5]^- + D^+ \rightarrow H_2 + \mu\text{-}6,9\text{-}D_6B_{10}H_8 \tag{e}$$

$$\mu\text{-}6,9\text{-}D_6B_{10}H_8 + 2\ H_2O \rightarrow 6,9\text{-}D_2B_{10}H_{12} \tag{f}$$

Electrophilic deuterium exchange may be effected[20] in $B_{10}H_{14}$ by xs DCl in CS_2 in the presence of $AlCl_3$ to give 1,2,3,4-$D_4B_{10}H_{10}$ after 36 h. Repetition four times allows 2,4,5,7,8,10-$D_6B_{10}H_8$ to be isolated[19]. A similar reaction allowed to proceed for several days affords[21] 1,2,3,4,5,7,8,10-$D_8B_{10}H_5$. When HCl is bubbled through 1,2,3,4,5,7,8,10-$D_8B_{10}H_6$ in CS_2, in the presence of $AlCl_3$ for 36 h, the species 5,7,8,10-$D_4B_{10}H_{10}$ may be isolated. If μ-$D_4B_{10}H_{10}$ is allowed to react with an xs of DCl in CS_2 in the presence of $AlCl_3$ at 25°C for 6 days, μ,1,2,3,4,5,7,8,10-$D_{12}B_{10}H_2$ may be isolated[21]. The labeling scheme for $B_{10}H_{14}$ is found in Fig. 1, §1.7.7.1.

(L. BARTON)

1. R. Maruca, J. D. Odom, R. Schaeffer, *Inorg. Chem.*, 7, 412 (1968).
2. J. E. Todd, W. S. Koski, *J. Am. Chem. Soc.*, 81, 2139 (1959).
3. A. D. Norman, R. Schaeffer, A. B. Bayliss, G. A. Pressley Jr., F. E. Stafford, *J. Am. Chem. Soc.*, 88, 2151 (1966).
4. R. Schaeffer, L. G. Sneddon, *Inorg. Chem.*, 11, 3098 (1972).
5. A. C. Bond, M. L. Pinsky, *J. Am. Chem. Soc.*, 92, 7585 (1970).
6. A. C. Bond, M. L. Pinsky, *Inorg. Chem.*, 12, 605 (1963).
7. A. D. Norman, R. Schaeffer, *Inorg. Chem.*, 4, 1225 (1965).
8. A. D. Norman, R. Schaeffer, *J. Am. Chem. Soc.*, 88, 1143 (1966).
9. T. P. Onak, R. E. Williams, *Inorg. Chem.*, 1, 106 (1962).
10. J. A. Heppert, D. F. Gaines, *Inorg. Chem.*, 22, 3155 (1983).
11. D. F. Gaines, T. V. Iorns, *J. Am. Chem. Soc.*, 89, 3375 (1967).
12. T. P. Onak, F. J. Gerhart, R. E. Williams, *J. Am. Chem. Soc.*, 85, 1754 (1963).
13. J. D. Odom, R. Schaeffer, *Inorg. Chem.*, 9, 2157 (1970).
14. P. C. Keller, *Inorg. Chem.*, 9, 75 (1970).
15. M. F. Hawthorne, J. J. Miller, *J. Am. Chem. Soc.*, 80, 754 (1958).
16. J. J. Miller, M. F. Hawthorne, *J. Am. Chem. Soc.*, 81, 4501 (1959).
17. M. Hillman, *J. Am. Chem. Soc.*, 82, 1096 (1960).
18. I. Shapiro, M. Lustig, R. E. Williams, *J. Am. Chem. Soc.*, 81, 838 (1959).
19. J. A. Dupont, M. F. Hawthorne, *J. Am. Chem. Soc.*, 81, 838 (1959).
20. J. A. Dupont, M. F. Hawthorne, *J. Am. Chem. Soc.*, 84, 1804 (1962).
21. F. Hanousek, B. Stibi, S. Hermanek, J. Plesek, *Coll. Czech. Chem. Commun.*, 38, 1312 (1973).
22. E. Hoel, M. F. Hawthorne, *J. Am. Chem. Soc.*, 97, 6388 (1973).
23. J. A. Slater, A. D. Norman, *Inorg. Chem.*, 10, 205 (1971).

1.8. Formation of Bonds between Hydrogen and Metals of Group IA (Li, Na, K, Rb, Cs, Fr) or IIA (Be, Mg, Ca, Sr, Ba, Ra)

1.8.1. Introduction

Compounds in which hydrogen is bonded to the alkali and alkaline-earth metals except beryllium are prepared by direct synthesis from the metals or amalgams. Beryllium hydride is prepared by pyrolysis or reduction of organic derivatives.

(J. J. ZUCKERMAN, ED.)

1.8.2. Alkali-Metal Hydrides

The alkali-metal hydrides are colorless solids that crystallize in the NaCl cubic system. They are saline or ionic hydrides containing an M^+ cation (M = Li, Na, K, Rb, Cs) and the H^- hydride anion, the dimensions of which are comparable to those of the fluoride ion[1].

Preparation is by direct synthesis. The H_2 must be free of O_2 and of H_2O, which react rapidly and irreversibly with the alkali metals and their hydrides. Although less reactive, the volatile hydrides present in H_2 (H_2S, NH_3, CH_4, C_2H_2) also should be eliminated. Finally, in the synthesis of LiH, the presence of N_2 must be avoided.

The conditions for preparing LiH are different from those of the other alkali hydrides because of its high stability, making it similar to the alkaline-earth hydrides. This hydride, therefore, is discussed separately.

(A. HEROLD, J.F. MARECHE)

1.8.2.1. Lithium Hydride

Lithium hydride is an industrial product utilized in organic synthesis, preparation of other hydrides, etc. The hydride, deuteride and tritide of Li play a role in the nuclear industry.

(i) Direct Synthesis. This method differs little from that first utilized[2] in 1896. The Li, contained in Fe, is heated under H_2. The reaction starts at RT but becomes rapid only above the mp of the metal. As shown by the fusion diagram of the Li–LiH system[3] (Fig. 1), the hydride is partially soluble in the liq metal, which helps the reaction. It is also favored by the density of LiH ($0.7-0.8$ g/cm^{-3}) which is greater than that of Li (~ 0.5 g/cm^{-3}), so that LiH falls to the bottom of the reactor and frees the surface of the metal. Nevertheless, to obtain complete hydrogenation, T must be higher than the mp of the hydride (689°C), but lower than the T of the monotectic plateau of the Li–LiH system.

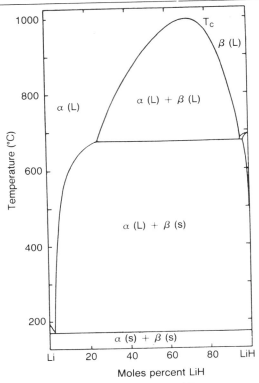

Figure 1. Diagram of the Li–LiH system.

At this T, the H_2 pressure for the pure hydride is only 23.7 torr (3.16×10^3 Pa) and, based on the equations of ref. 4, atm P is attained only at 889°C.

The hydride therefore, can be melted easily under H_2 without decomposing.

Industrially, the operation is carried out on 20–30 kg of metal at a time[3]. The metal is melted under vacuum; then the H_2 is progressively introduced. The heat given off by:

$$Li_{(l)} + \tfrac{1}{2} H_{2(g)} \rightarrow LiH_{(s)} \qquad (a)$$

is ca. 80 kJ/mol^{-1}, sufficient to maintain the T without external heating. The efficiency of converting the metal into the hydride is $\geq$ 98%.

Single crystals are prepared starting from the melted hydride[5-7], or also by growth in organic solvents, using three methods:

1. Vertical drawing from a bath of melted hydride using a metallic rod[5].
2. Vertical downward movement of a cylindroconical crucible containing the molten hydride from a hot zone to a cold one; the crystallization begins at the point of the cone and progresses throughout the whole crucible[6,7].
3. Growth in solvents, the best being CCl_4, which dissolves 2×10^{-4} g L^{-1} of hydride[8].

(ii) Other Syntheses. Complete hydrogenation of the metal can be obtained below 350°C in the presence of catalysts[9] such as WS_2 or MoS_2, but these introduce impurities into the product.

Lowering the T to below the mp of the metal by the use of an electric discharge[10] under low P is not a practical means of preparation.

More practical is the synthesis of the hydride under an H_2 current by heating an amalgam obtained by the electrolysis of a solution of chlorides with an Hg cathode[11]. The Li combines with the H_2, whereas the Hg is carried out and condensed. This method, which avoids preparation of the metal by high-T electrolysis, is only at the testing stage.

(iii) Lithium Deuteride and Tritide. Although the dissociation P of Li deuteride and tritide are different from those of the protide, the preparation conditions are the same[12,13].

The tritide is special because the isotope has a half-life of 12.26 y, and LiT can be obtained by isotopic exchange:

$$LiH + HT \rightarrow LiT + H_2 \tag{b}$$

or by neutron bombardment[14-16] of LiH.

(A. HEROLD, J.F. MARECHE)

1. C. B. Magee, in *Metal Hydrides*, W. M. Mueller, ed., Academic Press, New York, 1968, p. 165.
2. A. Guntz, *C. R. Hebd. Seances Acad. Sci. 122*, 244; *123*, 1273 (1896).
3. C. E. Messer, E. B. Damon, P. C. Maybury, J. Mellor, R. Seales, *J. Phys. Chem., 62*, 220 (1958).
4. C. E. Messer, *USAEC Report* NYO-9470, Tufts Univ., (1960). A report on LiH.
5. J. Tuffier, S. Bedere, *C. R. Hebd. Seances Acad. Sci., Ser. C, 280*, 337 (1975).
6. V. D. Pirogev, S. O. Cholakh, F. F. Gavilov, G. I. Philipenko, B. V. Schudgin, S. I. Somov, N. I. Kanunnikov, B. V. Vlasov, V. G. Ovechkin, *Fiz. Khim. Gidridov, 155*, 72 (1972); *Chem. Abstr., 83*, 185,69 (1975).
7. C. E. Holcombe, D. H. Johnson, *J. Cryst. Growth, 19*, 53 (1973).
8. R. Suchansko, K. Kulichi, W. Cetner, *Biul. Wojsk. Akad. Tech., 23*, 538 (1974); *Chem. Abstr., 82*, 205,257 (1975).
9. S. Landa, F. Petri, J. Vit, V. Prochazka, J. Mostecky, *Sb. Vys. Sk. Chem.-Technol. Praze, Oddil Fak. Anorg. Org. Technol.*, 495 (1958); *Chem. Abstr., 55*, 6225 (1961).
10. United Aircraft Corp., Br. Pat. 1,004,769 (1960); *Chem. Abstr., 63*, 15,893 (1965).
11. J. Novotny, M. Skolova, Czech. Pat. 86,588 (1957); *Chem. Abstr., 54*, 9229 (1960).
12. F. K. Neumann, O. N. Salmon, USAEC Report KA PL 1667, Knolls Atomic Power Laboratory, 1956; *Chem. Abstr., 51*, 9280 (1957).
13. E. Welekis, *J. Nucl. Mater., 79*, 20 (1979); *Chem. Abstr., 90*, 93,904 (1979).
14. V. A. Maroni, E. Velekis, E. H. Van Deventer, *Proc. Sympos. Tritium Technol., Relat. Fusion Reactor Systs.*, p. 120 (1974); *Chem. Abstr., 85*, 130,698 (1976).
15. L. A. Zaputryacva, O. N. Pavlov, A. T. Uverskaya, M. S. Fadecva, *Trudy Gos. Inst. Prikl. Khim., 45*, 97 (1960); *Chem Abstr., 56*, 2141 (1962).
16. M. Ziclinsky, *Nukleonika 7*, 789 (1962); *Chem. Abstr., 59*, 5015 (1963).

1.8.2.2. Sodium, Potassium, Rubidium and Cesium Hydrides

These four compounds form a homogeneous group, and their means of preparation are similar.

Sodium hydride is used industrially in organic synthesis; the annual production is several million tons. Utilized much less, KH is also commercial, whereas RbH and CsH are laboratory products.

TABLE 1. KINETICS OF THE REACTION OF H_2 WITH THE
ALKALI METALS (v = SPEED OF THE REACTION AND 100°C AND 1 ATM).

Metal	Na	K	Rb	Cs
v at 100°C	0.00061	0.0105	0.062	0.53
$E_t(kJ)$[a]	71.5	49.4	39.6	27.7

[a] Activation energy.

(i) M + $\frac{1}{2}$ H$_2$ → MH (M = Na, K, Rb, Cs). The combination of Na and K with
H$_2$ is reversible[1]. The data in Table 1 give the orders of magnitude of the rates when
dissociation of the hydrides is negligible[2]. At 100°C, the speed rapidly increases Na <
K < Rb < Cs, and the activation energy varies in the opposite order.

The four hydrides are dissociated easily by heat. Based on ref. 2, Table 2 gives the
enthalpies of formation of 1 mol of solid hydride starting from H$_2$ gas and the liq metal deduced from the equilibrium P. It also gives the T, θ, in °C, for which this P attains 1 atm. Less stable than LiH, these four hydrides dissociate without melting; they
are only slightly soluble in the metals at θ°C.

The system Na–NaH under P at higher T shows a miscibility of the two phases[3].
At sufficiently high P and T the hydrides can be melted[4]: 107.3 atm and 638°C for
NaH; 67.3 atm and 619°C for KH; 147.5 atm and 585°C for RbH and 31.5 atm and
528°C for CsH.

These data explain the difficulties encountered in completing the reaction between
H$_2$ and the alkali metals other than Li. Under atm P, the hydride, which does not wet
the metal or dissociate, slowly forms a superficial layer, the low permeability of which
stops the reaction. The synthesis of pure hydrides however, is possible by vapor- and
liquid-phase methods.

(ii) Direct Vapor-Phase Synthesis of the Hydrides. The homogeneous synthesis
can be carried out by heating the alkali metal above the hydride dissociation T under a
flow of H$_2$; the vapor is carried away by the gas and combines with the H$_2$ in a cooler
zone of the reactor, giving hydride particles that remain in suspension in the gas. Elec-
trostatic filtering[5] can be used to separate them, but it is more efficient to clean the gas
in a column using an organic liquid[6].

Pure hydrides can be obtained in the laboratory by reacting the metal vapor with
H$_2$ in contact with a wall and, therefore, in a heterogeneous phase[2]. The apparatus in
Fig. 1 is derived from that used to prepare pure hydrides[7]. The reactor is a horizontal
borosilicate glass tube, a few cm in diameter; the lower portion is heated by a half-cy-
lindrical oven. The T = T$_1$ of the metal contained in an Fe or Ni nacelle is lower than
the dissociation T of the hydride, so that the hydride covers the metal. The system is
not in equilibrium, however, because of the gradient between T$_1$ and T$_2$ of the upper
wall of the tube (T$_1$ > T$_2$). The hydride layer constantly gives off vapor, which com-

TABLE 2. THERMODYNAMIC DATA[2] FOR THE REACTION $M_{(L)}$ + $\frac{1}{2}$ $H_{2(G)}$ ⇌ $MH_{(S)}$

Metal	Na	K	Rb	Cs
$-\Delta H$ (kJ)	58.4	59.1	54.4	56.4
θ°C (1 atm)	420	427	364	389

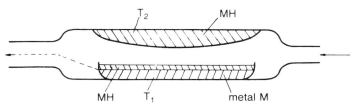

Figure 1. Direct synthesis of the alkali-metal hydrides by vapor-phase reaction in a T gradient.

bines with the H_2 on contact with the upper wall or with the hydride that covers it. At the same time that it dissociates in the upper portion, the hydride layer covering the metal is reformed from the latter, which can be transformed entirely into hydride deposited on the upper wall of the reactor. The nacelle can be withdrawn, e.g., with a magnet, and the tube sealed.

This process, also utilized for the synthesis of deuterides[8], allows small quantities of pure hydrides to be obtained in the laboratory. The reaction speed, low for Na, a metal of low volatility, increases with the T_1 of the metal. This can be increased without dissociating the hydride layer at the metal surface here (the hydride formed on the upper wall of the tube is mixed with metal) by increasing the P to 2 or 3 atm.

(iii) Direct Liquid-Phase Synthesis of the Hydrides. As with Li, the other melted alkali metals can be hydrogenated[9] above their mp, for NaH at 636–670°C and 68–100 atm; for CsH, 529–550°C, under 33–50 atm. The bulk hydride is obtained after cooling.

Below the mp of the hydride, the metal–gas contact surface, covered with an impermeable layer of hydride, must be renewed constantly for complete hydrogenation. This is done through agitation in the presence of a solid or liquid dispersant.

The best solid dispersant is the hydride itself. Liquid metal is introduced progressively into the agitated reactor containing the hydride from a previous preparation at 300°C under 20–30 atm H_2. At the end of the operation, half the hydride in the reactor is withdrawn in order to restart the operation[10-12]. The hydrogenation efficiency is 95–97%. This industrial technique furnishes hydride free of organic compounds.

The specific surface of the hydride can be increased[13] by mixing CO with the H_2 and adding[14] 0.1% MoS_2 and WS_2 lowers the reaction T.

The simplest and most used industrial synthesis of NaH and KH consists in reacting 2–3 atm of H_2 (at ca. 300°C) with a dispersion of 50% Na or 25% K (by wt) in an inert liquid (mineral oil or kerosene), also at ca. 300°C[11,15,16]. The conversion efficiency can reach 98%. The hydride can be conserved and manipulated in the form of a dispersion in the inert liquid. To obtain it dry, it is filtered and washed with a volatile solvent, such as petroleum ether. The synthesis can be accelerated by adding activating agents to the oil, such as anthracene or ortho-xylene, which act through the intermediate formation of π complexes with the alkali metal.

This method is employed for NaH in the laboratory at 200–220°C, using cyclic or noncyclic saturated hydrocarbons (C_7–C_9) as dispersing agents.

Hydrogen acts[17] under pressure on the amides MNHR (M = Na, K) derived from primary amines C_1–C_4 at ca. 90–100°C.

(iv) Sodium Hydride Synthesis Starting from an Amalgam. The industrial preparation of NaH starting from an amalgam[18,19] is not used because of the low concentration of Na in the liquid amalgams.

(v) Synthesis Starting from Carbonates or Hydroxides. The reactions[20,21]:

$$M_2CO_3 + 2 C + H_2 \rightarrow 2 MH + 3 CO \qquad (a)$$

$$2 NaOH + C + H_2 \rightarrow 2 NaH + CO + H_2O \qquad (b)$$

are not utilized.

(A. HEROLD, J.F. MARECHE)

1. J. L. Gay Lussac, L. J. Thenard, *Recherches Physicochim.*, *1*, 176 (1811).
2. A. Herold, *Ann. Chim. (Paris)*, *6*, 536 (1951).
3. M. D. Banus, J. J. McScharry, E. A. Sullivan, *J. Am. Chem. Soc.*, *77*, 2009, (1955).
4. O. A. Skuratov, O. N. Pavlov, V. I. Danilkin, I. V. Volkov, *Russ. J. Inorg. Chem. (Engl. Transl.)*, *21*, 2910 (1976).
5. G. F. Huttig, and F. Brodkorb, *Z. Anorg. Allg. Chem.*, *161*, 353 (1927).
6. H. Cooper, U.S. Pat. 3,495,938 (1970); *Chem. Abstr.*, *72*, 80,959 (1970).
7. H. Moissan, *C. R. Hebd. Seances Acad. Sci.*, *134*, 18 (1902); *Ann. Chim. Phys.* 27, 355 (1902).
8. L. Hackspill, Borocco, *Bull. Soc. Chim. Fr.*, *6*,91 (1939).
9. O. A. Skuratov, O. N. Pavlov, I. V. Volkov, V. I. Damilkin, USSR Pat. 519,393 (1976); *Chem. Abstr.*, *85*, 110,594 (1976); 79,738 (1976).
10. M. D. Banus, A. A. Hickley, *Handling and Use of Alkali Metals*, M. Sittig, ed., Advances in Chemistry Series, No. 19, American Chemical Society Washington, DC, 1957, p. 106.
11. T. R. P. Gibb Jr., *Prog. Inorg. Chem.*, *3*, 315 (1962).
12. A. Lenz, W. Rogler, Br. Pat. 1,309,806 (1973); *Chem. Abstr.*, *79*, 33,223 (1973).
13. H. Nawich, V. Prochazka, P. Kritz, S. Tlaskal, *Khim. Prom. (Moscow)*, *23*, 615 (1973); *Chem. Abstr.*, *80*, 64,248 (1974).
14. S. Landa, F. Petri, J. Mostecky, J. Vit, V. Prochazka, *Chem. Listy*, *52*, 1357 (1958); *Chem. Abstr.*, *58*, 8905 (1959).
15. V. N. Mikheeva, T. N. Dymova, M. M. Shkrabkina, *Russ. J. Inorg. Chem. (Engl. Transl.)*, *4*, 709 (1959).
16. M. D. Banus, E. R. Winiarczyk, H. W. Lambe, Br. Pat. 877,510 (1959); *Chem. Abstr.*, *56*, 11,216 (1962). U.S. Pat. 2,898,195 (1959); *Chem. Abstr.*, *53*, 20,720 (1959).
17. R. U. Lenieux, E. C. Sanford, J. F. Prescott, Ger. Offen. 2,613,113 (1976).
18. G. L. Cunningham, U.S. Pat. 2,829,950 (1958); *Chem. Abstr.*, *52*, 13,207 (1958).
19. G. L. Cunningham, U.S. Pat. 2,867,568, (1968); *Chem. Abstr.*, *53*, 8895 (1959).
20. G. F. Huff, U.S. Pat. 2,884,311 (1959); *Chem. Abstr.*, *53*, 14,438 (1959).
21. A. Lichte, S. Afr. Pat. 6,900,840 (1969); *Chem. Abstr.*, *72*, 102,209 (1970).

1.8.3. Alkaline-Earth Metal Hydrides

Calcium, Sr, Ba and presumably Ra react directly with H_2 to form ionic hydrides that have the orthorhombic $PbCl_2$ structure resembling the heavier alkaline-earth halides (e.g., $BaCl_2$, $BaBr_2$, BaI_2, $SrBr_2$).

Beryllium hydride is covalent, similar to hydrides formed by the group IB–VB elements. Magnesium hydride is a borderline case; it can be formed by direct reaction of Mg metal and H_2 as well as by the procedures used to prepare covalent hydrides (see below).

(G.G. LIBOWITZ)

1.8.3.1. Beryllium Hydride

Beryllium hydride cannot be formed by direct reaction with H_2. Techniques used to prepare solid, amorphous Be hydride never yield pure BeH_2. Pyrolysis of di-t-butylberyllium etherate yields a product containing 97 mol% BeH_2, the remainder being mostly ether[1]:

$$[(CH_3)_3C]_2Be \xrightarrow{200°C} BeH_2 + 2\ (CH_3)_2C=CH_2 \tag{a}$$

If ether-free di-t-butylberyllium is pyrolyzed, the remaining impurity contains unreacted t-butyl groups[2]. Beryllium hydride is also prepared by reacting $LiAlH_4$ with dimethylberyllium[3]:

$$LiAlH_4 + 2\ (CH_3)_2Be \xrightarrow{Et_2O} 2\ BeH_2 + LiAl(CH_3)_4 \tag{b}$$

and by removing BH_3 from $Be(BH_4)_2$ with triphenylphosphine[4]:

$$Be(BH_4)_2 + 2\ Ph_3P \rightarrow 2\ Ph_3PBH_3 + BeH_2 \tag{c}$$

Residual impurities consisting of solvent, unreacted species or products of side reactions are present, although there are claims of pure BeH_2 from processes using Eq. (c)[5].

Beryllium hydride prepared by these methods is an amorphous polymer, and IR spectra[4] indicate long-chain bridge bonding.

Crystalline BeH_2 is formed[6] by high-P compaction–fusion of amorphous BeH_2 (prepared by pyrolysis of di-t-butylberyllium etherate). Pressures ≥ 275 MPa (T $\geq$ 200°C) and temperatures $> 130°C$ (P ≥ 620 MPa) are required. Crystallization is catalyzed by 0.5–2.5 mol% Li. A hexagonal phase is observed below 195°C, and a phase of unknown structure is predominant above that T.

(G.G. LIBOWITZ)

1. G. E. Coates, F. Glockling, *J. Chem. Soc.*, 2526 (1954).
2. E. C. Head, C. E. Holley Jr., S. W. Rabideau, *J. Am. Chem. Soc.*, 79, 3687 (1957).
3. G. D. Barbaras, C. Dillard, A. E. Finholt, T. Wartik, K. E. Nilzbach, H. I. Schlesinger, *J. Am. Chem. Soc.*, 73, 4585 (1951).
4. L. Banford, G. E. Coates, *J. Chem. Soc.*, 5591 (1964).
5. R. H. Kratzer, K. J. L. Paciorek, U.S. Pat. 3,729,552 (1973); *Chem. Abstr.*, 79, 7578 (1973).
6. G. J. Brendel, E. M. Marlett, L. M. Niebylski, *Inorg. Chem.*, 17, 3589 (1978).

1.8.3.2. Magnesium Hydride

Magnesium hydride can be prepared by thermal decomposition of organometallic compounds[1] similar to reaction (a) (§1.8.3.1) for the preparation of BeH_2, e.g.:

$$Et_2Mg \xrightarrow{\sim175°C} MgH_2 + 2\ C_2H_4 \tag{a}$$

$$2\ EtMgBr \xrightarrow{220°C} MgH_2 + MgBr_2 + 2\ C_2H_4 \tag{b}$$

or by metathesis with $LiAlH_4$ similar to reaction (b) (§1.8.3.1)[2]:

$$LiAlH_4 + 2\ Mg(CH_3)_2 \xrightarrow{Et_2O} 2\ MgH_2 + LiAl(CH_3)_4 \tag{c}$$

but it is impossible to remove the last traces of ether without decomposing the hydride. The hydride may be prepared[3] by reacting acetaldehyde or benzaldehyde with Mg:

$$2\ C_6H_5CHO + Mg \rightarrow MgH_2 + C_6H_5COCOC_6H_5 \tag{d}$$

Direct reaction of Mg and H_2 at elevated T and P yields MgH_2. The rate is slow, and the process is difficult to carry to completion because the metal particles become coated with a dense layer of the hydride, through which the H_2 must diffuse. If the T is raised to increase the diffusion rate, the dissociation P is also increased and, for the reaction to proceed, the ambient H_2 P must exceed the dissociation P; e.g., 40 MPa and over 500°C are employed[1,4–6]. To increase the rate and facilitate complete reaction, various catalysts, such as I_2 [5], CCl_4 [6], allyl iodide[7] and MgI_2 [8] are used. Intimate mixing of $LaNi_5$ [9], Mg_2Cu [10] or Fe [11] with Mg metal decreases the time needed to complete hydride formation; e.g., the addition of 20% $LaNi_5$ increases the percentage completion after 15 min (at 345°C) from ca. 12% for pure Mg, to over 80% for the mixture[9]. These latter three catalysts provide an oxide-free surface[10] to facilitate dissociative adsorption of the H_2.

The addition of about 1% (or less) of a group III metal, such as Al, Ge or In, to the Mg increases the hydride-formation rate[12].

Homogeneous catalysts via the action of anthracene Mg in tetrahydrofuran (THF)[12] with Cr, Ti or Fe halides in THF allow the complete formation of MgH_2 in ca. 10 h at 8 MPa and only 60–70°C.

Magnesium powder of small particle size can be hydrided completely and rapidly without the use of a catalyst; e.g., under 3.5 MPa of H_2 complete conversion to MgH_2 occurs[14] within 15 min at 400°C, provided the particle size is < 100 μm.

Under high hydrostatic P (8000 MPa) and 800°C, normal MgH_2, in which the coordination number for Mg is six, is converted to a phase of higher density, in which the Mg has eight nearest-neighbor H atoms[15].

(G.G. LIBOWITZ)

1. E. Wiberg, R. Bauer, *Chem. Ber.,* 85, 593 (1952).
2. G. D. Barbaras, C. Dillard, A. E. Finholt, T. Wartik, K. E. Nilzbach, H. I. Schlesinger, *J. Am. Chem. Soc.,* 73, 4585 (1951).
3. M. Givelet, *C.R. Hebd. Seances Acad. Sci.,* 267, 881 (1968).
4. F. H. Ellinger, C. E. Holley Jr., B. B. McInteer, D. Pavone, R. M. Potter, E. Staritzky, W. H. Zachariasen, *J. Am. Chem. Soc.,* 77, 2647 (1955).
5. J. F. Stampfer Jr., *J. Am. Chem. Soc.,* 82, 3504 (1960).
6. T. N. Dymova, Z. K. Sterlyadkina, V. G. Safronov, *Russ. J. Inorg. Chem. (Engl. Transl.),* 6, 389 (1961).
7. J. P. Faust, E. D. Whitney, H. D. Batha, T. L. Heying, C. E. Fogle, *J. Appl. Chem. Biotechnol.,* 10, 187 (1960).
8. E. Wiberg, H. Goeltzer, R. Bauer, *Z. Naturforsch., Teil B,* 6, 394 (1951).
9. B. Tanguy, J. L. Soubeyroux, M. Pezat, J. Portier, P. Hagenmuller, *Mater. Res. Bull.,* 11, 1441 (1976).
10. J. Genossar, P. S. Rudman, *Z. Phys. Chem. (Leipzig),* 116, 215 (1979).
11. J. M. Welter, P. S. Rudman, *Scr. Metall.,* 16, 285 (1982).
12. M. H. Mintz, Z. Gavra, Z. Hadari, *J. Inorg. Nucl. Chem.,* 40, 765 (1978).
13. B. Bogdanovic, S. Liao, M. Schwickardi, P. Sikorsky, B. Spliethoff, *Angew. Chem., Int. Ed. Engl.,* 19, 818 (1980).
14. B. Vigeholm, J. Kjøller, B. Larsen, *J. Less-Common Met.,* 74, 341 (1980).
15. J. P. Bastide, B. Bonnetot, J. M. Létoffé, P. Claudy, *Mat. Res. Bull.,* 15, 1779 (1980).

1.8.3.3. Calcium, Strontium and Barium Hydrides

The hydrides of Ca, Sr and Ba are formed by direct reaction of the metal and H_2 at elevated T (see §1.12.1).

Because these metals are reactive with air and moisture, care must be taken to prevent oxidation, which would impede hydride formation. For the formation of CaH_2, 300–350°C is optimum[1]. Formation of SrH_2 and BaH_2 occurs[2] at elevated T. These hydrides react violently with H_2O, as do the alkali-metal hydrides.

(G.G. LIBOWITZ)

1. C. E. Messer, *Prep. Inorg. React., 1*, 203 (1964).
2. A. F. Zhigach, D. S. Stasinevitch, *Chemistry of Hydrides*, Khimiya, Leningrad, 1969.

1.9. Formation of Bonds between Hydrogen and Metals of Group IB (Cu, Ag, Au) or IIB (Zn, Cd, Hg)

1.9.1. Introduction

The reaction of transition metals with H_2 is distinct from the physical diffusion of H_2 through the solid metal. The group IB metals give no evidence of hydride formation, although they are important in the reaction[1-3] of alloys such as Pd-Cu and Pd-Ag with H_2.

There are many examples of borohydride compounds of these metals, e.g., Cu, Ag, Zn and Cd–BH_4 as neutral and anionic complexes in which the mode of bonding of BH_4 is dependent on the coordination number of the metal[4]. Higher borane anions also combine with Cu and Ag, yielding both neutral and anionic complexes. Although no borohydrides of Au are isolated, treatment of Au-halide complexes with, e.g., $NaBH_4$, is a standard method for the preparation of Au-cluster compounds[5]. Copper(I) hydride, first reported[6] in 1844, has the ZnS structure[7] [d(Cn–H) = 0.173 nm (1.73 Å); d(Cu–Cu) = 0.289 nm (2.89 Å)] and decomposes to the elements when heated. **At $\geq 100°C$ the decomposition is explosive.**

(F. GLOCKLING)

1. D. Fisher, D. M. Chisdes, T. B. Flanagan, *J. Solid-State Chem., 20,* 149 (1977).
2. B. Baranowski, *Z. Phys. Chem., Neue Folge, 114,* 59 (1979).
3. B. Baranowski, *Top. Appl. Phys., 29,* 157 (1978).
4. T. J. Marks, J. R. Kalb, *Chem. Rev., 77,* 263 (1977).
5. F. Cariati, L. Neldini, *Inorg. Chim. Acta, 5,* 172 (1971).
6. A. Wurtz, *Ann. Chim. Phys., 11,* 250 (1844).
7. T. P. R. Gibb, *Prog. Inorg. Chem., 3,* 492 (1962).

1.9.2. from the Elements.

Hydrogen gas diffuses through Cu at above 450°C, provided the metal is oxygen-free, the rate being proportional to $(P_{H_2})^{1/2}$, but CuH is not established under these conditions[1]. In the vapor phase at 1400°C, however, Cu and H_2 form[2] CuH.

The formation of Ag(I) and Au(I) hydrides at > 1000°C in H_2 is confirmed by spectroscopic studies on the metal hydrides and deuterides in the vapor phase[3]. Attempts to prepare an Au hydride–phosphine complex from H_2 in solution fail, although one is almost certainly formed as an intermediate[4]:

$$(Ph_3P)AuGePh_3 + H_2 \xrightarrow[3 \times 10^4 \text{ kPa (300 atm)}]{50°C} Ph_3P + Ph_3GeH + Au \qquad (a)$$

When the intermetallic alloy $CaAg_2$ is heated in H_2 at 575–600°C for 16–18 h, the

160

tertiary hydride CaAg$_2$H is formed, and this hydride also results from the solid-state reaction between Ag and Ca hydride at 650°C. It is a black, crystalline, nonvolatile solid, reactive to H$_2$O and slowly decomposed by air[5].

(F. GLOCKLING)

1. W. Eichenauer, A. Pebler, Z. Metallkd., 48, 373.
2. J. C. Warf, W. Keitnecht, Helv. Chim. Acta, 33, 613 (1950), and refs. therein.
3. K. M. Guggenheimer, Proc. Phys. Soc., 58, 456 (1946).
4. F. Glockling, M. D. Wilbey, J. Chem. Soc., 2168 (1968).
5. M. H. Mendelsohn, J. Tanaka, R. Lindsay, R. O. Moyer, Inorg. Chem., 14, 2911 (1975).

1.9.3. by Reaction with Hydrogen Atoms and Ions with Compounds of Group IB.

Atomic hydrogen in the presence of Cu gives[1,2] a black Cu(I) hydride at low T. If an electric discharge is passed between Ag electrodes in the presence of H$_2$, AgH can be identified spectroscopically.

The reaction of metals with energetic hydrogen or deuterium ions is important in nuclear reactors. Ion beams may be generated thermally and allowed to interact with the metal, and the reaction products then may be examined by matrix-isolation techniques. Alternatively, metal atoms are sputtered from a cathodic surface by a low-energy plasma. If H$_2$ or D$_2$ at low P is added to the discharge, then molecular species are formed by the interaction with the sputtered metal atoms. Applied to Cu this technique leads to the identification of CuH and CuD in Ar matrices by their IR spectra. The reacting species are believed to be atomic Cu and H or D formed in the hollow-cathode discharge[3].

(F. GLOCKLING)

1. J. C. Warf, W. Keitknecht, Helv. Chim. Acta, 33, 613 (1950).
2. F. A. McMahon, P. L. Robinson, J. Chem. Soc., 854 (1934).
3. R. B. Wright, J. K. Bates, D. M. Gruen, Inorg. Chem., 17, 2275 (1978).

1.9.4. by Hydride Ion Reduction

1.9.4.1. of Compounds of Group IB.

Alkali-metal hydrides are not used as a route to group IB-metal hydrides; complex metal hydrides are used instead.

(F. GLOCKLING)

1.9.4.2. of Compounds of Group IIB.

This is the best method for preparing Zn dihydride, because using equimol ratios of NaH and ZnCl$_2$ or LiH and ZnBr$_2$ gives Zn dihydride free from anionic species. Sodium hydride and ZnI$_2$ in a 2:1 ratio also give ZnH$_2$, but if the ratio of alkali-metal hydride to ZnX$_2$ is increased, then the ZnH$_2$ is contaminated with complex hydrides[1,2].

Zinc dialkyls are also reduced by alkali-metal hydrides; the products depend on the alkyl group and on the ratio of reactants. Di(s-butyl)zinc and KH yield an anionic hydride:

$$(\text{s-Bu})_2\text{Zn} + \text{KH} \rightarrow \text{K}_2\text{ZnH}_4 \qquad \text{(a)}$$

By contrast, dimethylzinc gives first a monohydride anion, which is insoluble in ether but dissolves in THF:

$$\text{Me}_2\text{Zn} + \text{KH} \rightarrow \text{KZnHMe}_2 \qquad \text{(b)}$$

Reaction of $KZnHMe_2$ with a further mol of Me_2Zn yields the unstable complex, $K[Zn_2HMe_4]$. Related arylzinc hydrides (and deuterides) such as $Na[ZnHPh_2]$ are prepared in the same way. The product from $(C_6F_5)_2Zn$ and NaH is ether soluble:

$$2 \ (\text{C}_6\text{F}_5)_2\text{Zn} + 2 \ \text{NaH} \xrightarrow[\text{0°C}]{\text{Et}_2\text{O}} \text{Na}_2[\text{ZnH}(\text{C}_6\text{F}_5)_2]_2 \qquad \text{(c)}$$

and on the basis of IR and mol wt measurements has a double hydrogen-bridged structure. Reduction of methylzinc iodide with Na hydride produces a mixed dimeric alkyl hydride:

$$2 \ \text{MeZnI} + 3 \ \text{NaH} \rightarrow 2 \ \text{NaI} + \text{Na}[\text{Zn}_2\text{H}_3\text{Me}_2] \qquad \text{(d)}$$

Anion structures with single- or triple-hydrogen bridges are proposed. It decomposes at $\geq$ RT in vacuo to $NaZnH_3$, which has greater thermal stability but is rapidly hydrolyzed[3]:

$$\text{Na}[\text{Zn}_2\text{H}_3\text{Me}_2](\text{THF})_n \rightarrow \text{NaZnH}_3 + \text{Me}_2\text{Zn} + n \ \text{THF} \qquad \text{(e)}$$

A neutral dimeric Zn-hydride complex is isolated as a white, crystalline solid having terminal rather than bridging Zn—H bonds[4]:

$$2 \ \text{ZnH}_2 + 2 \ \text{Me}_2\text{NCH}_2\text{CH}_2\text{NHMe} \rightarrow 2 \ \text{H}_2 + [\text{ZnH}(\text{NMeCH}_2\text{CH}_2\text{NMe}_2)]_2 \qquad \text{(f)}$$

(F. GLOCKLING)

1. E. C. Ashby, J. J. Watkins, *Inorg. Chem.*, *12*, 2493 (1973).
2. G. J. Kubas, D. F. Shriver, *J. Am. Chem. Soc.*, *92*, 1949 (1970).
3. D. F. Shriver, D. J. Kubas, J. A. Marshall, *J. Am. Chem. Soc.*, *93*, 5076 (1971).
4. N. A. Bell, G. E. Coates, *J. Chem. Soc., A*, 823 (1968).

1.9.5. by Neutral and Anionic Metal Hydride Reduction

1.9.5.1. of Compounds of Group IB.

Copper(I) chloride reacts with AlH_3 in ether at $-78°C$, and on the basis of elemental analysis and IR spectra, it appears that the reaction up to RT is:

$$2 \ \text{CuCl} + 2 \ \text{AlH}_3 \rightarrow \text{CuH} + \text{CuAlH}_4 + \text{HAlCl}_2 \qquad \text{(a)}$$

In the absence of strongly π-bonding ligands such as tertiary phosphines, complex hydrides, e.g., $[AlH_4]^-$ and $[BH_4]^-$ react with Cu, Ag and Au halides to form metal-hydrogen compounds, usually of low thermal stability. Uncomplexed CuH is unstable, even at $-80°C$.

1.9. Formation of Bonds between Hydrogen and Cu, Ag, Au or Zn, Cd, Hg   163
1.9.5. by Neutral and Anionic Metal Hydride Reduction
1.9.5.1. of Compounds of Group IB.

When $LiAlH_4$ is added to Cu(I) iodide in pyridine, a red-brown powder separates on addition of ether. This solid contains Cu(I) hydride together with LiI, CuI and pyridine, which stabilize it[1].

When $Li[CuMe_2]$ reacts with metal boro- or alumino-hydrides, complex hydrides of Cu are formed. Although not fully characterized, these can be formulated[2] as $MCuH_2$. Similarly, reduction of CuI by $K[n-Bu_3BH]$ yields[3] $KCuH_2$. By varying the experimental conditions, the reduction of $LiCuMe_2$ with $LiAlH_4$ yields a range of anionic Cu hydrides (e.g., $LiCuH_2$, Li_2CuH_3, Li_3CuH_4, Li_4CuH_5, Li_6CuH_6). In addition, this reaction leads to salts with more than one Cu atom in the anion (e.g., $LiCu_2H_3$, $Li_2Cu_3H_5$). Most of these complexes have transitory stability at RT; e.g., $LiCuH_2$ is stable at 0°C for several weeks. These hydrides are almost certainly formed by $[AlH_4]^-$ reduction of the preformed methylcopper anion, e.g.:

$$4 \text{ MeLi} + \text{CuI} \rightarrow Li_3CuMe_4 + \text{LiI} \tag{b}$$

$$Li_3CuMe_4 + 2 \text{ } LiAlH_4 \rightarrow Li_3CuH_4 + 2 \text{ } LiAlH_2Me_2 \tag{c}$$

These complexes are unsolvated. Their thermal decomposition by differential-thermal and thermal-gravimetric analyses (DTA and TGA) provides evidence for the conversion of one hydride complex into another with partial loss of Cu and H_2, e.g.:

$$3 \text{ } Li_2CuH_3 \xrightarrow{90°C} 2 \text{ } Li_3CuH_4 + \text{CuH} \tag{d}$$

$$4 \text{ } Li_3CuH_4 \xrightarrow{110°C} 3 \text{ } Li_4CuH_5 + \text{Cu} + \tfrac{1}{2} H_2 \tag{e}$$

$$\tfrac{3}{2} Li_4CuH_5 \xrightarrow{120°C} \tfrac{6}{5} Li_5CuH_6 + \tfrac{3}{10} \text{Cu} + \tfrac{3}{20} H_2 \tag{f}$$

With Ag no binary hydride is known, but several thermally unstable complex hydrides are isolated that contain H bonded to Ag. For example, $LiAlH_4$, $LiBH_4$ or $LiGaH_4$ react with Ag perchlorate in ether to give white or yellow precipitates of $AgAlH_4$, $AgBH_4$ and $AgGaH_4$. Thermal decomposition occurs below RT and is catalyzed[5] by Ag:

$$AgBH_4 \xrightarrow{-30°C} \text{Ag} + B_2H_6 + H_2 \tag{g}$$

$$AgGaH_4 \xrightarrow{-75°C} \text{Ag} + GaH_3 + H_2 \tag{h}[6]$$

Gold(III) also forms complex hydrides with $LiAlH_4$ and $LiBH_4$, stable only at low T. Analyses of the precipitated solids lead to such compositions as $Al_6Au_2H_{24}$, of unknown structure[7].

Stable molecular borohydride complexes of Cu, and to a lesser extent Ag and Au, are isolated, and each of the four possible bonding modes (I–IV) is identified[8]:

$M-H-BH_3$	M⟨H,H⟩BH$_2$	M⟨H,H⟩BH	$M[BH_4]$
(I)	(II)	(III)	(IV)

When Na borohydride in ethanol is added to Cu(I) chloride containing an electron-pair donor, reaction occurs at RT, with its course determined by the steric and

164 1.9. Formation of Bonds between Hydrogen and Cu, Ag, Au or Zn, Cd, Hg
     1.9.5. by Neutral and Anionic Metal Hydride Reduction
     1.9.5.1. of Compounds of Group IB.

donor–acceptor properties of the base. For triarylphosphines the reaction gives stable borohydride complexes such as $(Ph_3P)_2Cu(BH_4)$. With trialkylphosphines, Ph_3As, Ph_3Sb, or the sterically hindered tri-o-tolyphosphine, H_2 evolves and Cu is precipitated. The chelating diphosphine $(Ph_2PCH_2)_2$ (dppe) also yields a stable complex, (dppe)Cu(BH$_4$). Bis(triphenylphosphine)copper(I) borohydride is a crystalline complex in which both Cu and boron have a quasi-tetrahedral configuration (V):

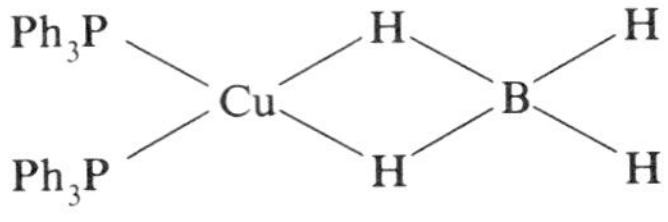

(V)

The deuterium analogue of (I) is also known[9,10]. When $(Ph_3P)_2CuBH_4$ is treated with HClO$_4$ in ethanol a soluble complex is first produced, which with more HClO$_4$ yields $[(Ph_3P)_2CuClO_4]_2$. The soluble complex, $[(Ph_3P)_4Cu_2BH_4]ClO_4$, is a white, diamagnetic solid; IR evidence suggests the structure[11] (VI):

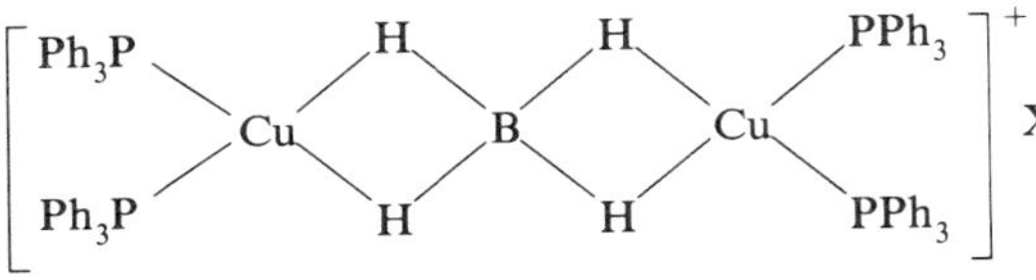

(VI)

Substitution reactions in which halide ion is displaced by $[BH_4]^-$ are also dependent on the halide; e.g., whereas neither $(Ph_2MeP)_3CuI$ nor $(PhMe_2P)_3CuI$ reacts with $[BH_4]^-$ at RT, reaction with the corresponding chloro complexes is rapid and quantitative. For both phosphines, a single Cu—H—B bridged structure results[12] with four-coordinated Cu:

$$(PhMe_2P)_3CuI + NaBH_4 \xrightarrow{\ CHCl_3-EtOH\ } (PhMe_2P)_3Cu-H-B{\overset{\displaystyle H}{\underset{\displaystyle H}{\big|}}}H + NaI \qquad (i)$$

Substituted borohydrides, $[BH_3X]^-$, are used to form Cu and Ag complexes[15]. The anion $[H_3BCO_2R]^-$ reacts with triphenylphosphine complexes of Cu(I) and Ag(I) in chloroform–ethanol, yielding $(Ph_3P)_nMH_3BCO_2R$ (where n = 2, 3 and R = H, Me, Et). If 3 mol of phosphine are coordinated to Cu or Ag, structure (VII) results, whereas in complexes with 2 mol of triphenylphosphine, two hydrogen bridges are formed (VIII):

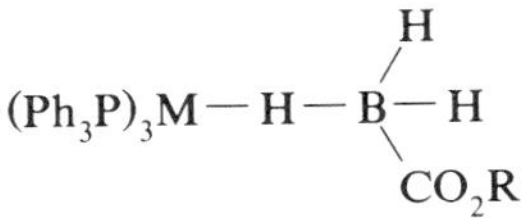

(VII) 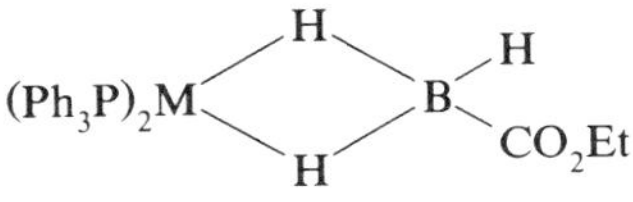   (VIII)

The structural evidence for these two forms of bonding comes from IR and NMR data[13].

1.9. Formation of Bonds between Hydrogen and Cu, Ag, Au or Zn, Cd, Hg   165
1.9.5. by Neutral and Anionic Metal Hydride Reduction
1.9.5.1. of Compounds of Group IB.

Monodentate bonding is confirmed for $(Ph_2MeP)_3CuBH_4$ by crystallography[14]. The Ag complex $(Ph_3P)_3AgH_3BCO_2R$ can also be obtained from:

$$2\ (Ph_3P)_4AgF + Cu(H_3BCO_2H)_2 \xrightarrow{H_2O-CHCl_3} (Ph_3P)_3AgH_3BCO_2H \qquad (j)$$

Silver nitrate–phosphine complexes are used to prepare borane derivatives[12]:

$$(Ph_2MeP)_3AgNO_3 + Na[H_3BCO_2Et] \rightarrow (Ph_2MeP)_3AgHBH_2CO_2Et \qquad (k)$$

The extent of neutral ligand dissociation follows the order: $L_3CuBH_4 > L_3AgBH_4 \geq L_3AgH_3BCO_2Et > L_3CuH_3BCO_2Et$.

The isocyanotrihydroborate ion, $[NCBH_3]^-$, coordinates to transition metals through nitrogen, and likewise with Cu(I), giving, e.g., $(Ph_3P)_3Cu(NCBH_3)$ which is weakly conducting in acetonitrile (Ag analogues of lower stability are also obtained). However, complexes with only two neutral ligands have different IR spectra, consistent with a hydrogen-bridged dimer (IX)[16], from a crystal structure[17] of $(Ph_3P)_2Cu(NCBH_3)_2$.

(IX)

A hydridocopper cluster is isolated as red, hexagonal plates from the reaction between $(Ph_3PCuCl)_4$ and Na trimethoxyborohydride in dimethylformamide (DMF). The reaction takes place over 30 min at RT yielding $Cu_6H_6(PPh_3)_6 \cdot DMF$ as a thermally stable solid that slowly oxidizes in air. X-Ray diffraction shows that the Cu_6^- cluster is distorted from regular O_h symmetry, and that one Ph_3P molecule is bonded to each Cu atom. The positions of the hydrogen atoms are not known nor are they detected by IR or ^1H NMR spectroscopy; their presence is demonstrated by decomposition of the complex using $PhCO_2D$ when HD and H_2 are formed, but not D_2. Moreover, the geometry of the skeleton [(a) and (b), Fig. 1] precludes the possibility of terminal Cu—H bonding and hence suggests that the six hydrogen atoms either bridge each edge of the large (or small) triangular faces or form triple bridges to the triangular faces. Assuming that the Ph_3P and H^- ligands each contribute two electrons, then this cluster has a noble-gas configuration[18].

Formation of Cu(I) borohydride complexes with bridging H at is not limited to the $[BH_4]^-$ ion; if $(Ph_3P)_2CuCl$ is treated with $Cs[B_3H_8]$ in acetone the crystalline complex $(Ph_3P)_2CuB_3H_8$ is isolated. A similar procedure leads to the triphenylarsine and -stibine analogues. The structure (XII) of solid $(Ph_3P)_2Cu(B_3H_8)$ is:

(XII)

166   1.9. Formation of Bonds between Hydrogen and Cu, Ag, Au or Zn, Cd, Hg
1.9.5. by Neutral and Anionic Metal Hydride Reduction
1.9.5.1. of Compounds of Group IB.

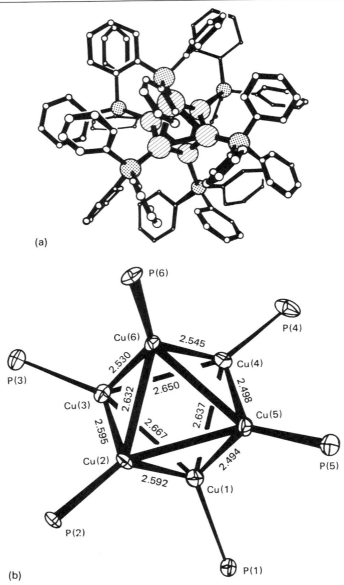

(a)

(b)

Figure 1. (a) Stereochemistry of $H_6Cu_6(PPh_3)_6$. (b) Cu-Cu distances within the $H_6Cu_6(PPh_3)_6$ cluster. Reproduced by permission of the authors and the Editor of the *Journal of the American Chemical Society.*

1.9. Formation of Bonds between Hydrogen and Cu, Ag, Au or Zn, Cd, Hg   167
1.9.5. by Neutral and Anionic Metal Hydride Reduction
1.9.5.1. of Compounds of Group IB.

In solution it is monomeric and a nonelectrolyte, and its structure involves bonding between Cu and terminal protons on two different boron atoms. In this and related complexes with two Ph_3As or $(PhO)_3P$ ligands bonded[19] to Cu, variable-T NMR studies show that the $[B_3H_8]^-$ ion rearranges, and that the activation energy for this scrambling or fluctional behavior depends on the effective positive charge on Cu: the greater the positive charge, the stronger the binding to $[B_3H_8]^-$.

In $(Ph_3P)_2CuB_3H_8$ the phosphine ligands are also labile (by ^{31}P NMR), and the labile Ph_3P molecules can be captured by the addition of B_2H_6 to Cu(I)-borane complexes[20]:

$$[(Ph_3P)_2Cu]_2B_{10}H_{10} + 2 B_2H_6 \rightarrow 4 Ph_3PBH_3 + Cu_2B_{10}H_{10} \tag{l}$$

$$(Ph_3P)_2CuB_3H_8 + B_2H_6 \rightarrow 2 Ph_3PBH_3 + CuB_3H_8 \tag{m}$$

The polymeric borane CuB_3H_8 is stable at RT in the absence of air; it is suggested that each Cu atom is bonded to more than one B_3H_8 unit through Cu—H—B bridges.

The capture of labile Ph_3P from $(Ph_3P)_2CuB_3H_8$ can be taken to the halfway stage:

$$(Ph_3P)_2CuB_3H_8 + \tfrac{1}{2} B_2H_6 \rightarrow Ph_3PBH_3 + Ph_3PCuB_3H_8 \tag{n}$$

Attempts to precipitate $Ph_3PCuB_3H_8$ fail because of its redistribution to $(Ph_3P)_2CuB_3H_8$ and CuB_3H_8. In a similar reaction one mol of Ph_3P may be removed from $(Ph_3P)_2$-$CuBH_4$, yielding $(Ph_3P)CuBH_4$, but this also is too unstable to isolate the solid. Its structure, based on vibrational and NMR spectroscopy, is consistent with a triple Cu—H—B bridge and four-coordinated Cu(I):

(XIII)

At above $-20°C$ Ph_3PCuBH_4 decomposes to a black solid, believed to be CuH, together with Ph_3PBH_3.

Trisphosphine silver complexes (L=phosphine) with $[B_3H_8]^-$ disproportionate:

$$2 L_3AgB_3H_8 \rightleftharpoons L_2AgB_3H_8 + [L_4Ag][B_3H_8] \tag{o}$$

Gold(I) complexes form only $[L_4Au][borane]$ salts on reaction with anionic boranes.

Copper and Ag complexes of the higher boranes are prepared by reaction of the metal-halide complex with a borane anion, including[21-23] B_5H_8, B_6H_9, $B_9H_{12}S$, B_9H_{14}, $B_{10}H_{13}$ and $B_{11}H_{14}$. Pentaborane(9) and hexaborane(10) form the Cu complexes $(Ph_3P)_2CuB_5H_8$ and $(Ph_3P)_2CuB_6H_9$; both are white, air-stable solids, formed in THF–CH_2Cl_2 from which they are precipitated by ether[24].

Most compounds have neutral (group VA or VIA) ligands bonded to Cu, but $CuCl_2$ and $[B_{10}H_{13}]^{2-}$ yield the complex, $(Et_4N)_2(Cl_2CuB_{10}H_{13})$. With $[B_9H_{14}]^-$, $[B_9H_{12}S]^-$ and $[B_{11}H_{14}]^-$ only ionic tetrakis(triphenylphosphine)Cu and -Ag salts are isolated. The decaborane $Cu_{12}B_{10}H_{10}$ is also prepared by:

$$(C_2H_3O_2)_2Cu \cdot H_2O + K_2B_{10}H_{10} \xrightarrow{H_2O} Cu_2B_{10}H_{10} \tag{p}$$

and separates as white crystals from acetonitrile. The corresponding deuteride is also known. X-Ray diffraction gives the geometry without revealing the positions of the H atoms; comparison of its IR spectrum and that of $Cu_2B_{10}D_{10}$ with the free ion $[B_{10}H_{10}]^{2-}$ suggests that the terminal $B_{10}H_{10}$ hydrogen atoms are involved in forming three-center $Cu—H—B$ bridge bonds, making the Cu(I) atoms sp^3 hybridized[2].

Bridge $Cu—H—B$ bonding is present in $[(Ph_3P)_2Cu]_2B_{10}H_{10}$ from solid-state studies, and solution IR studies also favor this view. The complex is made by the direct reaction[25,26]:

$$(Ph_3P)_2CuCl + [NH_4]_2B_{10}H_{10} \rightarrow [(Ph_3P)_2Cu]_2B_{10}H_{10} \qquad (q)$$

The pentaborane–Cu(I) complex $(Ph_3P)_2CuB_5H_8$ has the Cu atom bonded to a basal boron–site; it is nonfluctional and is prepared by standard techniques. The tendency toward coordination to Cu follows: $B_3H_8 > B_{10}H_{13} > B_9H_{12}S > B_9H_{14} \approx B_{11}H_{14}$.

(F. GLOCKLING)

1. J. A. Dilts, D. F. Shriver, *J. Am. Chem. Soc.*, 90, 5789 (1968).
2. E. C. Ashby, T. F. Korenowski, R. D. Schwarz, *J. Chem. Soc., Chem. Commun.*, 157 (1974).
3. T. Yoshida, E. Negishi, *J. Chem. Soc., Chem. Commun.*, 762 (1974).
4. E. C. Ashby, A. B. Goel, *Inorg. Chem.*, 16, 3043 (1977).
5. E. Wiberg, W. Henle, *Z. Naturforsch., Teil B*, 76, 250, 575, 576 (1952).
6. J. P. Chan, R. Hultgren, *J. Chem. Thermodyn.*, 1, 45 (1969).
7. E. Wiberg, H. Neumauer, *Inorg. Nucl. Chem. Lett.*, 1, 35 (1965).
8. S. J. Lippard, D. A. Ucko, *Inorg. Chem.*, 7, 1051 (1968).
9. F. Cariati, L. Naldini, *Gazz. Chim. Ital.*, 95, 3 (1965).
10. S. J. Lippard, K. M. Melmed, *J. Am. Chem. Soc.*, 89, 3929 (1967).
11. F. Cariati, L. Naldini, *J. Inorg. Nucl. Chem.*, 28, 2243 (1966).
12. J. C. Bommer, K. W. Morse, *Inorg. Chem.*, 19, 587, (1980).
13. J. C. Bommer, K. W. Morse, *Inorg. Chem.*, 18, 531 (1979).
14. C. Kutal, P. Grutsch, J. L. Atwood, R. D. Rogers, *Inorg. Chem.*, 17, 3558 (1978).
15. J. C. Bommer, K. W. Morse, *J. Am. Chem. Soc.*, 96, 6222 (1974).
16. S. J. Lippard, P. S. Welcker, *Inorg. Chem.*, 11, 6 (1972).
17. K. M. Melmed, T. Li, J. J. Mayerle, S. J. Lippard, *J. Am. Chem. Soc.*, 96, 69 (1974).
18. S. A. Bezman, M. R. Churchill, J. A. Osborn, J. Wormald, *J. Am. Chem. Soc.*, 93, 2063 (1971).
19. C. H. Bushweller, H. Beall, W. J. Dewkett, *Inorg. Chem.*, 15, 1739 (1976).
20. R. K. Hertz, R. Goetze, S. G. Shore, *Inorg. Chem.*, 18, 2813 (1979).
21. F. Clanberg, E. L. Muetterties, L. J. Guggenberger, *Inorg. Chem.*, 7, 2272 (1968).
22. E. L. Muetterties, W. G. Peet, P. A. Wegner, C. W. Abgranti, *Inorg. Chem.*, 9, 2447 (1970).
23. V. T. Brice, S. G. Shore, *J. Chem. Soc., Chem. Commun.*, 1312 (1970).
24. T. E. Paxson, M. F. Hawthorne, L. D. Brown, *Inorg. Chem.*, 13, 2772 (1974).
25. J. T. Gill, S. J. Lippard, *Inorg. Chem.*, 14, 751 (1975).
26. G. G. Outterson, V. T. Brice, S. G. Shore, *Inorg. Chem.*, 15, 1456 (1976).

1.9.5.2. of Compounds of Group IIB.

Although Hg hydrides are claimed, most studies relate to Zn and Cd, where polymeric binary compounds, $(MH_2)_n$, are isolated as well as anionic metal hydrides, e.g., $LiZnH_3$, Li_2ZnH_4, Li_3ZnH_5. In addition, borohydride complexes such as $Cd(BH_4)_2 \cdot$ THF are reported, some having more than one metal atom in the anion.

A white, solid, polymeric Zn dihydride is made by the action of LiAl hydride on Zn iodide or dimethylzinc:

$$2 ZnI_2 + LiAlH_4 \xrightarrow{-40°C, Et_2O} 2 ZnH_2 + LiI + AlI_3 \qquad (a)$$

1.9. Formation of Bonds between Hydrogen and Cu, Ag, Au or Zn, Cd, Hg   169
1.9.5. by Neutral and Anionic Metal Hydride Reduction
1.9.5.2. of Compounds of Group IIB.

$$Me_2Zn + LiAlH_4 \xrightarrow{0°C, Et_2O} ZnH_2 + Li[MeAlH_2] \qquad (b)$$

This polymeric Zn dihydride is converted into the borohydride by reaction with diborane:

$$ZnH_2 + B_2H_6 \rightarrow Zn(BH_4)_2 \qquad (c)$$

Cadmium dihydride, which is also polymeric, is prepared by the same methods[1-3] and HgH_2 is claimed from the reaction of HgI_2 with $LiAlH_4$ at low T.

Aluminium hydride reacts with anhyd Zn chloride or bromide to give a polymeric complex, Zn_2H_3X (X = Cl, Br), whereas Zn iodide gives an insoluble product, ZnI_2AlH_3. Under similar conditions $CdBr_2$ reacts[4] slowly to yield CdHBr. Structural information is limited to IR spectra because of their low solubility and reluctance to form crystals.

Complex Zn hydrides are isolated by procedures that are variations on earlier methods[5-7]:

$$2\ MgH_2 + Me_2Zn \xrightarrow{THF} MgZnH_4 + MgZnMe_4 \qquad (d)$$

$$2\ Me_2Mg + 2\ ZnH_2 \xrightarrow{THF} MgZnH_4 + MgZnMe_4 \qquad (e)$$

$$MgZnMe_4 + 2\ LiAlH_4 \xrightarrow{THF} MgZnH_4 + 2\ LiAlH_2Me_2 \qquad (f)$$

The complex $MgZnH_4$ forms a white, insoluble solid with two mol of THF, leaving the metal–methyl complex in solution.

Alkylzinc complexes including $LiZnMe_3$ and Li_2ZnMe_4 are reduced[6,7] with $LiAlH_4$, $NaAlH_4$ and AlH_3. These reactions, which yield complex hydrides of zinc, proceed in high yield, without the complications that result from ether-cleavage side reactions:

$$Li_2ZnMe_4 + LiAlH_4 \xrightarrow{100\%} Li_2ZnH_4 + LiAlMe_2H_2 \qquad (g)$$

$$LiZnMe_2H + LiAlH_4 \rightarrow Li_2ZnH_4 \qquad (h)$$

Although these reactions proceed rapidly, $LiZn_2Me_4H$ reacts with $LiAlH_4$ slowly, requiring five days. The reaction involves a slow exchange of a methyl group from $LiZn_2Me_4H$ to yield the intermediate $LiZn_2Me_3H_2$ which then disproportionates:

$$2\ LiZn_2Me_3H_2 \rightarrow Li_2ZnH_4 + 3\ Me_2Zn \qquad (i)$$

The reduction of Li_3ZnMe_5 with $LiAlH_4$ in diethyl ether at RT gives Li_3ZnH_5 quantitatively. This complex is not a physical mixture of $Li_3ZnH_4 + LiH$, but its insolubility limits structural studies. Similar limitations apply to $LiZnH_3$, formed by:

$$LiMe + ZnMe_2 \rightarrow LiZnMe_3 \qquad (j)$$

$$LiZnMe_3 + LiAlH_4 \rightarrow LiZnH_3 + LiAlMe_3H \qquad (k)$$

Its thermal decomposition to Li + Zn + H_2 proceeds through a disproportionation step:

$$2\ LiZnH_3 \rightarrow Li_2ZnH_4 + ZnH_2 \qquad (l)$$

X-ray powder diffraction demonstrates that complexes with the stoichiometry $LiZn_2H_5$ and $LiZn_3H_7$ are physical mixtures of $LiZnH_3$ and ZnH_2. However, the reaction between $KZnMe_2H$ or KZn_2Me_4H and Al hydride in THF proceeds according to:

170   1.9. Formation of Bonds between Hydrogen and Cu, Ag, Au or Zn, Cd, Hg
1.9.5. by Neutral and Anionic Metal Hydride Reduction
1.9.5.2. of Compounds of Group IIB.

$$2 \text{ KZnMe}_3\text{H} + 2 \text{ AlH}_3 \rightarrow \text{KZn}_2\text{H}_5 + \text{K[AlMe}_2\text{H}_2, \text{AlMe}_2\text{H]} \qquad \text{(m)}$$

$$\text{KZn}_2\text{Me}_4\text{H} + 2 \text{ AlH}_3 \rightarrow \text{KZn}_2\text{H}_5 + 2 \text{ AlMe}_2\text{H} \qquad \text{(n)}$$

X-Ray studies suggest that KZn_2H_5 is a genuine hydride complex; it decomposes thermally to $\text{Zn} + \text{K} + \text{H}_2$ via KZnH_3 and K_2ZnH_4. Changing the stoichiometry in the AlH_3 reaction leads to a solid having the composition KZn_3H_7, but this, like the lithium analogue, is a mixture of ZnH_2 and KZn_2H_5.

The complex $\text{Mg(ZnH}_3)_2 \cdot 0.5 \text{ Et}_2\text{O}$ is obtained by:

$$\text{Me}_2\text{Mg} + 2 \text{ Me}_2\text{Zn} \rightarrow \text{Mg(ZnMe}_3)_2 \qquad \text{(o)}$$

$$\text{Mg(ZnMe}_3)_2 + \text{LiAlH}_4 \rightarrow \text{Mg(ZnH}_3)_2 \qquad \text{(p)}$$

X-Ray powder examination shows that it is not a physical mixture of MgH_2 and ZnH_2, although when heated its thermal decomposition proceeds via dissociation into MgH_2 + 2 ZnH_2 after initial loss of ether.

Aluminum hydride and Li methylzinc hydrides form a further series of complexes:

$$\text{AlH}_3 + \text{LiZnMe}_2\text{H} \rightarrow \text{LiZnMe}_2\text{AlH}_4 \qquad \text{(q)}$$

$$\text{AlH}_3 + \text{LiZn}_2\text{Me}_4\text{H} \rightarrow \text{LiZn}_2\text{Me}_4\text{AlH}_4 \qquad \text{(r)}$$

Spectroscopic data suggest structures (I) and (II):

(I)

(II)

The borane complex, $[\text{NaOEt}_2][\text{Zn(BH}_4)_3]$, formed[8] by reaction of xs NaBH_4 on ZnCl_2 in ether is stable to 80°C. Anionic complexes are synthesized[9,10]:

$$\text{Zn(BH}_4)_2 + 2 \text{ LiBH}_4 \rightarrow \text{Li}_2\text{Zn(BH}_4)_4 \qquad \text{(s)}$$

Other starting materials may be used to give anionic-borane complexes:

$$\text{Na}_2\text{Zn(OMe)}_4 + \text{B}_2\text{H}_6 \rightarrow \text{Na}_2\text{Zn(BH}_4)_4 \qquad \text{(t)}$$

$$\text{NaZn(OMe)}_3 + \text{B}_2\text{H}_6 \rightarrow \text{NaZn(BH}_4)_3 \qquad \text{(u)}$$

Surprisingly, the nature of the isolated product depends on the cation, and trimetallic borohydride ions are obtained:

$$3 \text{ Zn(BH}_4)_2 + 2 \text{ KBH}_4 \rightarrow \text{K}_2\text{Zn}_3\text{(BH}_4)_8 \qquad \text{(v)}$$

1.9. Formation of Bonds between Hydrogen and Cu, Ag, Au or Zn, Cd, Hg   171
1.9.5. by Neutral and Anionic Metal Hydride Reduction
1.9.5.2. of Compounds of Group IIB.

$$3 \, Zn(BH_4)_2 + Ba(BH_4)_2 \rightarrow BaZn_3(BH_4)_8 \qquad (w)$$

Infrared and ^{11}B NMR evidence suggests the presence of ZnH_2BH_2 units with eight-co-ordinated Zn in the $[Zn(BH_4)]_4^{2-}$ anion.

The cleavage of dimethylzinc by borohydride can also yield the unsymmetrical compound[11], $MeZnBH_4$. Its IR spectrum shows a five-coordinated zinc polymer with ZnH_2BH_2Zn units:

(III)

Impure cadmium bis(tetrahydroborate), $Cd(BH_4)_2$, is obtained by:

$$CdCl_2 + 2 \, LiBH_4 \xrightarrow{Et_2O} Cd(BH_4)_2 \qquad (x)$$

It decomposes at ca. 25°C and is air sensitive[12]. The main product is the complex salt, $Li_2[Cd(BH_4)_4]$, isolated as an oily etherate. The analogous reaction using Na boro-hydride also yields a complex salt, $Na[Cd(BH_4)_3]$.

Neutral Cd borohydride complexes analogous to those of Zn also are isolated and are more thermally stable than $Cd(BH_4)_2$:

$$Cd(OMe)_2 + B_2H_6 \xrightarrow{THF} Cd(BH_4)_2 \cdot THF \qquad (y)$$

Pyridine or NH_3 may take the place of THF in these complexes[12–13].

Both Zn and Cd dimethyls (but not Me_2Hg) react with hexaborane(10) at 0°C lib-erating methane. From the Zn reaction the complex $Zn(THF)_2(B_6H_9)_2$ is isolated as a white solid, whereas the Cd analogue contains only a little coordinated THF. Both compounds are stereochemically nonrigid by ^1H NMR spectroscopy, with both the metal and bridging hydrogen atoms involved in the dynamic processes. However, at low T their structures result from insertion of Zn (or Cd) into a basal boron–boron bond in the anion[13]. Mercury carboranes are known, but their structures involve B—Hg—B bonding as in[14] $\mu,\mu\text{-}(Me_2B_4H_5)_2Hg$.

(F. GLOCKLING)

1. G. D. Barbaras, C. Dillard, A. E. Finholt, T. Wartik, K. E. Wilzbach, H. I. Schlesinger, *J. Am. Chem. Soc., 73*, 4585 (1951).
2. E. Wiberg, W. Henle, R. Bauer, *Z. Naturforsch., Teil B, 6*, 393 (1957).
3. E. Wiberg, W. Henle, *Z. Naturforsch., Teil B, 6*, 461 (1951).
4. E. C. Ashby, H. S. Prasad, *Inorg. Chem., 14*, 1608 (1975).
5. A. B. Gall, S. Gall, E. C. Ashby, *Inorg. Chem., 18*, 1433 (1969).
6. E. C. Ashby, K. C. Nainan, H. S. Prasad, *Inorg. Chem., 16*, 348 (1977).
7. E. C. Ashby, J. J. Watkins, *Inorg. Chem., 16*, 1445 (1977).
8. N. N. Maltseva, N. S. Kedrova, V. I. Mikleava, *Russ. J. Inorg. Chem. (Engl. Transl.), 18*, 1054 (1973).
9. H. Nöth, E. Wiberg, L. P. Winter, *Z. Anorg. Allg. Chem., 386*, 73 (1971).

10. N. N. Maltseva, N. S. Kedrova, V. V. Klinkova, N. A. Chumaevski, *Russ. J. Inorg. Chem.* (*Engl. Transl.*), *20,* 339 (1975).
11. J. W. Nibler, T. H. Cook, *J. Chem. Phys., 58,* 1596 (1973).
12. H. Nöth, L. P. Winter, *Z. Anorg. Allg. Chem., 389,* 225 (1972).
13. D. L. Denton, W. R. Clayton, M. Mangion, S. G. Shore, E. A. Meyers, *Inorg. Chem., 15,* 541 (1976).
14. N. S. Hosmane, R. N. Grimes, *Inorg. Chem., 18,* 2886 (1979).

1.9.6. by Other Methods.

When $CuSO_4$ is treated with hypophosphorous acid, $H_2P(O)OH$, in H_2O a red-brown solid separates[1,2] consisting mainly of CuH, although it also contains some Cu, Cu_2O and H_2O so that the ratio Cu:H varies from 1:0.97 to 1:0.65. One of the advantages of this method is its simplicity, and by using deuterohypophosphorous acid, the corresponding deuteride, CuD, may be obtained. Presumably, the impurities, including H_2O, stabilize Cu(I) hydride formed in this way.

Liquid NH_3 solns of CuI and an alkali-metal amide react with H_2 to give the solvated Cu hydride complexes[3], $K_2Cu_3H_5(NH_3)_x$ and $CsCuH_2(NH_3)_x$.

The chemisorption of H_2 on Zn oxide is, in part, fast and reversible at RT, and this system involves both Zn—OH and Zn—H bonds (by IR), formed by the dissociation of H_2 (or D_2) on Zn oxide pair sites[4].

Gas-phase HgH can be produced by the photosensitized decomposition of H_2 or alkanes[5]:

$$Hg(6^3P1_DO) + RH \rightarrow HgH + R^\cdot \qquad \text{(a)}$$

(F. GLOCKLING)

1. J. A. Goldkoop, A. F. Anderson, *Acta Crystallogr., 8,* 118 (1955).
2. J. C. Warf, W. Keitknecht, *Helv. Chim. Acta, 33,* 613 (1950).
3. K. A. Strom, W. L. Jolly, *J. Inorg. Nucl. Chem., 35,* 3445 (1973).
4. C. C. Chang, R. Kokes, *J. Am. Chem. Soc., 93,* 7107 (1971).
5. A. G. Vikis, D. J. Le Roy, *Can. J. Chem., 50,* 595 (1972).

1.10. Formation of Bonds between Hydrogen and Transition and Inner-Transition Metals

1.10.1. Introduction

Molecular transition-metal hydrides, known since the early 1930s, cover a variety of structural types involving mono- and bidentate hydrogen atoms. Their synthesis is described here. Reactions yielding reversible transition and inner transition-metal hydrides are discussed in §1.12.

(J.J. ZUCKERMAN)

1.10.2. from the Elements.

Few examples exist for preparing molecular transition-metal hydrides from the bulk metal. Hydride complexes of Co and Fe are prepared from the bulk metal, H_2 and CO or a phosphine ligand $[PF_3, $ or o-$C_6H_4(PEt_2)_2]$ at high T and P:

$$Co + 4\ CO + \tfrac{1}{2}H_2 \xrightarrow[\text{(25 MPa)}]{\text{250 atm, 180°C}} HCo(CO)_4 \qquad (a)^{1,2}$$

$$Co + 4\ PF_3 + \tfrac{1}{2}H_2 \xrightarrow[\text{(20 MPa)}]{\text{200 atm, 250°C}} HCo(PF_3)_4 \qquad (b),^3$$

$$Co + 2\ PF_3 + 2\ CO + \tfrac{1}{2}H_2 \xrightarrow[\text{(20 MPa)}]{\text{200 atm, 250°C}} HCo(PF_3)_{4-n}(CO)_n\ (n = 1-3) \qquad (c),^3$$

$$Fe + 2\ o\text{-}C_6H_4(PEt_2)_2 + H_2 \rightarrow HFe_2[o\text{-}C_6H_4(PEt_2)_2]_2 \qquad (d),^{4,5}$$

Cocondensation of metal-atom vapors and organic ligands yields transition-metal hydrides as well as other organometallics (see Table 1)[6]. The bulk metal is vaporized at a controlled rate by resistance heating, laser heating or electron bombardment under vacuum $[< 10^{-3}$ torr $(<0.1$ Pa)]. The atoms condense either on the cooled walls $(< -100°C)$ of the vacuum chamber with an excess of the organic compound or into a solution of the compound in an inert solvent. Yields of hydride often are based on the amount of metal evaporated.

(T. J. LYNCH)

1. W. Hieber, H. Schulten, B. Marin, *Z. Anorg. Allg. Chem.*, *240*, 261 (1939).
2. M. Orchin, *Acc. Chem. Res.*, *14*, 259 (1981).
3. T. Kruck, K. Baur, W. Lang, *Chem. Ber.*, *101*, 138 (1968).
4. J. Chatt, F. A. Hart, R. G. Hayter, *Nature (London)*, *187*, 55 (1960).
5. J. Chatt, F. A. Hart, D. T. Rosevear, *J. Chem. Soc.*, 5504 (1961).
6. P. L. Timms, T. W. Turney, *Adv. Organomet. Chem.*, *15*, 53 (1977).

TABLE 1. TRANSITION-METAL HYDRIDES SYNTHESIZED BY METAL-ATOM COCONDENSATION WITH ORGANIC LIGANDS

Metal hydride (yield in %)	Organic compound	Ref.
h^5-$C_8H_{11}CrH(PF_3)_3$ (2)	1,5-Cyclooctadiene, PF_3	7
h^5-$C_5H_5MoH_2$ (30–50)	Cyclopentadiene	8,9
$(h^5$-$C_9H_7)_2MoH_2$	Indene	9
h^6-$C_9H_8MoH(C_9H_7$-$h^5)$	Indene	9
$(h^5$-$C_5H_5)_2WH_2$ (40–60)	Cyclopentadiene	8,9
$C_6H_6WH(C_5H_5$-$h^5)$ (30)	Benzene, cyclopentadiene	9
$C_6H_5MeWH(C_5H_5$-$h^5)$ (55)	Toluene, cyclopentadiene	9
$(h^5$-$C_9H_7)_2WH_2$	Indene	9
h^6-$C_9H_8WH(C_9H_7$-$h^5)$	Indene	9
$H_2Fe(PF_3)_4$	Propene, PF_3	10
$(PMe_3)_3HFeCH_2PMe_2$ (6)	Trimethylphosphine	11
$(1,3$-$C_4H_6)_2CoH$	1,3-Butadiene, isobutane	12
$(1,3$-pentadiene$)_2CoH$	1,4-Pentadiene	12

7. E. A. K. von Gustorf, O. Jaenicke, O. Wolfbeis, C. R. Eady, *Angew. Chem., Int. Ed. Engl.*, *14*, 278 (1975).
8. M. T. D'Annello, E. K. Barefield, *J. Organomet. Chem.*, *76*, C50 (1974).
9. E. M. Van Dam, W. N. Brent, M. P. Silvon, P. S. Skell, *J. Am. Chem. Soc.*, *97*, 465 (1975).
10. C. Francis, B.Sc. Thesis, Univ. Bristol, England (1974); P. W. Taylor, B.Sc. Thesis, Univ. Bristol, England (1974).
11. P. L. Timms, T. W. Turney, unpublished results.
12. P. S. Skell, M. J. McGlinchey, *Angew. Chem., Int. Ed. Engl.*, *14*, 195 (1975).

1.10.3. by Hydrogenation

1.10.3.1. of Metal Oxides.

When OsO_4 (**Extremely toxic**[2]; **all manipulations should be carried out in a well-ventilated hood**) is treated with H_2 and CO in a ratio of 1:3 at high T and P, $H_2Os(CO)_4$ is formed in high yield[1]:

$$OsO_4 + H_2 + CO \xrightarrow[\text{heptane, 6 h}]{180 \text{ atm (18 MPa), }160°C} H_2Os(CO)_4 \text{ (100\%)} \qquad (a)$$

The polynuclear dihydride $H_2Os_3(CO)_{10}$ is prepared in two steps from OsO_4 by first preparing[2] $Os_3(CO)_{12}$ then treating this with H_2 at 1 atm[3].

$$OsO_4 + CO \xrightarrow[\text{methanol, 12 h}]{75 \text{ atm (7.6 MPa), }125°C} Os_3(CO)_{12} \text{ (75\%)} \qquad (b)$$

$$Os_3(CO)_{12} + H_2 \xrightarrow[\text{octane, 1.5 h}]{1 \text{ atm, (0.1 MPa), }120°C} H_2Os(CO)_4(73\%) \qquad (c)$$

(T. J. LYNCH)

1. F. L'Eplattenier, F. Calderazzo, *Inorg. Chem.*, *6*, 2092 (1967).
2. B. F. G. Johnson, J. Lewis, *Inorg. Synth.*, *13*, 93 (1972).

3. S. A. R. Knox, J. W. Koepke, M. A. Andrews, H. D. Kaesz, *J. Am. Chem. Soc.*, 97, 3942 (1975).

1.10.3.2. of Metal Salts.

Transition-metal salts are used to synthesize hydrides incorporating the trifluorophosphine ligand[1], PF_3 (Table 1). The reactions are performed at high T and at high P of H_2 and PF_3. Powdered Cu or Zn in the reaction mixture acts as a reducing agent and halogen acceptor. The yields are high:

$$MX_n + H_2 + PF_3 \xrightarrow{\text{Cu or Zn}} MH(PF_3)_n \qquad (a)$$

When transition-metal salts are combined with H_2, alkali-metal and free-ligand polyhydrides result, e.g., for the Ta hydride[2]:

$$TaCl_5 + H_2 + dmpe \xrightarrow[\text{12 h, K, THF}]{\text{10 atm (1 MPa), 115°C}} H_5Ta(dmpe) \ (<40\%) \qquad (b)$$

where dmpe = bis(dimethylphosphino)ethane and THF = tetrahydrofuran. The yields are variable, but when the Ta source is $[TaPh_6]^-$, reproducible yields of 30–40% are obtained. The hydride ligands can be exchanged for deuterium [60 atm, (6 MPa), 80°C], and this complex catalyzes the exchange of hydrogen on benzene.

Binary transition-metal acetates also form hydrides when treated with free ligand, H_2 and Na amalgam. The Mo dimer forms a tetrahydride[3]:

$$Mo_2(O_2CMe)_4 + H_2 + PMe_3 \xrightarrow[\text{24 h, Na(Hg)}]{\text{3 atm (0.3 MPa), 25°C}}$$

$$(Me_3P)_3HMo(\mu\text{-}H_2)MoH(PMe_3)_3 \ (80\%) \qquad (c)$$

TABLE 1. TRIFLUOROPHOSPHINE-HYDRIDE COMPLEXES PREPARED FROM TRANSITION-METAL SALTS

Metal salt	Conditions	Product (yield, %)
$ReCl_5$	Cu, PF_3 (250 atm, 25 MPa), H_2 (100 atm, 10 MPa), 300°C	$HRe(PF_3)_5$ (40)
FeI_2	Zn, PF_3 (300 atm, 30 MPa), H_2 (100 atm, 10 MPa), 250°C	$H_2Fe(PF_3)_4$ (trace)
$RuCl_3$	Cu, PF_3 (300 atm, 30 MPa), H_2 (100 atm, 10 MPa), 270°C	$H_2Ru(PF_3)_4$ (70)
$OsCl_3$	Cu, PF_3 (400 atm, 41 MPa), H_2 (100 atm, 10 MPa), 270°C	$H_2Os(PF_3)_4$ (80)
CoI_2	Cu, PF_3 (50–300 atm, 5–30 MPa), H_2 (100 atm, 10 MPa), 170°C	$HCo(PF_3)_4$ (100)
$RhCl_3$	Cu, PF_3 (90 atm, 9 MPa), H_2 (100 atm, 10 MPa), 70°C	$HRh(PF_3)_4$ (100)
$IrCl_3$	Cu, PF_3 (160 atm, 16 MPa), H_2 (45 atm, 46 MPa), 215°C	$HIr(PF_3)_4$ (100)

Niobium pentachloride is used in the synthesis of $NbH_3(C_5H_5-h^5)_2$; H_2 [800 atm (81 MPa)], $Na(C_5H_5)$, $NaBH_4$ and $NbCl_5$ form the trihydride in 20–30% yield. The initial source of the hydride ligand is the $NaBH_4$, but these are in fast exchange with H_2, for separate experiments demonstrate the loss of H_2 at 80°C and also hydride exchange with deuterium of benzene-d_6. The corresponding Ta, Mo and W hydrides, $(h^5-C_5H_5)_2TaH_3$ and $(h^5-C_5H_5)_2MH_2$, M = Mo, W, do not require H_2 in their syntheses but only $NaBH_4$, reflecting the higher stability of these complexes[5].

The Nb cluster Nb_6I_{11} absorbs H_2 at atm P above 420°C to form[6] HNb_6I_{11}.

Metal salts are used in the preparation of metal-carbonyl hydrides; e.g., $HCo(CO)_4$ is made[7] by heating CoS or CoX_2 (X = I, Br, Cl) with H_2 and CO at 200 atm (20 MPa) and 180°C in the presence of Cu. The Ru cluster, $H_4Ru_4(CO)_{12}$, is prepared by bubbling CO through $RuCl_3$ in refluxing EtOH for 4 h, then treating with CO [40 atm (4 MPa)] and H_2 [80 atm (8 MPa)] at 75–100°C for 3 days[8,9].

(T. J. LYNCH)

1. T. Kruck, Agnew. Chem., Int. Ed. Engl., 6, 53 (1967).
2. F. N. Tebbe, J. Am. Chem. Soc., 95, 5823 (1973).
3. K. W. Chiu, R. A. Jones, G. Wilkinson, J. Chem. Soc., Dalton Trans., 1892 (1981).
4. F. N. Tebbe, G. W. Parshall, J. Am. Chem. Soc., 93, 3793 (1971).
5. M. L. H. Green, J. A. McClemerty, L. Pratt, G. Wilkinson, J. Chem. Soc., 4854 (1961).
6. A. Simon, Z. Anorg. Allg. Chem., 355, 311 (1967).
7. W. Hieber, H. Shulten, B. Marin, Z. Anorg. Allg. Chem., 240, 261 (1939).
8. B. F. G. Johnson, R. D. Johnston, J. Lewis, B. H. Robinson, G. Wilkinson, J. Chem. Soc., A., 2856 (1968).
9. Only one isomer of $Ru_4H_4(CO)_{12}$ exists: C. R. Eady, B. F. G. Johnson, J. Lewis, J. Chem. Soc., Dalton Trans., 477 (1977).

1.10.3.3. of Metal Carbonyls.

Transition-metal carbonyls are a source of mono- and polynuclear metal-hydride complexes[1]. Osmium carbonyls are the most versatile, e.g., $Os(CO)_5$ and $Os_3(CO)_{12}$ react with H_2 at 80 atm (8.1 MPa) at 100°C in 6 h to form $H_2Os(CO)_4$. Whereas $H_2Ru(CO)_4$ can be observed only under high H_2 pressure on $Ru_3(CO)_{12}$ by IR spectroscopy, phosphine substitution for carbonyls stabilizes hydride products. Therefore, $RuH_2(CO)_2(PPh_3)_2$ (50%) can be prepared from the hydrogenation for 12 h of $Ru(CO)_3$-$(PPh_3)_2$ at 120 atm (12 MPa) and 130°C in tetrahydrofuran (THF)[3]. Likewise, H_3Fe_3-$(\mu_3COMe)(CO)_9$ is unstable relative to H_2 elimination, but $H_3Fe_3(\mu_3-COMe)(CO)_7$-$(SbPh_3)_2$ can be isolated in fair yield by the hydrogenation of $HFe_3(\mu-COMe)(CO)_{10}$ in the presence of $SbPh_3$.

The coordinatively unsaturated cluster $Os_3H_2(CO)_{10}$ is formed[5] at atm P:

$$Os_3(CO)_{12} + H_2 \xrightarrow[\text{1.5 h, 73\%}]{\text{octane, 120°C,}} H_2Os_3(CO)_{10} \qquad (a)$$

This method, which does not require high-P apparatus, is also applicable to the synthesis of $H_3Re_3(CO)_{12}$ (55%)[6,7] and $H_4Re_4(CO)_{12}$ (55%)[6] from $Re_2(CO)_{10}$; $H_4Ru_4(CO)_{12}$ (88%) from $Ru_3(CO)_{12}$; $H_4Os_4(CO)_{12}$ (29%) from $Os_3(CO)_{12}$; and $H_4FeRu_3(CO)_{12}$ (83%)[5] from $H_2FeRu_3(CO)_{13}$.

Hydrogenolysis of metal carbonyls, such as $Mn_2(CO)_{10}$ [250 atm (25 MPa), 200°C] or $Co_2(CO)_8$ [250 atm (25 MPa), 110°C], leads to metal–metal bond cleavage, forming $HMn(CO)_5$ [8] or $HCo(CO)_4$ [9]. The conversion of $Co_2(CO)_8$ to $HCo(CO)_4$ is the rate-determining step in Co-catalyzed hydroformylations at high P and T. Tertiary amines, nitrogen heterocycles, tertiary phosphorus bases or halide ions enhance the rate of $HCo(CO)_4$ formation[10]; e.g., pyridine can increase the rate of $HCo(CO)_4$ formation 300-fold at 40°C.

Metal dimers containing cyclopentadienyl ligands also can be separated to monomers with H_2 as for $(h^5-C_5H_5)_2Cr_2(CO)_6$ at 150 atm (15 MPa) and 70°C to yield[11] $HCr(C_5H_5-h^5)(CO)_3$. Mixed-metal hydride clusters may be synthesized from the H_2-assisted condensation of two different metal carbonyls. When $Os_3(CO)_{12}$ or $Ru_3(CO)_{12}$ reacts with $Ni_2(C_5H_5-h^5)_2(CO)_2$ in the presence of H_2 at atm P, the mixed-metal clusters $M_3(\mu-H)_3Ni(C_5H_5-h^5)(CO)_9$ [M = Ru (39%), Os (70%)] are formed[12].

The addition of H_2 to metal carbonyls also occurs with clusters containing noncarbonyl ligands, e.g., $HM_3(\mu-COMe)(CO)_{10}$ react with H_2 at atm P in refluxing hexane for 2 h to give $H_3M_3(\mu_3-COMe)(CO)_9$ [M = Ru (93%), Os (79%)][13]. The hydride ligands bridge the metal–metal bonds. The reaction[14] proceeds by dissociation of a carbonyl ligand prior to the rate-determining oxidative addition of H_2. Under severe conditions [CO, H_2, 34 atm (3.4 MPa), 130°C, 23 h, M = Ru] the methoxycarbyne ligand is hydrogenated fully to dimethyl ether, while the cluster is isolated as $Ru_3(CO)_{12}$.

Noncarbonyl ligands can participate more readily in the hydrogenation of the metal carbonyl[15] as for $Ru_3(CO)_{11}(CNBu-t)$. Bubbling H_2 through the cluster in cyclohexane at reflux for 1 h yields $HRu_3(\mu_3-HCNBu-t)(CO)_9$ (52%) by addition of H_2 to the coordinated isocyanide. These reactions require milder conditions than those used for the unfunctionalized $Ru_3(CO)_{12}$ with H_2.

Triruthenium dodecacarbonyl is functionalized with two edge-bridging bisphosphine or bisarsine ligands to form $Ru_3(CO)_8(L-L)_2$, where L-L = bis(diphenylphosphino)methane or bis(diphenylarsino)methane, and then treated with H_2 as with $H_2Os_3(CO)_{10}$ above. The imposed retention of the trinuclear Ru frame by the bidentate ligands could favor formation of the Ru analogue of $H_2Os_3(CO)_{10}$; however, concomitant with H_2 addition to the cluster is cleavage of the phenyl-phosphine or -arsine bond to yield the metal hydride, $Ru_3(\mu-H)_2[\mu-P(Ph)CH_2P(Ph_2)]_2(CO)_6$

(T. J. LYNCH)

1. H. D. Kaesz, *Chem. Ber.*, *9*, 344 (1973).
2. F. L'Eplattenier, F. Calderazzo, *Inorg. Chem.*, *6*, 2092 (1967).
3. F. L'Eplattenier, F. Calderazzo, *Inorg. Chem.*, *7*, 1290 (1968).
4. J. B. Keister, M. W. Payne, M. J. Muscatella, *Organometallics*, *2*, 219 (1983).
5. S. A. R. Knox, J. W. Koepke, M. A. Andrews, H. D. Kaesz, *J. Am. Chem. Soc.*, *97*, 3942 (1975).
6. H. D. Kaesz, S. A. R. Knox, J. W. Koepke, R. B. Saillant, *J. Chem. Soc., Chem. Comm.*, 477 (1971).
7. Another satisfactory route to this compound is described: M. A. Andrews, S. W. Kirtley, H. D. Kaesz, *Inorg. Synth.*, *17*, 66 (1977).
8. W. Hieber, G. Wagner, *Z. Naturforsch., Teil B, 13,* 339 (1958).
9. M. Orchin, *Acc. Chem. Res., 14,* 259 (1981).
10. A. Sisak, F. Ungváry, L. Markó, *Organometallics, 2,* 1244 (1983). The $Fe(CO)_5$-catalyzed hydrogenation of nitrogen heterocycles is similarly enhanced: T. J. Lynch, M. Banah, H. D. Kaesz, C. R. Porter, *J. Org. Chem., 49,* 1266 (1984).

11. E. O. Fischer, W. Hafner, H. O. Stahl, *Z. Anorg. Allg. Chem., 282*, 47 (1955).
12. G. Lavigne, F. Papageorgiou, C. Bergounhou, J. J. Bonnet, *Inorg. Chem., 22*, 2485 (1983).
13. J. B. Keister, M. W. Payne, M. J. Muscatella, *Organometallics, 2*, 219 (1983).
14. L. M. Bavaro, P. Montangero, J. B. Keister, *J. Am. Chem. Soc., 105*, 4977 (1983).
15. M. I. Bruce, R. C. Wallis, *Aust. J. Chem., 35*, 709 (1982).

1.10.4. by Oxidative Addition of Hydrogen

1.10.4.1. to Neutral, Coordinatively Unsaturated Species

1.10.4.1.1. Involving Iridium.

Oxidative addition of H_2 occurs to coordinatively unsaturated complexes of Ir(I):

$$IrCl(CO)(PPh_3)_2 + H_2 \text{ (1 atm)} \rightleftharpoons cis\text{-}H_2IrCl(CO)(PPh_3)_2 \qquad (a)^1$$

$$HIr(CO)(PPh_3)_2 + H_2 \text{ (1 atm)} \rightleftharpoons H_3Ir(CO)(PPh_3)_2 \qquad (b)^2$$

$$HIr(CO)(dppe) \xrightarrow{\;H_2 \text{ (or } D_2)\;} H_3Ir(CO)(dppe) \qquad (c)^3$$

where dppe = $Ph_2P(CH_2)_2PPh_2$. The kinetics of the addition of H_2 to square-planar Ir(I) complexes are known[4,5].

Oxidative addition of H_2 also occurs to the $[Ir(\mu\text{-SBu-t})(CO)(PR_3)]_2$ (R = Me, Ph, NMe_2, OMe) thiolato-bridged dinuclear Ir complexes. These react irreversibly with H_2 to give quantitative yields of the thiolato-bridged dihydridoiridium complexes $[Ir(H)(\mu\text{-SBu-t})(CO)(PR_3)]_2$. This complex is diamagnetic and contains an Ir—Ir single bond. The dihydrido complex can be protonated. The interaction between H_2 and $[Ir(\mu\text{-SBu-t})(CO)(PR_3)]_2$ can be interpreted as a one-electron oxidative addition to each metal atom. An electron-pair coupling interaction between the two Ir^{2+} (d^7) is necessary for each Ir to attain a closed-shell configuration and is in accord with the diamagnetic character of these complexes. In the dihydrido complex, $[Ir(H)(\mu\text{-SBu-t})(CO)P(OMe)_3]_2$, each Ir is in a rectangular-pyramidal environment with an Ir–Ir distance of 267.3 pm. The formation of the dihydrido complex occurs through an Ir^{3+}–Ir^+ intermediate in which the two hydride ligands are bound to the same metal atom[6,7].

Iridium complexes containing 1,2- and 1,7-dicarba-closo-dodecarborane(12) (carb) are formed[8,9] through a metal–carbon σ bond:

$$trans\text{-}(PR_3)_2Ir(CO)Cl + Li(carb) \rightarrow trans\text{-}(PR_3)_2Ir(CO)(\sigma\text{-carb}) \qquad (d)$$

These complexes are isoelectronic and isostructural with the bis-phosphine Ir carbonyl chloride complex, but they differ in two respects: (a) the reaction with H_2 is irreversible and (b) the oxidative addition of H_2 yields three different cis isomers as a result of a solvent dependence for this addition. The complexes $trans\text{-}(PR_2R')_2Ir(CO)(\sigma\text{-carb})$ (R = C_6H_5; R' = C_6H_5, CH_3) react with H_2 or D_2, giving the $(PR_2R')_2IrH_2$ (or D_2)(CO)(σ-carb) dihydrides or dideuterides. These complexes are colorless crystalline compounds, nonelectrolytes in solution and stable with respect to thermal loss of either carborane or H_2; however, they are light sensitive.

$$
\begin{array}{ccc}
\underset{L}{\overset{\displaystyle OC}{\diagdown}}\!\!\underset{\displaystyle L}{\overset{\displaystyle H}{\underset{\displaystyle |}{Ir}}}\!\!\underset{}{\overset{\displaystyle H}{\diagup}} Y & \underset{OC}{\overset{\displaystyle L}{\diagdown}}\!\!\underset{\displaystyle Y}{\overset{\displaystyle H}{\underset{\displaystyle |}{Ir}}}\!\!\underset{}{\overset{\displaystyle H}{\diagup}} L & \underset{OC}{\overset{\displaystyle L}{\diagdown}}\!\!\underset{\displaystyle L}{\overset{\displaystyle H}{\underset{\displaystyle |}{Ir}}}\!\!\underset{}{\overset{\displaystyle H}{\diagup}} Y \\
(\mathbf{I}) & (\mathbf{II}) & (\mathbf{III}) \\
(\sim 60\%) & (\sim 40\%) & \text{(small amounts)}
\end{array}
$$

$$
\underset{L}{\overset{\displaystyle CO}{\diagdown}} Ir \underset{Y}{\overset{\displaystyle L}{\diagup}} \xrightarrow{\;H_2,\;\text{nonpolar solvent}\;}
$$

(via polar solvent H₂ / CH₃CN and CH₂Cl₂ or C₂H₄Cl₂)

$$
\begin{array}{cc}
\underset{OC}{\overset{\displaystyle L}{\diagdown}}\!\!\underset{\displaystyle L}{\overset{\displaystyle H}{\underset{\displaystyle |}{Ir}}}\!\!\underset{}{\overset{\displaystyle H}{\diagup}} Y & \underset{OC}{\overset{\displaystyle L}{\diagdown}}\!\!\underset{\displaystyle Y}{\overset{\displaystyle H}{\underset{\displaystyle |}{Ir}}}\!\!\underset{}{\overset{\displaystyle H}{\diagup}} L \\
(\mathbf{III}) & (\mathbf{II}) \\
(100\%) & (100\%)
\end{array}
$$

where L = PPh$_3$, PMePh$_2$; Y = [2-H-1,2-B$_{10}$C$_2$H$_{10}$]$^-$, [7-R-1,7-B$_{10}$C$_2$H$_{10}$]$^-$ (R = H, CH$_3$, C$_6$H$_5$). Solid 1-[(PPh$_3$)$_2$Ir(CO)](2-CH$_3$-1,2-B$_{10}$C$_2$H$_{10}$) is insoluble but reacts with H$_2$. The reaction is complete in a few seconds at RT under 1 atm of H$_2$, yielding a single dihydridocarborane iridium(III) isomer (configuration **III**) that is soluble and stable in organic solvents. Steric effects, resulting from the presence of the large carborane and phosphine ligands, determine the stereochemistry of the oxidative addition reactions of the carborane–iridium(I) complexes.

Trans-[Ir(CO)(PPh$_3$)$_2$(7-C$_6$H$_5$-1,7-B$_{10}$C$_2$H$_{10}$)] reacts with RCN (R = CH$_3$, C$_6$H$_5$) at RT to give monophosphino nitrile Ir(I) complexes, [Ir(CO)(RCN)(PPh$_3$)Y] (Y = 7-C$_6$H$_5$-1,7-B$_{10}$C$_2$H$_{10}$). When [Ir(CO)(RCN)(PPh$_3$)Y] is treated with the bidentate ligand dppe, facile and irreversible displacement of both RCN and PPh$_3$ ligands occurs to give the four-coordinated complex [Ir(CO)(dppe)Y]. This complex reacts with H$_2$ in a benzene suspension to give the white dihydro complex[9]:

$$
\underset{P}{\overset{\displaystyle P}{\diagdown\!\!\!\diagup}}\!\!\underset{\displaystyle CO}{\overset{\displaystyle H}{\underset{\displaystyle |}{Ir}}}\!\!\underset{}{\overset{\displaystyle H}{\diagup}} Y
$$

The reactions of $[Ir(CO)(RCN)(PPh_3)Y]$ are:

$$[Ir(CO)_3(PPh_3)Y] \xrightarrow{H_2} \begin{array}{c} H \\ Ph_3P \diagdown \mid \diagup H \\ Ir \\ OC \diagup \mid \diagdown Y \\ CO \end{array}$$

$$\Big\uparrow CO \qquad\qquad \Big\downarrow CO \qquad\qquad (f)$$

$$[Ir(CO)(RCN)(PPh_3)Y] \xrightarrow{H_2} \begin{array}{c} H \\ Ph_3P \diagdown \mid \diagup H \\ Ir \\ RCN \diagup \mid \diagdown Y \\ CO \end{array}$$

The ^1H NMR spectrum for $[H_2Ir(CO)(RCN)(PPh_3)(7-C_6H_5-1,7-B_{10}C_2H_{10})$ shows two doublets that have chemical shift and coupling constant values consistent with a structure in which the two mutually cis-hydride ligands are trans to CO and RCN, respectively.

Bis(3-diphenylphosphinopropyl)phenylphosphinechloroiridium(I) is prepared from $[IrCl(C_8H_{14})_2]_2$ and the triphosphine ligand. This complex reacts with H_2 in toluene[10]:

$$\begin{array}{c} PPh_2 \\ \mid \\ PhP - Ir - Cl \\ \mid \\ PPh_2 \end{array} \xrightarrow[\text{toluene}]{H_2} \begin{array}{c} PPh_2 \diagdown \begin{array}{c}H \\ \mid\end{array} \diagup H \\ Ir \\ PhP \diagup \mid \diagdown PPh_2 \\ Cl \end{array} \qquad (g)$$

IR and NMR spectroscopy indicate that the dihydrido complex is a cis-mer O_h.

Section 1.10.4 presents enthalpy data for the oxidative addition of substrates to Ir(I) complexes. Table 1 summarizes thermochemical data[11] for the oxidative addition of H_2 to Ir(I) complexes.

(J. TOPICH)

1. L. Vaska, R. E. Rhodes, *J. Am. Chem. Soc.*, 87, 4970 (1965).
2. L. Malatesta, G. Caglio, M. Angoletta, *J. Chem. Soc.*, 6974 (1965).
3. B. J. Fisher, R. Eisenberg, *Organometallics*, 2, 764 (1983).
4. P. B. Chock, J. Halpern, *J. Am. Chem. Soc.*, 88, 3511 (1966).
5. J. Halpern, *Acc. Chem. Res.*, 3, 386 (1970).
6. A. Thorez, A. Maisonnat, R. Poilblanc, *J. Chem. Soc., Chem. Commun.*, 518 (1977).
7. J. J. Bonnet, A. Thorez, A. Maisonnat, J. Galy, R. Poilblanc, *J. Am. Chem. Soc.*, 101, 5940 (1979).
8. B. Longato, F. Morandini, S. Bresadola, *Inorg. Chem.*, 15, 650 (1976).
9. B. Longato, S. Bresadola, *Inorg. Chim. Acta*, 33, 189 (1979).
10. E. Arpac, L. Dahlenburg, *Z. Naturforsch., Teil B*, 36, 672 (1981).
11. J. U. Mondal, D. M. Blake, *Coord. Chem. Rev.*, 47, 205 (1982).
12. L. Vaska, *Acc. Chem. Res.*, 1, 335 (1968).
13. W. Strohmeier, *J. Organomet. Chem.*, 32, 137 (1971).
14. W. Strohmeier, F. J. Muller, *Z. Naturforsch., Teil B*, 24, 931 (1969).
15. L. Vaska, M. F. Werneke, *Trans. N.Y. Acad. Sci.*, 70 (1971).

1.10. Formation of Bonds between Hydrogen and Transition Metals   181
1.10.4. by Oxidative Addition of Hydrogen
1.10.4.1. to Neutral, Coordinatively Unsaturated Species

TABLE 1. THERMOCHEMICAL DATA FOR THE REACTION OF TRANS-$[IrXL'L_2]$ WITH H_2

L	X	L'	$-\Delta H$ (kJ mol^{-1})	T (°C)	Solvent	Ref.
PPh$_3$	Cl	CO	62.3	30	Chlorobenzene	12
			66.1	80	Toluene	14
	Br		71.1	30	Chlorobenzene	15
			34.7	80		14
	I		79.5	30		12
			11.3	80	Toluene	12
P(Tol-4)$_3$	Cl		55.2			14
P(C$_6$H$_{11}$)$_3$			50.2	30		15
			44.8	80	Toluene	13, 14
	Br		51.9			13, 14
	I		75.3			13, 14
P(OPh)$_3$	Cl		51.5			13, 14
	Br		40.6			13, 14
	I		11.3			13, 14
P(Pr-i)$_3$	Cl		44.4			13, 14
	Br		54.8			13, 14
	I		17.2			13, 14
PEt$_3$	Cl		55.6		Chlorobenzene	15
P(Bu-t)Ph$_2$			48.1		Toluene	13, 14
P(CH$_2$Ph)$_3$			62.3			13, 14

1.10.4.1.2. Involving Platinum.

Two-coordinated 14-electron complexes of Pt, Pt(PR$_3$)$_2$, are sources of Pt(0). Preparation of Pt[P(C$_6$H$_{11}$)$_3$]$_2$ and Pt[PMe(Bu-t)$_2$]$_2$ proceeds from Pt(COD)$_2$ (COD = 1,5-cyclooctadiene) with 2 equivs of the tertiary phosphine[1]. Both complexes are air sensitive and dissolve in organic solvents; however, Pt[PMe(Bu-t)$_2$]$_2$ decomposes after a few days when stored as a solid under N$_2$. Bis(tricyclohexylphosphine)Pt in toluene reacts rapidly with H$_2$ at RT to give white crystals[2,3] of trans-H$_2$[P(C$_6$H$_{11}$)$_3$]$_2$Pt.

Phosphine complexes[4,5] of Pt(0) and Pd(0) react with small molecules. Not all PdL$_2$ complexes [where L = PEt$_3$, P(Bu-t)$_3$, PPh(Bu-t)$_2$, P(C$_6$H$_{11}$)$_3$ and P(Pr-i)$_3$] react with H$_2$. Strongly electron donating phosphines are necessary for PtL$_2$ to react with H$_2$ under normal pressure at RT in benzene:

$$Pt(PEt_3)_2 + H_2 \rightarrow \text{trans-}H_2Pt(PEt_3)_2 \tag{a}$$

$$Pt[P(Bu\text{-}t)_3]_2 + H_2 \rightarrow \text{no reaction} \tag{b}$$

$$Pt[PPh(Bu\text{-}t)_2]_2 + H_2 \rightarrow \text{no reaction} \tag{c}$$

$$Pt[P(C_6H_{11})_3]_2 + H_2 \rightarrow \text{trans-}H_2Pt[P(C_6H_{11})]_2 \tag{d}$$

$$Pt[P(Pr\text{-}i)_3]_2 + H_2 \rightarrow \text{trans-}H_2Pt[P(Pr\text{-}i)_3]_2 \tag{e}$$

The reactivity of the Pt phosphine complexes toward H$_2$ increases in the order P(Bu-t)$_3$ << PPh(Bu-t)$_2$ < P(C$_6$H$_{11}$)$_3$ < P(Pr-i)$_3$. The cone angles of the phosphines[6] increase in the order P(Pr-i)$_3$ (160 $\pm$ 10°) < P(C$_6$H$_{11}$)$_3$ (179 $\pm$ 10°) $\leq$ PPh(Bu-t)$_2$ (170 $\pm$ 2°) < P(Bu-t)$_3$ (182 $\pm$ 2°), suggesting that the reactivity of the two-coordinated Pt(0) complexes is governed by the size of the tertiary phosphine. Electronic properties also may be important. The basicity of the phosphines[7] increases in the order $PPh(Bu-t)_2 < P(Pr-i)_3 < P(C_6H_{11})_3 < P(Bu-t)_3$.

cis-Dihydride diphosphine complexes of Pt^{2+} can be prepared[8]. Four-coordinated, planar d^8-metal complexes with cis-hydrido ligands are rare, but stable trans-dihydride complexes are isolated with bulky tertiary phosphines[4,5], e.g., $[Pt(Bu-t)_2P(CH_2)_3P(Bu-t)_2]_2$ is prepared from $PtCl_2[(t-Bu)_2P(CH_2)_3P(Bu-t)_2]$ and Na/Hg (1%) in tetrahydrofuran (THF) at RT. Cis-$H_2Pt[(t-Bu)_2P(CH_2)_3P(Bu-t)_2]$ is prepared from $[Pt(t-Bu)_2P(CH_2)_3$-$P(Bu-t)_2]_2$ in toluene by bubbling H_2 through it for 30 min. The cis-dihydride also can be prepared by reducing $PtCl_2[(t-Bu)_2P(CH_2)_3P(Bu-t)_2]$ with Na/Hg under H_2. The reaction of this dimer with D_2 or $CHCl_3$ is accompanied by formation of a Pt—H-containing species, indicating hydrogen abstraction upon Pt—Pt bond cleavage. Although direct evidence for dissociation to a mononuclear species, $Pt[(t-Bu)_2P(CH_2)_3$-$P(Bu-t)_2]$, is lacking, the instability in solution suggests Pt—Pt bond cleavage. The mass spectrum[5] shows the mononuclear ion (m/e 527) along with the parent ion (m/e 1054).

Cis-$H_2Pt[Men(Ph)P(CH_2)_2P(Ph)Men]$ (Men = 1-menthyl) and cis-$H_2Pt[(t-Bu)$-$(Ph)P(CH_2)_2P(Ph)(Bu-t)]$ are prepared[8] from the corresponding $PtCl_2(diphos)$ complex by reduction with Na/Hg (1%) in THF under H_2. These two complexes are stable crystalline solids, and the dihydrogen coordination is reversible in solution.

$$\text{(f)}$$

where R = t-Bu or Men. This reaction can be followed by ^1H NMR spectroscopy, and the dimeric nature of the product inferred. The large $|^1J(^{195}Pt-^1H)|$ and high $\nu(Pt-H)$ IR frequencies of cis-dihydrides compared with[9] trans-PtH_2L_2 suggest stronger Pt—H bonding, consistent with the stronger trans influence of the hydride compared with that of the phosphine.

The stability of a cis-dihydride depends on the nature of the diphosphine and increases: $Ph_2P(CH_2)_2PPh_2 < Men(Ph)P(CH_2)_2P(Ph)Men < (t-Bu)(Ph)P(CH_2)_2P(Ph)$-$(Bu-t) < Men_2P(CH_2)_2PMen_2 \approx (t-Bu)_2P(CH_2)_nP(Bu-t)_2$ (n = 2, 3).

(J. TOPICH)

1. J. Fornies, M. Green, J. L. Spencer, F. G. A. Stone, J. Chem. Soc., Dalton Trans., 1006 (1977).
2. B. L. Shaw, M. F. Uttley, J. Chem. Soc., Chem. Commun., 918 (1974).
3. A. Immirzi, A. Musco, G. Carturan, U. Belluco, Inorg. Chim. Acta, 12, L23 (1975).
4. T. Yoshida, S. Otsuka, J. Am. Chem. Soc., 99, 2134 (1977).
5. S. Otsuka, J. Organomet. Chem., 200, 191 (1980).
6. C. A. Tolman, J. Am. Chem. Soc., 92, 2956 (1970).
7. C. A. Tolman, J. Am. Chem. Soc., 92, 2953 (1970).
8. T. Yoshida, T. Yamagata, T. H. Tulip, J. A. Ibers, S. Otsuka, J. Am. Chem. Soc., 100, 2063 (1978).
9. S. Otsuka, T. Yoshida, J. Am. Chem. Soc., 99, 2108 (1977).

1.10. Formation of Bonds between Hydrogen and Transition Metals 183
1.10.4. by Oxidative Addition of Hydrogen
1.10.4.1. to Neutral, Coordinatively Unsaturated Species

1.10.4.1.3. Involving Rhodium.

The equilibrium constant[1] in toluene at 25°C:

$$H_2 + RhClL_3 \xrightleftharpoons{K} H_2RhClL_3 \tag{a}$$

is 18 atm^{-1} for L = PPh$_3$, and 40 atm^{-1} for L = [P(tolyl-4)$_3$]. The equilibrium for oxidative addition of H$_2$ is favored for the complex with the better electron-donating phosphine. The equilibrium constant for the sluggish reaction of $\{RhCl[P(tolyl-4)_3]_2\}_2$ in toluene at 25°C is 11 atm^{-1}. The equilibrium constant for the p-tolylphosphine dimer is 3.6 times less than that of the analogous monomer, reflecting a reduced electron density at Rh in the Cl-bridged dimer. The absence of a dimer tetrahydride, even under forcing conditions, indicates that the electron-withdrawing nature of the hydride can be transmitted through the chlorine bridge to the other end of the dimer.

The reaction[2]:

$$H_2 + RhCl[P(tolyl-4)_3]_2B \rightarrow H_2RhCl[P(tolyl-4)_3]_2B \tag{b}$$

[where B = P(tolyl-4)$_3$, pyridine (py), tetrahydrothiophene (THTP) and N-methylimidazole (MeIm)] gives dihydride complexes containing trans phosphines and nonequivalent cis hydrides:

$$\begin{array}{c}
PR_3 \\
Cl \diagdown \ | \diagup H \\
Rh \\
B \diagup \ | \diagdown H \\
PR_3
\end{array}$$

For H$_2$ oxidative addition with RhCl[P(tolyl-4)$_3$]$_3$, K (30.1°C) = 29.9 atm^{-1} and ΔH = −46.0 kJ mol^{-1}, and with RhCl(THTP)[P(tolyl-4)$_3$]$_2$, K (30.0°C) = 21.5 atm^{-1} and ΔH = −48.5 kJ mol^{-1}. Metal–hydrogen bond dissociation energies of 241.0 and 242.3 kJ mol^{-1} result for the P(tolyl-4)$_3$ and THTP adducts, respectively, of RhH$_2$ClB-[P(tolyl-4)$_3$]$_2$. Hence the average metal–hydrogen bond energy is insensitive to differences in these two ligands.

Bubbling H$_2$ through solutions of HRh[P(Pr-i)$_3$]$_2$ and HRh[P(Bu-t)$_3$]$_2$ yields H$_3$Rh[P(Pr-i)$_3$]$_2$ and H$_3$Rh[P(-Bu-t)$_3$]$_2$, respectively[3].

Soluble (chlorosilyl)phosphine complexes of Rh(I) can be polymerized into poly(siloxyphosphine) Rh(I) species[4]. In addition, soluble siloxyphosphine–Rh(I) complexes, L$_3$RhCl and L$_4$Rh$_2$Cl$_2$, where L = [(CH$_3$)$_3$SiO]$_2$(CH$_3$)Si(CH$_2$)$_2$PPh$_2$, serve as models for their polymeric counterparts.

$$RhClL_3 + H_2 \text{ (1 atm)} \rightleftharpoons \begin{array}{c} H \\ Cl \diagdown \ | \diagup L \\ Rh \\ L \diagup \ | \diagdown H \\ L \end{array} \tag{c}$$

$$\begin{array}{c}
L \diagdown Cl \diagdown L \\
Rh Rh \\
L \diagup Cl \diagup L
\end{array} + H_2 \text{ (1 atm)} \rightleftharpoons \begin{array}{c}
 L \\
L \diagdown Cl \diagdown \ | \diagup H \\
Rh Rh \\
L \diagup Cl \diagup \ | \diagdown H \\
 L
\end{array} \tag{d}$$

The uptake of H$_2$ by RhClL$_3$ is reversible at 25°C, as is the uptake by L$_4$Rh$_2$Cl$_2$. The two siloxyphosphine complexes catalyze the hydrogenation of styrene. The hydrogenation rate is higher for the tris(phosphine) than for the tetrakis(phosphine) complex.

(J. TOPICH)

184   1.10. Formation of Bonds between Hydrogen and Transition Metals
1.10.4. by Oxidative Addition of Hydrogen
1.10.4.1. to Neutral, Coordinatively Unsaturated Species

1. C. A. Tolman, P. Z. Meakin, D. L. Linder, J. P. Jesson, *J. Am. Chem. Soc.*, 96, 2762 (1974).
2. R. S. Drago, J. G. Miller, M. A. Hoselton, R. D. Farris, M. J. Desmond, *J. Am. Chem. Soc.*, 105, 444 (1983).
3. T. Yoshida, T. Okano, D. L. Thorn, T. H. Tulip, S. Otsuka, J. A. Ibers, *J. Organomet. Chem.*, 181, 183 (1979).
4. Z. C. Brzezinska, W. R. Cullen, *Inorg. Chem.*, 18, 3132 (1979).

1.10.4.1.4. Involving Fe—Ir and Ta=Ta Clusters.

Heterobimetallic complexes, $FeIr(\mu\text{-}PPh_2)(CO)_x(PPh_3)_2$ (x = 4–6), are synthesized[1] from $Li[Fe(CO)_4(PPh_2)]$ with trans-$IrCl(CO)(PPh_3)_2$. In the parent complex (I):

$$Ph\quad Ph$$
$$\backslash\;/$$
$$P$$
$$/$$
$$(CO)_3(PPh_3)Fe—Ir(CO)_2(PPh_3)$$

(I)

both metal centers are saturated coordinatively with 18 valence electrons (with Fe as a two-electron donor), although cleavage of the Fe—Ir bond leads to a vacent coordination site on Ir. Addition of H_2 (1 atm, 25°C) proceeds rapidly and quantitatively to cleave the Fe—Ir bond. Addition of H_2 to the Ir center is indicated by the nuclear magnetic resonance (NMR) of PPh_3 on Ir, which is a doublet of triplets at $\delta = 1.9$ ppm:

(a)

(b)

Complex (I) also loses CO when heated at 110°C or irradiated ($\lambda = 366$ nm) under N_2. This product (II) also adds H_2 and a tetrahydride complex (III) is formed. The ^1H NMR spectrum (-40°C) shows four distinct hydride resonances (1 : 1 : 1 : 1). Selective decoupling NMR experiments show that each of the terminal PPh_3 ligand couples to two inequivalent hydrides, suggesting structure (III).

The Ta^{3+} dimer, $Ta_2Cl_6(PMe_3)_4$, reacts readily and irreversibly with H_2 under mild conditions (25°C, 1 atm) to form the Ta^{4+} dimer[2], $Ta_2Cl_6(PMe_3)_4H_2$, an example of H_2 addition to a metal–metal multiple bond:

The dihydrido product can be isolated in quantitative yield. Oxidative additions at multiply metal–metal-bonded centers are accompanied by gross structural reorganizations; however the Ta–Ta dimer remains intact.

(J. TOPICH)

1. M. J. Breen, M. R. Duttera, G. L. Geoffroy, G. C. Novotnah, D. A. Roberts, P. M. Shulman, G. R. Steinmetz, *Organometallics, 1,* 1008 (1982).
2. A. P. Sattelberger, R. B. Wilson Jr., J. C. Huffman, *Inorg. Chem., 21,* 4179 (1982).

1.10.4.2. to Cationic, Coordinatively Unsaturated Species of Rhodium, Ruthenium and Iridium.

Cationic transition-metal complexes activate small molecules and serve as homogeneous catalysts, especially for selective hydrogenation of unsaturated organics. Section 1.10.4.2 deals with coordinatively unsaturated complexes in which the ligands remain coordinated after the oxidative addition of H_2. In §1.10.4.4 the complexes may or may not be coordinatively unsaturated; complexes that are not must lose ligand(s) before oxidative addition of H_2. Coordinatively unsaturated complexes are described that exchange ligands either before or after the oxidative addition of H_2.

Bis(ditertiaryphosphine)Rh(I) complexes add H_2 reversibly[1]:

$$[Rh(L_2)_2]BF_4 + H_2 \rightarrow [cis\text{-}H_2Rh(L_2)_2]BF_4 \qquad (a)^1$$

where L_2 is either 1,3-bis(diphenylphosphino)propane (dppp) or 2,3-O-isopropylidene-2,3-dihydroxy-1,4-bis(diphenylphosphino)butane [(+)-diop]. The $[H_2Rh(diop)_2]^+$ cation is known as the $[ClO_4]^-$ and $[PF_6]^-$ salts[2]. When L_2 is bis(diphenylphosphino)methane (dppm) or bis(diphenylphosphino)ethane (dppe), the Rh complex cations are unreactive toward H_2, whereas a mixture of hydrides is formed when L_2 is bis(diphenylphosphino)butane (dppb).

The cations $[cis\text{-}H_2Rh(dppp)_2]^+$ and $[cis\text{-}H_2Rh[(+)\text{-}diop]_2]^+$ are similar to the known[3] $[cis\text{-}H_2Rh[P(CH_3)_3]_4]^+$ and show a $\upsilon(Rh–H)$ stretch in the IR, a pair of multiplets in the high-field nuclear magnetic resonance (NMR) at 25°C and two resonances in the ^{31}P NMR. The lower field resonance for each system is attributed to the mutually trans-P atoms, and that at higher field to the P atoms trans to the hydrides. The presence of two ^{31}P resonances at all temperatures rules out a cis–trans rearrangement. Both $[cis\text{-}H_2Rh(dppp)_2]^+$ and $[cis\text{-}H_2Rh[(+)\text{-}diop]_2]^+$ lose H_2 on dissolving in solution under Ar.

With similar bidentate phosphines, as well as $PMePh_2$, Ru^{2+} hydrides react[4] with H_2 at RT and ambient P:

$$[HRu(L_2)_2]^+ \underset{-H_2}{\overset{+H_2}{\rightleftharpoons}} [H_3Ru(L_2)_2]^+ \underset{+H^+}{\overset{-H^+}{\rightleftharpoons}} H_2Ru(L_2)_2 \qquad (b)$$

where $L_2 = (PMePh_2)_2$, dppp, dppb.

The salts $[H_3Ru(L_2)_2]^+$ lose 1 equiv of H_2 on warming in alcohol under Ar. When $L_2 = $ dppp prolonged heating is required for H_2 loss.

Cationic coordinatively unsaturated Ir^+ complexes react with H_2 in homogeneous hydrogenation reactions.[5-7]:

$$[Ir(COD)L_2]^+ + H_2 \xrightarrow{-80°C} [cis-H_2Ir(COD)L_2]^+ \qquad (c)$$

where $L = 1/2$ diop, $1/2$ dppm, $P(Bu-n)_3$, PPh_3, $PMePh_2$; $COD = 1,5$-cyclooctadiene;

$$[Ir(COD)L(py)]^+ + COD + 2 H_2 \xrightarrow[0°C]{CH_2Cl_2} [cis-H_2Ir(COD)L(py)]^+ + COE \qquad (d)$$

where $L = P(Pr-i)_3$, $P(C_6H_{11})_3$; $COE = $ cyclooctene;

$$[Ir(COD)(py)_2]^+ + H_2 \rightarrow \text{no reaction} \qquad (e)$$

$$[Ir(COD)_2]^+ + H_2 \xrightarrow{-80°C} [cis-H_2Ir(COD)_2]^+ \qquad (f)$$

$$[Ir(olefin)_2L_2]^+ + H_2 \rightarrow [cis,cis,trans-H_2Ir(olefin)_2L_2]^+ \qquad (g)$$

where olefin $= C_2H_4$, $PhCH=CH_2$, COE; $L = PMePh_2$.

Reaction (c) is reversible. On warming, H_2 is lost and the parent complex is recovered in high yield. If H_2 is bubbled through the solution [Eq. (c)] as it is warmed to RT, the Ir complex is completely hydrogenated to give a polynuclear Ir hydride, $[HL_2IrH_3IrHL_2]^+$. Reaction (d) involves $[Ir(COD)L(py)]^+$, the most active hydrogenation catalyst among the Ir complexes in reactions (c)–(g). Reaction (f) shows that electron-withdrawing ligands, such as olefins, do not deactivate the metal toward addition of H_2. In fact, ^{13}C-NMR evidence[7,8] supports the reductive character of the addition. An upfield shift in the vinyl-carbon resonances of $[Ir(COD)_2]^+$ on addition of H_2 suggests that the electron density at the Ir increases.

Electron-donating ligands normally enhance the oxidative addition of H_2 to a metal center, but addition of H_2 to Ir complexes can be promoted by electron-withdrawing ligands and inhibited by electron-donating ones. The addition of H_2 to these complexes may be not oxidative, but reductive in character. This unusual reactivity pattern allows the catalytic hydrogenation systems derived from them to be less susceptible to oxidizing poisons.

(J. TOPICH)

1. B. R. James, D. Mahajan, *Can. J. Chem.*, *58*, 996 (1980).
2. D. Sinuo, H. B. Kagan, *J. Organomet. Chem.*, *114*, 325 (1976).
3. R. R. Schrock, J. A. Osborn, *J. Am. Chem. Soc.*, *93*, 2397 (1971).
4. T. V. Ashworth, E. Singleton, *J. Chem. Soc., Chem. Commun.*, 705 (1976).
5. R. Crabtree, *Acc. Chem. Res.*, *12*, 331 (1979).
6. R. H. Crabtree, H. Felkin, T. Fillebeen-Khan, G. E. Morris, *J. Organomet. Chem.*, *168*, 183 (1979).
7. R. H. Crabtree, J. M. Quirk, *J. Organomet. Chem.*, *199*, 99 (1980).
8. R. H. Crabtree, G. G. Hlatky, *Inorg. Chem.*, *19*, 571 (1980).

1.10.4.3. to Neutral Species with Replacement of Coordinated Ligands

In contrast with §1.10.4.1, which deals with the oxidative addition of H_2 to neutral coordinatively unsaturated species, §1.10.4.3 describes the action of H_2 with neu-

1.10. Formation of Bonds between Hydrogen and Transition Metals   187
1.10.4. by Oxidative Addition of Hydrogen
1.10.4.3. to Neutral Species

tral transition-metal complexes that lose ligand(s) during the reaction. The loss of a coordinated ligand can be photoinduced.

(J. TOPICH)

1.10.4.3.1. Involving Niobium, Tantalum and Zirconium.

In the presence of H_2, $(h^5\text{-}C_5H_5)_2M(CO)H$ (M = Nb and Ta) can be photolyzed[1]:

$$(h^5\text{-}C_5H_5)_2M(CO)H + H_2 \xrightarrow{h\nu} (h^5\text{-}C_5H_5)_2MH_3 + CO \qquad (a)$$

Only 50% of the $(h^5\text{-}C_5H_5)_2Nb(CO)H$ is converted to the trihydride, whereas $(h^5\text{-}C_5H_5)_2$-Ta(CO)H converts completely. The reverse of reaction (a) occurs if the photolysis of $(h^5\text{-}C_5H_5)_2NbH_3$ and $(h^5\text{-}C_5H_5)_2TaH_3$ is carried out in the presence of CO. Both $(h^5\text{-}C_5H_5)_2NbH_3$ and $(h^5\text{-}C_5H_5)_2TaH_3$ lose H_2 when photolyzed in benzene. The 16-electron species $(h^5\text{-}C_5H_5)_2NbH$ and $(h^5\text{-}C_5H_5)_2TaH$ are capable of activating carbon–hydrogen bonds in benzene and catalyzing the H–D exchange between D_2 and benzene.

Stirring $[(h^5\text{-}C_5Me_5)_2ZrN_2]_2N_2$ in toluene with 1 atm of H_2 yields $(h^5\text{-}C_5Me_5)_2ZrH_2$ quantitatively[2,3]:

$$[(h^5\text{-}C_5Me_5)_2ZrN_2]_2N_2 + 2 H_2 \rightarrow 2 (h^5\text{-}C_5Me_5)_2ZrH_2 + 3 N_2 \qquad (b)$$

The product is soluble in hydrocarbons and ethers as a monomer, in contrast to the polymeric $[(h^5\text{-}C_5H_5)_2ZrH_2]_x$. In toluene at $-80°C$ the $(h^5\text{-}C_5Me_5)_2ZrH_2$ hydrogens exchange with D_2. At higher T (ca. 50°C), all 30 methyl hydrogens of the two $h^5\text{-}C_5Me_5$ rings also exchange with D_2.

The evolution of 1 equiv of free CO in Eq. (c) suggests that $(h^5\text{-}C_5Me_5)_2Zr(CO)$ is an intermediate[3,4]:

$$(h^5\text{-}C_5Me_5)_2Zr(CO)_2 + 2 H_2 \xrightarrow{110°C} (h^5\text{-}C_5Me_5)_2ZrH(OCH_3) + CO \qquad (c)$$

Oxidative addition of H_2 to $(h^5\text{-}C_5Me_5)_2Zr(CO)$ to generate the dihydride is plausible in light of:

$$(h^5\text{-}C_5Me_5)_2ZrH_2 + CO \rightleftharpoons (h^5\text{-}C_5Me_5)_2ZrH_2(CO) \qquad (d)$$

(J. TOPICH)

1. D. F. Foust, R. D. Rogers, M. D. Rausch, J. L. Atwood, *J. Am. Chem. Soc., 104,* 5646 (1982).
2. J. E. Bercaw, in *Transition-Metal Hydrides,* R. Bau, ed., Advances in Chemistry Series No. 167, American Chemical Society, Washington, DC, 1978, p. 136.
3. P. T. Wolczanski, J. E. Bercaw, *Acc. Chem. Res., 13,* 121 (1980).
4. J. Manriquez, D. R. McAlister, R. D. Sanner, J. E. Bercaw, *J. Am. Chem. Soc., 98,* 6733 (1976).

1.10.4.3.2. Involving Molybdenum and Rhenium.

Reaction of $Mo(N_2)_2L_2$, where L is a bidentate tertiary phosphine ligand, occurs with H_2:

$$Mo(N_2)_2L_2 + 2 H_2 \rightarrow MoH_4L_2 + 2 N_2 \qquad (a)[1]$$

where L = $Ph_2PCH_2CH_2PPh_2$(dppe), (3-tolyl)$_2PCH_2CH_2P$(tolyl-3)$_2$(dmtpe), (4-tolyl)$_2$-PCH_2CH_2P(tolyl-4)$_2$ (dptpe) yields a tetrahydride, and not a dihydride as previously claimed[2]. Reaction (a) can be reversed, although not quantitatively, if N_2 is bubbled through MoH_4L_2 in cyclohexane. Irradiation of MoH_4(dppe)$_2$ under N_2, however, re-

188   1.10. Formation of Bonds between Hydrogen and Transition Metals
       1.10.4. by Oxidative Addition of Hydrogen
       1.10.4.3. to Neutral Species

sults in $Mo(N_2)_2(dppe)_2$ quantitatively[3,4]. Variable-T ^1H nuclear magnetic resonance (NMR) shows that $MoH_4(dmtpe)_2$ is stereochemically nonrigid and undergoes an intramolecular exchange in which all the hydride ligands are coupled equally to all ^{31}P nuclei[1].

Photolysis of polyhydride Re complexes under H_2 produces[4,6]:

$$ReH_3L_4 + H_2 \xrightarrow{h\nu} ReH_5L_3 + L \qquad (b)$$

$$ReH_5L_3 + H_2 \xrightarrow{h\nu} ReH_7L_2 + L \qquad (c)$$

where $L = PMe_2Ph, PMePh_2, PPh_3$. The ReH_7L_2 product converts thermally and photochemically to the dimer[6], $Re_2H_8L_4$. In the absence of H_2, ReH_5L_3 loses the phosphine ligand, L, rather than H_2, on photolysis[4,5,7].

Hydrogenation[8] of $ReH_3(PPh_3)_2(h^4\text{-}2,3\text{-dimethylbutadiene})$ leads to $ReH_7(PPh_3)_2$, which is thermally unstable. 2,3-Dimethylbutane and 2,3-dimethylbut-1-ene also are observed as products:

$$(PPh_3)_2ReH_3 \xrightarrow[50°C]{H_2} (PPh_3)_2ReH_7 + CH_3-\underset{\underset{CH_3}{|}}{CH}-\underset{\underset{CH_3}{|}}{CH}-CH_3$$

$$+ CH_3-\underset{\underset{CH_3}{|}}{CH}-\underset{\underset{CH_3}{|}}{C}=CH_2 \qquad (d)$$

Photolysis[9] of $(h^5\text{-}C_5H_5)Re(CO)_3$ in the presence of H_2 yields $(h^5\text{-}C_5H_5)ReH_2(CO)_2$.

(J. TOPICH)

1. L. J. Archer, T. A. George, *Inorg. Chem.*, 18, 2079 (1979).
2. M. Hidai, K. Tominari, Y. Uchida, *J. Am. Chem. Soc.*, 94, 110 (1972).
3. G. L. Geoffroy, M. G. Bradley, R. Pierantozzi, in *Transition-Metal Hydrides*, R. Bau, ed., Advances in Chemistry Series No. 167, American Chemical Society, Washington, DC, 1978; p. 181.
4. G. L. Geoffroy, in *Transition-Metal Hydrides*, R. Bau, ed., Advances in Chemistry Series No. 198, American Chemical Society, Washington, DC, 1982, p. 347.
5. D. A. Roberts, G. L. Geoffroy, *J. Organomet. Chem.*, 214, 221 (1981).
6. J. Chatt, R. S. Coffey, *J. Chem. Soc, A*, 1963 (1969).
7. M. A. Green, J. C. Huffman, K. G. Caulton, *J. Am. Chem. Soc.*, 103, 695 (1981).
8. D. Baudry, M. Ephritikhine, H. Felkin, *J. Organomet. Chem.*, 224, 363 (1982).
9. J. K. Hoyano, W. A. G. Graham, *J. Am. Chem. Soc.*, 104, 3722 (1982).

1.10.4.3.3. Involving Ruthenium and Mixed Metal–Ruthenium Clusters.

The Ru complex $Ru(COD)(COT)$ (COD = 1,5-cyclooctadiene; COT = 1,3,5-cyclooctatriene) is a precursor to Ru–hydride phosphine complexes[1]:

$$Ru(COD)(COT) + 2\ P(Pr\text{-}i)_3 \xrightarrow{H_2} \{H_4Ru[P(Pr\text{-}i)_3]_2\}_2 \qquad (a)$$

$$Ru(COD)(COT) + 2\ P(C_6H_{11})_3 \xrightarrow{H_2} H_6Ru[P(C_6H_{11})_3]_2 \qquad (b)$$

$$Ru(COD)(COT) + 2\ dppe \xrightarrow{H_2} H_2Ru(dppe)_2 \qquad (c)$$

$$Ru(COD)(COT) + 2\ dppm \xrightarrow{H_2} [H_2Ru(dppm)_2]_3 \qquad (d)$$

where dppe $= Ph_2P(CH_2)_2PPh_2$; dppm $= Ph_2PCH_2PPh_2$.

The Ru-carbonyl clusters react with H_2; e.g., $H_4Ru_4(CO)_{12}$ is prepared[2] by bubbling H_2 through $Ru_3(CO)_{12}$ in n-octane at reflux for 1 h. Likewise, $D_4Ru_4(CO)_{12}$ is prepared[2] using D_2 gas, and $H_2D_2Ru_4(CO)_{12}$ results from $H_2Ru_4(CO)_{13}$ and D_2 in refluxing hexane. In contrast to the analogous Os system, no hydrogen exchange with the hydrocarbon solvent occurs with $H_2D_2Ru_4(CO)_{12}$. Reaction of $H_2Ru_4(CO)_{13}$ under H_2 also yields $H_4Ru_4(CO)_{12}$ when this solution is stirred for 24 h at 25°C. The reaction is accelerated by 355-nm irradiation[3,4].

Irradiation (UV) of $Ru_2(CO)_3(\mu\text{-CHR})(C_5H_5\text{-}h^5)_2$ (R = H, Me, CO_2Et) in toluene (25°C, 3 d) under H_2 produces $H_3Ru_3(CO)_3(C_5H_5\text{-}h^5)$ in 40–60% yield[5]:

(e)

Hydrogen reacts[6] with $Ru_3(CO)_8(dppm)_2$ and $Ru_3(CO)_8(dpam)_2$ (dpam $= Ph_2AsCH_2AsPh_2$):

$$Ru_3(CO)_8(dppm)_2 + 2\ H_2 \xrightarrow{85°C} (\mu\text{-H})_2Ru_3(CO)_6(\mu\text{-PPhCH}_2PPh_2)_2$$
$$+\ 2\ CO + 2\ C_6H_6 \quad \text{(f)}$$

$$Ru_3(CO)_8(dpam)_2 + 2\ H_2 \xrightarrow{85°C} (\mu\text{-H})Ru_3(CO)_7(\mu\text{-AsPhCH}_2AsPh_2)(dpam)$$
$$+\ CO + C_6H_6 \quad \text{(g)}$$

$$(\mu\text{-H})Ru_3(CO)_7(\mu\text{-AsPhCH}_2AsPh_2)(dpam) + H_2 \xrightarrow{120°C}$$

$$(\mu\text{-H})_2Ru_3(CO)_6(\mu\text{- AsPhCH}_2AsPh_2)_2 + CO + C_6H_6 \quad \text{(h)}$$

The Ru cluster $HRu_3(\mu\text{-COMe})(CO)_{10}$ reacts[7,8] with H_2 to produce $H_3Ru_3\text{-}(\mu_3\text{-COMe})(CO)_9$.

$$HRu_3(\mu\text{-COMe})(CO)_{10} + H_2 \rightleftharpoons H_3Ru_3(\mu_3\text{-COMe})(CO)_9 + CO \quad \text{(i)}$$

The equilibrium constant for reaction (i) is 2.6 in the range 60–80°C in decane[9].

Hydrogen (1 atm) reacts[10] with $Ru_3(CO)_{11}[CN(Bu\text{-t})]$ in refluxing cyclohexane for 1 h to produce $HRu_3[\mu_3\text{-HCN(Bu-t)}](CO)_9$, along with $HRu_3[\mu\text{-HCN(Bu-t)}](CO)_8\text{-}[CN(Bu\text{-t})]$ and $H_4Ru_4(CO)_{12\text{-}n}[CN(Bu\text{-t})]_n$ (n = 0, 1, 2). Higher yields of the latter two products are obtained by hydrogenation of $Ru_3(CO)_{10}[CN(Bu\text{-t})]$.

Reaction of H_2 with $FeRu_2(CO)_{12}$ produces[2] $H_4Ru_4(CO)_{12}$. However, $H_4FeRu_3\text{-}(CO)_{12}$ can be prepared from $H_2FeRu_3(CO)_{13}$ and H_2 in refluxing hexane[2] or by photolysis[3] of $H_2FeRu_3(CO)_{13}$ in a hydrocarbon under H_2, and $HCoRu_3(CO)_{13}$ reacts with H_2 in refluxing hexane to produce $H_3CoRu_3(CO)_{12}$ in 75% yield[11]. Stirring $Ru_2Co_2(CO)_{13}$ in n-hexane with H_2 for 3 h at 50°C produces $H_2Ru_2Co_2(CO)_{12}$ in 78% yield[12].

(J. TOPICH)

1. B. Chaudret, G. Commenges, R. Poilblanc, *J. Chem. Soc., Chem. Commun.*, 1388 (1982).

190   1.10. Formation of Bonds between Hydrogen and Transition Metals
       1.10.4. by Oxidative Addition of Hydrogen
       1.10.4.3. to Neutral Species

2. S. A. R. Knox, J. W. Keopke, M. A. Andrews, H. D. Kaesz, *J. Am. Chem. Soc., 97*, 3942 (1975).
3. H. C. Foley, G. L. Geoffroy, *J. Am. Chem. Soc., 103*, 7176 (1981).
4. J. L. Graff, M. S. Wrighton, *Inorg. Chim. Acta, 63*, 63 (1982).
5. N. J. Forrow, S. A. R. Knox, M. J. Morris, A. G. Orpen, *J. Chem. Soc., Chem. Commun.,* 234 (1983).
6. G. Lavigne, N. Lugan, J. J. Bonnett, *Organometallics, 7*, 1040 (1982).
7. J. B. Keister, *J. Chem. Soc., Chem. Commun.,* 214 (1979).
8. J. B. Keister, M. W. Payne, M. J. Muscatella, *Organometallics, 2*, 219 (1983).
9. L. M. Bavaro, P. Montangero, J. B. Keister, *J. Am. Chem. Soc., 105*, 4977 (1983).
10. M. I. Bruce, R. C. Wallis, *Aust. J. Chem., 35*, 709 (1982).
11. W. L. Gladfelter, G. L. Geoffroy, J. C. Calabrese, *Inorg. Chem., 19*, 2569 (1980).
12. E. Roland, H. Vahrenkamp, *Organometallics, 2*, (1983).

1.10.4.3.4. Involving Iron, Osmium, Rhodium, Iridium and Platinum.

The clusters $HM_3(\mu\text{-COMe})(CO)_{10}$ (M = Fe, Os) react[1,2] with H_2 in refluxing hexane to give $H_3M_3(\mu_3\text{-COMe})(CO)_9$. This process can be reversed by reacting the trihydrides with CO. Although $H_3Fe_3(\mu_3\text{-COMe})(CO)_9$ is unstable under ambient conditions, the substituted derivative, $H_3Fe_3(\mu_3\text{-COMe})(CO)_7(SbPh_3)_2$, can be obtained by carrying out the hydrogenation in the presence of $SbPh_3$.

The cluster $H_2FeOs_3(CO)_{13}$ adds H_2 and loses CO when irradiated[3] under H_2 to produce $H_4FeOs_3(CO)_{12}$. The product is photosensitive in the presence of H_2, decomposing to a complex mixture of unidentified compounds.

Hydrogen reacts[4] with $Os_3(CO)_{12}$ in n-octane to produce $H_2Os_3(CO)_{10}$ and $H_4Os_4(CO)_{12}$:

$$Os_3(CO)_{12} + H_2 \xrightarrow[1.5\ h]{120°C} H_2Os_3(CO)_{10} + 2\ CO \tag{a}$$

$$Os_3(CO)_{12} + H_2 \xrightarrow[41\ h]{120°C} H_4Os_4(CO)_{12} \tag{b}$$

When D_2 is bubbled through $Os_3(CO)_{12}$ for 2 h, exchange occurs with hydrogen from the hydrocarbon solvent.

Photochemically generated 16-electron intermediates are highly reactive. Photolysis[5] of $(h^5\text{-}C_5Me_5)Os(CO)_2H$ under H_2 produces:

Reaction[6] of $[(\mu_2\text{-H})_2Os_3(CO)_9(\mu_3\text{-NCH}_2CF_3)]$ with H_2 produces $[H(\mu_2\text{-H})_3Os_3$-$(CO)_8(\mu_3\text{-NCH}_2CF_3)]$ in an oxidative addition of H_2 involving CO loss with incorporation of a terminal and a μ_2-H atom into the cluster.

Hydridodinitrogentrialkylphosphine complexes of Rh(I) react[7]:

$$HRh(N_2)[P(Bu\text{-}t)_3]_2 + H_2 \rightarrow H_3Rh[P(Bu\text{-}t)_3]_2 + N_2 \tag{d}$$

$HRh(N_2)[P(Pr\text{-}i)_3]_2$ does not form under the same conditions as its $P(Bu\text{-}t)_3$ analogue, $HRh[P(Pr\text{-}i)_3]_3$ being formed instead. This Rh(I) hydride will add H_2 with loss of phosphine:

$$HRh[P(Pr\text{-}i)_3]_3 + H_2 \rightarrow H_3Rh[P(Pr\text{-}i)_3]_2 + P(Pr\text{-}i)_3 \tag{e}$$

Cis addition is the stereochemical course of the oxidative addition[8,9] of H_2 to Ir(I) complexes, e.g.[8]:

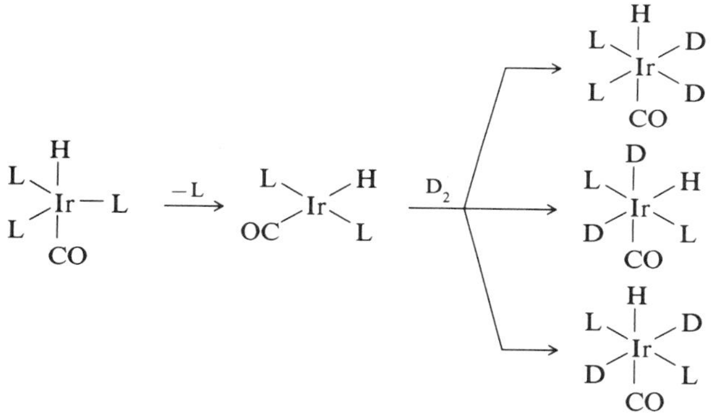

where $L = PPh_3$. Trans addition of D_2 could have ocurred, but the H_2 or D_2 addition is cis with H or D scrambling occurring, possibly through a dihydrido-bridged Ir dimer[10]:

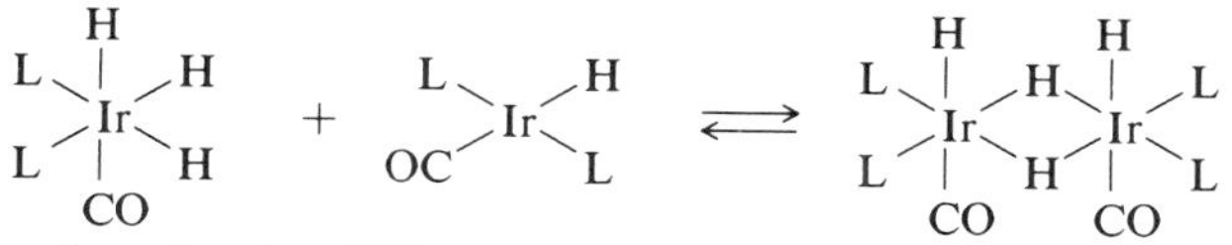

In the photochemistry[11,12] of the Ir hydride complexes, the primary photochemical process for $H_3Ir(PPh_3)_3$ is loss of H_2 to form $HIr(PPh_3)_3$. However, irradiation under H_2 suppresses this pathway and allows another to be observed:

$$H_3Ir(PPh_3)_3 \xrightarrow[H_2]{h\nu} H_5Ir(PPh_3)_2 + PPh_3 \tag{h}$$

Hydrogen reacts[13] with $1\text{-}[Ir(CO)_3(PPh_3)](7\text{-}C_6H_5\text{-}1,7\text{-}B_{10}C_2H_{10})$ in benzene, losing 1 equiv of CO to produce $1\text{-}[H_2Ir(CO)_2(PPh_3)](7\text{-}C_6H_5\text{-}1,7\text{-}B_{10}C_2H_{10})$.

Iridium tetrahydrides can be prepared[14]:

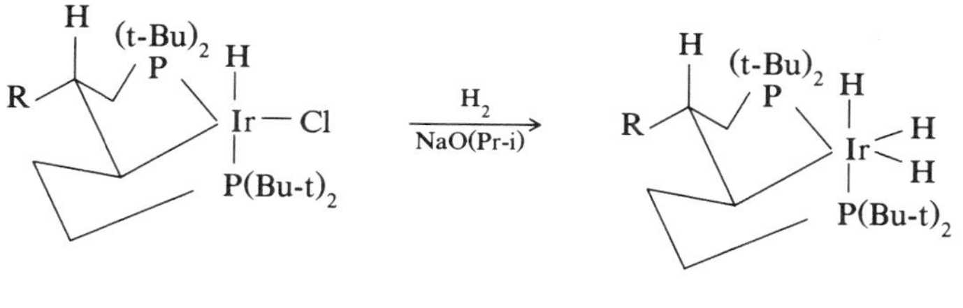

---

where R = H, Me.

Hydrogen adds[15] to $Pt[P(Pr-i)_3]_3$ to yield $trans$-$H_2Pt[P(Pr-i)_3]_2$ with dissociation of 1 equiv of $P(Pr-i)_3$. Cis- and $trans$-H_2PtL_2 (L = PEt_3, PMe_3) can be prepared by bubbling H_2 (1 atm, 25°C) through $PtL_2(C_2H_4)$ solutions[16].

$$PtL_2(C_2H_4) + H_2 \rightarrow H_2PtL_2 + C_2H_4 \qquad (j)$$

The ratio of cis and trans isomers formed depends on L and the solvent. The cis isomer has a dipole moment and is better solvated in polar media; hence the relative concentration of cis isomer increases on going from toluene to acetone. In a given solvent the cis geometry is more favored for L = PMe_3 than for L = PEt_3.

A source of the $Pt(PR_3)$ fragment is $Pt(C_2H_4)_2PR_3$. The products of the reaction of $Pt(C_2H_4)_2PR_3$ with H_2 depend on the phosphine[17,18]:

$$Pt(C_2H_4)_2[P(Bu-t)_3] \xrightarrow[\text{(30 MPa)}]{H_2, \ 300 \text{ atm}} H_6Pt_3[P(Bu-t)_3]_3 \qquad (k)$$

$$Pt(C_2H_4)_2[PPh(Pr-i)_2] \xrightarrow[\text{(30 MPa)}]{H_2, \ 300 \text{ atm}} H_8Pt_4[PPh(Pr-i)_2]_4 \qquad (l)$$

$$Pt(C_2H_4)_2[PPh(Bu-t)_2] \xrightarrow[\text{(30 MPa)}]{H_2, \ 300 \text{ atm}} H_8Pt_5[PPh(Bu-t)_2]_5 \qquad (m)$$

(J. TOPICH)

1. J. B. Keister, *J. Chem. Soc., Chem. Commun.*, 214 (1979).
2. J. B. Keister, M. W. Payne, M. J. Muscatella, *Organometallics, 2*, 219 (1983).
3. H. C. Foley, G. L. Geoffroy, *J. Am. Chem. Soc., 103*, 7176 (1981).
4. S. A. R. Knox, J. W. Koepke, M. A. Andrews, H. D. Kaesz, *J. Am. Chem. Soc., 97*, 3942 (1975).
5. J. K. Hoyano, W. A. G. Graham, *J. Am. Chem. Soc., 104*, 3722 (1982).
6. J. Banford, Z. Dawoodi, K. Henrick, M. J. Mays, *J. Chem. Soc., Chem. Commun.*, 554 (1982).
7. T. Yoshida, T. Okano, D. L. Thorn, T. H. Tulip, S. Otsuka, J. A. Ibers, *J. Organomet. Chem., 181*, 183 (1979).
8. J. F. Harrod, G. Hamer, W. Yorke, *J. Am. Chem. Soc., 101*, 3987 (1979).
9. M. Drouin, J. F. Harrod, *Inorg. Chem. 22*, 999 (1983).
10. R. Crabtree, *Acc. Chem. Res., 12*, 331 (1979).
11. G. L. Geoffroy, R. Pierantozzi, *J. Am. Chem. Soc., 98*, 8054 (1976).
12. G. L. Geoffroy, M. G. Bradley, R. Pierantozzi, in *Transition-Metal Hydrides*, R. Bau, ed., Advances in Chemistry Series No. 167, American Chemical Society, Washington, DC, 1978, p. 181.
13. B. Longato, S. Bresadola, *Inorg. Chim. Acta, 33*, 189 (1979).
14. R. J. Errington, B. L. Shaw, *J. Organomet. Chem., 238*, 319 (1982).
15. T. Yoshida, S. Otsuka, *J. Am. Chem. Soc., 99*, 2134 (1977).
16. R. S. Paonessa, W. C. Trogler, *J. Am. Chem. Soc., 104*, 1138 (1982).
17. D. Gregson, J. A. K. Howard, M. Murray, J. L. Spencer, *J. Chem. Soc., Chem. Commun.*, 716, (1981).
18. P. W. Frost, J. A. K. Howard, J. L. Spencer, D. G. Turner, *J. Chem. Soc., Chem. Commun.*, 1104 (1981).

1.10.4.4. to Cationic Species of Iridium, Rhodium and Platinum with Replacement of Coordinated Ligands.

Oxidative addition of H_2 to cationic metal complexes can occur with replacement of coordinated ligands[1]; e.g., $[Ir(CO)_3L_2]BPh_4$ (L = tertiary phosphine or arsine) are prepared[2] by passing CO through $trans$-$[IrCl(CO)L_2]$ solutions in the presence of

1.10. Formation of Bonds between Hydrogen and Transition Metals   193
1.10.4. by Oxidative Addition of Hydrogen
1.10.4.4. to Cationic Species of Ir, Rh and Pt.

NaBPh$_4$. The cationic dihydrides are obtained in 70–100% yields by bubbling H$_2$ through $[Ir(CO)_3L_2]BPh_4$ in acetone–methanol at RT:

$$
\begin{array}{ccc}
\underset{\textbf{(I)}}{\underset{\displaystyle L}{\overset{\displaystyle L}{OC-Ir\begin{smallmatrix} CO^+ \\ \\ CO \end{smallmatrix}}}}
& \xrightarrow[-CO]{+H_2} &
\underset{\textbf{(II)}}{\underset{\displaystyle L}{\overset{\displaystyle L}{\underset{OC}{H}\; Ir \; \overset{H^+}{CO}}}}
\end{array}
\tag{a}
$$

where L = PPh$_3$, AsPh$_3$, PMePh$_2$, PEtPh$_2$, PEt$_2$Ph, PEt$_3$, P(C$_6$H$_{11}$)$_3$, P(Pr-i)$_3$. This re-action is reversible, and H$_2$ is liberated and (I) regenerated by passing CO through the solution. Initial loss of CO occurs for $[Ir(CO)_3L_2]^+$, where L = P(C$_6$H$_{11}$)$_3$ or P(Pr-i)$_3$, and this is the rate-determining step with H$_2$ adding to the resulting four-coordinated complex. In the reductive elimination[3] of H$_2$ from $[IrH_2(CO)_2L_2]BPh_4$, the rate de-creases: L = PPh$_3$ > AsPh$_3$ > PMePh$_2$ > PEtPh$_2$ > PEt$_2$Ph > PEt$_3$ > P(C$_6$H$_{11}$)$_3$ ≈ P(Pr-i)$_3$. The breaking of the metal–hydrogen bonds is the rate-determining step and is related to the strength of the metal–hydrogen bond in each complex. The Ir(III) complexes are stabilized toward reductive elimination by the more basic phosphines.

Cationic Rh hydrides[4] can be isolated:

$$[Rh(NBD)(PPh_3)_2]^+ \xrightarrow[S]{H_2} [H_2Ph(PPh_3)_2S_2]^+ \tag{b}$$

where S = acetone, ethanol, CH$_3$CN; NBD = norbornadiene:

$$[Rh(NBD)Cl]_2 \xrightarrow{\substack{\text{(i) AgPF}_6,\ S \\ \text{(ii) PPh}_2C_6H_{11} \\ \text{(iii) H}_2}} [H_2Rh[PPh_2(C_6H_{11})]_2S_2]^+ \tag{c}$$

where S = acetone, CH$_3$CN;

$$[Rh(NBD)(AsPh_3)_2]^+ \xrightarrow[S]{H_2} [H_2Rh(AsPh_3)_2S_2]^+ \tag{d}$$

where S = acetone, CH$_3$CN;

$$[Rh(NBD)L_2]^+ \xrightarrow{H_2,\ 2.2'\text{-bipy}} [H_2RhL_2(bipy)]^+ \tag{e}$$

where L = PPh$_3$, PMePh$_2$, AsPh$_3$ and 2,2'-bipyridine (bipy). Reaction of $[Rh(NBD)L_2]^+$ with H$_2$ yields a transient intermediate, $[H_2Rh(NBD)L_2]^+$. Hydride transfer yields norbornene, which may dissociate before it is reduced to norbornane. On reduction of NBD, solvent or other ligand occupies the vacated coordination sites.

Cationic olefin complexes of Ir(I) are also susceptible to loss of coordinated li-gands on oxidative addition of H$_2$; e.g., $[IrH_2S_2(PPh_3)_2]BF_4$ (S = solvent) is prepared[5] from $[Ir(COD)(PPh_3)_2]BF_4$ (COD = 1,5-cyclooctadiene) and H$_2$ at 0–20°C in Me$_2$CO, MeCN, tetrahydrofuran (THF), MeOH, EtOH, i-PrOH, t-BuOH and H$_2$O. The ease of solvent displacement from $[IrH_2S_2(PPh_3)_2]^+$ follows the order[5]: H$_2$O ≈ THF > t-BuOH > i-PrOH > Me$_2$CO > EtOH > MeOH > MeCN.

For binuclear Pt complexes[6,7]:

$$\text{(f)}$$

where $\widehat{(PP}$ $=$ $Ph_2PCH_2PPh_2)$. Displacement of H_2 from (IV) by CO is readily reversible, so treatment of (V) with xs H_2 produces (IV) in good yield.

(J. TOPICH)

1. H. D. Kaesz, R. B. Saillant, *Chem. Rev., 72,* 231 (1972).
2. M. J. Church, M. J. Mays, R. N. F. Simpson, F. P. Stefanini, *J. Chem. Soc., A,* 2909 (1970).
3. M. J. Mays, R. N. F. Simpson, F. P. Stefanini, *J. Chem. Soc., A,* 3000 (1970).
4. R. R. Schrock, J. A. Osborn, *J. Am. Chem. Soc., 98,* 2134 (1976).
5. R. H. Crabtree, P. C. Demou, D. Eden, J. M. Mihelcic, C. A. Parnell, J. M. Quirk, G. E. Morris, *J. Am. Chem. Soc., 104,* 6994 (1982).
6. M. P. Brown, J. R. Fisher, A. J. Mills, R. J. Puddephatt, M. Thomson, *Inorg. Chim. Acta,* 44, L271 (1980).
7. J. R. Fisher, A. J. Mills, S. Sumner, M. P. Brown, M. A. Thomson, R. J. Puddephatt, A. A. Frew, L. Manojlovic-Muir, K. W. Muir, *Organometallics, 1,* 1421 (1982).

1.10.5. by Oxidative Addition to Metal Complexes

Reviews of oxidative addition to d^8-transition metals[1] and transition metals in general[2], including the formation of hydrogen–metal bonds[3], are available. Hydrogen–element bonds add to metals predominantly to the left in the transition-metal series. This is a formal 2 e^- oxidation of the metal by transfer of electron density to the proton. Therefore, the reaction is oxidative addition and the reverse reaction is reductive elimination:

$$M^nL_m + H\!-\!Z \rightleftharpoons HM^{n+2}(Z)L_m \qquad \text{(a)}$$

with no mechanistic information implied[4]. Mechanisms may be concerted or stepwise with either the hydrogen or the element adding first[5]. Oxidative addition to a metal complex requires nonbonding electron density on the metal, two vacant coordination sites and a metal with oxidation states separated by two units. Both the hydrogen and element bond are coordinated to the metal in the product such that the overall charge on the metal complex is unchanged. This is in contrast to the protonation of metal complexes discussed in §1.10.6. The tendency of a metal complex to undergo oxidative addition parallels its basicity. Basicity increases on descending a transition-metal triad, by decreasing metal oxidation state and with the increase of electron-donating ligands in the complex.

(T. J. LYNCH)

1. J. P. Collman, W. R. Roper, *Adv. Organomet. Chem.*, *7*, 53 (1968).
2. J. Halpern, *Acc. Chem. Res.*, *3*, 386 (1970).
3. H. D. Kaesz, R. B. Saillant, *Chem. Rev.*, *72*, 231 (1972).
4. J. S. Bradley, D. E. Connor, D. Dolphin, J. A. Labinger, J. A. Osborn, *J. Am. Chem. Soc.*, *94*, 4043 (1972).
5. R. H. Crabtree, J. M. Quirk, *J. Organomet. Chem.*, *181*, 203 (1979).

1.10.5.1. of Hydrogen Halides.

Oxidative addition of hydrogen halides is most extensively studied with Ir complexes. Stereochemical, mechanistic and thermochemical[1,2] data are known. The trans-$IrCl(CO)(PR_3)_2$ compounds demonstrate the clearest examples of oxidative addition because the products $HIr(X)Cl(CO)(PR_3)_2$ are stable O_h complexes. The analogous Rh complexes are unstable with respect to the reductive elimination of hydrogen halide.

When anhyd HCl is bubbled through trans-$IrCl(CO)(PPh_3)_2$ in ether, the O_h product $HIr(Cl)_2(CO)(PPh_3)_2$ is formed rapidly and quantitatively. In nonpolar solvents or in the solid state the cis isomer is obtained. In polar solvents mixtures of cis and trans adducts form[3]. When the phosphine is tri-o-tolylphosphine, HX addition to trans-$IrCl(CO)(PR_3)_2$ is slow because the apical sites are blocked[4]. When the sterically bulky but strongly basic tricyclohexylphosphine is incorporated into trans-$IrCl(CO)(PR_3)_2$, the HCl adduct is formed easily[5]. Moreover, when the phosphine of the complex is PMe_2(o-$MeOC_6H_4$), interaction of the methoxy oxygen with the Ir center increases its nucleophilicity and so facilitates oxidative addition[6].

With non-phosphorus-containing ligands, e.g., $Ir(Oq)(CO)(PPh_3)$ (Oq = 8-oxyquinolate), when dry HCl is added in 1 min $HIrCl(Oq)(CO)(PPh_3)$ (82%) forms[7]. Addition of HCl, HBr and HI to the carborane—Ir complexes, trans-Ir(carborane)-$(CO)PR_3$, occurs in both the solid state and solution[8]. The stereochemistry of the adducts $HIrX$(carborane)$(CO)PR_3$ depends on the medium and the carborane ligand. The adducts reductively eliminate H-carborane on heating in benzene. Indirect addition of HCl to an Ir complex occurs between $[Ir(dppe)_2]Cl$ where dppe is $1,2[(C_6H_5)_2P]_2CH_2$-CH_2 and $B_{10}H_{13}X$ (X = H or 6-Cl) which yields $[IrHCl(dppe)_2][B_{10}H_{12}X]$ (74%)[9]. The cyclometallated tetrahydride H_4Ir(t-$Bu_2PCH_2CH_2CHCH_2CH_2PBu$-$t_2$) is formally reduced by HBr, producing $HIrBr$(t-$Bu_2PCH_2CH_2CHCH_2CH_2PBu$-$t_2$)[10].

Reviews of synthetic methods to hydrido-Ni, -Pd and -Pt, including oxidative addition of hydrogen halides, are available[11,12].

Two-, three- and four-coordinated Pd(0) complexes react with HCl to form trans-$HPdCl(PR_3)_2$. The two-coordinated complexes, $Pd(PBu$-$t_3)_2$ and $Pd(PPhBu$-$t_2)_2$, react[13] with HCl at RT quantitatively, whereas the higher coordinated complexes, $Pd(PMe_3)_4$, $Pd(PMePh_2)_4$ and $Pd(Pi$-$Pr_3)_3$, react[14] at −50°C. Similarly, $HPtCl[PPh(Bu$-$t_2)]_2$ is prepared almost quantitatively at RT from HCl and $Pt[PPh(Bu$-$t_2)]_2$ in n-hexane[13]. The complex H_2Pt $[As(Bu-t)_3]_2$, a Pt dihydride not containing any phosphorus donor ligands, does not add HX oxidatively but yields[15] the hydride-halogen exchange product, $HPtX[As(Bu-t)_3]_2$ (X = Cl, 90%; Br, 88%), in benzene in 0.5 h.

Other transition metals undergo oxidative addition with hydrogen halides[16], e.g., the group VIA metals. An Mo dimer reacts with HCl:

$$Mo_2(O_2CMe)_4 + 8 \ HCl \rightarrow [Cl_3Mo(\mu\text{-}Cl_2)(\mu\text{-}H)MoCl_3]^{3-} + 3 \ H^+ + 4 \ MeCO_2H \quad \text{(b)}$$

in an oxidative addition to an M–M multiple bond. The yield is quantitative when carried out[17] above 60°C and in the absence of O_2. Treatment of another quadruply bonded Mo dimer, $Mo_2(mhp)_4$ (mhp = 2-hydroxy-6-methylpyridine anion), in EtOH with HCl gas in the presence of either PEt_3 or $P(Pr-n)_3$ affords the same anion, $[R_3PH]_3$-$[Mo_2Cl_8H]$ $[PEt_3$ (41%), P(Pr-n)$_3$ (26%)][18].

The labile cluster $Os_3(NCMe)_2(CO)_{10}$ adds HCl or HBr to form the hydride- and halide-bridged compound $Os_3HX(CO)_{10}$ under mild conditions[19]. A similar cluster, $Os_3(NCMe)(CO)_{11}$, adds HCl and HBr to afford[20] first $Os_3HX(CO)_{11}$ in 50% yields and then, with further loss of CO, $HOs_3X(CO)_{10}$. When HI is passed through $Os_3(NCMe)(CO)_{11}$ in CH_2Cl_2, $HOs_3I(CO)_{10}$ is isolated[20].

(T. J. LYNCH)

1. J. U. Mondal, D. M. Blake, *Coord. Chem. Rev., 47*, 205 (1982).
2. G. Yoneda, D. M. Blake, *Inorg. Chem., 20,* 67 (1981).
3. The stereochemistry of these reactions is known: H. D. Kaesz, R. B. Saillant, *Chem. Rev., 72*, 238 (1972).
4. R. Brady, W. H. DeCamp, B. R. Flynn, M. L. Schneider, J. D. Scott, L. Vaska, M. F. Werneke, *Inorg. Chem., 14*, 2669 (1975).
5. F. G. Moers, J. A. M. De Jong, P. M. H. Beaumont, *J. Inorg. Nucl. Chem., 35,* 1915 (1973).
6. E. M. Miller, B. L. Shaw, *J. Chem. Soc., Dalton Trans.,* 480 (1974). Rates are available for the oxidative addition of MeI.
7. R. Usón, L. A. Oro, M. A. Ciriano, R. Gonzalez, *J. Organomet. Chem., 205,* 259 (1981).
8. B. Longato, F. Morandini, S. Bresadola, *Inorg. Chim. Acta, 39,* 27 (1980).
9. N. N. Greenwood, W. S. McDonald, D. Reed, J. Staves, *J. Chem. Soc., Dalton Trans.,* 1339 (1979).
10. J. Errington, B. L. Shaw, *J. Organomet. Chem., 238,* 319 (1982).
11. D. M. Roundhill, *Adv. Organomet. Chem., 13,* 276 (1975).
12. D. M. Roundhill, in *Transition Metal Hydrides,* R. Bau, ed., Advances in Chemistry Series No. 167, American Chemical Society, Washington, DC, 1978, p. 160.
13. T. Yoshida, S. Otsuka, *J. Am. Chem. Soc., 99,* 2134 (1977).
14. H. Werner, W. Bertleff, *Chem. Ber., 116,* 823 (1983).
15. R. G. Goel, W. O. Ogini, R. C. Srivastava, *Inorg. Chem., 21,* 1627 (1982).
16. J. Halpern, *Acc. Chem. Res., 3,* 386 (1970).
17. A. Bino, F. A. Cotton, *Angew. Chem., Int. Ed. Engl., 18,* 332 (1979).
18. J. L. Pierce, D. DeMarco, R. A. Walton, *Inorg. Chem., 22,* 9 (1983).
19. M. Tachikawa, J. R. Shapley, *J. Organomet. Chem., 124,* C19 (1977).
20. B. F. G. Johnson, J. Lewis, D. A. Pippard, *J. Chem. Soc., Dalton Trans.,* 407 (1981).

1.10.5.2. of Hydrogen–Boron Bonds.

The chemistry of metalloborane compounds is vast[1,2], although most are not prepared by oxidative addition of the B—H bond to a metal complex; neither do they contain metal–hydrogen bonds. The synthesis of metalloborane hydrides involves many routes, the mechanisms of which are obscure. Reactions between borohydrides and transition-metal complexes involve hydride transfer without the formation of a metal–boron bond (see §1.10.7).

The most explicit oxidative additions of B—H bonds to metal centers involve Ir complexes, e.g., the o-metallation of a B—H bond obtained[3] with $[Ir(C_8H_{14})_2Cl]_2$ and xs $1-(Me_2P)-1,2-C_2B_{10}H_{11}$ in cyclohexane at reflux for 2 h under N_2 in quantitative yield. The product is $HIrCl(C_2B_{10}H_{10}PMe_2)(C_2B_{10}H_{11}PMe_2)_2$. The hydride ligand originates from a boron of the carborane cage (by deuterium labeling). The intermolecular conversion occurs[4] between $IrCl(PPh_3)_2$ and $1,2-C_2B_{10}H_{12}$ at reflux in cyclohexane for 1 h. The five-coordinated complex, $3-[HIrCl(PPh_3)_2]-1,2-C_2B_{10}H_{11}$, is isolated in $>90\%$ yield. The 1,12-carborane isomer of the above Ir complex catalyzes deuterium–hydrogen exchange in $1,7-C_2B_{10}H_{12}$. The isolated metallocarboranes are, therefore, models for intermediates in this exchange. Pentaborane, B_5H_9, is oxidatively added[5] to $IrCl(CO)(PMe_3)_2$ but not to $IrCl(CO)(PPh_3)_2$. This demonstrates the increased basicity of the trimethylphosphine-Ir complex, which decomposes slowly at 25°C but can be kept indefinitely at $-20°C$.

Several Ir—B bonds form in the same complex[6] from $Tl(B_3H_8)$ with trans-$IrCl(CO)(PPh_3)_2$ when stirred in benzene at 20°C for 48 h. The product is an O_h-borallyl compound (by ^1H and ^{11}B NMR) containing one two-electron, two-center Ir—B bond and one two-electron, three-center Ir—B_2 bond, $\mu^3-B_3H_7IrH(CO)(PPh_3)_2$.

Reactions of the anion $[B_3H_8]^-$ with other transition-metal complexes do not lead to metal–boron bonds. The borane anion acts as a bidentate ligand with bonding to the metal only through hydrogen[7], but this type of metal–hydrogen–boron interaction can transform to direct metal–boron bonding. When $Fe_2(B_2H_6)(CO)_6$ is treated with $Fe_2(CO)_9$ in pentane at 25°C for 7 h, the diborane moiety oxidatively adds to yield $HFe_4(BH_2)(CO)_{12}$, which consists of an Fe_4 butterfly structure and a BH_2 fragment bonded to all four Fe atoms[8].

When $CoCl_2$, NaC_5H_5 and B_5H_9 react at $-78°C$ overnight, metalloboranes result, but only in low yield[9]. Two of these contain both Co—H and Co—B bonds[10].

A less straightforward example of oxidative addition of a B—H bond to a metal is the reaction[11] of $Ru_3(CO)_{12}$ with $NaBH_4$. One of the five products isolated is formulated as $HRu_4(BH_2)(CO)_{12}$.

(T. J. LYNCH)

1. C. E. Housecroft, T. P. Fehlner, *Adv. Organomet. Chem.*, *21*, 57 (1982).
2. N. N. Greenwood, *Pure Appl. Chem.*, *55*, 1415 (1983).
3. E. L. Hoel, M. F. Hawthorne, *J. Am. Chem. Soc.*, *95*, 2712 (1973).
4. E. L. Hoel, M. F. Hawthorne, *J. Am. Chem. Soc.*, *96*, 6770 (1974).
5. M. R. Churchill, J. J. Hackbarth, A. Davison, D. D. Traficante, S. S. Wreford, *J. Am. Chem. Soc.*, *96*, 4041 (1974).
6. N. N. Greenwood, J. D. Kennedy, D. Reed, *J. Chem. Soc., Dalton Trans.*, 196, (1980).
7. D. F. Gaines, S. J. Hildebrandt, *Inorg. Chem.*, *17*, 794 (1978).
8. K. S. Wong, W. R. Scheidt, T. P. Fehlner, *J. Am. Chem. Soc.*, *104*, 1111 (1982).
9. V. R. Miller, R. Weiss, R. N. Grimes, *J. Am. Chem. Soc.*, *99*, 5646 (1977).
10. J. R. Pipal, R. N. Grimes, *Inorg. Chem.*, *16*, 3251, 3255 (1977).
11. C. R. Eady, B. F. G. Johnson, J. Lewis, *J. Chem. Soc., Dalton Trans.*, 477, (1977).

1.10.5.3. of Hydrogen–Carbon Bonds.

Oxidative addition of H—C bonds to transition metals, which also is referred to as an activation of the H—C bond, is of scientific and technological importance.[1-7] (See §1.10.2 for activation of H—C bonds by metal-atom vapors.) Intramolecular oxidative addition of H—C bonds (cyclometallation) occurs, but organic functional

198   1.10. Formation of Bonds between Hydrogen and Transition Metals
1.10.5. by Oxidative Addition to Metal Complexes
1.10.5.3. of Hydrogen–Carbon Bonds.

groups undergo intermolecular H—C oxidative additions to metal complexes less commonly.

The rates of intramolecular oxidative addition of aromatic H—C bonds[8]:

$$\text{IrCl(PPh}_3)_3 \longrightarrow \qquad\qquad\qquad\qquad\qquad \text{(a)}$$

(cyclometallation) can be increased by $Li\text{-}2H\text{-}1,2\text{-}B_{10}C_2H_{10}$ [9] or Li alkyls or aryls[10]. Aliphatic H—C bonds also are activated toward intramolecular oxidative addition when bound to a phosphine ligand. When $RhCl_3 \cdot 3\ H_2O$ is treated with the long-chain diphosphine $t\text{-}Bu_2PCH_2CH_2CH_2CH_2CH_2PBu\text{-}t_2$, the product[11] is $[HRhCl[t\text{-}Bu_2PCH_2\text{-}CH_2CHCH_2CH_2PBu\text{-}t_2]$ formed by oxidative addition of the middle H—C bond. Further dehydrogenation leads to $[RhCl[t\text{-}Bu_2PCH_2CH_2CH=CHCH_2PBu\text{-}t_2]$, in which the olefin and trans-phosphorus atoms are bound to the Rh center. A second H—C bond cleavage from the same middle carbon atom of a similar Ir complex forms an Ir-coordinated carbene[12]. In related complexes with similar ligands the cleavage of three H—C bonds yields π-allyl complexes[13]. Polynuclear Os complexes induce H—C bond oxidative addition with organic ligands[14-16].

Intermolecular oxidative addition of H—C usually involves activated H—C bonds. The weak acid HCN reacts with transition-metal complexes; e.g., HCN and NiL_4 lead to the hydride complexes $HNi(CN)L_3$ (L = various phosphorus ligands)[17]. The versatile complex $IrCl(CO)(PPh_3)_2$ adds HCN cleanly in CH_2Cl_2 at RT to form[18] $HIr(CN)(Cl(PPh_3)_2$. The zero-valent complexes $Pt(PPh_3)_4$ or $Pt(PPh_3)_3$ also add HCN to yield[19] $HPt(CN)(PPh_3)_2$. Reactions of $HMNp(dmpe)_2$ (M = Fe, Ru, Os; Np = 2-naphthyl; dmpe = $Me_2PCH_2CH_2PMe_2$) with HCN and terminal acetylenes give $HMR(dmpe)_2$ that contain new M—C bonds (R = $-CN$, $-C_2R'$)[20].

Terminal alkyne H—C bonds add to $Pt(PPh_3)_3$; e.g., the reaction with 1-ethynylcyclohexanol gives[21] the trans-dihydride, $H_2Pt(CCR)_2(PPh_3)_2$. Metal vapors of Yb, Sm and Er react with 1-hexyne, and the first step is oxidative addition of the terminal H—C bond[22].

Through the oxidative addition of aldehydes, hydridoformyl and -acyl compounds are formed. Formaldehyde adds to the reactive complex, $[Ir(PMe_3)_4]PF_6$, to afford[23] $[HIr(CHO)(PMe_3)_4]PF_6$. Cyclometallation of an aldehyde H—C bond results from treating $RhCl(PPh_3)_3$ with 8-quinolinecarboxaldehyde in CH_2Cl_2 in 10 min to yield (95%) $HRh(CRO)Cl(PPh_3)_2$ (R = 8-carboxyquinoline)[24].

The sterically crowded $Pt[P(C_6H_{11})_3]_2$ reacts with fluorobenzenes such as C_6HF_5 to afford[25] $HPt(C_6F_5)[P(C_6H_{11})_3]_2$.

Cyclopentadiene reacts with transition metals by oxidative addition of one H—C bond at the saturated carbon. Metal-atom vapors of Mo and W codeposit with cyclopentadiene to form[26] $(h^5\text{-}C_5H_5)_2MH_2$. Irradiation of $Mo(CO)_6$ with cyclopentadiene in isooctane yields[27] $(h^5\text{-}C_5H_5)Mo(CO)_2C_5H_7\text{-}h^3)$[28] that involves overall H transfer between ligands.

The H—C bond cleavage in saturated hydrocarbons is only realized in low-T matrices and in homogeneous solution. Reactions of methane and ethane with Ni are calculated theoretically[28]. Metals (Mn, Fe, Co, Cu, Zn, Ag and Au) in CH_4 matrices insert into the H—C bonds when irradiated at 15K to form[29] $HMCH_3$.

1.10. Formation of Bonds between Hydrogen and Transition Metals   199
1.10.5. by Oxidative Addition to Metal Complexes
1.10.5.3. of Hydrogen–Carbon Bonds.

Cyclopentane reacts with $[IrH_2S_2(PPh_3)]^+$ $[S = H_2O, (CH_3)_2CO]$ at 80°C in the presence of 3,3-dimethyl-1-butene, which acts as a hydrogen acceptor, to yield[30] $[IrH(C_5-H_5-h^5)(PPh_3)_2]^+$. Another Ir complex, $H_2Ir(C_5Me_5-h^5)(PMe_3)$, when irradiated in cyclohexane or neopentane at RT yields $HIrR(C_5Me_5-h^5)(PMe_3)$ (R = cyclohexyl, neopentyl, respectively)[31]. Irradiation of $Ir(C_5Me_5-h^5)(CO)_2$ in cyclohexane or neopentane at RT forms the hydridoalkyl addition products $HIrR(C_5Me_5-h^5)(CO)$ (R = cyclohexyl, neopentyl)[32]. The mechanism and thermodynamics of alkane and arene H—C bond activation[33] for the alkyl- and aryl-hydride complexes, $HRhR(C_5Me_5-h^5)-(PMe_3)$ (R = Me, Ph) are known.

Chlorination of CH_4 is catalyzed by supported Rh complexes involving H—C activation of CH_4 methane by the metal[34]. Methane exchanges with $La-CH_3$ complexes[35]

(T. J. LYNCH)

1. E. L. Muetterties, *Chem. Soc. Rev.*, *11*, 283 (1982).
2. G. W. Parshall, *Acc. Chem. Res.*, *8*, 113 (1975).
3. D. E. Webster, *Adv. Organomet. Chem.*, *15*, 147 (1977).
4. G. W. Parshall, D. L. Thorn, T. H. Tulip, *ChemTech.*, 571 (1982).
5. R. Dagani, *Chem. Eng. News*, *60*, 59 (18 Jan., 1982).
6. J. P. Collman, L. S. Hegedus, *Principles and Applications of Organotransition Metal Chemistry*, University Science Books, Mill Valley, CA, 1980, p. 211.
7. F. A. Cotton, G. Wilkinson, *Advanced Inorganic Chemistry*, 4th ed., Wiley, New York, 1980, p. 1244.
8. M. A. Bennett, D. L. Milner, *J. Am. Chem. Soc.*, *91*, 6983 (1969).
9. S. Bresadola, B. Longato, F. Morandini, *Inorg. Chim. Acta*, *25* L135 (1977).
10. K. V. Deuten, L. Dahlenburg, *Cryst. Struct. Commun.*, *9*, 421 (1980).
11. C. Crocker, R. J. Errington, R. Markham, C. J. Moulton, K. J. Odell, B. J. Shaw, *J. Am. Chem. Soc.*, *102*, 4373 (1980).
12. H. D. Empsall, E. M. Hyde, R. Markham, W. S. McDonald, M. C. Norton, B. L. Shaw, B. Weeks, *J. Chem. Soc., Chem. Commun.*, 589 (1977).
13. M. A. Bennett, H. Neumann, *Aust. J. Chem.*, *33*, 1251 (1980).
14. J. R. Shapley, D. E. Samkoff, C. Bueno, M. R. Churchill, *Inorg. Chem.*, *21*, 634 (1982), and refs. therein.
15. C. C. Yin, A. J. Deeming, *J. Organomet. Chem.*, *133*, 123 (1977).
16. A. J. Deeming, M. Underhill, *J. Chem Soc., Dalton Trans.*, 2727 (1973).
17. J. D. Druliner, A. D. English, J. P. Jesson, P. Meakin, C. A. Tolman, *J. Am. Chem. Soc.*, *98*, 2156 (1976).
18. H. Singer, G. Wilkinson, *J. Chem. Soc., A*, 2516 (1968).
19. F. Cariati, R. Ugo, F. Bonati, *Inorg. Chem.*, *5*, 1128 (1966).
20. S. D. Ittel, C. A. Tolman, A. D. English, J. P. Jesson, *J. Am. Chem. Soc.*, *100*, 7577 (1978).
21. D. M. Roundhill, in *Transition Metal Hydrides*, R. Bau, ed., Advances in Chemistry Series No. 167, American Chemical Society, Washington, DC, 1978, p. 160.
22. W. J. Evans, S. C. Engerer, K. M. Coleson, *J. Am. Chem. Soc.*, *103*, 6672 (1981).
23. D. L. Thorn, *Organometallics*, *1*, 197 (1982).
24. J. W. Suggs, *J. Am. Chem. Soc.*, *100*, 640 (1978).
25. M. Fornies, M. Green, J. L. Spencer, F. G. A. Stone, *J. Chem. Soc., Dalton Trans.*, 1006 (1977).
26. E. M. Van Dam, W. N. Brent, M. P. Silvon, P. S. Skell, *J. Am. Chem. Soc.*, *97*, 465 (1975).
27. W. C. Mills, III, M. S. Wrighton, *J. Am. Chem. Soc.*, *101*, 5830 (1979).
28. M. R. A. Blomberg, U. Brandemark, P. E. M. Siegbahn, *J. Am. Chem. Soc.*, *105*, 5557 (1983).
29. W. E. Billups, M. M. Konarski, R. H. Hauge, J. L. Margrave, *J. Am. Chem. Soc.*, *102*, 7393 (1980).
30. R. H. Crabtree, M. F. Mellea, J. M. Mihelcic, J. M. Quirk, *J. Am. Chem. Soc.*, *104*, 107 (1982).
31. A. H. Janowicz, R. G. Bergman, *J. Am. Chem. Soc.*, *104*, 352 (1982).
32. J. K. Hoyano, W. A. G. Graham, *J. Am. Chem. Soc.*, *104*, 3723 (1982).
33. W. D. Jones, F. J. Feher, *J. Am. Chem. Soc.*, *106*, 1650 (1984).

34. N. Kitajima, J. Schwartz, *J. Am. Chem. Soc., 106,* 2220 (1984).
35. P. L. Watson, *J. Am. Chem. Soc., 105,* 6491 (1983).

1.10.5.4. of Hydrogen–Other Group IVB Element Bonds.

Oxidative addition of the H—Si bond is the most studied of the group IVB elements with the exception of carbon because of its relation to the industrially important catalytic hydrosilation of alkenes, alkynes and ketones[1]. Compounds containing the Si—M bond are stable. While they are also synthesized by routes other than oxidative addition of the H—Si bond[2,3]. H—Si, H—Ge and H—Sn add to transition metals:

$$Fe(CO)_5 + SiHPh_3 \longrightarrow FeH(SiPh_3)(CO)_4 + CO \qquad (a)^4$$

$$Pt(C_2H_4)(PCy_3) + GeHMe_3 \longrightarrow \underset{Cy_3P}{\overset{Me_3Ge}{>}}Pt\underset{H}{\overset{H}{<}}Pt\underset{GeMe_3}{\overset{PCy_3}{<}} \qquad (b)^5$$

$$Pt(PPr-i_3)_2 + SnHPh_3 \longrightarrow trans\text{-}PtH(SnPh_3)(PPr-i_3)_2 \qquad (c)^6$$

where Cy = cyclohexyl. The stereochemistry of the oxidative addition at Si using an optically active silane proceeds with retention of configuration[7,8]:

$$Mn(C_5H_4Me)(CO)_3 + R(+)\text{-}SiHMePh(1\text{-}Np) \rightarrow$$
$$S(-)\text{-}MnH[SiMePh(1\text{-}Np)](C_5H_4Me)(CO)_2 \qquad (d)$$

where Np = napthyl.

Oxidative addition of H—Si bonds to metals with concomitant loss of CO is performed photochemically as in Eq. (d). The silyl hydride, cis-HRe(SiPh_3)(C_5H_5-h^5)-(CO)_2, is obtained in 15% yield by UV irradiation[9] of (h^5-C_5H_5)Re(CO)_3 with HSiPh_3 in cyclohexane for 7 h. Triphenylsilane with Rh(C_5H_5-h^5)(CO)_2 yields HRh(SiPh_3)-(C_5H^5-h^5)(CO) also under irradiation in nonpolar solvents[10]. The anion $[Mn(CO)_5]^-$, when photolyzed in the presence of HSiPh_3, results in high yields of the oxidative-addition product,[11] $[HMn(SiPh_3)(CO)_4]^-$. A formal Ir^{3+} to Ir^{5+} oxidative addition occurs in the reaction of $Ir_2(\mu\text{-}H_2)(C_5Me_5\text{-}h^5)_2Cl_2$ with two SiHEt_3 to form[12] two $H_2Ir(SiEt_3)$-(C_5Me_5-h^5)Cl. When the Ru or Os clusters, $M_3(CO)_{12}$, are treated with HSiCl_3 at 75°C or 135°C, respectively, in hexane, three equiv of the silane add to the clusters to afford $M_3(\mu\text{-}H)_3(SiCl_3)_3(CO)_9$ (M = Ru, 48%; Os, 62%)[13].

(T. J. LYNCH)

1. J. L. Speier, *Adv. Organomet. Chem., 17,* 407 (1979).
2. C. S. Cundy, B. M. Kingston, M. F. Lappert, *Adv. Organomet. Chem., 17,* 407 (1979).
3. R. J. P. Corriu, E. Colomer, *Ann. Chim. (Paris), 8,* 121 (1983).
4. W. Jetz, W. A. G. Graham, *Inorg. Chem., 10,* 4 (1971).
5. M. Green, J. A. K. Howard, J. Proud, J. L. Spencer, F. G. A. Stone, C. A. Tsipis, *J. Chem. Soc., Chem. Commun.,* 671 (1976).
6. H. C. Clark, A. B. Goel, C. Billard, *J. Organomet. Chem., 182,* 431 (1979).
7. E. Colomer, R. J. P. Corriu, A. Vioux, *J. Chem. Soc., Chem. Commun.,* 175 (1976).
8. E. Colomer, R. J. P. Corriu, A. Vioux, *Inorg. Chem., 18,* 695 (1979).
9. D. F. Dong, J. K. Hoyano, W. A. G. Graham, *Can. J. Chem., 59,* 1455 (1981).
10. A. J. Oliver, W. A. G. Graham, *Inorg. Chem., 10,* 1 (1971).
11. R. A. Faltynek, M. S. Wrighton, *J. Am. Chem. Soc., 100,* 2702 (1978).

12. M. J. Fernandez, P. M. Maitlis, *Organometallics, 2,* 164 (1983).
13. G. N. van Buuren, A. C. Willis, F. W. B. Einstein, L. K. Peterson, R. K. Pomeroy, D. Sutton, *Inorg. Chem., 20,* 4361 (1981).

1.10.5.5. of Hydrogen – Group VB Element Bonds

1.10.5.5.1. Involving Hydrogen – Nitrogen Bonds.

Ammonia undergoes H—N oxidative addition to transition-metal complexes; e.g., the Ti complex μ-(h^1:h^5-cyclopentadienyl)-tris(h^5-cyclopentadienyl)dititanium (Ti–Ti), which has a bridged structure, reacts with NH_3 and primary and secondary amines in toluene at low T, generating[1] 1 mol of H_2. The product from the NH_3 reaction is $[Ti(C_5H_5$-$h^5)_2]_2(NH)_2H$, a binuclear Ti complex bridged by two nitrogen atoms; all four atoms in Ti_2N_2 are coplanar. The three hydrogens could not be located in the structure. One may bridges the Ti—Ti bond, and the other two a nitrogen and a Ti atom.

A highly reactive Ni(0) complex, $Ni[P(C_6H_{11})_3]_2$, oxidatively adds pyrrole at RT in toluene to form[2] $HNi(NC_4H_4)[P(C_6H_{11})_3]_2$.

When an Os trinuclear cluster contains weakly coordinated ligands, H—N bonds add under mild conditions. The clusters $Os_3(CO)_{10}(C_8H_{14})_2$ (C_8H_{14} = cyclooctene) and $Os_3(CO)_{10}(NCMe)_2$ react with H—Y compounds including aniline, which yields $HOs_3(NHPh)(CO)_{10}$ where the hydride and -NHPh group bridge the same Os—Os bond[3].

Under more forcing conditions, aniline reacts with the parent cluster carbonyl to yield successively four H—N oxidative-addition products[4]. In refluxing aniline Os_3-$(CO)_{12}$ yields a cluster with two coordinated aniline ligands, one of which has oxidatively added a H—N and a C—H bond: $H_2Os_3(NHC_6H_4)(CO)_8(H_2NPh)$. When treated with CO, this compound loses the aniline ligand to form $H_2Os_3(NHC_6H_4)(CO)_9$, and addition of a second CO yields $Os_3H(NHPh)(CO)_{10}$. A fourth compound, H_2Os_3-$(NPh)(CO)_9$, is isolated on refluxing in nonane; this product contains two bridging hydrides and the -NPh group bonded to all three Os atoms of one face of the cluster through the nitrogen. A minor product[5] in the reaction of $Os_3(CO)_{12}$ with NMe_3 is $HOs_3(NH_2)(CO)_{10}$, arising from NH_3 impurity in the NMe_3.

The reaction of $Pt(PPh_3)_4$ with succinimide, phthalimide or saccharin gives[6] the H—N oxidative-addition Pt(II) product, HPt(imide)$(PPh_3)_2$, in good yields.

(T. J. LYNCH)

1. J. N. Armor, *Inorg. Chem., 17,* 203 (1978).
2. K. Jonas, G. Wilke, *Angew. Chem., Int. Ed. Engl., 8,* 519 (1969).
3. M. Tachikawa, J. R. Shapley, *J. Organomet. Chem., 124,* C19 (1977).
4. C. C. Yin, A. J. Deeming, *J. Chem. Soc., Dalton Trans.,* 1013, (1974).
5. C. C. Yin, A. J. Deeming, *J. Organomet. Chem., 133,* 123 (1977).
6. D. M. Roundhill, *J. Chem. Soc., Chem. Commun.,* 567 (1969).

1.10.5.5.2. Involving Hydrogen–Phosphorus Bonds.

Whereas diphenylphosphine $(HPPh_2)$ does not add oxidatively across the H—P bond to metal centers, $HP(CF_3)_2$ does so with either $Fe(CO)_5$ or $Fe_3(CO)_{12}$. When $HP(CF_3)_2$ is heated for 8 h in a sealed tube with a slight xs of either of the Fe carbonyls at 100°C in the absence of solvent, sublimation of the residue gives[1] 15% of a metal dimer, $H(CO)_3Fe[\mu$-$P(CF_3)_2]_2FeH(CO)_3$. The hydrides occupy axial sites, and the product exists as a mixture of cis and trans isomers.

The phosphine oxide OPHPh$_2$ readily adds to Pt(PPh$_3$)$_3$, forming[2] HPt[P(O)Ph$_2$]-[P(OH)PPh$_2$]PPh$_3$. Because OPHPh$_2$ and HOPPh$_2$ exist as an equilibrium mixture of tautomers, the overall addition of H—P to Pt may take an indirect route though the latter species.

(T. J. LYNCH)

1. R. C. Dobbie, D. Whittaker, *J. Chem. Soc., Chem. Commun.*, 796 (1970).
2. D. M. Roundhill, in *Transition Metal Hydrides*, R. Bau, ed., Advances in Chemistry Series No. 167, American Chemical Society, Washington DC, 1978, p. 160.

1.10.5.6. of Hydrogen–Group VIB Element Bonds

1.10.5.6.1. Involving Hydrogen–Oxygen Bonds.

The H—O bonds of water, alcohols and carboxylic acids oxidatively add to metal complexes. Water reacts with the zero-valent Pt complex, Pt[P(Pr-i)$_3$]$_2$; with an xs of H$_2$O in tetrahydroheran (THF), HPt(OH)[P(Pr-i)$_3$]$_2$ is formed in 0.5 h in 18% yield[1]. The compound is thermally unstable but can be stored for several weeks at < 0°C. Heating Os$_3$(CO)$_{12}$ and H$_2$O to 230°C in a sealed and evacuated vessel leads to polynuclear complexes, one of which is the double edge-bridged cluster HOs$_3$(OH)(CO)$_{10}$ in 2% yield[2,3]. A 71% yield of this compound[3] is isolated by column chromatography by treating Os$_3$(NCMe)$_2$(CO)$_{10}$ with xs H$_2$O in THF for 0.5 h. The Ru analogue, Ru$_3$-(CO)$_{12}$ forms only hydrides without the -OH group[2]. The product from electrochemical reduction[4] of aq [trans-[Rh(en)$_2$Cl$_2$]]$^+$ (where en = ethylenedianine) under Ar is [trans-[Rh(en)$_2$H(OH)]]$^+$.

The photochemical reaction of Re$_2$(CO)$_{10}$ with H$_2$O leads to the incorporation of fragments of H$_2$O into separate complexes: HRe(CO)$_5$ and Re$_4$(OH)$_4$(CO)$_{12}$. Transition-metal complexes that contain the -H and -OH ligands are prepared by routes other than oxidative addition of H$_2$O. These include [(C$_5$H$_5$)Mo]$_2$(μ-OH)(μ-H)(μ-C$_5$H$_4$-C$_5$H$_4$)$^{2+}$ [6], Ru$_2$(μ-OH)(μ-H)(PMe$_3$)$_6$ [7] and HRu(OH)(PPh$_3$)$_2$(H$_2$O) [8].

Alcohols are used in the synthesis of metal hydrides[9] where metal alkoxides are intermediates[10]. Isolation of hydridometal alkoxides from the oxidative addition of alcohols to metal centers is not common. Hydrocarbon solns of W$_2$(NMe$_2$)$_6$ and 2-propanol generate[11] NHMe$_2$ and H$_2$ at RT. The product results from alcoholysis, oxidative addition of i-PrOH and condensation to the tetramer, H$_2$W$_4$(OPr-i)$_{14}$. Both Os$_3$(CO)$_{12}$[3] and Os$_3$(NCMe)$_2$(CO)$_{10}$[12] oxidatively add HOR (R = Me, Et), forming HOs$_3$(OR)(CO)$_{10}$. Phenol adds[13] to Ni[P(C$_6$H$_{11}$)$_3$]$_2$ under mild conditions to yield HNi(OPh)[P(C$_6$H$_{11}$)$_3$]$_2$.

Carboxylic and sulfonic acids react with low-valent metal complexes, forming metal hydrides. Reactions of the two-coordinated complexes PdL$_2$ [L = P(Bu-t)$_3$, PPh-(Bu-t)$_2$] and Pt[PPh(Bu-t)$_2$] with HOCOCF$_3$ produces trans-HM(OCOCF$_3$)L$_2$ (M = Pd, Pt)[14]. The three-coordinated Pd(PPr-i$_3$)$_3$ also reacts with HOCOCF$_3$ to yield trans-HPd(OCOCF$_3$)(PPr-i$_3$)$_2$. The complex HPt(OCOCF$_3$)[As(Bu-t$_3$)]$_2$ is prepared in 96% yield[15] in 40 min by combining HPt$_2$[As(Bu-t$_3$)]$_2$ and HOCOCF$_3$ in hexane at RT. The sulfonic acids, HOSO$_2$CF$_3$ and HOSO$_2$C$_4$F$_{11}$ react with Ir(L)Cl(PPh$_3$)$_2$ (L = CO, N$_2$) leading to the O$_h$ HIr(X)(L)Cl(PPh$_3$)$_2$ (X = OSOCF$_3$, OSOC$_4$F$_{11}$)[16].

Formic acid adds to the lightly stabilized cluster, Os$_3$(CO)$_{10}$(h^2-C$_8$H$_{14}$)$_2$ (C$_8$H$_{14}$ = cyclooctene to form a double edge-bridged cluster, Os$_3$(μ-H)(μ-O$_2$CH)(CO)$_{10}$, with the formate ligand bridging through the oxygen atoms[17].

1.10. Formation of Bonds between Hydrogen and Transition Metals   203
1.10.5. by Oxidative Addition to Metal Complexes
1.10.5.6. of Hydrogen–Group VIB Element Bonds

The fragments of carbonic acid, in the form of CO_2 and H_2O, add to Rh complexes, such as $HRh(PPr-i_3)_3$, which, in THF at RT in 2h, yield[18] 85% of the six-coordinated $H_2Rh(O_2COH)(PPr-i_3)_2$.

(T. J. LYNCH)

1. T. Yoshida, T. Matsuda, T. Okano, T. Kitani, S. Otsuda, *J. Am. Chem. Soc.*, *101*, 2027 (1979).
2. C. R. Eady, B. F. G. Johnson, J. Lewis, *J. Chem. Soc., Dalton Trans.*, 838 (1977).
3. The compound $Os_3H(OH)(CO)_{10}$ is formed in 14% yield from $Os_3(CO)_{12}$ and base in MeOH at RT and atm P: B. F. G. Johnson, J. Lewis, P. A. Kilty, *J. Chem. Soc., A.*, 2859 (1968). High-yield synthesis (71%): N. L. Jones, private communication, 1984.
4. B. F. G. Johnson, J. Lewis, T. L. Odiaka, P. R. Raithby, *J. Organomet. Chem.*, *216*, C56 (1981).
5. R. D. Gillard, B. T. Heaton, D. H. Vaughan, *J. Chem. Soc., A*, 3126 (1970).
6. D. T. Gard, T. L. Brown, *J. Am. Chem. Soc.*, *104*, 6340 (1982).
7. N. J. Cooper, M. L. H. Green, C. Couldwell, K. Prout, *J. Chem. Soc., Chem. Commun.*, 145 (1977).
8. R. A. Jones, G. Wilkinson, I. J. Colquohoun, W. McFarlane, A. M. R. Galas, M. B. Hursthouse, *J. Chem. Soc., Dalton Trans.*, 2480 (1980).
9. B. N. Chaudret, D. J. Cole-Hamilton, R. S. Nohr, G. Wilkinson, *J. Chem. Soc., Dalton Trans.*, 1546 (1977).
10. H. D. Kaesz, R. B. Saillant, *Chem. Rev.*, *72*, 236 (1972).
11. M. Akiyama, D. Little, M. H. Chisholm, D. A. Haitko, F. A. Cotton, M. W. Extine, *J. Am. Chem. Soc.*, *101*, 2504 (1979).
12. M. Tachikawa, J. R. Shapley, *J. Organomet. Chem.*, *124*, C19 (1977).
13. K. Jonas, G. Wilke, *Angew. Chem. Int. Ed. Engl.*, *8*, 519 (1969).
14. T. Yoshida, S. Otsuka, *J. Am. Chem. Soc.*, *99*, 2134 (1977).
15. R. G. Goel, W. O. Ogini, R. C. Srivastava, *Inorg. Chem.*, *21*, 1627 (1982).
16. B. Olgemöller, H. Bauer, W. Beck, *J. Organomet. Chem.*, *213*, C57 (1981).
17. J. R. Shapley, G. M. St. George, M. R. Churchill, *Inorg. Chem.*, *21*, 3295 (1982).
18. T. Yoshida, D. L. Thorn, T. Okano, J. A. Ibers, S. Otsuka, *J. Am. Chem. Soc.*, *101*, 4212 (1979).

1.10.5.6.2. Involving Hydrogen–Sulfur Bonds.

The hydrogen–sulfur bonds of H_2S, RSH (R = alkyl, aryl) and RC(O)SH add to low-valent transition-metal complexes. Unsaturated Rh, Ir and Pt complexes add[1] H_2S. The complexes $RhCl(PPh_3)_3$, $IrCl(CO)PPh_3$ and $Pt(PPh_3)_3$ in benzene at RT react with H_2S to form $HRh(SH)Cl(PPh_3)_2$ (82%)[2], $HIr(SH)Cl(PPh_3)_2$ (100%)[2] and $HPt(SH)(PPh_3)_2$ (80%)[3], respectively. Thiophenol also adds to the former complexes, giving high yields of the analogous oxidative-addition products[2,3].

The mixed thioether–thiol, $HSCH_2CH_2SMe$, when treated with $Pt(PPh_3)_3$ forms[4] the hydride, $HPt(SCH_2CH_2SMe)(PPh_3)$ in which both the thiolate and the thioether are coordinated to Pt. The compound $HSCH_2CH_2SCH_2CH_2CH_2SMe$ transforms $Pt(PPh_3)_3$ to yield $HPt(SCH_2CH_2SCH_2CH_2CH_2SMe)(PPh_3)$, which has a four-coordinated, planar structure with the terminal thioether group uncoordinated.

When HSMe is bubbled through the binuclear complex, $[HPt_2(CO)(\mu\text{-dppm})_2]PF_6$, in CH_2Cl_2, a pale yellow precipitate of $[H_2Pt(\mu\text{-SMe})(\mu\text{-dppm})_2]PF_6$ is obtained in 72% yield[5] [dppm = bis(diphenylphosphino)methane].

The cis-diazido Pt complex, $Pt(N_3)_2(PPh_3)_2$, reacts with HSC(O)Me to yield[6] trans-$HPt[SC(O)Me](PPh_3)_2$. Thioacids also add to $Pt(PPh_3)_4$ to form $HPt[SC(O)R](PPh_3)_2$ (R = Me, Ph)[7].

The cluster $Ru_3(CO)_{12}$ can add EtSH or n-BuSH oxidatively in C_6H_6 at reflux in 15 min to form $HRu_3(SR)(CO)_{10}$ in 72 or 50% yield, respectively[8]. The Os analogues

$HOs_3(SR)(CO)_{10}$ are obtained in 56 and 38% yields in toluene reflux after longer reaction times (7 and 4 h, respectively). The hydride and thiolate ligands form a double bridge between two of the metal centers. The same Os product is made[9] under milder conditions starting with $Os_3(NCMe)_2(CO)_{10}$. The reaction of $Os_3(CO)_{12}$ with H_2S in n-octane at 125°C leads to $H_2Os_3(S)(CO)_9$ through the double oxidative addition of both H—S bonds[10]. The sulfur atom bridges one Os_3 face. The Ru analogue is also known. The Fe trimer $Fe_3(CO)_{12}$ loses an additional CO in contrast to the Ru and Os analogues in its reaction with alkanethiols RSH ($R = i\text{-}C_3H_7$, $s\text{-}C_4H_9$, $t\text{-}C_4H_9$). In refluxing C_6H_6, $HFe_3(SR)(CO)_9$ is formed ($>60\%$) where the -SR group bridges all three Fe atoms[11]. Further reaction yields the iron dimers, $Fe_2(SR)_2(CO)_6$.

(T. J. LYNCH)

1. R. A. Schunn, in *Transition Metal Hydrides*, E. L. Muetterties, ed., Marcel Dekker, New York, 1971, p. 216.
2. H. Singer, G. Wilkinson, *J. Chem. Soc. A*, 2516 (1968).
3. R. Ugo, G. LaMonica, S. Cenini, A. Segre, F. Conti, *J. Chem. Soc. A*, 522 (1971).
4. D. M. Roundhill, in *Transition Metal Hydrides*, R. Bau, ed., Advances in Chemistry Series No. 167, American Chemical Society, Washington, DC, 1978, p. 160.
5. J. R. Fisher, A. J. Mills, S. Sumnes, M. P. Brown, M. A. Thomson, R. J. Puddephatt, A. A. Frew, L. Manojlovic-Muir, K. W. Muir, *Organometallics, 1,* 1421 (1982).
6. P. H. Kreutzer, K. T. Schorpp, W. Beck, *Z. Naturforsch. Teil B, 30,* 544 (1975).
7. D. M. Roundhill, P. B. Tripathy, B. W. Renoe, *Inorg. Chem., 10,* 727 (1971).
8. G. R. Crooks, B. F. G. Johnson, J. Lewis, I. G. Williams, *J. Chem. Soc. A,* 797 (1969).
9. M. Tachikawa, J. R. Shapley, *J. Organomet. Chem., 124,* C19 (1977).
10. A. J. Deeming, M. Underhill, *J. Organomet. Chem., 42,* C60 (1972).
11. J. A. De Beer, R. J. Haines, *J. Organomet. Chem., 24,* 757 (1970).

1.10.6. by Protonation

1.10.6.1. of Neutral and Cationic Complexes

1.10.6.1.1. Which Are Mononuclear: Scope.

Transition-metal complexes can serve as proton acceptors. Protonation, like the addition of covalent hydrogen–element bonds (§1.10.5), is a formal oxidative addition and so requires a formal d-electron configuration of d^2 or higher. The reaction is common for complexes of early as well as late transition metals[1-3]. Table 1 lists examples for even d-electron configurations.

Protonation of d^0 complexes, such as $(h^5\text{-}C_5H_5)_2TaH_3$, does not give stable hydrides. For less-apparent reasons, protonation of some d^{10} complexes such as those of Au(I) and Hg(II) also does not give stable hydrides. Transient species, such as $[Me_2HgH]^+$, are observed in the gas phase[24], and hydride bridges to Au are known[25,26].

With many neutral complexes the protonated cation can be identified in acid [by observation of an upfield resonance in the ^1H nuclear magnetic resonance (NMR)] and, therefore, is frequently prepared in situ. However, isolation can be difficult because protonation generally is reversible. (In Table 1 no counterion is written when the cationic hydride is prepared only in solution; an anion is shown where a solid product is isolated.) For example, ferrocene is protonated[5] in aq BF_3, but isolation is difficult and is accomplished only for HCl and $AlCl_3$ in CH_2Cl_2 (giving the $[AlCl_4]^-$ salt[27]) and for HX and BX_3 in benzene (giving the $[BX_4]^-$ salts[9]). Oxidation to the ferricenium cation and loss of HX to give ferrocenyldihaloboranes are complications.

1.10.6. by Protonation
1.10.6.1. of Neutral and Cationic Complexes
1.10.6.1.1. Which Are Mononuclear: Scope.

205

TABLE 1. FORMATION OF CATIONIC HYDRIDES BY PROTONATION
OF NEUTRAL TRANSITION-METAL COMPLEXES

Neutral complex (d-electron configuration)	Acid	Cationic hydride (d-electron configuration)	Refs.
$(h^5-C_5H_5)_2WH_2$ (d^2)	aq HCl	$[(h^5-C_5H_5)_2WH_3]Cl$ (d^0)	4
$(h^5-C_5H_5)_2ReH$ (d^4)	aq HCl	$[(h^5-C_5H_5)_2ReH_2]Cl$ (d^2)	5
$h^5-C_5H_5Nb(CO)_4$ (d^4)	$CF_3CO_2H-CH_2Cl_2$	$[h^5-C_5H_5Nb(CO)_4H]^+$ (d^2)	6
$(h^5-C_5H_5)_2Fe$ (d^6)	$BX_3-HX-C_6H_6$	$[(h^5-C_5H_5)_2FeH][BX_4]$ (d^4)	7–9
$h^5-C_5H_5Re(CO)_2PPh_3$ (d^6)	$CF_3CO_2H-CHCl_3$	$[h^5-C_5H_5Re(CO)_2(PPh_3)H]^+$ (d^4)	6, 10
$h^5-C_5H_5Mn(CO)_2PPhMe_2$ (d^6)	$CF_3SO_3H-CD_2Cl_2$	$[h^5-C_5H_5Mn(CO)_2(PPhMe_2)H]^+$ (d^4)	11, 12
$h^6-C_6H_3Me_3Cr(CO)_3$ (d^6)	FSO_3H-SO_2	$[h^6-C_6H_3Me_3Cr(CO)_3H]^+$ (d^4)	12–16
$Os(CO)_3(PPh_3)_2$ (d^8)	$HClO_4$ (or HPF_6 or HBF_4) in EtOH	$[Os(CO)_3(PPh_3)_2H][ClO_4]$ (d^6)	17
$h^5-C_5Me_5Ir(CO)_2$ (d^8)	HBF_4–ether	$[h^5-C_5Me_5Ir(CO)_2H][BF_4]$ (d^6)	18
$h^5-C_5Me_5Co(PMe_3)_2$ (d^8)	$MeOH-NH_4PF_6$	$[h^5-C_5Me_5Co[PMe_3]_2H][PF_6]$ (d^6)	19
$Fe[P(OMe)_3]_5$ (d^8)	NH_4PF_6–THF	$[HFe[P(OMe)_3]_5]^+$ (d^6)	20
$Pt(PPh_3)_3$ (d^{10})	CF_3CO_2H	$[Pt(PPh_3)_3H][(CF_3CO_2)_2H]$ (d^8)	21, 22
$Pd(PEt_3)_4$ (d^{10})	EtOH–NaBPh$_4$–HCl	$[Pd(PEt_3)_3H][BPh_4]$ (d^8)	23

Competition for protons between the metal and the ligand is common; e.g., dienes and other unsaturated ligands undergo protonation, as do electronegative ligand atoms, such as acyl oxygens[28]:

$$h^5-C_5H_5[(C_6H_{11})_3P](OC)Fe\overset{O}{\overset{\|}{C}}CH_3 \xrightarrow[\text{MeOH-H}_2O]{HBF_4}$$

$$\left[h^5-C_5H_5[(C_6H_{11})_3P](OC)Fe=C\begin{smallmatrix}OH\\CH_3\end{smallmatrix} \right] [BF_4] \qquad (a)$$

Protonation at the latter is favored kinetically because it can occur with little electronic rearrangement. Initial protonation at the metal may give a cation in which the hydride is exchanged between the metal and various ligand sites; e.g., protonation of $h^5-C_5H_5Rh$(1,3-cyclohexadiene) gives a hydride cation (which can be isolated as a $[PF_6]^-$ salt):

$$(b)$$

When CF_3CO_2D is used as the acid proton, nuclear magnetic resonance (NMR) shows rapid and stereospecific exchange between the Rh hydride and the two endo positions on the six-membered ring, reflecting the reversible transfer of the Rh hydride to endo sites at the ends of the diene system[29].

Protonation of organometallic complexes containing metal–carbon σ bonds leads to cleavage of those bonds—a reaction assumed to proceed by formation of a cationic

206   1.10. Formation of Bonds between Hydrogen and Transition Metals
         1.10.6. by Protonation
         1.10.6.1. of Neutral and Cationic Complexes

hydride followed by C—H elimination. However, protonation by strong acids of $CH_3Mn(CO)_5$ in the gas phase gives an observable cation (by protonation at the metal to form $[CH_3Mn(CO)_5H]^+$), whereas weak acids give immediate methane loss without intermediate formation of the cation[30]. Protic cleavage of a metal–carbon σ bond may occur without formation of a M—H bond, presumably by direct electrophilic attack on the electron pair in the σ bond.

Although stable cationic polyhydrides are formed on protonation of some hydrides, e.g., the first two entries in Table 1 and:

$$h^6\text{-}C_6H_6Mo(PEt_3)_3 \xrightarrow[\text{NH}_4\text{PF}_6]{\text{2 N HCl-EtOH}} [h^6\text{-}C_6H_6Mo(PEt_3)_3H][PF_6] \xrightarrow[\text{NH}_4\text{PF}_6]{\text{conc HCl}}$$

$$[h^6\text{-}C_6H_6Mo(PEt_3)_3H_2][PF_6]_2 \qquad (c)^{31}$$

$$h^6\text{-}Me_6C_6RuPPh_3H_2 \xrightarrow[\text{THF, } -78°C]{\text{NH}_4\text{PF}_6} [h^6\text{-}Me_6C_6RuPPh_3H_3][PF_6] \qquad (d)^{32}$$

$$WH_4(PMePh_2)_4 \xrightarrow[\text{THF, } -10°C]{\text{HPF}_6\text{-MeOH}} [WH_5(PMePh_2)_4][PF_6] \qquad (e)^{33}$$

$$HRh[P(CHMe_2)_3]_3 \xrightarrow{\text{py-H}_2\text{O}} [H_2Rh[P(CHMe_2)_3]_2(py)_2]^+ \qquad (f)^{34,35}$$

many cationic dihydrides are unstable to H_2 evolution. The protonation of $h^5\text{-}C_5H_5\text{-}W(CO)_3H$ gives a cationic dihydride[13] that evolves H_2 within several hours at RT:

$$h^5\text{-}C_5H_5W(CO)_3H \xrightarrow{\text{BF}_3\text{-H}_2\text{O-CF}_3\text{CO}_2\text{H}} [h^5\text{-}C_5H_5W(CO)_3H_2]^+ \xrightarrow{-\text{H}_2}$$

$$h^5\text{-}C_5H_5W(CO)_3O_2CCF_3 \qquad (g)$$

The protonation of $HCo[P(OCHMe_2)_3]_4$ gives a cationic dihydride that evolves[39] H_2 within 1 day at 20°C:

$$HCo[P(OCHMe_2)_3]_4 \xrightarrow[\text{ether}]{\text{HPF}_6} [H_2Co[P(OCHMe_2)_3]_4]^+ \xrightarrow{-\text{H}_2} [Co[P(OCHMe_2)_3]_4]^+ \text{ (h)}$$

Care must be taken in protonations to avoid further protonation and H_2 evolution; e.g., although clean monoprotonation of $Fe[P(OMe)_3]_5$ can be effected by NH_4PF_6 (see Table 1), CH_3CO_2H decomposes the cationic hydride initially formed[20,40]:

$$Fe[P(OMe)_3]_5 \xrightarrow[\text{THF}]{\text{NH}_4\text{PF}_6} [HFe[P(OMe)_3]_5][PF_6] \xrightarrow{\text{CH}_3\text{CO}_2\text{H}} [Fe[P(OMe)_3]_5]^{2+} \qquad (i)$$

Careful protonation of $Fe(CNCMe_3)_5$ at $-40°C$ allows the isolation of $[HFe(CNCMe_3)_5]^+$:

$$Fe(CNCMe_3)_5 \xrightarrow[\text{THF, } -40°C]{\text{HBF}_4\cdot2 \text{ Et}_2\text{O}} [HFe(CNCMe_3)_5][BF_4] \qquad (j)$$

but protonation with xs acid at RT gives loss of H_2 and formation of a dication[41]:

$$Fe(CNCMe_3)_5 \xrightarrow[\text{Et}_2\text{O, } 25°C]{\text{HBF}_4\cdot\text{H}_2\text{O}} [Fe(CNCMe_3)_5OH_2][BF_4]_2 \qquad (k)$$

If the addition of aq HBF_4 to $h^5\text{-}C_5H_5Co(PMe_3)_2$ is not carefully controlled, the heat from hydrolysis of the propionic anhydride solvent causes H_2 evolution and formation of a dication[42]:

$$h^5\text{-}C_5H_5Co(PMe_3)_2 \xrightarrow[\text{acetone}]{NH_4PF_6} [h^5\text{-}C_5H_5Co(PMe_3)_2H][PF_6] \quad (1)$$

$$h^5\text{-}C_5H_5Co(PMe_3)_2 \xrightarrow[\text{propionic anhydride}]{HBF_4\text{-}H_2O,\ 95°C} [h^5\text{-}C_5H_5Co(PMe_3)_3][BF_4]_2 \quad (m)$$

Although $Pt(PPh_3)_4$ becomes $[Pt(PPh_3)_3H]^+$ between pH $= 2$ and an acidity function, $H_0 \approx -10$, it reacts[38] and generates H_2 at $H_0 < -10$. Protic cleavage of hydride ligands may occur by direct electrophilic attack on the electron pair in the existing $M\text{—}H$ σ bond without the formation of a new $M\text{—}H$ σ bond (in analogy with the process discussed above for cleavage of $M\text{—}C$ σ bonds).

Cationic complexes can be protonated, e.g., in Eq. (c), and in §1.10.6.1.2, where the equilibrium constant for protonation is discussed. However, in:

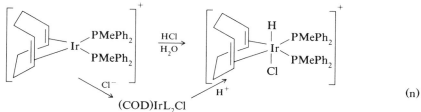

where L $=$ PMe_2Ph, both kinetic evidence[41] and isolation of the intermediate shown[42] establish that the anion, X^-, of a coordinating acid must be added, forming a neutral intermediate, before protonation can occur. It seems likely that this pattern will prove common for cationic complexes.

(J. R. NORTON)

1. J. C. Kotz, D. G. Pedrotty, *Organomet. Chem. Rev., A,* 4, 479 (1969).
2. D. F. Shriver, *Acc. Chem. Res.,* 3, 231 (1970).
3. R. A. Schunn, in *Transition-Metal Hydrides,* E. L. Muetterties, ed., Marcel Dekker, New York, 1971, p. 203.
4. M. L. H. Green, J. A. McCleverty, L. Pratt, G. Wilkinson, *J. Chem. Soc.,* 4854 (1961).
5. M. L. H. Green, L. Pratt, G. Wilkinson, *J. Chem. Soc.,* 3916 (1958).
6. B. V. Lokshin, A. A. Pasinsky, N. E. Kolobova, K. N. Anisimov, T. V. Makarov, *J. Organomet. Chem.,* 55, 315 (1973).
7. T. J. Curphey, J. O. Santer, M. Rosenblum, J. H. Richards, *J. Am. Chem. Soc.,* 82, 5249 (1960).
8. T. E. Bitterwolf, A. C. Ling, *J. Organomet. Chem.,* 40, 197 (1972).
9. W. Siebert, W. Ruf, K.-J. Schaper, T. Renk, *J. Organomet. Chem.,* 128, 219 (1977).
10. G. A. Panosyan, P. V. Petrovskii, N. I. Pyshnograyeva, N. E. Kolobova, V. N. Setkina, E. I. Fedin, *J. Organomet. Chem.,* 108, 209 (1976).
11. B. V. Lokshin, A. G. Ginzburg, V. N. Setkina, D. N. Kursanov, I. B. Nemirovskaya, *J. Organomet. Chem.,* 37, 347 (1972).
12. T. C. Flood, E. Rosenberg, A. Sarhangi, *J. Am. Chem. Soc.,* 99, 4334 (1977).
13. A. Davison, W. McFarlane, L. Pratt, G. Wilkinson, *J. Chem. Soc.,* 3653 (1962).
14. D. N. Kursanov, V. N. Setkina, P. V. Petrovskii, V. I. Zdanovich, N. K. Baranetskaya, I. D. Rubin, *J. Organomet. Chem.,* 37, 339 (1972).
15. G. A. Olah, S. H. Yu, *J. Org. Chem.,* 41, 717 (1976).
16. C. P. Lillya, R. A. Sahatjian, *Inorg. Chem.,* 11, 889 (1972).
17. K. R. Laing, W. R. Roper, *J. Chem. Soc., A,* 1889 (1969).
18. J. Plank, D. Riedel, W. A. Herrmann, *Angew. Chem., Int. Ed. Engl.,* 19, 937 (1980).
19. H. Werner, B. Heiser, B. Klingert, R. Dölfel, *J. Organomet. Chem.,* 240, 179 (1982).
20. T. V. Harris, J. W. Rathke, E. L. Muetterties, *J. Am. Chem. Soc.,* 100, 6966 (1978).
21. F. Cariati, R. Ugo, F. Bonati, *Inorg. Chem.,* 5, 1128 (1966).

208   1.10. Formation of Bonds between Hydrogen and Transition Metals
        1.10.6. by Protonation
        1.10.6.1. of Neutral and Cationic Complexes

22. D. M. Roundhill, in *Transition-Metal Hydrides*, R. Bau, ed., Advances in Chemistry Series No. 167, American Chemical Society, Washington, DC, 1978, p. 160.
23. R. A. Schunn, *Inorg. Chem., 15*, 208 (1976).
24. J. A. Stone, J. R. M. Camicioli, M. C. Baird, *Inorg. Chem., 19*, 3128 (1980).
25. H. Lehner, D. Matt, P. S. Pregosin, L. M. Venanzi, A. Albinati, *J. Am. Chem. Soc., 104*, 6825 (1982).
26. M. Green, A. G. Orpen, I. D. Salter, F. G. A. Stone, *J. Chem. Soc., Chem. Commun.*, 813 (1982).
27. M. Rosenblum, J. O. Santer, W. G. Howells, *J. Am. Chem. Soc., 85*, 1450 (1963).
28. M. L. H. Green, L. C. Mitchard, M. G. Swanwick, *J. Chem. Soc., A.*, 794 (1971).
29. B. F. G. Johnson, J. Lewis, D. J. Yarrow, *J. Chem. Soc., Dalton Trans., 2084 (1972).*
30. A. E. Stevens, J. L. Beauchamp, *J. Am. Chem. Soc., 101*, 245 (1979).
31. M. L. H. Green, L. C. Mitchard, W. E. Silverthorn, *J. Chem. Soc., Dalton Trans.*, 1361 (1974).
32. H. Werner, H. Kletzin, *J. Organomet. Chem., 243*, C59 (1983).
33. E. Carmona-Guzman, G. Wilkinson, *J. Chem. Soc., Dalton Trans.*, 1716 (1977).
34. T. Yoshida, T. Okano, Y. Ueda, S. Otsuka, *J. Am. Chem. Soc., 103*, 3411 (1981).
35. R. F. Jones, D. J. Cole-Hamilton, *J. Chem. Soc., Chem. Commun.*, 58 (1981).
36. E. L. Muetterties, P. L. Watson, *J. Am. Chem. Soc., 100*, 6978 (1978).
37. E. L. Muetterties, J. W. Rathke, *J. Chem. Soc., Chem. Commun.*, 850 (1974).
38. J.-M. Bassett, L. J. Ferugia, F. G. A. Stone, *J. Chem. Soc., Dalton Trans.*, 1789 (1980).
39. H. Werner, W. Hofmann, *Chem. Ber., 110*, 3481 (1977).
40. P. Foley, G. M. Whitesides, *Inorg. Chem., 19*, 1402 (1980).
41. V. Ashworth, J. E. Singleton, D. J. A. de Waal, E. Singleton, E. van der Stok, *J. Chem. Soc., Dalton Trans.*, 340 (1978).
42. R. H. Crabtree, J. M. Quirk, T. Fillebeen-Khan, G. E. Morris, *J. Organomet. Chem., 181*, 203 (1979).

1.10.6.1.2. Which Are Mononuclear:
Position of the Protonation Equilibria.

Few quantitative data are available. The pK_a of $[(h^5\text{-}C_5H_5)_2ReH_2]^+$ in 60% aq dioxane is[1] 8.5. The equilibrium constants for protonation of NiL_4 by H_2SO_4 in MeOH at 0°C:

$$H^+ + NiL_4 \overset{K}{\rightleftharpoons} [HNiL_4]^+ \tag{a}$$

are known for phosphite ligands and for $L_2 = Ph_2PCH_2CH_2PPh_2$. Values of K range from 410 M^{-1} for $Ni(Ph_2PCH_2CH_2PPh_2)_2$ to 34 M^{-1} for $Ni[P(OR)_3]_4$ (R=Me, or Et) to small values for phosphites with electron-withdrawing substituents[2]. Under the same conditions the equilibrium constant for protonation of $Ni(CO)_4$ is only[2] 10^{-7} M^{-1}. In acetonitrile, $Ni[P(OEt)_3]_4$ is fully protonated by 1 equiv of $HClO_4$ or H_2SO_4 and completely unprotonated by CH_3CO_2H; $[Ni(P(OEt)_3)_4H]^+$ has the acid strength[3] of trifluoroacetic acid in CH_3CN. In the same solvent the pK_a values[4] of $[M[P(OMe)_3]_4H]^+$ are 18.5, 8.0 and 12.3 for M = Pt, Pd and Ni, respectively; in MeOH they are 10.2, 1.0 and 1.5.

In MeOH or in acetonitrile the d^8 cation $[Rh(Ph_2PCH_2CH_2PPh_2)]^+$ is protonated by HBF_4, HPF_6 or $HClO_4$ to give a dicationic hydride:

$$[Rh(Ph_2PCH_2CH_2PPh_2)]^+ + H^+ \overset{K}{\rightleftharpoons} \begin{bmatrix} \begin{smallmatrix} Ph_2 & H \\ P & \vert & S \\ & Rh & \\ P & \vert & S \\ Ph_2 & S \end{smallmatrix} \end{bmatrix}^{2+} \tag{b}$$

$$S = CH_3CN$$

The equilibrium constant in MeOH is[5] 11 M^{-1}. For $IrCl(CO)(AsPh_3)_2$ in MeOH at

1.10. Formation of Bonds between Hydrogen and Transition Metals   209
1.10.6. by Protonation
1.10.6.1. of Neutral and Cationic Complexes

RT, the protonation equilibrium constant is 203 M^{-1}; for $RhCl(CO)(AsPh_3)_2$ it is 75 M^{-1}. Similar K values are found for Ir and Rh complexes with phosphine and halide ligands.[6] Finally, ruthenocene is half-protonated in H_2SO_4 with an acidity function, H_0, of -5.7, whereas half-protonation of ferrocene requires[7] an H_0 of -7.7.

From qualitative knowledge about protonation equilibria, basicity tends to increase down a periodic column: ruthenocene (see above) is protonated in less acidic media than is ferrocene[7], cis-$OsH_2(PMe_3)_4$ is protonated in less acidic media than is its Fe analogue[8], h^5-$C_5H_5Re(CO)_2PPh_3$ is protonated in less acidic media than is[9] h^5-C_5H_5-$Mn(CO)_2PPh_3$ and h^5-$C_5H_5Nb(CO)_3PPh_3$ is protonated in less acidic media than is[9] h^5-$C_5H_5V(CO)_3PPh_3$. However, the quantitative data above show that basicity decreases from Ni to Pd for $M[P(OMe)_3]_4$. The replacement of π-acceptor by σ-donor ligands also increases basicity: h^6-$C_6H_6Cr(CO)_2PPh_3$ requires less acidic media for protonation than does[10] h^6-$C_6H_6Cr(CO)_3$, and base strength increases in the order[11] h^5-$C_5H_5Co(CO)_2 < h^5$-$C_5H_5Co[P(OR)_3]_2 < h^5$-$C_5H_5Co(PMe_3)_2$.

The presence of sufficiently powerful σ-donor ligands can make a metal a proton acceptor. The trialkylphosphine ligands in[12] $Pt(PEt_3)_3$, and[13] h^5-$C_5H_5Co(PMe_3)_2$ make the metals sufficiently basic to deprotonate EtOH and H_2O:

$$Pt(PEt_3)_3 + EtOH \rightarrow [HPt(PEt_3)_3]^+ + [EtO]^- \qquad (c)$$

and $Ni(PEt_3)_4$, $Pd(PEt_3)_3$ and $Pd(PEt_3)_4$ also form hydride cations by deprotonating[14] EtOH.

Tricarbonyl complexes are more basic than their dinitrosyl analogues[15]; e.g., $[HMoL(CO)_3][BF_4]$ (L = 1,4,7-triazacyclononane) is a strong aqueous acid, whereas $[HMoL(NO)_2][ClO_4]$ is weak[16].

(J. R. NORTON)

1. M. L. H. Green, L. Pratt, G. Wilkinson, *J. Chem. Soc.*, 3916 (1958).
2. C. A. Tolman, *Inorg. Chem.*, *11*, 3128 (1972).
3. C. A. Tolman, *J. Am. Chem. Soc.*, *92*, 4217 (1970).
4. R. G. Pearson, personal communication, 1985.
5. J. Halpern, D. P. Riley, A. S. C. Chan, J. J. Pluth, *J. Am. Chem. Soc.*, *99*, 8055 (1977).
6. R. G. Pearson, C. T. Kresge, *Inorg. Chem.*, *20*, 1878 (1981).
7. G. Cerichelli, G. Illuminati, G. Ortaggi, A. M. Giuliani, *J. Organomet. Chem.*, *127*, 357 (1977).
8. H. Werner, J. Gotzig, *Organometallics*, *2*, 547 (1983).
9. B. V. Lokshin, A. A. Pasinsky, N. E. Kolobova, K. N. Anisimov, Y. V. Makarov, *J. Organomet. Chem.*, *55*, 315 (1973).
10. D. N. Kursanov, V. N. Setkina, P. V. Petrovskii, V. I. Zdanovich, N. K. Baranetskaya, I. D. Rubin, *J. Organomet. Chem.*, *37*, 339 (1972).
11. H. Werner, *Pure Appl. Chem.*, *54*, 177 (1982).
12. D. H. Gerlach, A. R. Kane, G. W. Parshall, J. P. Jesson, E. L. Muetterties, *J. Am. Chem. Soc.*, *93*, 3543 (1971).
13. H. Werner, W. Hofmann, *Angew. Chem., Int. Ed. Engl.*, *17*, 464 (1978).
14. R. A. Schunn, *Inorg. Chem.*, *15*, 208 (1976).
15. B. E. Bursten, M. G. Gatter, *J. Am. Chem. Soc.*, *106*, 2554 (1984).
16. P. Chaudhuri, K. Wieghart, Y.-H. Tsai, C. Krüger, *Inorg. Chem.*, *23*, 427 (1984).

1.10.6.1.3. Which Are Mononuclear: Rates.

Even fewer quantitative data are available on protonation rates than on equilibria. For the protonation[1]:

$$Ni[P(OEt)_3]_4 + H^+ \underset{k_{-1}}{\overset{k_1}{\rightleftharpoons}} [HNi[P(OEt)_3]_4]^+ \qquad (a)$$

210   1.10. Formation of Bonds between Hydrogen and Transition Metals
1.10.6. by Protonation
1.10.6.1. of Neutral and Cationic Complexes

at 25°C in MeOH, $k_1 = 1550$ M^{-1} s^{-1} and $k_{-1} = 45$ s^{-1}.

The rate of protonation of $IrCl(CO)(PPh_3)_2$ in MeOH reflects the rate of prior solvent coordination[2].

The rate constants for protonation at the metal of (6-dimethylamino-fulvene)$M(CO)_3$ increase[3] in the order Cr < Mo < W; quantitative results are given[3], but they appear unreliable. Proton transfer between protonated and unprotonated forms of (arene)$Cr(CO)_3$ is fast on the NMR time scale and occurs by rate-determining proton removal by the conjugate base ($[FSO_3]^-$) of the acid employed[4] (FSO_3H).

(J. R. NORTON)

1. C. A. Tolman, *J. Am. Chem. Soc.*, 92, 4217 (1970).
2. R. G. Pearson, C. T. Krege, *Inorg. Chem.*, 20, 1878 (1981).
3. V. N. Setkina, B. N. Strunin, D. N. Kursanov, *J. Organomet. Chem.*, 186, 325 (1980).
4. C. P. Lillya, R. A. Sahatjian, *Inorg. Chem.*, 11, 889 (1972).

1.10.6.1.4. Which Are Polynuclear.

Polynuclear complexes are frequently protonated in strong acid. In Table 1 the anion is shown when the hydride cation is isolated; if it is omitted, the hydride cation is made only in situ. Often the location (terminal or bridging and, if the latter, among which metal atoms) of the hydrogen atom is not known, although an upfield shift in the ^1H NMR identifies the hydrogen as attached to the metal framework.

Substituent effects on protonation equilibria are similar to those for mononuclear compounds. The metal–metal-bonded dimers, $[(\mu-SCH_3)Fe(CO)_2L]_2$, are stronger proton bases[12] when L is a strong donor phosphine such as PMe_3, than when L is PPh_3. The pK_b of $[h^5-C_5H_5Fe(CO)_2]_2$ in acetic acid is 7.5; both the phosphite-substituted dimer $(h^5-C_5H_5)_2Fe_2(CO)_3P(OMe)_3$ and the second-row analogue, $[h^5-C_5H_5Ru(CO)_2]_2$, are stronger proton bases[4]. Proton-transfer rates involving $(h^5-C_5H_5)_2Fe_2(CO)_3P(OMe)_3$ and its protonated analogue are faster than those involving the $[h^5-C_5H_5Fe(CO)_2]_2$ system and its Ru analogue[4], and $Os_3(CO)_{12}$ is ca. five times more basic[13] than $Ru_3(CO)_{12}$.

As is seen in Table 1, protonation occurs on the metal–metal-bonded framework despite the presence of hydride, sulfur ligands, or organic fragments. Diprotonation may result in protonation of a carbon ligand, but only after initial protonation on the metal core; e.g., $HRu_3(CO)_9C\equiv CCMe_3$ undergoes monoprotonation on a Ru—Ru bond in CF_3CO_2H; in HSO_3Cl or H_2SO_4 a dication is formed, with the second proton going onto carbon[14].

TABLE 1. CATIONIC HYDRIDES RESULTING FROM THE PROTONATION OF POLYNUCLEAR COMPLEXES

Polynuclear complex	Acid	Cationic hydride	Refs.
$[h^5-C_5H_5W(CO)_3]_2$	$HF-PCl_5$	$[[h^5-C_5H_5W(CO)_3]_2H][PF_6]$	1
$Re(CO)_4(\mu-Ar_2PC_5H_4)$-$Mo(CO)_3$ [a]	$CF_3SO_3H-CD_2Cl_2$	$[Re(CO)_4(\mu-Ar_2PC_5H_4)(\mu-H)$-$Mo(CO)_3]^+$ [a]	2
$[h^5-C_5H_5Fe(CO)_2]_2$	$HCl-PF_5$	$[[h^5-C_5H_5Fe(CO)_2]_2H][PF_6]$	3, 4
$Os_3(CO)_{12}$	$H_2SO_4-NH_4PF_6$	$[HOs_3(CO)_{12}][PF_6]$	5, 6
$H_2Os_3(CO)_9C\equiv CH_2$	$CF_3CO_2H-CDCl_3$	$[H_3Os_3(CO)_9C\equiv CH_2]^+$	7
$HOs_3(CO)_9(PEt_3)SPh$	$H_2SO_4-NH_4PF_6$	$[H_2Os_3(CO)_9(PEt_3)SPh][PF_6]$	8
$(h^5-C_5H_5)_2Rh_2(CO)_3$	$HBF_4-H_2O-Et_2O$	$[(h^5-C_5H_5)_2Rh_2(CO)_3H][BF_4]$	9
$(h^5-C_5H_5CoPMe_2)_2$	$CF_3CO_2H-NH_4PF_6$	$[(h^5-C_5H_5CoPMe_2)_2H][PF_6]$	10
$(\mu-Ph_2PCH_2PPh_2)_2Pt_2Cl_2$	$HCl-CHCl_3-NH_4PF_6-$ MeOH	$[(\mu-Ph_2PCH_2PPh_2)_2Pt_2Cl_2H][PF_6]$	11

[a] Ar = p-tolyl.

1.10.6. by Protonation
1.10.6.1. of Neutral and Cationic Complexes
1.10.6.1.4. Which Are Polynuclear.

211

Preferential protonation of an organic ligand rather than a metal–metal bond occurs when the latter is sterically inaccessible, as for $Ru_2(\mu\text{-}CH_2)_3(PMe_3)_6$ where the Ru—Ru bond is already triply bridged[15]. In contrast, protonation of $\mu\text{-}CH_2[h^5\text{-}C_5H_5\text{-}Rh(CO)]_2$ occurs at the metal–metal bond, although deuterium labeling shows rapid exchange of the hydride and methylene protons and suggests that the methyl tautomer of the observed structure is accessible. Addition of a coordinating anion allows isolation of a dimer with a methyl ligand[9]. Protonation labilizes the Rh—Rh bond and leads to the formation of a $\mu_3\text{-}CH$ trimer:

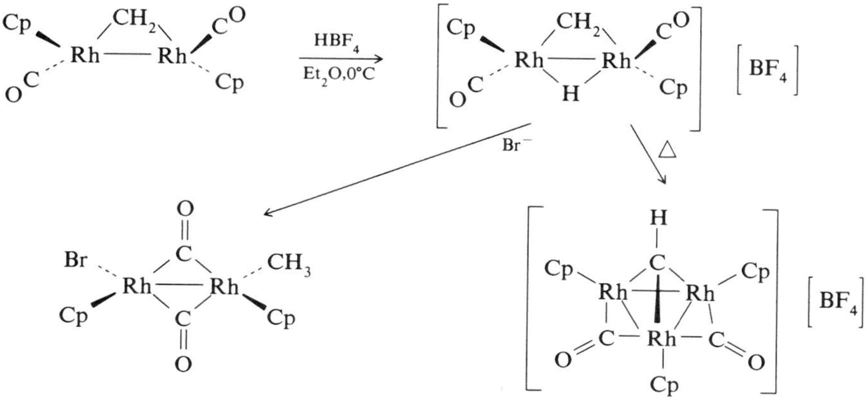

(a)

just as protonation of a $(h^5\text{-}C_5H_5)_2Rh_2(CO)_3$ (as in Table 1) leads to formation[9] of a $\mu\text{-}CH_3$ trimer. The reverse process is also possible: protonation of a mononuclear complex can form a protonated metal–metal bond[16]:

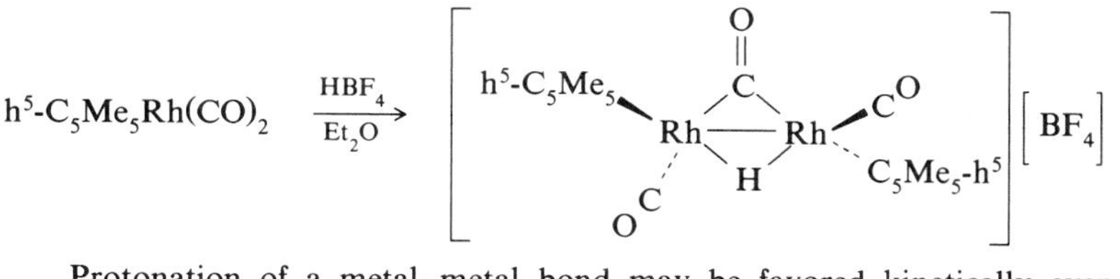

(b)

Protonation of a metal–metal bond may be favored kinetically even when it is disfavored thermodynamically. The diplatinacyclobutene shown initially forms an isolable bridged cationic hydride which rearranges to a bridged vinyl complex[17] on standing for 1 h at RT:

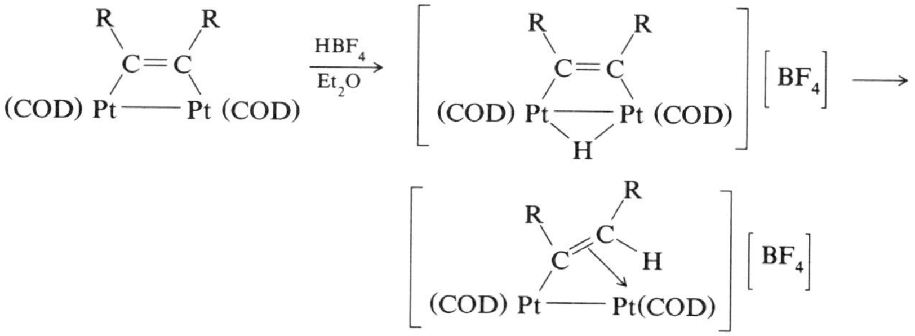

(c)

where $R = 4\text{-}MeOC_6F_4$.

(J. R. NORTON)

1. A. Davison, W. McFarlane, L. Pratt, G. Wilkinson, *J. Chem. Soc.*, 3653 (1962).
2. C. P. Casey, R. M. Bullock, *J. Organomet. Chem.*, *251*, 245 (1983).
3. D. A. Symon, T. C. Waddington, *J. Chem. Soc., A*, 953 (1971).
4. D. C. Harris, H. B. Gray, *Inorg. Chem.*, *14*, 1215 (1975).
5. J. Knight, M. J. Mays, *J. Chem. Soc., A*, 711 (1970).
6. A. J. Deeming, B. F. G. Johnson, J. Lewis, *J. Chem. Soc., A*, 2967 (1970).
7. A. J. Deeming, S. Hasso, M. Underhill, A. J. Canty, B. F. G. Johnson, W. G. Jackson, J. Lewis, T. W. Matheson, *J. Chem. Soc., Chem. Commun.*, 807 (1974).
8. A. J. Deeming, B. F. G. Johnson, J. Lewis, *J. Chem. Soc., A*, 2517 (1970).
9. W. A. Herrmann, J. Plank, D. Riedel, M. L. Ziegler, K. Weidenhammer, E. Guggloz, B. Balbach, *J. Am. Chem. Soc.*, *103*, 63 (1981).
10. H. Werner, W. Hofmann, *Angew. Chem., Int. Ed. Engl.*, *18*, 158 (1979).
11. M. P. Brown, R. J. Puddephatt, M. Rashidi, K. R. Seddon, *J. Chem. Soc., Dalton Trans.*, 516 (1978).
12. K. Fauvel, R. Mathieu, R. Poilblanc, *Inorg. Chem.*, *15*, 976 (1976).
13. B. Delley, M. C. Manning, D. E. Ellis, J. Berkowitz, W. C. Trogler, *Inorg. Chem.*, *21*, 2247 (1982).
14. C. Barner-Thorsen, E. Rosenberg, G. Saatjian, S. Aime, L. Milone, D. Osella, *Inorg. Chem.*, *20*, 1592 (1981).
15. M. B. Hursthouse, R. A. Jones, K. M. Abdul Malik, G. Wilkinson, *J. Am. Chem. Soc.*, *101*, 4128 (1979).
16. J. Plank, D. Riedel, W. A. Herrmann, *Angew. Chem., Int. Ed. Engl.*, *19*, 937 (1980).
17. N. M. Boag, M. Green, F. G. A. Stone, *J. Chem. Soc., Chem. Commun.*, 1281 (1980).

1.10.6.2. of Metal Anions

1.10.6.2.1. Which are Mononuclear: Scope.

This widely used method is, just like the protonation of neutral complexes, a formal oxidative addition requiring a d-electron configuration between d^2 and d^{10}. In practice, most applications are to d-electron configurations between d^6 and d^{10} (see Table 1).

Most transition-metal anions give hydrides by protonation at the metal, but acyl anions generally give hydroxycarbene complexes instead of acyl hydrides[18-21]:

$$[h^5\text{-}C_5H_5Re(CO)_2\overset{O}{\overset{\|}{C}}CH_3]^- \xrightarrow[H_2O/Et_2O]{H_2SO_4} h^5\text{-}C_5H_5(OC)_2Re=C\overset{OH}{\underset{CH_3}{<}} \tag{a}$$

$$[Re(CO)_4Br(\overset{O}{\overset{\|}{C}}CH_3)]^- \xrightarrow[Et_2O]{HCl} (OC)_4BrRe=C\overset{OH}{\underset{CH_3}{<}} \tag{b}$$

Because the charge on the acyl anion is localized at the acyl oxygen, protonation at that oxygen can occur with little electronic rearrangement, and O-protonation is favored kinetically. It is not clear whether hydroxycarbene complexes are more stable than acyl hydrides[22].

Just as further protonation of a cationic hydride can lead to H_2 evolution and formation of a dication, futher protonation of a neutral hydride can lead to H_2 loss and formation of a monocation. Although CF_3SO_3H in THF gives[23] $HMn(CO)_5$ from $[Mn(CO)_5]^-$:

$$[Mn(CO)_5]^- \xrightarrow{CF_3SO_3H\text{-}THF} HMn(CO)_5 \tag{c}$$

1.10.6. by Protonation
1.10.6.2. of Metal Anions
1.10.6.2.1. Which Are Mononuclear: Scope.

213

TABLE 1. FORMATION OF HYDRIDES BY PROTONATION OF ANIONIC TRANSITION-METAL COMPLEXES

Anion (d-electron configuration)	Acid	Hydride (d-electron configuration)	Refs.
$K_5[W(CN)_7]$ (d^4)	HOAc	$K_4[W(CN)_7H]$ (d^2)	1
$Na[h^5\text{-}C_5H_5Mo(CO)_3]$ (d^6)	HOAc	$h^5\text{-}C_5H_5Mo(CO)_3H$ (d^4)	2–5
$Na_2[h^5\text{-}C_5H_5Nb(CO)_3]$ (d^6)	CH_3CN	$Na[h^5\text{-}C_5H_5Nb(CO)_3H$ (d^4)	6
$Na_3V(CO)_5$ (d^8)	$EtOH^a$	$Na_2[HV(CO)_5]$ (d^6)	7
$Na[Re(CO)_5]$ (d^8)	H_3PO_4	$HRe(CO)_5$ (d^6)	5, 8, 9
$Na[h^5\text{-}C_5H_5Fe(CO)_2]$ (d^8)	HOAc	$h^5\text{-}C_5H_5Fe(CO)_2H$ (d^6)	10
$K[h^5\text{-}C_5Me_5Ru(CO)_2$ (d^8)	HOAc	$h^5\text{-}C_5Me_5Ru(CO)_2H$ (d^6)	11
$K[HFe(CO)_4]$ (d^8)	H_2SO_4	$H_2Fe(CO)_4$ (d^6)	12
$[Co(py)_6][Co(CO)_4]_2$ (d^{10})	H_2SO_4	$HCo(CO)_4$ (d^8)	13, 14
$K[Rh(PF_3)_4]$ (d^{10})	H_2SO_4 or H_3PO_4	$HRh(PF_3)_4$ (d^8)	15
$Na_2Os(CO)_4$ (d^{10})	H_3PO_4	$H_2Os(CO)_4$ (d^6)	16, 17

^{a} In liq NH_3 at $-50°C$.

further reaction occurs[24] in neat CF_3SO_3H:

$$[Mn(CO)_5]^- \xrightarrow{CF_3SO_3H} [Mn(CO)_5][O_3SCF_3] \qquad \text{(d)}$$

(J. R. NORTON)

1. A. M. Soares, P. M. Kiernan, D. J. Cole-Hamilton, W. P. Griffith, J. Chem. Soc., Chem. Commun., 84 (1981).
2. T. S. Piper, G. Wilkinson, J. Inorg. Nucl. Chem., 3, 104 (1956).
3. R. B. King, F. G. A. Stone, Inorg. Synth., 7, 107 (1963).
4. E. O. Fischer, Inorg. Synth., 7, 136 (1963).
5. R. B. King, Organometallic Synthesis, Vol. 1, Academic Press, New York, 1965.
6. F. Näumann, D. Rehder, V. Pank, J. Organomet. Chem., 240, 363 (1982).
7. G. F. P. Warnock, S. B. Philson, J. E. Ellis, J. Chem. Soc., Chem. Commun., 893 (1984).
8. W. Beck, W. Hieber, G. Braun, Z. Anorg. Allg. Chem., 308, 23 (1961).
9. B. H. Byers, T. L. Brown, J. Am. Chem. Soc., 98, 2527 (1977).
10. S. B. Ferguson, L. J. Sanderson, T. A. Shakelton, M. C. Baird, Inorg. Chim. Acta, 83, L45 (1984).
11. A. Stasunik, D. R. Wilson, W. Malisch, J. Organomet. Chem., 270, C18 (1984).
12. L. Vancea, W. A. G. Graham, J. Organomet. Chem., 134, 219 (1971).
13. H. W. Sternberg, I. Wender, M. Orchin, Inorg. Synth., 5, 192 (1957).
14. R. J. Clark, S. E. Whiddon, R. E. Serfass, J. Organomet. Chem., 11, 637 (1968).
15. T. Kruck, W. Lang, N. Derner, M. Stadler, Chem. Ber., 101, 3816 (1968).
16. R. D. George, S. A. R. Knox, F. G. A. Stone, J. Chem. Soc., Dalton Trans., 972 (1973).
17. J. Evans, J. R. Norton, J. Am. Chem. Soc., 96, 7577 (1974).
18. E. O. Fischer, A. Riedel, Chem. Ber., 101, 156 (1968).
19. E. O. Fischer, G. Kreiss, F. R. Kreissl, J. Organomet. Chem., 56, C37 (1973).
20. J. R. Moss, M. Green, F. G. A. Stone, J. Chem. Soc., Dalton Trans., 975 (1973).
21. K. P. Darst, C. M. Lukehart, J. Organomet. Chem., 171, 65 (1979).
22. M. J. Breen, P. M. Schulman, G. L. Geoffroy, A. L. Rheingold, W. C. Fultz, Organometallics, 3, 782 (1984).
23. J. A. Gladysz, W. Tam, G. M. Williams, D. L. Johnson, D. W. Parker, Inorg. Chem., 18, 1163 (1979).
24. W. C. Trogler, J. Am. Chem. Soc., 101, 6459 (1979).

1.10.6.2.2. Which Are Mononuclear: Position of the Protonation Equilibria.

The available pK_a data derive from a variety of solvents. Values for representative hydride complexes are given in Table 1.

As can be seen by comparing $HCo(CO)_4$ with $HCo(CO)_3L$ and $HV(CO)_6$ with $HV(CO)_5L$, σ-donor ligands decrease the acidity of hydrides; the data[18] on $HCo(CO)_3L$ and $HCoL_4$ show that phosphines decrease the acidity more than do phosphites. Comparison of $H_2Os(CO)_4$ and $H(CH_3)Os(CO)_4$ suggests that replacement of a hydride by an alkyl also decreases acidity.

Acidity usually decreases down a periodic column, e.g., the acidity of $h^5C_5H_5$-$M(CO)_3H$ decreases Cr > Mo > W; $H_2Os(CO)_4$ is a weaker acid than $H_2Fe(CO)_4$ and $HRe(CO)_5$ cannot be deprotonated[19] under the conditions employed for the titration and pK_a determination of $HMn(CO)_5$. However, there is little difference in pK_a between the Co and Rh dimethylglyoxime complexes (see Table 1), and IR observations

TABLE 1. ACIDITY OF NEUTRAL AND ANIONIC TRANSITION-METAL HYDRIDES

Hydride	pK_a	Solvent[a]	Refs.
$HMn(CO)_5$	7.1	H_2O (20°C)	1
$HMn(CO)_5$	15.1	CH_3CN	2
$HRe(CO)_5$	ca. 22	CH_3CN	2
$H_2Fe(CO)_4$	4.00 (pK_1)	H_2O (20°C)	7
	12.68 (pK_2)		
$H_2Fe(CO)_4$	5.88 (pK_1)	70% aq MeOH	4
$H_2Fe(CO)_4$	11.4 (pK_1)	CH_3CN	2
$H_2Os(CO)_4$	15.2	MeOH	5
$H_2Os(CO)_4$	20.8	CH_3CN	6
$H(CH_3)Os(CO)_4$	23.0	CH_3CN (0°C)	6
$HCo(CO)_4$	Strong acid	H_2O	7
$HCo(CO)_4$	Strong acid	MeOH (0°C)	7
$HCo(CO)_4$	8.4	CH_3CN	8
$HCo(CO)_3P(OPh)_3$	4.95	H_2O	9
$HCo(CO)_3P(OPh)_3$	11.4	CH_3CN	8
$HCo(CO)_3PPh_3$	6.96	H_2O	9
$HCo(CO)_3PPh_3$	15.4	CH_3CN	8
$HV(CO)_6$	Strong acid	H_2O	10, 11
$HV(CO)_5PPh_3$	6.8	H_2O	10
$h^5-C_5H_5Cr(CO)_3H$	13.3	CH_3CN	6
$h^5-C_5H_5Mo(CO)_3H$	13.9	CH_3CN (0°C)	6
$h^5-C_5H_5W(CO)_3H$	16.1	CH_3CN	6
$h^5-C_5H_5W(CO)_3H$	9.0	MeOH	12
$h^5-C_5H_5W(CO)_2(PMe_3)H^b$	26.6	CH_3CN	2
$h^5-C_5H_5Fe(CO)_2H$	19.4	CH_3CN	2
$h^5-C_5Me_5Fe(CO)_2H$	26.3	CH_3CN	2
$h^5-C_5H_5Ru(CO)_2H$	20.2	CH_3CN	2
$HCo(DMGH)_2PBu_3{}^c$	10.5	$H_2O-MeOH$	13
$HRh(DMGH)_2PPh_3{}^c$	9.5	$H_2O-MeOH$	14, 15
$[HCo(CN)_5]^{3-}$	20	H_2O	16, 17

[a] Temperature 25°C unless otherwise stated.
[b] Rapidly equilibrating mixture of cis- and trans-isomers.
[c] DMGH = dimethylglyoximato monoanion.

under high CO pressure in solution[20] suggest the order of basicities $[Rh(CO)_4]^-$ < $[Co(CO)_4]^-$ < $[Ir(Co)_4]^-$. Conductivity-based estimates[21] of the basicity of the PF_3 analogues in acetone give the order $[Co(PF_3)_4]^- \approx [Rh(PF_3)_4]^-$ < $[Ir(PF_3)]^-$. The more acidic behavior in H_2O of $[Rh(CN)_5H]^{3-}$ in comparison with $[Co(CN)_5H]^{3-}$ is not a true pK_a difference but a reflection of the greater tendency of $[Rh(CN)_5]^{4-}$ to dissociate[22] to the four-coordinated complex, $[Rh(CN)_4]^{3-}$.

Some hydrides are strong acids, and the protonation of their anions is correspondingly difficult. The equilibria in Table 1 indicate that acetic acid is not strong enough to protonate some anions (the pK_a of acetic acid[23] in acetonitrile is 22.3), and it should be avoided except where experience shows that it is satisfactory. The hydrides $HM(PF_3)_4$ (M = Co, Rh, Ir) are all strong acids in H_2O–acetone, and their preparation from the corresponding anions requires 50% sulfuric or phosphoric acid.[21]

Some group IVB ligands cause hydrides to become acidic; e.g., h^5-C_5H_5-$Fe(CO)(SiCl_3)_2H$ has[24] a pK_a of 2.6 in CH_3CN. Less strong but still deprotonated by Et_3N in CH_2Cl_2 are $H(Cl_3Si)Fe(CO)_4$, $H(Ph_3Si)Fe(CO)_4$ and $H(Cl_3Si)Mn(CO)_2C_5H_5$-h^5; their acidity decreases in the order given,[24] and $H(Ph_3Ge)Fe(CO)_4$ is sufficiently acidic to be deprotonated by Cl^- in CH_2Cl_2. Protonation of $[Et_4N][Ph_3GeFe(CO)_4]$ nevertheless can be accomplished[25] by HCl in Et_2O–THF because of the low solubility of the resulting Et_4NCl.

(J. R. NORTON)

1. W. Hieber, G. Wagner, Z. Naturforsch., Teil B, 13, 339 (1958).
2. J. M. Sullivan, J. R. Norton, unpublished work, 1985.
3. F. Galembeck, P. Krumholz, J. Am. Chem. Soc., 43, 1939 (1971).
4. R. G. Pearson, H. Mauermann, J. Am. Chem. Soc., 104, 500 (1983).
5. H. W. Walker, R. G. Pearson, P. C. Ford, J. Am. Chem. Soc., 105, 1179 (1983).
6. R. F. Jordan, J. R. Norton, J. Am. Chem. Soc., 104, 1255 (1982).
7. W. Hieber, W. Hubel, Z. Elektrochem., 57, 235 (1953).
8. E. J. Moore, J. R. Norton, unpublished work, 1985.
9. W. Hieber, E. Lindner, Chem. Ber., 94, 1417 (1961).
10. W. Hieber, E. Winter, and E. Schubert, Chem. Ber., 95, 3070 (1962).
11. F. Calderazzo, G. Pompaloni, D. Vitali, Gazz. Chim. Ital., 111, 455 (1981).
12. C. Amman, R. G. Pearson, unpublished work, quoted in R. G. Pearson, Chem. Rev., 85, 41 (1985).
13. G. N. Schrauzer, R. J. Holland, J. Am. Chem. Soc., 93, 1505 (1971).
14. J. H. Weber, G. N. Schrauzer, J. Am. Chem. Soc., 92, 726 (1970).
15. T. Ramasami, J. H. Espenson, Inorg. Chem., 19, 1846 (1980).
16. H. S. Lim, F. C. Anson, Inorg. Chem., 10, 103 (1971).
17. G. D. Venerable II, J. Halpern, J. Am. Chem. Soc., 93, 2176 (1971).
18. H.-F. Klein, Angew. Chem., Int. Ed. Engl., 19, 362 (1980).
19. W. Beck, W. Hieber, G. Braun, Z. Anorg. Allg. Chem., 308, 23 (1961).
20. J. L. Vidal, W. E. Walker, Inorg. Chem., 20, 249 (1981).
21. T. Kruck, W. Lang, N. Derner, M. Stadler, Chem. Ber., 101, 3816 (1968).
22. J. Halpern, R. Cozens, L.-Y. Goh, Inorg. Chim. Acta, 12, L35 (1975).
23. I. M. Kolthoff, M. K. Chantooni, S. Bhowmik, J. Am. Chem. Soc., 90, 23 (1968).
24. W. Jetz, W. A. G. Graham, Inorg. Chem., 10, 1647 (1971).
25. E. R. Isaacs, W. A. G. Graham, J. Organomet. Chem., 85, 237 (1975).

1.10.6.2.3. Which Are Mononuclear: Rates.

Proton transfers to metal anions are much slower (see Table 1) than proton transfers to other bases of comparable thermodynamic strength, reflecting the extent of the electronic rearrangement that must occur on protonation of a metal anion[3,4]. A kinetic-

TABLE 1. RATES OF PROTONATION OF MONONUCLEAR ANIONS AT 25°C

Anion	Acid	Solvent	k	Refs.
$[Co(CN)_5]^{4-}$	H_2O	H_2O	1.1×10^5 s^{-1} (20°C)	1
$[Rh(DMGH)_2PPh_3]^-$ ^{a}	$[H_3O]^+$	H_2O–MeOH	36 M^{-1} s^{-1}	2
$[HOs(CO)_4]^-$	CH_3OH	CH_3OH	0.8 s^{-1}	3
$[HOs(CO)_4]^-$	$[Et_3NH]^+$	CH_3CN	1.0×10^5 M^{-1} s^{-1}	4
$[HOs(CO)_4]^-$	$H_2Os(CO)_4$	CH_3CN	0.075 M^{-1} s^{-1} (30°C)	4
$[CH_3Os(CO)_4]^-$	$[Et_3NH]^+$	CH_3CN	1×10^6 M^{-1} s^{-1}	4
$[h^5\text{-}C_5H_5W(CO)_3]^-$	$[morH]^+$ ^{b}	CH_3CN	5.6×10^4 M^{-1} s^{-1}	4
$[h^5\text{-}C_5H_5W(CO)_3]^-$	$h^5\text{-}C_5H_5W(CO)_3H$	CH_3CN	650 M^{-1} s^{-1}	4, 5
$[h^5\text{-}C_5H_5Mo(CO)_3]^-$	$h^5\text{-}C_5H_5Mo(CO)_3H$	CH_3CN	2.5×10^3 M^{-1} s^{-1}	4, 5
$[h^5\text{-}C_5H_5Cr(CO)_3]^-$	$h^5\text{-}C_5H_5Cr(CO)_3H$	CH_3CN	1.8×10^4 M^{-1} s^{-1}	4, 5

^{a} DMGH = dimethylglyoximato monoanion.
^{b} mor = morpholine.

isotope effect of 5.8 is observed[1] on replacing H_2O by D_2O in the protonation of $[Co(CN)_5]^{4-}$.

(J. R. NORTON)

1. G. D. Venerable II, J. Halpern, J. Am. Chem. Soc., 93, 2176 (1971).
2. T. Ramasami, J. H. Espenson, Inorg. Chem., 19, 1846 (1980).
3. H. W. Walker, R. G. Pearson, P. C. Ford, J. Am. Chem. Soc., 105, 1179 (1983).
4. R. F. Jordan, J. R. Norton, J. Am. Chem. Soc., 104, 1255 (1982).
5. J. M. Sullivan, J. R. Norton, unpublished work, 1985.

1.10.6.2.4. Which Are Polynuclear.

Polynuclear anions are generated readily, and some polynuclear hydrides that can be made by this method[1-3] are listed[4-12] in Table 1. Although these anions are isolated and characterized, polynuclear hydrides frequently are made from anions prepared in situ. As with the cationic polynuclear hydrides discussed previously, the location of the H atom may be unknown, even though the ^1H NMR specifies that it is attached to the metal core.

In the hexagonally close-packed Rh anions[13-15], $[Rh_{13}(CO)_{24}H_{5-n}]^{n-}$, where n = 2–4, protonation, instead of occurring at any specific Rh atoms or Rh—Rh bond, inserts an additional interstitial hydrogen atom:

$$[Rh_{13}(CO)_{24}H_2]^{3-} \xrightarrow{H^+} [Rh_{13}(CO)_{24}H_3]^{2-} \qquad (a)$$

Similar interstitial hydrogens are obtained on protonation[16] of $[Co_6(CO)_{15}]^{2-}$:

$$K_2[Co_6(CO)_{15}] \xrightarrow[H_2O]{HCl} K[HCo_6(CO)_{15}] \qquad (b)$$

and $[Ru_6(CO)_{18}]^{2-}$:

$$[(Ph_3P)_2N]_2[Ru_6(CO)_{18}] \xrightarrow[THF\text{-}MeOH]{H_2SO_4} [(Ph_3P)_2N][HRu_6(CO)_{18}] \qquad (c)^{17}$$

A large kinetic-isotope effect on protonation rates of polynuclear anions indicate

1.10.6. by Protonation
1.10.6.2. of Metal Anions
1.10.6.2.4. Which Are Polynuclear.

217

TABLE 1. FORMATION OF POLYNUCLEAR HYDRIDES BY PROTONATION OF POLYNUCLEAR ANIONS

Anion	Acid	Hydride	Refs.
$Na_2[Fe_2(CO)_8]$	HOAc–EtOH–$[(Ph_3P)_2N]$Cl	$[(Ph_3P)_2N][HFe_2(CO)_8]$	4, 5
$[Et_4N]_2[Cr_2(CO)_{10}]$	HCl–THF–Et_2O	$[Et_4N][HCr_2(CO)_{10}]$	6
$K_2[Ru_4(CO)_{13}]$	HCl (1 equiv)–THF	$K[HRu_4(CO)_{13}]$	7
$K_2[Ru_4(CO)_{13}]$	H_2SO_4–THF	$H_2Ru_4(CO)_{13}$	7
$K_4[Ru_4(CO)_{12}]$	H_2SO_4	$H_4Ru_4(CO)_{12}$	7
$K[CoRu_3(CO)_{13}]$	H_2SO_4–hexane	$HCoRu_3(CO)_{13}$	8
$[Me_4N][MnOs_2(CO)_{12}]$	H_3PO_4–acetone	$HMnOs_2(CO)_{12}$	9
$[Et_4N][Fe_4N(CO)_{12}]$	H_2SO_4–toluene	$HFe_4N(CO)_{12}$	10
$[Et_4N][Fe_5N(CO)_{14}]$	H_2SO_4–toluene	$HFe_5N(CO)_{14}$	10
$[Mo_2Cl_8]^{4-}$	HCl–H_2O	$[HMo_2Cl_8]^{3-}$	11, 12

that the proton is becoming an interstitial hydrogen[18], but such kinetic-isotope effects occur generally in the protonation of polynuclear complexes. Reactions (d), (e) and (f) show k_H/k_D values of 16.8, 8.3 and 16.2, respectively[18-20]:

$$[Co(Me_2CO)_6][FeCo_3(CO)_{12}]_2 \xrightarrow[H_2O]{HCl} HFeCo_3(CO)_{12} \qquad (d)$$

$$[Et_4N][FeCo_3(CO)_{11}P(OCHMe_2)_3] \xrightarrow[H_2O]{HCl} HFeCo_3(CO)_{11}P(OCHMe_2)_3 \qquad (e)$$

$$Na[OsCo_3(CO)_{12}] \xrightarrow[H_2O]{HCl} HOsCo_3(CO)_{12} \qquad (f)$$

Some polynuclear anions change the structure of their metal cores on protonation. For example, the dianion $[Os_6(CO)_{18}]^{2-}$ has O_h symmetry; the first proton bridges an O_h face, and the second causes the metal framework to rearrange to a monocapped square pyramid in which the exact location of the hydrogen ligands is not clear[21,22]:

$$[Os_6(CO)_{18}]^{2-} \xrightarrow[THF]{H_2SO_4} [HOs_6(CO)_{18}] \xrightarrow[CH_3CN]{H_2SO_4} H_2Os_6(CO)_{18}$$

(g)

Protonation of $[Fe_4(CO)_{13}]^{2-}$ changes the core geometry from T_d to a butterfly arrangement in which a carbonyl group behaves as a four-electron donor[23,24]:

$$[Fe[py]_6][Fe_4[CO]_{13}] \xrightarrow[H_2O]{HCl} [pyH][HFe_4[CO]_{13}]$$

(h)

A similar rearrangement occurs when $[(Ph_3P)_2N][Fe_4(CO)_{12}COMe]$ is protonated, giving[25,26] $HFe_4(CO)_{12}COMe$.

Such structural differences between anion and hydride may be a factor in the high acidity of some polynuclear hydrides and the consequent difficulty of forming them by protonation of their anions. However, a more general cause is probably the delocalization of charge over several metal centers and π-acceptor ligands; e.g., $[HCo_6(CO)_{15}]^-$ is deprotonated by THF, H_2O, acetone and alcohols (although equilibration in alcohols requires[16] about an hour). Its formation in reaction (b), therefore, requires strong acid; there is no evidence for diprotonation. Similarly, the second protonation in reaction (g) succeeds only when a solvent (CH_3CN) is used in which the neutral dihydride is insoluble[21]. Other examples of anions for which attempts at isolating a protonated derivative have failed are: $[Cr_2(CO)_{10}]^{2-}$ [6], $[Rh_{12}(CO)_{30}]^{2-}$ [2] and $[Re_4(CO)_{16}]^{2-}$ [27]. The proton basicity of anions increases markedly with increasing charge per metal: the pK_a of $H_3Re_3(CO)_{12}$ in H_2O is 3, that of $[H_2Re_3(CO)_{12}]^-$ is 10, and that of $[HRe_3(CO)_{12}]^{2-}$ is[28] 25.

The same factors (structural rearrangement on protonation and delocalization of negative charge) that produce the low proton basicity of polynuclear anions are probably also responsible for the slow rate at which protonation equilibria are established with polynuclear systems[28-30] (see Table 2). If $[HOs(CO)_4Os(CO)_4]^-$ is compared with the mononuclear Os anions in the previous two subsections, it is clear that the second $Os(CO)_4$ unit has decreased the rate of protonation as well as the proton basicity.

The interplay of these kinetic and thermodynamic factors determines the site of protonation (i.e., metal core or ligand) of polynuclear anions. The Fe_2H hydrogen in $HFe_4(CH-h^2)(CO)_{12}$ is both kinetically and thermodynamically less acidic than the h^2-CH one[31]. The reaction of CH_3CN with Fe-carbonyl anion systems under protic conditions occurs by initial kinetically controlled protonation at nitrogen, forming $[Fe_3(CO)_9CH_3C=NH]^-$; this is followed by protonation of the weakly basic Fe core and movement of the ligand proton from nitrogen to the thermodynamically more basic carbon to form[32] $HFe_3(N=CHCH_3)(CO)_9$. Kinetic vs. thermodynamic control of the site of protonation is illustrated by[33]:

$$[Bu_4N]\,[HRu_3[\mu_2\text{-}CO]\,(CO)_{10}] \xrightarrow[CH_2Cl_2]{CF_3SO_3H,\ -60°C}$$

$$\text{(i)}$$

The oxygen of the μ_2-CO is the kinetically preferred site of protonation (the analogous Fe_3 compound can be isolated[34]), but protonation of the Ru_3 core is favored thermodynamically [the analogous Os_3 compound is obtained[35] on protonation of $[HOs_3$-$(\mu_2\text{-}CO)(CO)_{10}]^-$ at RT]. Similarly, $[Ru_3(CO)_{10}NO]^-$ undergoes kinetic protonation (e.g., by the strong acid CF_3SO_3H) on the nitrosyl oxygen, but thermodynamic protonation (e.g., by the weaker acid, CF_3CO_2H) in the presence of an anion that catalyzes the attainment of equilibrium on the Ru_3 core[36]:

1.10.6. by Protonation
1.10.6.2. of Metal Anions
1.10.6.2.4. Which Are Polynuclear.

219

TABLE 2. RATE AND EQUILIBRIUM DATA FOR PROTONATION OF POLYNUCLEAR ANIONS

Anion	pK_a (solvent), 25°C	Acid	Protonation rate, 25°C	Refs.
$[H_3Ru_4(CO)_{12}]^-$	11.7 (MeOH)	MeOH	0.08 s^{-1}	29
$[H_3FeRu_3(CO)_{12}]^-$	11.8 (MeOH)	MeOH	0.1 s^{-1}	29
$[HRu_4(CO)_{13}]^-$	11.1 (MeOH)	MeOH	0.02 s^{-1}	29
$[HFeRu_3(CO)_{13}]^-$	11.2 (MeOH)	MeOH	0.05 s^{-1}	29
$[H_3Os_4(CO)_{12}]^-$	12.3 (MeOH)	MeOH	0.1 s^{-1}	29
$[HOs_2(CO)_8]^-$	20.4 (CH$_3$CN)	$[Et_3NH]^+$	$8 \times 10^2 \text{ M}^{-1} \text{ s}^{-1}$	30

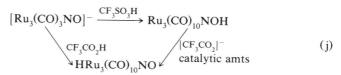

$$(j)$$

(J. R. NORTON)

1. A. P. Humphries, H. D. Kaesz, *Prog. Inorg. Chem.*, 25, 145 (1979); a comprehensive review of polynuclear hydrides, including a section on their formation by protonation of polynuclear anions.
2. P. Chini, G. Longoni, V. G. Albano, *Adv. Organomet. Chem.*, 14, 285 (1976); although confined to clusters containing five or more metal atoms, a good general discussion of synthetic methods for polynuclear carbonyl anions.
3. D. A. Roberts, G. L. Geoffroy, in *Comprehensive Organometallic Chemistry*, G. Wilkinson, F. G. A. Stone, E. W. Abel, eds., Pergamon Press, Oxford, 1982, Ch. 40; although confined to heterometallic systems, a good general discussion of synthetic methods for polynuclear carbonyl anions.
4. J. P. Collman, R. G. Finke, P. L. Matlock, R. Wahren, R. G. Komoto, J. I. Brauman, *J. Am. Chem. Soc.*, 100, 1119 (1978).
5. H. B. Chin, R. Bau, *Inorg. Chem.*, 17, 2314 (1978).
6. R. G. Hayter, *J. Am. Chem. Soc.*, 88, 4376 (1966).
7. C. C. Nagel, S. G. Shore, *J. Chem. Soc., Chem. Commun.*, 530 (1980).
8. P. C. Steinhardt, W. L. Gladfelter, A. D. Harley, J. R. Fox, G. L. Geoffrey, *Inorg. Chem.*, 19, 332 (1980).
9. J. Knight, M. J. Mays, *J. Chem. Soc., Dalton Trans.*, 1022 (1972).
10. M. Tachikawa, J. Stein, E. L. Muetterties, R. G. Teller, M. A. Beno, E. Gebert, J. M. Williams, *J. Am. Chem. Soc.*, 102, 6649 (1980).
11. F. A. Cotton, P. C. W. Leung, W. J. Roth, A. J. Schultz, J. M. Williams, *J. Am. Chem. Soc.*, 106, 117 (1984).
12. S. S. Miller, A. Haim, *J. Am. Chem. Soc.*, 105, 5624 (1983).
13. V. G. Albano, A. Ceriotti, P. Chini, G. Chiani, S. Martinengo, W. M. Anker, *J. Chem. Soc., Chem. Commun.*, 859 (1975).
14. V. G. Albano, G. Ciani, S. Martinengo, A. Sironi, *J. Chem. Soc., Dalton Trans.*, 978 (1979).
15. G. Ciani, A. Sironi, S. Martinengo, *J. Chem. Soc., Dalton Trans.*, 519 (1981).
16. D. W. Hart, R. G. Teller, C.-Y. Wei, R. Bau, G. Longoni, S. Campanella, P. Chini, T. F. Koetzle, *J. Am. Chem. Soc.*, 103, 1458 (1981).
17. C. R. Eady, P. F. Jackson, B. F. G. Johnson, J. Lewis, M. C. Malatesta, M. McPartlin, W. J. H. Nelson, *J. Chem. Soc., Dalton Trans.*, 383 (1980).
18. M. J. Mays, R. N. F. Simpson, *J. Chem. Soc., A*, 1444 (1968).
19. J. Knight, M. J. Mays, *J. Chem. Soc., A*, 711 (1970).
20. C. G. Cooke, M. J. Mays, *J. Organomet. Chem.*, 74, 449 (1974).
21. C. R. Eady, B. F. G. Johnson, J. Lewis, *J. Chem. Soc., Chem. Commun.*, 302 (1976).
22. M. McPartlin, C. R. Eady, B. F. G. Johnson, J. Lewis, *J. Chem. Soc., Chem. Commun.*, 883 (1976).

23. W. Hieber, R. Werner, *Chem. Ber.*, *90*, 286 (1957).
24. M. Manassero, M. Sansoni, G. Longoni, *J. Chem. Soc., Chem. Commun.*, 919 (1976).
25. E. M. Holt, K. H. Whitmire, D. F. Shriver, *J. Organomet. Chem.*, *213*, 125 (1981).
26. E. M. Holt, K. H. Whitmire, D. F. Shriver, *J. Am. Chem. Soc.*, *104*, 5621 (1982).
27. R. Bau, B. Fontal, H. D. Kaesz, M. R. Churchill, *J. Am. Chem. Soc.*, *89*, 6374 (1967).
28. H. D. Kaesz, *Chem. Br.*, *9*, 344 (1973).
29. H. W. Walker, R. G. Pearson, P. C. Ford, *J. Am. Chem. Soc.*, *105*, 1179 (1983).
30. R. F. Jordan, J. R. Norton, *J. Am. Chem. Soc.*, *104*, 1255 (1982).
31. M. Tachikawa, E. L. Muetterties, *J. Am. Chem. Soc.*, *102*, 4541 (1980).
32. M. A. Andrews, H. D. Kaesz, *J. Am. Chem. Soc.*, *101*, 7238 (1979).
33. J. B. Keister, *J. Organomet. Chem.*, *190*, C36 (1980).
34. H. A. Hodali, D. F. Shriver, *Inorg. Chem.*, *18*, 1236 (1979).
35. C. R. Eady, B. F. G. Johnson, J. Lewis, M. C. Malatesta, *J. Chem. Soc., Dalton Trans.*, 1358 (1978).
36. R. E. Stevens, W. L. Gladfelter, *J. Am. Chem. Soc.*, *104*, 6454 (1982).

1.10.7. by Reduction of Metal Complexes

1.10.7.1. with Borohydride.

This is a common method for preparing transition-metal hydrides (see Table 1)[1-7].

Borohydride reduction is complex, e.g., $Ru_3(CO)_{12}$ with $NaBH_4$ in THF gives over a dozen products[41], but a few useful mechanistic generalizations can be offered. Syntheses of metal hydrides proceed through intermediate borohydride complexes. Electron-pair bases (an ethereal solvent may be sufficient) are then necessary to complex and remove BH_3. In a few examples the borohydride complexes are observable:

$$(h^5\text{-}C_5H_5)_2ZrCl_2 \xrightarrow[Et_2O]{LiBH_4} (h^5\text{-}C_5H_5)_2Zr(BH_4)_2 \qquad (a)^{42}$$

$$(h^5\text{-}C_5H_5)_2Zr(BH_4)_2 \xrightarrow[C_6H_6]{Me_3N} (h^5\text{-}C_5H_5)_2Zr(H)BH_4 \xrightarrow[C_6H_6]{Me_3N} [(h^5\text{-}C_5H_5)_2ZrH_2]_n \quad (b)^{43}$$

$$(h^5\text{-}C_5H_5)_2NbCl_2 + 4\ NaBH_4 \xrightarrow{MeOCH_2CH_2OMe} (h^5\text{-}C_5H_5)_2NbBH_4 \qquad (c)^{44}$$

$$(h^5\text{-}C_5H_5)_2NbBH_4 \xrightarrow[C_6H_6]{PMe_2Ph} (h^5\text{-}C_5H_5)_2Nb(H)PMe_2Ph \qquad (d)^{44}$$

$$(h^5\text{-}C_5H_5)V(CO)_4 \xrightarrow[h\nu,\ THF]{[(Ph_3P)_2N]\ [BH_4]} [h^5\text{-}C_5H_5V(CO)_3BH_4]^- \qquad (e)^{45}$$

$$\swarrow{-CO} \qquad \searrow{-BH_3}$$

$$[(Ph_3P)_2N][h^5\text{-}C_5H_5V(CO)_2BH_4] \qquad [(Ph_3P)_2N][h^5\text{-}C_5H_5V(CO_3)H]$$

In Eq. (e), an intermediate, coordinatively saturated monodentate borohydride complex, $[h^5\text{-}C_5H_5V(CO)_3BH_4]^-$, is formed, which can either lose CO to give a bidentate borohydride complex or lose BH_3 to give an anionic hydride[45]. Similarly, the borohydride reduction of $Mo(CO)_6$ in THF proceeds through an $[Mo(CO)_5(BH_4)]^-$ intermediate that can either lose CO to form $[Mo(CO)_4(BH_4)]^-$ or lose BH_3 to form $[HMo_2(CO)_{10}]^-$. Diethyl ether (a weaker electron-pair donor) instead of THF favors

Initial metal complex	Reagents and conditions	Product hydride	Refs.
WCl_6 or $MoCl_5$	NaC_5H_5–THF–$NaBH_4$ (initial mixing below $-100°C$ followed by warming and brief reflux)	$(h^5\text{-}C_5H_5)_2MH_2$ ($M = W$ or Mo)	8–10
$TaCl_5$	NaC_5H_5–THF–$NaBH_4$ (initial mixing at 0°C followed by warming and reflux)	$h^5\text{-}C_5H_5TaH_3$	8
$ReCl_5$	NaC_5H_5–THF (0°C), $NaBH_4$ (50°C)	$(h^5\text{-}C_5H_5)_2ReH$	9, 11
$W(CO)_6$	$NaBH_4$–THF	$[HW_2(CO)_{10}]^-$	12, 13
$W(CO)_6$	$[Et_4N][BH_4]$–THF (reflux for several days)	$[H_2W_2(CO)_8]^{2-}$	14
$Re_2(CO)_{10}$	$NaBH_4$–THF–H_3PO_4	$H_3Re_3(CO)_{12}$	15, 16
$Re_2(CO)_{10}$	$NaBH_4$–THF (reflux overnight followed by several days at RT)	$[H_6Re_4(CO)_{12}]^{2-}$	17
$Os_6(CO)_{18}$	$NaBH_4$–THF	$[HOs_6(CO)_{18}]^-$	18
$WCl_4(PMe_2Ph)$	$NaBH_4$–EtOH	$H_6W(PMe_2Ph)_3$	19
$MoCl_4(PMePh_2)_2$	$NaBH_4$–EtOH–$PMePh_2$	$H_4Mo(PMePh_2)_4$	20, 21
$MoCl(h^3\text{-}C_3H_5)PPh_3.h^6\text{-}C_6H_6$	$NaBH_4$–EtOH–PPh_3	$H_2Mo(PPh_3)_2C_6H_6\text{-}h^6$	22
$ReCl_3(PMe_2Ph)_3$	$NaBH_4$–EtOH	$H_2Re(PMe_2Ph)_3$	23
$Na_2[OsCl_6]\cdot 6\,H_2O$	$NaBH_4$–EtOH–PPh_3	$H_4Os(PPh_3)_3$	24
$RuCl_2[P(OEt)_3]_4$	$NaBH_4$–EtOH–$P(OEt)_3$	$H_2Ru[P(OEt)_3]_4$	25
$FeCl_2$	EtOH–$Ph_2PCH_2CH_2PPh_2$–$NaBH_4$	$HFeCl(Ph_2PCH_2CH_2PPh_2)_2$	26
$h^5\text{-}C_5H_5Ru(PPh_3)_2Cl$	$NaBH_4$–THF	$h^5\text{-}C_5H_5Ru(PPh_3)_2H$	27
$Co[NO_3]_2\cdot 6\,H_2O$	$NaBH_4$–EtOH–$P(OPh)_3$	$CoH[P(OPh)_3]_4$	28–30
$CoCl_2$	$NaBH_4$–EtOH–PPh_3	$CoH(PPh_3)_3$	31
$Na_2[IrCl_6]$	EtOH–Ph_3P–$NaBH_4$	$IrH_3(PPh_3)_3$ (both isomers)	32
$trans$-$Ir(PPh_3)_2(CO)Cl$	EtOH–PPh_3–$NaBH_4$	$Ir(PPh_3)_3(CO)H$	33
$RhCl_3\cdot 3H_2O$	EtOH–PPh_3–$NaBH_4$	$HRh(PPh_3)_4$	34
$NiCl_2[P(C_6H_{11})_3]_2$	$NaBH_4$–Et_2O–C_6H_6–EtOH	$trans$-$HNiCl[P(C_6H_{11})_3]_2$	35
$[Pt_2Me_2(\mu\text{-}Cl)(\mu\text{-}Ph_2P(CH_2)_2PPh_2)_2][PF_6]$	$NaBH_4$	$[Pt_2Me_2(\mu\text{-}H)(\mu\text{-}Ph_2P(CH_2)_2PPh_2)_2][PF_6]$	36
$ClCo(DMGH)_2PBu_3$[a]	H_2O–MeOH–$NaBH_4$	$HCo(DMGH)_2\cdot 2PBu_3$[a]	37, 38
$ClRh(DMGH)_2PPh_3$	H_2O–MeOH $NaBH_4$	$HRh(DMGH)_2PPh_3$	39
$[cis\text{-}Ru(bipy)_2(CO)Cl][ClO_4]$	H_2O–EtOH–$NaBH_4$–NH_4PF_6	$[cis\text{-}Ru(bipy)_2(COH][PF_6]$	40

[a] DMGH = dimethylglyoximato monoanion.

221

222   1.10. Formation of Bonds between Hydrogen and Transition Metals
       1.10.7. by Reduction of Metal Complexes
       1.10.7.1. with Borohydride.

loss of the volatile BH_3 and $[HMo_2(CO)_{10}]^-$ formation[46]. The reduction of $Fe[Ph_2P-(CH_2)_2PPh_2]_2Cl_2$ with $[BH_4]^-$ affords a borohydride-containing species[47] that decomposes to $Fe[Ph_2P(CH_2)_2PPh_3]H_2$ on standing in THF.

Sodium borohydride can be used to synthesize reactive intermediates at low T, e.g., to prepare an unstable allyl hydride[48]:

$$[h^3\text{-}C_3H_5PtP(CMe_3)_3(Me_2O)][BF_4] \xrightarrow[-60°C, \text{ THF}]{NaBH_4, \text{ 48 h}} h^3\text{-}C_3H_5PtHP(CMe_3)_3 \qquad (f)$$

(J. R. NORTON)

1. R. A. Schunn, in *Transition Metal Hydrides*, E. L. Muetterties, ed., Marcel Dekker, New York, 1971, p. 203.
2. H. D. Kaesz, R. B. Saillant, *Chem. Rev., 72*, 231 (1972).
3. D. Giusto, *Inorg. Chim. Acta Rev., 6*, 91 (1972).
4. D. M. Roundhill, *Adv. Organomet. Chem., 13*, 273 (1975).
5. H. D. Kaesz, *Inorg. Synth., 17*, 52 (1977); a listing of all earlier preparations of transition-metal hydrides published in *Inorg. Synth.* as well as an introduction to newer preparations. Many of the preparations involve $NaBH_4$.
6. A. P. Humphries, H. D. Kaesz, *Prog. Inorg. Chem., 25*, 145 (1979); a comprehensive review of all aspects of the chemistry of polynuclear hydrides.
7. G. L. Geoffroy, J. R. Lehman, *Adv. Inorg. Chem. Radiochem., 20*, 189 (1977).
8. M. L. H. Green, J. A. McCleverty, L. Pratt, G. Wilkinson, *J. Chem. Soc.*, 4854 (1961).
9. R. B. King, *Organometallic Syntheses*, Vol. 1, Academic Press, New York, 1965.
10. M. L. H. Green, P. J. Knowles, *J. Chem. Soc., Perkin Trans. 1*, 989 (1973).
11. M. L. H. Green, L. Pratt, G. Wilkinson, *J. Chem. Soc.*, 3916 (1958).
12. R. G. Hayter, *J. Am. Chem. Soc., 88*, 4376 (1966).
13. D. C. Harris, H. B. Gray, *J. Am. Chem. Soc., 97*, 3073 (1975).
14. M. R. Churchill, S. W.-Y. N. Chang, M. L. Berch, A. Davison, *J. Chem. Soc., Chem. Commun.*, 691 (1973).
15. D. K. Huggins, W. Fellmann, J. M. Smith, H. D. Kaesz, *J. Am. Chem. Soc., 86*, 4841 (1964).
16. M. A. Andrews, S. W. Kirtley, H. D. Kaesz, *Inorg. Synth., 17*, 66 (1977).
17. H. D. Kaesz, B. Fontal, R. Bau, S. W. Kirtley, M. R. Churchill, *J. Am. Chem. Soc., 91*, 1021 (1969).
18. C. R. Eady, B. F. G. Johnson, J. Lewis, *J. Chem. Soc., Chem. Commun.*, 302 (1976).
19. J. R. Moss, B. L. Shaw, *J. Chem. Soc., Dalton Trans.*, 1910 (1972).
20. F. Pennella, *Inorg. Synth., 15*, 43 (1974).
21. P. Meakin, L. J. Guggenberger, W. G. Peet, E. L. Muetterties, J. P. Jesson, *J. Am. Chem. Soc., 95*, 1467 (1973).
22. W. E. Silverthorn, *Inorg. Synth., 17*, 57 (1977).
23. P. G. Douglas, B. L. Shaw, *Inorg. Synth., 17*, 64 (1977).
24. N. Ahmad, J. J. Levison, S. D. Robinson, M. F. Uttley, *Inorg. Synth., 15*, 56 (1974).
25. W. G. Peet, D. H. Gerlach, *Inorg. Synth., 15*, 41 (1974).
26. P. Gianoccaro, A. Sacco, *Inorg. Synth., 17*, 69 (1977).
27. T. Blackmore, M. I. Bruce, F. G. A. Stone, *J. Chem. Soc., A*, 2376 (1971).
28. J. J. Levison, S. D. Robinson, *Inorg. Synth., 13*, 105 (1972).
29. D. Titus, A. A. Orio, H. B. Gray, *Inorg. Synth., 13*, 118 (1972).
30. E. L. Muetterties, F. J. Hirsekorn, *J. Am. Chem. Soc., 96*, 7920 (1974).
31. A. Sacco, M. Rossi, *Inorg. Synth., 12*, 19 (1970).
32. N. Ahmad, S. D. Robinson, M. F. Uttley, *J. Chem. Soc., Dalton Trans.*, 843 (1972).
33. G. Wilkinson, *Inorg. Synth., 13*, 126 (1972).
34. J. J. Levison, S. D. Robinson, *J. Chem. Soc., A*, 2947 (1970).
35. T. Saito, H. Munakata, H. Imoto, *Inorg. Synth., 17*, 84 (1977).
36. M. P. Brown, S. J. Cooper, A. A. Frew, L. Manojlović-Muir, K. W. Muir, R. J. Puddephatt, M. A. Thompson, *J. Organomet. Chem., 198*, C33 (1980).
37. G. N. Schrauzer, R. J. Holland, *J. Am. Chem. Soc., 93*, 1505 (1971).
38. T.-H. Chao, J. H. Espenson, *J. Am. Chem. Soc., 100*, 129 (1978).
39. T. Ramasami, J. H. Espenson, *Inorg. Chem., 19*, 1846 (1980).

40. J. M. Kelly, J. G. Vos, *Angew. Chem., Int. Ed. Engl.,* 21, 628 (1982).
41. C. R. Eady, B. F. G. Johnson, J. Lewis, *J. Chem. Soc., Dalton Trans.,* 477 (1977).
42. R. K. Nanda, M. G. H. Wallbridge, *Inorg. Chem.,* 3, 1798 (1964).
43. B. D. James, R. K. Nanda, M. G. H. Wallbridge, *Inorg. Chem.,* 6, 1979 (1967).
44. C. R. Lucas, *Inorg. Synth.,* 16, 109 (1976).
45. R. J. Kinney, W. D. Jones, R. G. Bergman, *J. Am. Chem. Soc.,* 100, 7902 (1978).
46. S. W. Kirtley, M. A. Andrews, R. Bau, G. W. Grynkewich, T. J. Marks, D. L. Tipton, B. R. Whittlesey, *J. Am. Chem. Soc.,* 99, 7154 (1977).
47. M. V. Baker, L. D. Field, *J. Chem. Soc., Chem. Commun.,* 996 (1984).
48. G. Carturan, A. Scrivanti, F. Morandini, *Angew. Chem., Int. Ed. Engl.,* 20, 112 (1981).

1.10.7.2. with Trialkylborohydride and Other Substituted Borohydrides.

Trialkylborohydride reducing agents differ from borohydride in their ability to transfer hydride directly to a carbonyl ligand without prior substitution in the coordination sphere. They are used to synthesize formyl complexes[1-5]. When formyl complexes lose CO and undergo hydride migration from the formyl ligand to the metal, a transition-metal hydride results. The process is formally similar to nucleophilic attack by $[OH]^-$ on a carbonyl ligand, followed by loss of CO_2 and formation of a transition-metal hydride[6]. Examples of hydride syntheses via formyl complexes are:

$$[Ir(CO)_3(PPh_3)_2][PF_6] \xrightarrow[CH_2Cl_2, THF]{Li[Et_3BH], -60°C} Ir(CO)_2(PPh_3)_2CHO \xrightarrow[-CO]{-40°C} HIr(CO)_2(PPh_3)_2 \quad (a)^7$$

$$Re_2(CO)_{10} \xrightarrow[THF]{K[HB(OCHMe_2)_3]} [Re_2(CO)_9CHO]^- \xrightarrow[[Et_4N]Br]{hv, 50°C, -CO} [Et_4N][Re_2(CO)_9H] \quad (b)^8$$

$$Ir_4(CO)_{12} \xrightarrow[THF]{Li[Et_3BH], 0°C} [Ir_4(CO)_{11}CHO]^- \xrightarrow[-CO]{40°C} [HIr_4(CO)_{11}]^- \quad (c)^{9,10}$$

$$Os_3(CO)_{12} \xrightarrow[THF]{K[HB(OCHMe_2)_3], 0°C} [Os_3(CO)_{11}CHO]^- \xrightarrow[25°C]{[(Ph_3P)_2N]Cl}$$
$$[(Ph_3P)_2N][HOs_3(CO)_{11}] \quad (d)^{9-11}$$

However, formyl complexes undergo reactions other than decomposition to transition-metal hydrides[5,7]; e.g., h^5-$C_5H_5Re(CO)(NO)(CHO)$ disproportionates[12] to h^5-C_5H_5-$(CO)(NO)ReCO_2CH_2Re(CO)(NO)C_5H_5$-$h^5$. Thus, while trialkylborohydride reduction is the method of choice for formyl complexes and generates transition-metal anions for protonation to hydrides[13,14], it is not a general method for metal hydrides. Trialkylborohydride reduction generates an unstable allyl hydride at low T when no carbonyl ligands are present[15]:

$$h^3\text{-}C_3H_5Ni(PPh_3)Br \xrightarrow[-135°C]{NaHBMe_3} h^3\text{-}C_3H_5Ni(PPh_3)H \quad (e)$$

Lithium triethylborohydride is more effective than $NaBH_4$ for preparing Mo and W tetrahydrides[16], e.g.:

$$WCl_6 + 4\ PMePh_2 \xrightarrow[THF]{Li[Et_3BH]} WH_4(PMePh_2)_4 \quad (f)$$

The yield is 70%, but it is only 4% from $NaBH_4$.

Although cyanoborohydride ($[BH_3CN]^-$) is a selective reagent for ligand reductions[17,18], it also reduces some carbonyl cations to the corresponding hydrides in good yield[18]:

$$[h^5\text{-}C_5H_5Mo(CO)_4][PF_6] \xrightarrow[\text{MeOH}]{\text{NaBH}_3\text{CN}} h^5\text{-}C_5H_5Mo(CO)_3H \qquad (g)$$

$$[Mn(CO)_5(PPh_3)][PF_6] \xrightarrow[\text{MeOH}]{\text{NaBH}_3\text{CN}} HMn(CO)_4PPh_3 \qquad (h)$$

(J. R. NORTON)

1. C. P. Casey, S. M. Neumann, *J. Am. Chem. Soc.*, *98*, 5395 (1976).
2. S. R. Winter, G. W. Cornett, E. W. Thompson, *J. Organomet. Chem.*, *133*, 339 (1977).
3. J. C. Selover, M. Marsi, D. W. Parker, J. A. Gladysz, *J. Organomet. Chem.*, *206*, 317 (1981).
4. K. P. Darst, C. M. Lukehart, *J. Organomet. Chem.*, *171*, 65 (1979).
5. J. A. Gladysz, *Adv. Organomet. Chem.*, *20*, 1 (1982).
6. H. D. Kaesz, *J. Organomet. Chem.*, *200*, 145 (1980).
7. W. Tam, G.-Y. Lin, J. A. Gladysz, *Organometallics*, *1*, 525 (1982).
8. C. P. Casey, S. M. Neumann, *J. Am. Chem. Soc.*, *100*, 2544 (1978).
9. R. L. Pruett, R. C. Schoening, J. L. Vidal, R. A. Fiato, *J. Organomet. Chem.*, *182*, C57 (1979).
10. R. C. Schoening, J. L. Vidal, R. A. Fiato, *J. Organomet. Chem.*, *206*, C43 (1981).
11. G. R. Steinmetz, E. D. Morrison, G. L. Geoffroy, *J. Am. Chem. Soc.*, *106*, 2559 (1984).
12. C. P. Casey, M. A. Andrews, D. R. McAlister, J. E. Rinz, *J. Am. Chem. Soc.*, *102*, 1927 (1980).
13. J. A. Gladysz, G. M. Williams, W. Tam, D. L. Johnson, D. W. Parker, J. C. Selover, *Inorg. Chem.*, *18*, 553 (1979).
14. J. A. Gladysz, W. Tam, G. M. Williams, D. L. Johnson, D. W. Parker, *Inorg. Chem.*, *18*, 1163 (1979).
15. H. Bönnemann, *Angew. Chem., Int. Ed. Engl.*, *9*, 736 (1970).
16. R. H. Crabtree, G. G. Hlatky, *Inorg. Chem.*, *21*, 1273 (1982).
17. T. Bodnar, G. Coman, S. LaCroce, C. Lambert, K. Menard, A. Cutler, *J. Am. Chem. Soc.*, *103*, 2471 (1981).
18. T. Bodnar, E. Coman, K. Menard, A. Cutler, *Inorg. Chem.*, *21*, 1275 (1982).

1.10.7.3. with Tetrahydroaluminate.

Lithium aluminum hydride is used almost as frequently as $NaBH_4$ (see §1.10.7.1), but even less is known about its mechanism of action, especially whether it reacts via an intermediate $[AlH_4]^-$ ligand. Complexes containing that ligand are known[1-4].

Examples of $LiAlH_4$ use are found in reviews[5-10] and in[11-18] Table 1. Although $LiAlH_4$ is a more powerful reducing agent than $NaBH_4$, the latter is used frequently in EtOH, while $LiAlH_4$ requires ether. Comparison of Table 1 with the corresponding table in §1.10.7.1 shows that transition-metal hydrides can be prepared by the use of either reagent.

(J. R. NORTON)

1. P. C. Wailes, H. Weigold, *J. Organomet. Chem.*, *24*, 405 (1970).
2. T. J. McNeese, S. S. Wreford, B. M. Foxman, *J. Chem. Soc., Chem. Commun.*, 500 (1978); a critical evaluation of claims of tetrahydroaluminate complexes.
3. G. S. Girolami, G. Wilkinson, M. Thornton-Pett, M. B. Hursthouse, *J. Am. Chem. Soc.*, *105*, 6752 (1983); earlier claims of $[AlH_4]^-$ complexes are listed.
4. E. B. Lovkovskii, G. L. Soloveichik, A. I. Sisov, B. M. Bulychev, A. I. Gusev, N. I. Kirillova, *J. Organomet. Chem.*, *265*, 167 (1984).

TABLE 1. PREPARATIONS OF TRANSITION-METAL HYDRIDES BY $[AlH_4]^-$ REACTIONS

Initial metal complex	Reagents	Product hydride	Refs.
$TaCl_5$	$LiC_5H_4Me-THF-C_6H_6-LiAlH_4$	$(h^5-C_5H_4Me)_2TaH_3$	11
$WCl_4(PPh_2Et)_2$	$PPh_2Et-THF-LiAlH_4$	$H_4W(PPh_2Et)_4$	12
$WCl_4(PMe_3)_3$	$LiAlH_4$	$H_6H(PMe_3)_3$	13
$ReOCl_3(PPhEt_2)_2$	$LiAlH_4-THF$	$H_7Re(PPhEt_2)_2$	7, 14, 15
$ReCl_3(NPh)(PMe_3)_3$	$LiAlH_4$	$HRe(NHPh)(PMe_3)_2$	13
trans-$ReCl_3(PPhEt_2)_3$	$LiAlH_4-THF$	$H_5Re(PPhEt_2)_3$	7, 14, 15
$FeCl_2(Et_2PCH_2CH_2PEt_2)_2$	$LiAlH_4-THF$	$HFeCl(Et_2PCH_2CH_2PEt_2)_2$	16
$[Me_3PH][IrCl_4(PMe_3)_2]$	$LiAlH_4-THF$	$H_5Ir(PMe_3)_2$	15, 17
$CuCl(PPh_3)_3$	$LiAlH_4-Et_2O$	$HCuPPh_3$	18

5. R. A. Schunn, in *Transition Metal Hydrides*, E. L. Muetterties, ed., Marcel Dekker, New York, 1971, p. 203.
6. H. D. Kaesz, R. B. Saillant, *Chem. Rev., 72*, 231 (1972).
7. D. Giusto, *Inorg. Chim. Acta Rev., 6*, 91 (1972).
8. H. D. Kaesz, in *Inorg. Synth., 17*, 52 (1977); a listing of all earlier preparations of transition-metal hydrides published in *Inorg. Synth.* as well as an introduction to newer preparations.
9. A. P. Humphries, H. D. Kaesz, *Prog. Inorg. Chem., 25*, 145 (1979); a comprehensive review of all aspects of the chemistry of polynuclear hydrides.
10. G. L. Geoffroy, J. R. Lehman, *Adv. Inorg. Chem. Radiochem., 20*, 189 (1977).
11. U. Klabunde, G. W. Parshall, *J. Am. Chem. Soc., 94*, 9081 (1972).
12. P. Meakin, L. J. Guggenberger, W. G. Peet, E. L. Muetterties, J. P. Jesson, *J. Am. Chem. Soc., 95*, 1467 (1973).
13. M. B. Hursthouse, D. Lyons, M. Thorton-Pett, G. Wilkinson, *J. Chem. Soc., Chem. Commun.*, 476 (1983).
14. J. Chatt, R. S. Coffey, *J. Chem. Soc., A*, 1963 (1969).
15. R. Bau, W. E. Carroll, D. W. Hart, R. G. Teller, T. F. Koetzle, in *Transition Metal Hydrides*, R. Bau, ed., American Chemical Society, Advances in Chemistry Series No. 167, Washington, DC, 1978, p. 73.
16. M. J. Mays, B. E. Prater, *Inorg. Synth., 15*, 23 (1974).
17. E. K. Barefield, *Inorg. Synth., 15*, 36 (1974).
18. R. D. Stephens, *Inorg. Synth., 19*, 87 (1979).

1.10.7.4. with Alkoxyaluminum Hydrides.

The solubility of $Na[AlH_2(OCH_2CH_2OCH_3)_2]$ in aromatic hydrocarbons makes it unique among the main-group hydride reagents used to form transition–metal hydrogen bonds. It is the reagent of choice for:

$$(h^5-C_5H_5)_2NbCl_2 \xrightarrow[C_6H_6]{Na[AlH_2(OCH_2CH_2OCH_3)_2]} (h^5-C_5H_5)_2NbH_3 \qquad (a)^{1,2}$$

$$(h^5-C_5H_5)_2ZrCl_2 \xrightarrow[C_6H_6]{Na[AlH_2(OCH_2CH_2OCH_3)_2]} (h^5-C_5H_5)_2ZrHCl \qquad (b)^{3,4}$$

$$(h^5-C_5H_5)_2W(CH_3)I \xrightarrow[C_6H_6]{Na[AlH_2(OCH_2CH_2OCH_3)_2]} (h^5-C_5H_5)_2W(H)CH_3 \qquad (c)^5$$

$$(h^5-C_5H_5)_2W(O_2CPh)CH_3 \xrightarrow[C_6H_6]{Na[AlH_2(OCH_2CH_2OCH_3)_2]} (h^5-C_5H_5)_2W(H)CH_3 \qquad (d)^5$$

$$h^5\text{-}C_5H_5W(NO)_2Cl \xrightarrow[\text{toluene}]{Na[AlH_2(OCH_2CH_2OCH_3)_2]} (h^5\text{-}C_5H_5)W(NO)_2H \quad (e)^6$$

$$WCl_4(PMe_2Ph)_3 \xrightarrow[\text{THF-}C_6H_6]{Na[AlH_2(OCH_2CH_2OCH_3)_2]} WH_6(PMe_2Ph)_3 \quad (f)^7$$

$$h^5\text{-}MeC_5H_4Mn(CO)(PPh_3)I \xrightarrow[C_6H_6]{Na[AlH_2(OCH_2CH_2OCH_3)_2]} h^5\text{-}MeC_5H_4Mn(CO)(PPh_3)H \quad (g)^8$$

One such reaction results in the isolation of a stable aluminohydride complex[9]:

$$TaCl_2(Me_2PCH_2CH_2PMe_2)_2 \xrightarrow[Et_2O]{Na[AlH_2(OCH_2CH_2OCH_3)_2]}$$

$$\{Ta[H_2Al(OCH_2CH_2OCH_3)_2](Me_2PCH_2CH_2PMe_2)_2\}_2 \quad (h)$$

(J. R. NORTON)

1. J. A. Labinger, in *Transition Metal Hydrides*, R. Bau, ed., American Chemical Society, Advances in Chemistry Series No. 167, Washington, DC, 1978, p. 149.
2. J. A. Labinger, K. S. Wong, *J. Organomet. Chem., 170*, 373 (1979).
3. P. C. Wailes, H. Weigold, *Inorg. Synth., 19*, 226 (1979).
4. D. B. Carr, J. Schwartz, *J. Am. Chem. Soc., 101*, 3521 (1979).
5. N. J. Cooper, M. L. H. Green, R. Mahtab, *J. Chem. Soc., Dalton Trans.*, 1557 (1979).
6. P. Legzdins, D. T. Martin, *Inorg. Chem., 18*, 1250 (1979).
7. R. H. Crabtree, G. G. Hlatky, *Inorg. Chem., 23*, 2388 (1984).
8. B. W. Hames, P. Legzdins, J. C. Oxley, *Inorg. Chem., 19*, 1565 (1980).
9. T. J. McNeese, S. S. Wreford, B. M. Foxman, *J. Chem. Soc., Chem. Commun.*, 500 (1978).

1.10.8. by Hydrogen Transfer to the Metal

1.10.8.1. from Alcohols in Basic Media.

This method, most frequently employing EtOH, is used widely because the reaction is easy to perform. Examples are found in reviews[1-5] and in[6-17] Table 1.

β-Hydrogen is eliminated from a coordinated alkoxide ligand:

$$(a)$$

as is implied by the incorporation of a deuterium label[18]:

$$K_2IrCl_6 \xrightarrow[Ph_3P]{CH_3CD_2OH\text{-}H_2O} IrDCl_2(PPh_3)_3 \quad (b)$$

and proved by observation of an intermediate methoxy complex[19]:

$$[\text{trans-Pt(MeOH)Ph(PEt}_3)_2][BF_4] \xrightarrow[\text{or NaOH}]{NaOMe} \text{trans-Pt(OMe)Ph(PEt}_3)_2 \longrightarrow$$

$$\text{trans-PtHPh(PEt}_3)_2 \quad (c)$$

1.10. Formation of Bonds between Hydrogen and Transition Metals   227
1.10.8 by Hydrogen Transfer to the Metal
1.10.8.1. from Alcohols in Basic Media.

TABLE 1. PREPARATION OF TRANSITION-METAL HYDRIDES FROM ALCOHOLS

Initial metal complex	Reagents	Hydride	Refs.
$RuCl_3 \cdot x\ H_2O$	$PPh_3-HCHO-KOH-EtOH$	$H_2Ru(CO)(PPh_3)_3$	6, 7
$Na_2OsCl_6 \cdot 6\ H_2O$	$PPh_3-HCHO-MeOCH_2-$ CH_2OH	$HOsCl(CO)(PPh_3)_3$	6, 7
$RhCl_3 \cdot x\ H_2O$	$PPh_3-KOH-EtOH$	$HRh(PPh_3)_4$	6, 7
$Na_2IrCl_6 \cdot 6\ H_2O$	$PPh_3-HCHO-KOH-$ $MeOCH_2CH_2OH$	$HIr(CO)(PPh_3)_3$	6
$Ru(NO)Cl_3$	$PPh_3-KOH-EtOH$	$HRu(NO)(PPh_3)_3$	8
$RuCl_2(PPh_3)_3$	$NaOAc-MeOH$	$HRu(OAc)(PPh_3)_3$	9
$[Ru_2Cl_3(PEt_2Ph)_6]Cl$	$KOH-EtOH$	$HRuCl(CO)(PEt_2Ph)_3$	10
$cis\text{-}RuClMe(PMe_3)_4$	$NaOMe-THF$	$cis\text{-}HRuMe(PMe_3)_4$	11
$RhCl_3 \cdot x\ H_2O$	$EtOH-PEtPh_2$	$HRhCl_2(PEtPh_2)_3$	12, 13
$H_2[IrCl_6]$	$EtOH-COD$	$[HIrCl_2(C_8H_{12})]_2$	14
$[Ir(COD)Cl]_2$	$EtOH-PEt_2Ph-NaBPh_4$	$[H_2Ir(PEt_2Ph)_4][BPh_4]$	15
$[cis\text{-}Os(bipy)_2(CO)Cl][PF_6]$	$PPh_3-HO(CH_2)OH$	$[cis\text{-}Os(bipy)_2(CO)H][PF_6]$	16
$[cis\text{-}Ir(bipy)_2(OSO_2CF_3)_2]\text{-}$ $[CF_3SO_3]$	$PPh_3-HO(CH_2)_2OH$	$[cis\text{-}Ir(bipy)_2(PPh_3)H][PF_6]$	17

In favorable cases like Eq. (b), the addition of strong base is not required; either a weak base, such as PPh_3, can deprotonate sufficient alcohol to enable the reaction to proceed, or β-hydrogen elimination can occur even when the alcohol instead of the alkoxide is coordinated. The organic product is the aldehyde or ketone (confirmed in only a few cases)[2,20,21]:

$$IrCl_3(PEt_3)_3 \xrightarrow[EtOH]{KOH} HIrCl_2(PEt_3)_3\ +\ CH_3CHO \qquad (d)$$

$$cis\text{-}PtCl_2(PEt_3)_2 \xrightarrow[EtOH]{KOH} trans\text{-}HPtCl(PEt_3)_2\ +\ CH_3CHO \qquad (e)$$

$$cis\text{-}PtCl_2(PEt_3)_2 \xrightarrow[Me_2CHOH]{KOH} trans\text{-}HPtCl(PEt_3)_2\ +\ Me_2CO \qquad (f)$$

The alkoxide of a tertiary alcohol cannot generate a hydride by this mechanism. The importance of the type of alcohol employed is illustrated by the contrast between Eqs. (g) and (h)[22]:

$$[OsH(CO)_2(PPh_3)_3]^+ \xrightarrow{[OEt]^-} OsH_2(CO)(PPh_3)_3 \qquad (g)$$

$$[OsH(CO)_2(PPh_3)_3]^+ \xrightarrow[{[OMe]^-}]{[OCMe_3]^-\ or} Os(CO)_2(PPh_3)_3 \qquad (h)$$

Whereas treatment with ethoxide generates another hydride ligand, treatment with t-butoxide results in deprotonation. Similarly, whereas treatment of $Re(PPh_3)_2\text{-}(NC_6H_4Me)Cl_3$ with isopropoxide generates a hydride ligand:

$$Re(PPh_3)_2(N\text{-}p\text{-}C_6H_4Me)Cl_3 \xrightarrow[Me_2CHOH]{NaOCHMe_2} Re(PPh_3)_2(N\text{-}p\text{-}C_6H_4Me)(H)Cl_2 \qquad (i)$$

t-butoxide gives no reaction[23] at 100°C in t-BuOH.

228   1.10. Formation of Bonds between Hydrogen and Transition Metals
       1.10.8 by Hydrogen Transfer to the Metal
       1.10.8.1. from Alcohols in Basic Media.

As Eq. (h) illustrates, $[MeO]^-$ can fail to generate a hydride ligand; a stable methoxo complex is formed[23] when MeOH is used in Eq. (i). Complications may occur with other primary alcohols as well; EtOH, n-PrOH and n-BuOH are oxidized to carboxylato bridges[24] when substituted for i-PrOH, e.g.:

$$[(h^5\text{-}Me_5C_5)_2Rh_2(\mu\text{-}OH)_3][PF_6] \xrightarrow{\text{i-PrOH}} [(h^5\text{-}Me_5C_5)_3Rh_3(\mu\text{-}H)_3O][PF_6] \qquad (j)$$

In contrast, sec-BuOH gives satisfactory results.

Reaction (h) illustrates a limitation of the use of alkoxides for generating hydride ligands. A vacant coordination site must be generated so that the alcohol or alkoxide ion can attack the metal rather than any ligand [such as the hydrogen ligand in Eq. (h) or the carbonyl ligand in Eq. (k)[25]] subject to nucleophilic attack:

$$[Ir(CO)_3(PPh_3)_2][ClO_4] \xrightarrow[\text{MeOH}]{\text{KOH}} Ir(CO)_2(PPh_3)_2CO_2Me \qquad (k)$$

Carbonyl ligands with high $v(C\!-\!O)$ stretching frequencies are particularly susceptible to attack[26].

Sometimes either attack on the metal (leading to hydride formation) or attack on a carbonyl group can occur, depending on reaction conditions[27]:

$$[Ir(CO)(Me_2PCH_2CH_2PMe_2)_2]Cl \Big\langle \begin{array}{l} \xrightarrow[\text{or MeOH}]{\text{cold EtOH}} Ir(Me_2PCH_2CH_2PMe_2)_2CO_2R \\[2ex] \xrightarrow[\text{reflux}]{\text{EtOH}} [IrHCl(Me_2PCH_2CH_2PMe_2)_2]^+ \end{array} \qquad (l)$$

The formation of a carboalkoxy ligand, although kinetically favored, is reversible, whereas ligand dissociation and hydride formation are irreversible.

When side reactions do not interfere, the use of alcohol and base conveniently yields metal hydride. Each of the first four reactions in Table 1 prepares a complex transition-metal hydride from a metal salt in a few minutes, but only with triphenylphosphine.

The formaldehyde in some of these reactions is not only the source of the carbonyl ligands, but is also a potential alternative to the alcohol solvent as source of the hydride ligands. A similar ambiguity exists in the many hydride preparations in which H_2O and $[OH]^-$ are present as well as ROH and $[RO]^-$, e.g., in[28,29]:

$$Ir_4(CO)_{12} \Big\langle \begin{array}{l} \xrightarrow[\text{MeOH, CO}]{\text{K}_2\text{CO}_3} [HIr_4(CO)_{11}]^- \\[2ex] \xrightarrow[\text{EtOH}]{\text{KOH}} [H_2Ir_4(CO)_{10}]^{2-} \end{array} \qquad (m)$$

the hydrides may arise from $[OH]^-$ attack on carbonyl ligands (see §1.10.9), rather than from coordinated alkoxide.

(J. R. NORTON)

1. R. A. Schunn, in *Transition Metal Hydrides,* E. L. Muetterties, ed., Marcel Dekker, New York, 1971, p. 203.

2. H. D. Kaesz, R. B. Saillant, *Chem. Rev., 72*, 231 (1972).
3. H. D. Kaesz, *Inorg. Synth., 17*, 52 (1977); a listing of all earlier preparations of transition-metal hydrides published in *Inorg. Synth.* as well as an introduction to newer preparations.
4. A. P. Humphries, H. D. Kaesz, *Prog. Inorg. Chem., 25*, 145 (1979); a comprehensive review of all aspects of the chemistry of polynuclear hydrides.
5. G. L. Geoffroy, J. R. Lehman, *Adv. Inorg. Chem. Radiochem, 20*, 189 (1977).
6. N. Ahmad, S. D. Robinson, M. F. Uttley, *J. Chem. Soc., Dalton Trans.*, 843 (1972).
7. N. Ahmad, J. J. Levison, S. D. Robinson, M. F. Uttley, *Inorg. Synth., 15*, 45 (1974).
8. J. S. Bradley, G. Wilkinson, *Inorg. Synth., 17*, 73 (1977).
9. R. Young, G. Wilkinson, *Inorg. Synth., 17*, 79 (1977).
10. J. Chatt, B. L. Shaw, A. E. Field, *J. Chem. Soc.*, 3466 (1964).
11. J. A. Statler, G. Wilkinson, M. Thornton-Pett, M. B. Hursthouse, *J. Chem. Soc., Dalton Trans.*, 1731 (1984).
12. G. M. Intille, *Inorg. Chem., 11*, 695 (1972).
13. C. Masters, B. L. Shaw, *J. Chem. Soc., A*, 3679 (1971).
14. S. D. Robinson, B. L. Shaw, *J. Chem. Soc.*, 4997 (1965).
15. L. M. Haines, E. Singleton, *J. Chem. Soc., Dalton Trans.*, 1891 (1972).
16. J. V. Caspar, B. P. Sullivan, T. J. Meyer, *Organometallics., 2*, 551 (1983).
17. B. P. Sullivan, T. J. Meyer, *J. Chem. Soc., Chem. Commun.*, 403 (1984).
18. L. Vaska, J. W. DiLuzio, *J. Am. Chem. Soc., 84*, 4989 (1962).
19. D. P. Arnold, M. A. Bennett, *Inorg. Chem., 23*, 2110 (1984).
20. J. Chatt, R. S. Coffey, B. L. Shaw, *J. Chem. Soc.*, 7391 (1965).
21. J. Chatt, B. L. Shaw, *J. Chem. Soc.*, 5075 (1962).
22. B. E. Cavit, K. R. Grundy, W. R. Roper, *J. Chem. Soc., Chem. Commun.*, 60 (1972).
23. G. LaMonica, S. Cenini, F. Porta, *Inorg. Chim. Acta, 48*, 91 (1981).
24. A. Nutton, P. M. Bailey, P. M. Maitlis, *J. Organomet. Chem., 213*, 313 (1981).
25. L. Malatesta, G. Caglio, M. Angoletta, *J. Chem. Soc.*, 6974 (1965).
26. R. J. Angelici, *Acc. Chem. Res., 5*, 335 (1972).
27. S. D. Ibekwe, K. A. Taylor, *J. Chem. Soc., A*, 1 (1970).
28. M. Angoletta, L. Malatesta, G. Caglio, *J. Organomet. Chem., 94*, 99 (1975).
29. G. Ciani, M. Manassero, V. G. Albano, F. Canziani, G. Giordano, S. Martinengo, P. Chini, *J. Organomet. Chem., 150*, C17 (1978).

1.10.8.2. from Hydrazine.

Although one of the first discovered[1] for transition–metal hydrogen bonds, this method is now seldom used[2]. It gives the $(Et_3P)_2Pt$ hydridochloride in excellent yield[3,4]:

$$\text{cis-PtCl}_2(\text{PEt}_3)_2 \xrightarrow[\text{H}_2\text{O}]{\text{N}_2\text{H}_4} \text{trans-HPtCl(PEt}_3)_2 \qquad \text{(a)}$$

but it forms Pt(0) complexes when other phosphines are present[5]. Isolation of an intermediate mixture of $[Pt(PPh_3)_2(\mu\text{-}N_2H)]_2^{2+}$ and $[Pt(PPh_3)_2(\mu\text{-}NH_2)]_2^{2+}$ is claimed[6,7].

Hydrazine is also successfully employed in:

$$\text{Rh(PPh}_3)_2(\text{CO})\text{Cl} \xrightarrow[\text{PPh}_3]{\text{N}_2\text{H}_4} \text{HRh(CO)(PPh}_3)_3 \qquad \text{(b)[8]}$$

$$\text{OsCl}_3(\text{PBu}_2\text{Ph})_3 \xrightarrow[\text{EtOH-H}_2\text{O}]{\text{N}_2\text{H}_4} \text{HOsCl}_2(\text{PBu}_2\text{Ph})_3 \qquad \text{(c)[9]}$$

(J. R. NORTON)

1. J. Chatt, L. A. Duncanson, B. L. Shaw, *Proc. Chem. Soc.*, 343 (1957).
2. R. A. Schunn, in *Transition Metal Hydrides*, E. L. Muetterties, ed., Marcel Dekker, New York, 1971, p. 203.
3. J. Chatt, B. L. Shaw, *J. Chem. Soc.*, 5075 (1962).

4. G. W. Parshall, *Inorg. Synth.*, *12*, 28 (1970).
5. D. M. Roundhill, *Adv. Organomet. Chem.*, *13*, 273 (1975).
6. G. C. Dobinson, R. Mason, G. B. Robertson, R. Ugo, F. Conti, D. Morelli, S. Cenini, F. Bonati, *J. Chem. Soc., Chem. Commun.*, 739 (1967).
7. H. D. Kaesz, R. B. Saillant, *Chem. Rev.*, *72*, 231 (1972).
8. S. S. Bath, L. Vaska, *J. Am. Chem. Soc.*, *85*, 3500 (1963).
9. J. Chatt, G. J. Leigh, R. J. Paske, *J. Chem. Soc., Chem. Commun.*, 671 (1967).

1.10.8.3. from Aluminum Alkyls.

Aluminum alkyls, although inconvenient to handle **because of their spontaneous inflammability**, are powerful reducing agents and are applied to the synthesis of transition-metal hydrides[1,2]. They, and the organomagnesium halide reagents to be discussed in §1.10.8.4, function by initial alkylation of the transition metal (e.g., for Et_3Al):

$$M-X + (CH_3CH_2)_3Al \rightarrow M-CH_2CH_3 + (CH_3CH_2)_2AlX \qquad (a)$$

followed by formation of a hydride ligand by β-hydrogen elimination:

$$M-CH_2CH_3 \rightarrow M-H + C_2H_4 \qquad (b)$$

Reaction (a) requires that the metal be less electropositive than Al—a requirement met by most transition elements. Reaction (b) requires that a vacant coordination site be available (the olefin produced by β-hydrogen elimination must be at least transiently coordinated[3]), and that the alkyl group possess at least one β-hydrogen[4,5].

Organoaluminums are soluble in noncoordinating organic solvents, which are used in:

$$Zr(OCMe_3)_4 + COT + Et_2AlH \xrightarrow{\text{toluene}} (COT)ZrH_2 \qquad (c)^6$$

$$Mo(acac)_3 + Ph_2PCH_2CH_2PPh_2 + AlEt_3 \xrightarrow[\text{Ar}]{\text{toluene}} HMo(acac)(Ph_2PCH_2CH_2PPh_2)_2 \quad (d)^7$$

$$RuHCl(PPh_3)_3 + AlEt_3 + N_2 \xrightarrow{\text{Et}_2O} H_2Ru(N_2)(PPh_3)_3 \qquad (e)^{8,9}$$

$$Co(acac)_3 + PPh_3 + Al(i\text{-Bu})_3 + N_2 \xrightarrow{\text{toluene}} HCo(N_2)(PPh_3)_3 \qquad (f)^{10,11}$$

$$Co(acac)_3 + PPhMe_2 + Al(i\text{-Bu})_2OEt \xrightarrow{\text{Et}_2O} HCo(PPhMe_2)_4 \qquad (g)^{12,13}$$

$$Rh(PPh_3)_3Cl + Al(i\text{-Pr})_3 \xrightarrow{\text{Et}_2\text{O-hexane}} HRh(PPh_3)_3 \qquad (h)^{14}$$

(J. R. NORTON)

1. R. A. Schunn, in *Transition Metal Hydrides*, E. L. Muetterties, ed., Marcel Dekker, New York, 1971, p. 203.
2. H. D. Kaesz, R. B. Saillant, *Chem. Rev.*, *72*, 231 (1972).
3. D. E. Reger, E. C. Culbertson, *J. Am. Chem. Soc.*, *98*, 2789 (1976).
4. P. J. Davidson, M. F. Lappert, R. Pearce, *Chem. Rev.*, *76*, 219 (1976); an excellent review, including references to earlier reviews of decomposition mechanisms for σ-bonded organotransition-metal complexes.
5. R. R. Schrock, G. W. Parshall, *Chem. Rev.*, *76*, 243 (1976); a review of early transition-metal alkyls and their decomposition mechanisms, including β-hydrogen elimination.

6. H.-J. Kablitz, G. Wilke, *J. Organomet. Chem.*, *51*, 241 (1973).
7. T. Ito, A. Yamamoto, *Inorg. Synth.*, *17*, 61 (1977).
8. T. Ito, S. Kitazume, A. Yamamoto, S. Ikeda, *J. Am. Chem. Soc.*, *92*, 3011 (1970).
9. W. H. Knoth, *Inorg. Synth.*, *15*, 31 (1974).
10. A. Yamamoto, S. Kitazume, L. S. Pu, S. Ikeda, *J. Am. Chem. Soc.*, *93*, 371 (1971).
11. A. Misono, *Inorg. Synth.*, *12*, 12 (1970).
12. J. Lorberth, H. Nöth, P. V. Rinze, *J. Organomet. Chem.*, *16*, P1 (1969).
13. T. Ikariya, A. Yamamoto, *J. Organomet. Chem.*, *116*, 239 (1976).
14. K. C. Dewhirst, W. Keim, C. A. Reilly, *Inorg. Chem.*, *7*, 546 (1968).

1.10.8.4. from Alkylmagnesium Halides.

Although their use is becoming less common, organomagnesium halide reagents convert metal halides into hydrides[1-3]. Like the Al alkyls discussed in §1.10.8.3, organomagnesium halide reagents function by initial alkylation of the transition metal [e.g., in Eq. (a) for EtMgBr]:

$$M—X + CH_3CH_2MgBr \rightarrow M—CH_2CH_3 + MgXBr \qquad (a)$$

followed by formation of a hydride ligand by β-hydrogen elimination:

$$M—CH_2CH_3 \rightarrow M—H + C_2H_4 \qquad (b)$$

Reaction (a) requires that the metal be less electropositive than Mg—a requirement met by most transition elements. Reaction (b) requires that a vacant coordination site be available (the olefin produced by β-hydrogen elimination must be at least transiently coordinated[4]), and that the alkyl group possess at least one β-hydrogen. Such eliminations are easier for secondary and tertiary alkyl ligands than for the more stable primary ones[5,6].

Isopropylmagnesium halide reagents, therefore, are effective at forming transition-metal hydrides; methyl, phenyl and other magnesium halide reagents without β-hydrogens give the corresponding organometallic complexes[7,8]:

$$[Ir(COD)Cl]_2 + Ph_3P \xrightarrow[C_6H_6-Et_2O]{MeMgI} Ir(COD)(PPh_3)_2CH_3 \qquad (c)$$

$$[Ir(COD)Cl]_2 + Ph_3P \xrightarrow[C_6H_6-Et_2O]{i\text{-}PrMgBr} HIr(COD)(PPh_3)_2 \qquad (d)$$

$$h^5\text{-}Me_5C_5Ru(PMe_3)_2Cl \xrightarrow[\text{toluene-}Et_2O]{Me_3SiCH_2MgCl} h^5\text{-}Me_5C_5Ru(PMe_3)_2CH_2SiMe_3 \qquad (e)$$

$$h^5\text{-}Me_5C_5Ru(PMe_3)_2Cl \xrightarrow[\text{toluene-}Et_2O]{i\text{-}PrMgCl} h^5\text{-}Me_5C_5Ru(PMe_3)_2H \qquad (f)$$

Even an n-propylmagnesium halide reagent forms a hydride less readily than the corresponding isopropyl reagent, cf.[8]:

$$\text{trans-}PtCl_2(PEt_3)_2 \xrightarrow[Et_2O]{n\text{-}PrMgCl} \text{trans-}Pt(n\text{-}Pr)Cl(PEt_3)_2 \qquad (g)$$

$$\text{trans-}PtCl_2(PEt_3)_2 \xrightarrow[Et_2O]{i\text{-}PrMgCl} \text{trans-}HPtCl(PEt_3)_2 \qquad (h)$$

232   1.10. Formation of Bonds between Hydrogen and Transition Metals
1.10.8. by Hydrogen Transfer to the Metal
1.10.8.4. from Alkylmagnesium Halides.

Another example [related to Eq. (d)] of the use of an isopropylmagnesium halide for the synthesis of a transition-metal hydride is[10,11]:

$$[Ir(COD)Cl]_2 + 1,3\text{-cyclohexadiene} \xrightarrow[\text{Et}_2\text{O}]{\text{i-PrMgBr, } h\nu} HIr(COD)(1,3\text{-cyclohexadiene-h}^4) \qquad (i)$$

As a secondary organomagnesium halide reagent, cyclohexylmagnesium bromide may also be useful for the synthesis of metal hydrides[8,14,15]:

$$cis\text{-PtCl}_2(\text{PEt}_3)_2 + \text{C}_6\text{H}_{11}\text{MgBr} \xrightarrow{\text{Et}_2\text{O}} trans\text{-PtHBr(PEt}_3)_2 \qquad (j)$$

but over half of the hydride ligand arises from hydrolysis instead of from β-hydrogen elimination[3,13]. Tertiary organomagnesium halide reagents or other tertiary organometallics may be more effective for hydride synthesis[8,14,15]:

$$\text{h}^5\text{-C}_5\text{H}_5\text{Mo(C}_4\text{Ph}_4\text{-h}^4)\text{Cl} \xrightarrow[\text{Et}_2\text{O}]{\text{Me}_3\text{CMgCl}} \text{h}^5\text{-C}_5\text{H}_5\text{Mo(C}_4\text{Ph}_4\text{-h}^4)\text{H} \qquad (k)$$

$$[(\text{Me}_3\text{Si})_2\text{N}]_3\text{ThCl} \xrightarrow[\text{pentane}]{\text{Me}_3\text{CLi}} [(\text{Me}_3\text{Si})_2\text{N}]_3\text{ThH} \qquad (l)$$

$$\text{h}^5\text{-Me}_5\text{C}_5\text{Ru(PMe}_3)_2\text{Cl} \xrightarrow[\text{Et}_2\text{O}]{\text{Me}_3\text{CMgCl}} \text{h}^5\text{-Me}_5\text{C}_5\text{Ru(PMe}_3)_2\text{H} \qquad (m)$$

Organomagnesium halide reagents and organolithiums attack carbonyl ligands, particularly when the latter have high $\nu(\text{C}-\text{O})$ stretching frequencies and are susceptible to nucleophilic attack[16,17]. These organometallic reagents are, therefore, best used with complexes that do not contain carbonyls.

(J. R. NORTON)

1. R. A. Schunn, in *Transition Metal Hydrides*, E. L. Muetterties, ed., Marcel Dekker, New York, 1971, p. 203.
2. H. D. Kaesz, R. B. Saillant, *Chem. Rev.*, 72, 231 (1972).
3. D. M. Roundhill, *Adv. Organomet. Chem.*, 13, 273 (1975).
4. D. E. Reger, E. C. Culbertson, *J. Am. Chem. Soc.*, 98, 2789 (1976).
5. P. J. Davidson, M. F. Lappert, R. Pearce, *Chem. Rev.*, 76, 219 (1976); an excellent review, including references to earlier reviews of decomposition mechanisms for σ-bonded organotransition-metal complexes.
6. R. R. Schrock, G. W. Parshall, *Chem. Rev.*, 76, 243 (1976); a review of early transition-metal alkyls and their decomposition mechanisms, including β-hydrogen elimination.
7. H. Yamazaki, M. Takesada, N. Hagihara, *Bull. Chem. Soc. Jpn.*, 42, 275 (1969).
8. T. D. Tilley, R. H. Grubbs, J. E. Bercaw, *Organometallics*, 3, 274 (1984).
9. J. Chatt, R. S. Coffey, A. Gough, D. T. Thompson, *J. Chem. Soc., A*, 190 (1968).
10. J. Müller, H. Menig, P. V. Rinze, *J. Organomet. Chem.*, 181, 387 (1979).
11. J. Müller, H. Mehnig, J. Pickardt, *Angew. Chem., Int. Ed. Engl.*, 20, 401 (1981).
12. J. Chatt, B. L. Shaw, *J. Chem. Soc.*, 5075 (1962).
13. R. J. Cross, F. Glockling, *J. Organomet. Chem.*, 3, 253 (1964).
14. R. B. King, A. Efraty, *J. Chem. Soc., Chem. Commun.*, 1370 (1970).
15. H. W. Turner, S. J. Simpson, R. A. Andersen, *J. Am. Chem. Soc.*, 101, 2782 (1979).
16. D. J. Darensbourg, M. Y. Darensbourg, *Inorg. Chem.*, 9, 1691 (1970).
17. M. Y. Darensbourg, H. L. Conder, D. J. Darensbourg, C. Hasday, *J. Am. Chem. Soc.*, 95, 5919 (1973).

1.10.9. by Decarboxylation of Hydroxycarbonyl or Formate Complexes.

Hydride synthesis by metal-carbonyl hydrolysis is used widely, and mechanistic details are emerging. The reaction begins with nucleophilic attack by H_2O or $[OH]^-$ on a carbonyl ligand:

$$M-CO + H_2O \rightarrow [M-\overset{\overset{\displaystyle O}{\|}}{C}OH]^- + H^+ \qquad (a)$$

$$M-CO + [OH]^- \rightarrow [M-\overset{\overset{\displaystyle O}{\|}}{C}OH]^- \qquad (b)$$

$$[M-CO]^+ + [OH]^- \rightarrow M-\overset{\overset{\displaystyle O}{\|}}{C}OH \qquad (c)$$

The loss of CO_2 in some form leaves a metal hydride:

$$M-\overset{\overset{\displaystyle O}{\|}}{C}OH \longrightarrow CO_2 + M-H \qquad (d)$$

Overall, the reaction converts a cationic carbonyl to a neutral hydride:

$$[M-CO]^+ + [OH]^- \rightarrow CO_2 + M-H \qquad (e)$$

or a neutral carbonyl to an anionic hydride:

$$M-CO + [OH]^- \rightarrow CO_2 + [M-H]^- \qquad (f)$$

Several hydroxycarbonyl are isolable and convertable to hydrides[1,2]. The product of reaction (g) is stable when extracted into benzene immediately after formation[3]:

$$[h^5\text{-}C_5H_5FePPh_3(CO)_2]Cl \xrightarrow[C_6H_6\text{-}H_2O]{KOH} h^5\text{-}C_5H_5FePPh_3(CO)CO_2H \qquad (g)$$

It is amphoteric, being deprotonated by xs KOH while dissociating hydroxyl ion [i.e., reversing Eq. (g)] in solvents of high dielectric constant such as formamide. The neutral hydroxycarbonyl complex loses CO_2 on warming, whereas its potassium salt is stable in solution even at 100°C:

$$h^5\text{-}C_5H_5(Ph_3P)(OC)Fe-\overset{\overset{\displaystyle O}{/\!/}}{\underset{\underset{\displaystyle H-O}{|}}{C}} \longrightarrow h^5\text{-}C_5H_5(Ph_3P)(OC)FeH + CO_2 \qquad (h)$$

$$[h^5\text{-}C_5H_5(Ph_3P)(OC)Fe(CO_2)]^- \xrightarrow{\;\;\;\nparallel\;\;\;} [h^5\text{-}C_5H_5(Ph_3P)(OC)Fe]^- + CO_2 \qquad (i)$$

Concerted elimination of CO_2 is, therefore, faster than decarboxylation of the anion, perhaps because of the high energy of the basic anion that would be generated if reaction (i) occurred[3].

Similarly, treatment of trans-$[PtCl(CO)(PEt_3)_2]^+$ in acetone with H_2O gives a hydroxycarbonyl complex; the addition of $HClO_4$ reverses the process:

$$\text{trans-}[PtCl(CO)(PEt_3)_2]^+ \xrightarrow[\text{acetone}]{H_2O} \text{trans-}PtCl(COOH)(PEt_3)_2 + H^+ \qquad (j)$$

The solid hydroxycarbonyl complex loses CO_2 only slowly[2,4]:

$$\text{trans-}PtCl(COOH)(PEt_3)_2 \xrightarrow{170°C} \text{trans-}PtHCl(PEt_3)_2 + CO_2 \qquad (k)$$

A different conclusion of the decarboxylation step is offered[5,6] by $h^5\text{-}C_5H_5\text{-}Re(CO)(NO)CO_2H$:

$$[h^5\text{-}C_5H_5Re(CO)_2(NO)][PF_6] \xrightarrow[Et_2O\text{-}H_2O]{NaOH} h^5\text{-}C_5H_5Re(CO)(NO)CO_2H \qquad (l)$$

Although this hydroxycarbonyl complex is stable as a solid to 100°C, treatment with catalytic Et_3N or other bases[6] causes decarboxylation to the hydride:

$$h^5\text{-}C_5H_5Re(CO)(NO)CO_2H \xrightarrow[\text{acetone-}H_2O]{Et_3N} [Et_3NH][h^5\text{-}C_5H_5Re(CO)(NO)CO_2] \longrightarrow$$

$$CO_2 + [Et_3NH][h^5\text{-}C_5H_5Re(CO)(NO)] \longrightarrow Et_3N + h^5\text{-}C_5H_5Re(CO)(NO)H \qquad (m)$$

Treatment of $[h^5\text{-}C_5H_5Re(CO)_2(NO)]^+$ with aq Et_3N is the standard method for preparing[5,6] $h^5\text{-}C_5H_5Re(CO)(NO)H$.

The stability of $h^5\text{-}C_5H_5Re(CO)(NO)CO_2H$ arises from its high barrier to CO dissociation, whereas PPh_3 dissociation in $h^5\text{-}C_5H_5Fe(CO)(PPh_3)CO_2H$ is facile and provides a vacant coordination site as required for decarboxylation. The ease of hydroxycarbonyl complex decarboxylation varies greatly[7,8].

Hydride complexes also are available from the decarboxylation of complexes containing formate, an isomer of the hydroxycarbonyl ligand[9-11]:

$$[\overset{\overset{\textstyle O}{\|}}{H}COCr(CO)_5]^- \xrightarrow{25°C} [HCr(CO)_5]^- + CO_2 \qquad (n)$$

However, formate and hydroxycarbonyl complexes do not interconvert intramolecularly[10]; therefore, neither species is an intermediate in the decarboxylation of the other. Formate ion is useful for the generation[12] of thermally unstable hydrides at low T:

$$[PtMe(MeOH)dppe][BF_4] \xrightarrow{NaO_2CH} Pt(O_2CH)Me(dppe)$$
$$\downarrow 25°C \qquad\qquad (o)$$
$$Pt(H)Me(dppe)$$

The chiral complex $h^5\text{-}C_5H_5Re(NO)(PPh_3)O_2CH$ decarboxylates[12] to $h^5\text{-}C_5H_5Re(NO)\text{-}(PPh_3)H$ without PPh_3 dissociation and with retention at Re.

The synthetic limitations of metal-carbonyl hydrolysis arise from the requirement for initial nucleophilic attack on a carbonyl ligand. The susceptibility of carbonyl to such attack varies inversely with the extent to which it is serving as a π acceptor and can be predicted from its IR carbonyl stretching frequency and force constant[14,15]. Whereas some electron-poor cationic carbonyls can be attacked by H_2O, as in Eq. (j), neutral carbonyls require $[OH]^-$; for electron-rich carbonyls with CO stretching frequencies lowered by substantial π backbonding, strongly basic conditions may be required (see Table 1). The more nucleophilic NaSH [which leads to COS elimination after formation of an MC(O)SH intermediate] often gives[23] better results than NaOH:

$$[(Ph_3P)Mn(CO)_5][PF_6] \xrightarrow[CH_3CN]{NaSH} HMn(CO)_4PPh_3 \tag{p}$$

TABLE 1. FORMATION OF HYDRIDES BY BASE HYDROLYSIS OF METAL–CARBONYL COMPLEXES

Carbonyl complex	Base and conditions	Product	Refs.
$Cr(CO)_6$	$KOH-H_2O-MeOH-THF$, 50°C	$K[HCr_2(CO)_{10}]$	16
$Fe(CO)_5$	$NaOH-H_2O-MeOH-[(Ph_3P)_2N]Cl$	$[(Ph_3P)_2N][HFe(CO)_4]$	17
$Ru_3(CO)_{12}$	H_2O, 135°C	$H_4Ru_4(CO)_{12}$	18
$[Mn(CO)_6][BF_4]$	H_2O-CH_3CN	$HMn(CO)_5$	19, 20
$[PtCl(CO)(PEt_3)_2][BF_4]$	$H_2O-MeOH$, 110°C	$PtHCl(PEt_3)_2$	21, 22

(J. R. NORTON)

1. D. J. Darensbourg, *Isr. J. Chem.*, *15*, 247 (1977).
2. J. Halpern, *Comments Inorg. Chem.*, *1*, 3 (1981).
3. N. Grice, S. C. Kao, R. Petit, *J. Am. Chem. Soc.*, *101*, 1627 (1979); includes a brief but comprehensive historical survey with refs. to proposed MCO_2H intermediates.
4. M. Catellani, J. Halpern, *Inorg. Chem.*, *19*, 566 (1980); includes an excellent survey of MCO_2H complexes postulated as intermediates.
5. C. P. Casey, M. A. Andrews, J. E. Rinz, *J. Am. Chem. Soc.*, *101*, 741 (1979).
6. J. R. Sweet, W. A. G. Graham, *Organometallics*, *1*, 982 (1982).
7. W. Tam, G.-Y. Lin, W.-K. Wong, W. A. Kiel, V. K. Wang, J. A. Gladysz, *J. Am. Chem. Soc.*, *104*, 141 (1982).
8. D. H. Gibson, K. Owens, T.-S. Ong, *J. Am. Chem. Soc.*, *106*, 1125 (1984).
9. D. J. Darensbourg, R. A. Kudaroski, *Adv. Organomet. Chem.*, *22*, 129 (1983).
10. D. J. Darensbourg, A. Rokicki, *Organometallics*, *1*, 1685 (1982).
11. J. V. Caspar, B. P. Sullivan, T. J. Meyer, *Organometallics*, *2*, 551 (1983).
12. D. P. Arnold, M. A. Bennett, *J. Organomet. Chem.*, *199*, C17 (1980).
13. J. H. Merrifield, J. A. Gladysz, *Organometallics*, *2*, 782 (1983).
14. D. J. Darensbourg, M. Y. Darensbourg, *Inorg. Chem.*, *9*, 1691 (1970).
15. D. J. Darensbourg, B. J. Baldwin, J. A. Froelich, *J. Am. Chem. Soc.*, *102*, 4688 (1980).
16. M. D. Grillone, B. B. Kedzia, *J. Organomet. Chem.*, *140*, 161 (1977).
17. M. Y. Darensbourg, D. J. Darensbourg, H. L. C. Barros, *Inorg. Chem.*, *17*, 297 (1978).
18. J. A. Froelich, D. J. Darensbourg, *Inorg. Chem.*, *16*, 960 (1977).
19. C. R. Eady, B. F. G. Johnson, J. Lewis, *J. Chem. Soc., Dalton Trans.*, 838 (1977).
20. W. Hieber, T. Kruck, *Z. Naturforsch., Teil B*, *16*, 709 (1961).
21. D. J. Darensbourg, J. A. Froelich, *J. Am. Chem. Soc.*, *99*, 4726 (1977).
22. H. C. Clark, K. R. Dixon, W. J. Jacobs, *J. Am. Chem. Soc.*, *91*, 1346 (1967).
23. H. C. Clark, W. J. Jacobs, *Inorg. Chem.*, *9*, 1229 (1970).

1.11. Formation of Bonds between Hydrogen and Elements of Group 0

There are no compounds that contain hydrogen–group 0 element (G) bonds. The cations $[GH]^+$ however, are generated in ion–molecule reactions[1]:

$$G^+_{(g)} + H_{2(g)} \rightarrow [GH]^+_{(g)} + H_{(g)} \tag{a}$$

$$G + H^+_{2(g)} \rightarrow [GH]^+_{(g)} + H_{(g)} \tag{b}$$

The proton affinities [i.e., $-\Delta H$ for $G_{(g)} + H^+_{(g)}$] from such studies are listed in Table 1. As may be appreciated from the thermochemical cycle below, the electron affinity of $[GH]^+_{(g)}$ is equal to the electron affinity of the proton ($+13.6$ eV) less the proton affinity of G, if ΔH [$GH_{(g)} \rightarrow G_{(g)} + H(g)$] is zero. Because the last condition is likely to be approximately true for all of the gases, the electron affinities of the $[GH]^+$ cations decrease down the series, and $E[XeH]^+ \leq +7.6$ eV. Thus hydroxenonium salts may be preparable. To stabilize $[GH]^+$ in a salt, $[GH]^+X^-_{(s)}$, it is essential to have a combined lattice energy and electron affinity for X that exceeds the electron affinity of $[GH]^+$, otherwise the electron will simply transfer from X^- to $[GH]^+$. The problem of the generation of G—H bonds in cations $[GH]^+$ is to find an anion X^- that fulfills not only those conditions but also does not abstract the proton from the cation:

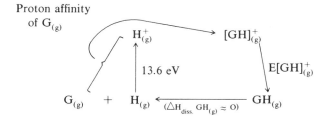

TABLE 1. PROTON AFFINITIES (PA) OF NOBLE GASES[a]

	He	Ne	Ar	Kr	Xe
PA (g) (eV)	1.8	2.2	3.0	≥ 4	≥ 6

[a] From Ref. 1.

(N. BARTLETT)

1. G. von Bünau, *Fortschr. Chem. Forsch.*, 5, 374 (1965).

1.12. Reversible Formation of Metal Hydrides by Direct Reaction of Hydrogen

1.12.1. Introduction

This chapter deals with metal–hydrogen compounds formed directly from metal and H_2:

$$M + \frac{x}{2} H_2 \rightleftharpoons MH_x \tag{a}$$

where M may be a group IA–VIII metal (including lanthanide or actinide), alloy or intermetallic compound. The reaction is spontaneous, exothermic and easily reversible.

Although many metals dissolve large amounts of hydrogen to form solid solutions, the hydride, MH_x shown in Eq. (a) refers only to a new phase, different from the hydrogen-saturated metal. The definite hydride phase may be detected either by a change in crystal structure, by discontinuous change in lattice parameters on hydride formation or by thermodynamic data indicating new phase formation. The latter technique is illustrated in Fig. 1, which is a pressure–composition isotherm of a metal–hydrogen system. As hydrogen is dissolved in a metal, the H_2 pressure in equilibrium with the solid solution increases with hydrogen concentration. When the solubility limit is reached at point y, the nonstoichiometric hydride phase, MH_x, is formed. Because there is now an additional phase in the system, the number of degrees of freedom decreases in accordance with the phase rule, and the H_2 pressure remains constant across the concentration range from y to x. Therefore, the appearance of a plateau pressure is indicative of new phase (hydride) formation. The plateau pressure represents the dissociation P of the hydride at the particular T of the isotherm. As H_2 is added to the system in this concentration range (y to x), hydrogen-saturated metal is converted to nonstoichiometric hydride, MH_x. After the metal is converted to hydride, further addition of H_2 beyond the composition x results in an increase in H_2 pressure as the composition approaches the stoichiometric value, s. Therefore, the stoichiometry range of the hydride, x to s, also can be obtained from such isotherms. Most hydrides discussed in this chapter are nonstoichiometric, and the deviation from the stoichiometric composition (s–x) may be large.

In the initial solution of H_2 in metals, the solubility is proportional to the square root of the H_2 pressure. This suggests that hydrogen is dissolved in the atomic rather than the molecular state. Structural investigations support and confirm this notion. Consequently, the heat of dissociation of H_2, E_D, which is 436 kJ/mol H_2, represents an energy barrier to the reaction and influences the rate of the reaction. Reaction (a) can be considered a three-stage process, consisting of H_2 adsorption on the metal surface, solution into the bulk and nucleation of the hydride phase. For the first stage, adsorption of H_2 on a clean metal surface[1], potential-energy curves for a metal and an H_2 molecule and for a metal and two H atoms can be calculated[2] as a function of distance from the metal surface. The resulting curves, shown schematically in Fig. 2, reveal two adsorption types, physi- and chemisorption. Physisorption, represented by curve 1, re-

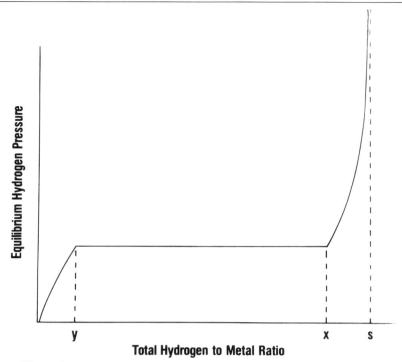

Figure 1. Typical pressure–composition isotherm for a metal–hydrogen system.

sults from attraction by dispersion forces of an H_2 molecule to the metal surface, whereas chemisorption, curve 2, arises from chemical bonding between the H atom and the metal. The minimum in the physisorption curve corresponds to the equilibrium distance of the H_2 molecule from the metal surface and is[1] ca. 0.2–0.3 nm; the heat of adsorption, E_{ad}, is[3] < 30 kJ/mol H_2. By contrast, in the chemisorption curve, the equilibrium distance of the H atom from the metal atom in the surface is[1] 0.05–0.1 nm (i.e., chemical bond length distances), and the heat of chemisorption, E_c, is[3] larger, > 100 kJ/mol H_2.

Hydrogen molecules striking a metal surface can be physi- or chemisorbed depending on their kinetic energies and the height of the activation barrier, E_{act}. The activation barrier is determined by the point of intersection of the physisorption curve and the chemisorption curve in relation to the zero-energy level. Intersection above the zero-energy level corresponds to activated chemisorption (as depicted in Fig. 2), whereas intersection below corresponds to nonactivated chemisorption. In the example of Fig. 2, H_2 molecules with kinetic energies < E_{ad} can be adsorbed physically, whereas molecules with energies > E_{act} can move along curve 1 to the intersection with curve 2, change over to curve 2 at the intersection, dissociate into H atoms and become chemisorbed. In the second stage[1] of the process, the chemisorbed H atoms penetrate the surface, a step that may or may not require an activation energy[3], and are subse-

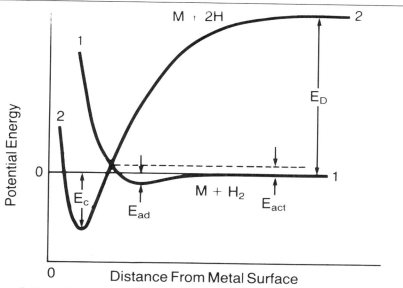

Figure 2. Potential energy as a function of metal–hydrogen distance for hydrogen molecules (curve 1) and for hydrogen atoms (curve 2).

quently dissolved in the bulk through diffusion. The diffusion process also requires an energy of activation, E_{diff}. The final stage is nucleation of the new phase; associated with this stage is an activation energy of nucleation, E_{nucl}. The rate-determining step in this entire process depends on the magnitude of E_{act}, E_c, E_{diff}, E_{nucl}, etc.[3].

The shapes of the potential-energy curves shown in Fig. 2 are very dependent on the condition of the metal surface; e.g., roughness, cleanliness and crystal orientation[3] influence rates. In the practical preparation of metal hydrides according to Eq. (a), surface contamination of the metal is the most serious problem. Most metals after exposure to air do not react with H_2 gas at RT without some form of activation. The activation process consists of the mechanical removal of surface layers (oxides, CO, sulfides, etc.), or by heating to high T in ultrahigh vacuum to evaporate or dissolve the surface-contaminating species. Gaseous H_2 is then admitted to the metal sample without exposure to air. The absorption is carried out at elevated T to increase rate.

The hydrides formed in reaction (a) may be classified as (1) saline or ionic hydrides, (2) metallic hydrides and (3) covalent hydrides. The saline hydrides include the hydrides of the alkali and alkaline-earth metals, except BeH_2, which is covalent. Transition metals form binary compounds with hydrogen that are classified as metallic hydrides including rare-earth and actinide hydrides. Intermetallic compound hydrides, such as $TiFeH_2$ and $LaNi_5H_6$, may be thought of as pseudobinary metallic hydrides.

The crystal lattices of the saline hydrides consist of hydrogen anions and metal cations, but not exclusively; e.g., in LiH, calculations[4] and diffraction experiments[5] suggest that electron transfer from Li to H is 0.8–1 e, implying a strong ionic bond with covalent character. Magnesium hydride occupies a special position. Although classified here as a saline hydride, its properties are intermediate between the ionic hydrides and covalent BeH_2.

The nature of chemical bonding in metallic hydrides is controversial[6-10], with two opposing models: the protonic and the ionic. In the former[11], hydrogen is assumed to donate its electron to the d band of the transition metal, forming an alloy with the metal. Hydrogen, therefore, exists as protons, partially screened by the conduction electrons in the metal sublattice. The opposing view asserts[12] that H accepts an electron from the metal to form a hydride anion and a metal cation, i.e., a saline hydride type. The protonic model is supported by the metallic conductivity of metallic hydrides. However, the rare-earth trihydrides become semiconductors with electronic properties better explained by the ionic model[13]. The enthalpies of formation of metallic hydrides resemble those of the saline hydrides. However, results from nuclear gamma-ray resonance fluorescence, positron annihilation, magnetic susceptibility and nuclear magnetic resonance (NMR) as well as theoretical considerations are not conclusive, although they are interpreted in terms of one of these two models[10].

Energy-band calculations[14-22] can resolve these differences; e.g., in the metallic hydride structure the 1s orbitals of H and the metal-band states mix and hybridize, and in such monohydrides as PdH with H in O_h sites, the sp metal bands mix with the H bonding orbitals to form a modified band, lowered in energy. The energy states of this band are filled below the energy E_F in the metal (where E_F is the electrochemical potential of electrons in the metal), and the added electrons from hydrogen go into the empty metallic states above E_F, and so the hydrogen donates its electron to the metal, i.e., it has the appearance of the proton model. The stability of the hydride is determined by the extent to which the states in the modified band are empty, and by the amount the energy of the modified band is lowered. In dihydrides and trihydrides new low-lying states, associated with the H atoms, are formed, and the additional electrons occupy these states, i.e., as in the ionic model. The energies of the new bands are dependent on the H–H distance, which in turn is determined by the sites occupied by H in the metal sublattice and the size of the metal atom. The positions of the energy bands determine the formation and stability of these hydrides.

(G. G. LIBOWITZ, A. J. MAELAND)

1. R. Speiser, in *Metal Hydrides*, W. M. Muller, J. P. Blackledge, G. G. Libowitz, eds., Academic Press, New York, 1968, p. 51.
2. J. F. Lennard-Jones, *Trans. Faraday Soc., 28*, 333 (1932).
3. L. Schlapbach, A. Seiler, F. Stucki, H. C. Siegmann, *J. Less-Common Met., 73*, 145 (1980).
4. H. Shull, *J. Appl. Phys., 33*, 292 (1962).
5. R. S. Calder, W. Cochran, D. Griffiths, R. D. Lowde, *J. Phys. Chem. Solids, 23*, 621 (1962).
6. T. R. P. Gibb, Jr., *Prog. Inorg. Chem., 3*, 315 (1962).
7. G. G. Libowitz, *The Solid-State Chemistry of Binary Metal Hydrides*, W. A. Benjamin, New York, 1965; good introduction to the subject.
8. J. P. Blackledge, in *Metal Hydrides*, W. M. Muller, J. P. Blackledge, G. G. Libowitz, eds., Academic Press, New York, 1968, p. 1.
9. K. M. Mackay, *Hydrogen Compounds of the Metallic Elements*, E. & F. N. Spon, London, 1966.
10. G. G. Libowitz, in *MTP Internat. Rev. Sci.*, Vol. 10, *Solid-State Chemistry*, L. E. J. Roberts, ed., Butterworths, London, 1972, p. 79.
11. N. F. Mott, H. Jones, *The Theory of the Properties of Metals and Alloys*, Dover Publications, New York, 1936.
12. G. G. Libowitz, T. R. P. Gibb, Jr., *J. Phys. Chem., 60*, 510 (1956).
13. G. G. Libowitz, *Ber. Bunsenges. Phys. Chem., 76*, 837 (1972).
14. A. C. Switendick, *Solid-State Commun. 8*, 1463 (1970).
15. A. C. Switendick, *Int. J. Quant. Chem., 5*, 459 (1971).
16. A. C. Switendick, in *Hydrogen in Metals*, Vol. I, G. Alefeld, J. Völk, eds., Springer-Verlag, Berlin, 1978, p. 101; review article.

17. A. C. Switendick, *Z. Phys. Chem. N. F.*, (*Frankfurt-am-Main*) 117, 447 (1979); review article.
18. A. C. Switendick, *J. Less-Common Met.*, *101*, 191 (1984).
19. D. A. Papaconstantopoulos, in *Metal Hydrides*, G. Bambakidis, ed., Plenum Press, New York, 1981, p. 215; review article.
20. N. I. Kulikov, V. N. Borzunov, A. D. Zvonkov, *Phys. Stat. Sol.*, *B*, *86*, 83 (1978).
21. V. I. Savin, R. A. Andriveskii, V. I. Potorocha, V. Ya. Markin, *Inorg. Mater.* (*Engl. Transl.*), *14*, 1254 (1979).
22. M. Gupta, in *Metal Hydrides*, G. Bambakidis, ed., Plenum Press, New York, 1981, p. 255.

1.12.2. with Alkali Metals.

The reactions of H_2 with the alkali metals are discussed in §1.8.2. The hydrides thus formed are ionic and consist of alkali-metal cations and hydrogen anions, H^-. They resemble the alkali halides, because hydrogen can be considered as the first halogen.

(G. G. LIBOWITZ, A. J. MAELAND)

1.12.3. with Alkaline-Earth Metals.

Direct reaction of Ca, Sr, Ba and presumably Ra with H_2 forms hydrides that are ionic, like the alkali-metal hydrides. Direct reaction with Mg also forms a hydride which may be synthesized by other methods. Alkaline-earth hydrides are discussed in §1.8.3.

(G. G. LIBOWITZ, A. J. MAELAND)

1.12.4. with Group IIIA Transition Metals

The group IIIA transition metals include the lanthanides and actinides, as well as Sc and Y; however, the hydrides of Sc and Y are so similar to those of the lanthanides that they are included with the rare-earth hydrides.

(G. G. LIBOWITZ, A. J. MAELAND)

1.12.4.1. Involving Lanthanides.

The lanthanides form dihydrides and trihydrides with H_2. The lanthanide hydrides are divided into three groups based on their crystal structures.

The first group consists of hydrides of La, Ce, Pr and Nd. The dihydrides are cubic, with H atoms occupying the T_d sites of a fcc metal sublattice (fluorite structure). The trihydrides of the first group are also cubic and form solid solutions with the dihydrides. The third H per formula unit occupies an O_h site in the fcc metal sublattice. Hydrogen atoms occupy O_h sites before all the T_d sites are filled, so that at the stoichiometric composition, MH_2, there is an equal concentration of vacant T_d sites and occupied O_h sites[1].

The second subgroup, Sc, Y, Sm, Gd, Tb, Dy, Ho, Er, Tm and Lu, also forms fluorite-type dihydrides, but their trihydrides are hexagonal and the dihydride undergoes a phase change on formation of the trihydride. Hexagonal scandium trihydride, $ScH_{2.65}$, can be prepared[2] only at > 30 MPa H_2.

The third subgroup consists of Eu and Yb dihydrides, which have an orthorhombic structure and resemble the alkaline-earth hydrides, CaH_2, SrH_2 and BaH_2 (see §1.8.3.3).

The rare-earth hydrides are prepared by heating the metal ($> 200°C$) under H_2 to initiate the reaction; after equilibrium is reached the sample is slowly cooled to RT. Alternatively, the metal is first heated in vacuum, followed by admission of H_2 gas and slow cooling to RT. Finely divided metal can react with H_2 even at RT. Initial heating removes surface oxidation[3] and other surface impurities.

The hydrides are pyrophoric, particularly when finely divided.

Single crystals of Ce hydrides are grown[4] by forming the compounds from saturated melts (see Fig. 1). A boat containing molten Ce is placed under a thermal gradient such that the lowest T is above the peritectic (ca. 1010°C), and H_2 is slowly dissolved in the melt. When the Ce at the cool end of the boat attains the H content corresponding to the liquidus line between the two-phase (liquid and hydride) region, solid nonstoichiometric Ce hydride, $CeH_{2-\delta}$, forms (where δ is determined by the composition at the solidus line for the hydride). As further H_2 is added, molten Ce at a higher T solidifies into hydride, so the solid–liquid interface moves toward hotter portions of the melt until completely solidified. In this manner single crystals of Ce hydride may be grown. The rate of growth is controlled by the rate H_2 is introduced into the system.

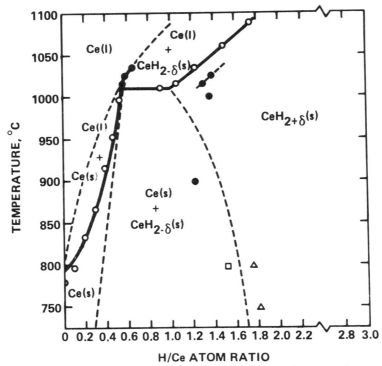

Figure 1. Partial phase diagram for the Ce–H system (From ref. 4).

Neodymium falls into the first group in forming a cubic trihydride. However, if the hydride is prepared at $< 350°C$, a hexagonal Nd trihydride is formed[5], similar to those in the second group. This trihydride is metastable, however; heating $> 350°C$ transforms it to the cubic phase, and subsequent cooling $< 350°C$ does not reform the hexagonal phase.

Although Eu forms only a dihydride, a higher hydride of Yb can be prepared[6] under high H_2 pressure. The Yb is slowly heated to ca. 200°C under 2.2 MPa H_2 and then slowly cooled to RT. The resulting $YbH_{2.55}$ has a cubic structure similar to that of the first-group trihydrides. Single crystals of Yb dihydride are prepared[7] by vapor transport, whereby YbH_2 is heated to 900°C for 5–10 h in an evacuated Mo tube. Crystals, 0.01–0.5 mm, deposit in the cooler (550–700°C) portions of the tube. Erbium dihydride may also be prepared[8] by vaporizing Er with high-energy laser pulses under 0.1 MPa of H_2. The ErH_2 condenses as chains of globular particles, ca. 10.0–12.5 nm.

Rare-earth hydrides are also prepared by methods other than direct reaction with H_2; e.g., rare-earth dihydrides can be formed from the metal with H_2O vapor at 100–150°C:

$$M + H_2O \rightarrow MH_2 + \tfrac{1}{2} O_2 \tag{a}$$

for Nd [9], Gd [10,11] and Dy [9,11]. Equation (a) is thermodynamically unfavorable, but the product may be stabilized by the presence of oxygen incorporated into the solid hydride phase[11].

Scandium dihydride may be prepared[12] by chemical-vapor deposition:

$$ScCl_{3(g)} + H_{2(g)} + 3 Na_{(g)} \rightarrow ScH_{2(s)} + 3 NaCl_{(g)} \tag{b}$$

or:

$$2 ScCl_{3(g)} + 2 H_{2(g)} + 3 Mg_{(g)} \rightarrow 2 ScH_{2(s)} + 3 MgCl_{2(g)} \tag{c}$$

Hydrogen alone is not a strong enough reducing agent, so Na or Mg must be used. The ScH_2 is deposited as brittle, black crystals sufficiently downstream from the chloride deposits that the hydride is free of chloride.

(G. G. LIBOWITZ, A. J. MAELAND)

1. G. G. Libowitz, A. J. Maeland, in *Handbook on the Physics and Chemistry of Rare Earths,* K. A. Gschneidner, L. Eyring, eds., North-Holland, Amsterdam, 1979, p. 299.
2. I. O. Bashkin, E. G. Ponyatovskii, M. E. Kost, *Phys. Status Solidi B, 87,* 369 (1978).
3. L. C. Beavis, R. S. Blewer, J. W. Guthrie, E. J. Nowak, W. G. Perkins, in *Proc. Hydrogen Economy, Miami Energy Conf.,* T. N. Veziroglu, ed., Univ. of Miami, Coral Gables, FL, 1974, p. S4-38.
4. G. G. Libowitz, J. G. Pack, in *Proc. Internat. Conf. Crystal Growth,* H. S. Peiser, ed., Pergamon Press, Oxford, 1967, p. 129; *Inorg. Synth., 4,* 184 (1973).
5. M. H. Mintz, Z. Hadari, M. Bixon, *J. Less-Common Met., 37,* 331 (1974).
6. J. C. Warf, K. Hardcastle, *J. Am. Chem. Soc., 83,* 2206 (1961); *Inorg. Chem., 5,* 1728 (1966).
7. J. M. Haschke, M. R. Clark, *High-Temp. Sci., 7,* 152 (1975).
8. H. Oesterreicher, H. Bittner, B. Kothari, *J. Solid-State Chem., 26,* 97 (1978).
9. J. Dexpert-Ghys, C. Loier, Ch. H. Blanchetais, P. E. Caro, *J. Less-Common Met., 41,* 105 (1975).
10. H. Oesterreicher, H. Bittner, K. Shuler, *J. Solid-State Chem., 29,* 191 (1979).
11. H. K. Smith, A. G. Moldovan, R. S. Craig, W. E. Wallace, S. G. Sankar, *J. Solid-State Chem., 32,* 239 (1980).
12. T. Kobayashi, H. Takei, *J. Cryst. Growth, 45,* 29 (1978).

1.12.4.2. Involving Actinides.

Actinides react directly with H_2 to form hydrides of different stoichiometries. Thorium forms a tetragonal dihydride, related to the fluorite structure, and a higher hydride, Th_4H_{15}, which has a complex bcc structure. The trihydrides of U and Pa have the bcc β-W structure, and U also forms a low-T trihydride with a different bcc structure. Hydrides similar to the third group of the lanthanides (see §1.12.4.1), i.e., an fcc dihydride and a hexagonal trihydride, are formed by Np through Bk.

The hydrides are prepared like the lanthanide hydrides, i.e., by heating the metal to react with H_2 gas followed by slow cooling. Some finely powdered metals react with hydrogen at RT.

The direct synthesis of AcH_2 is unknown. It is present only in samples of Ac metal prepared by reducing[1] $AcCl_3$ with hydrogen, which may have come from residual hydrogen in the reaction of $Ac(OH)_3$ and NH_4Cl to form the $AcCl_3$.

Structural changes and volume expansion on forming the hydride from the metal produce powdered hydride. However, Th_4H_{15} is prepared in massive form by preparing the hydride[2] at 850°C and 75 MPa.

The low-T form of UH_3 can be prepared by cathodic charging of U metal in a solution of $HClO_4$ or NaOH below 20°C, in addition to direct reaction with H_2 gas at -40°C.

(G. G. LIBOWITZ, A. J. MAELAND)

1. J. D. Farr, A. L. Giorgi, M. G. Bowman, R. K. Money, *J. Inorg. Nucl. Chem., 18*, 42 (1961).
2. C. B. Satterthwaite, D. T. Peterson, *J. Less-Common Met., 26*, 361 (1972).
3. R. Caillat, H. Coriou, P. Perio, *C. R. Hebd. Seances Acad. Sci., 237*, 812 (1953).

1.12.5. with Group IVA Transition Metals

Nonstoichiometric hydrides that have cubic CaF_2-type structures or a tetragonally distorted version form when Ti, Zr and Hf react with H_2[1]. The hydride phases have wide composition ranges that depend on T. The limiting composition is MH_2.

(G. G. LIBOWITZ, A. J. MAELAND)

1. G. G. Libowitz, *The Solid-State Chemistry of Binary Metal Hydrides*, W. A. Benjamin, New York, 1965.

1.12.5.1. Involving Titanium.

Titanium hydride is prepared by heating Ti in H_2 at atmospheric P. The reaction is slow below 300°C but increases[1] with T; optimum is 400–500°C. Because surface impurities impede the reaction, an initial outgassing step to 1000°C in vacuum is helpful in dissolving the surface oxide or other impurities. After equilibration with H_2 at the reaction T, the sample is cooled slowly to RT in H_2 to achieve maximum hydrogen content, TiH_2.

The hydride is also prepared electrochemically[2,3] by cathodic charging of Ti in 1 N H_2SO_4. Hydride formation, however, is limited to a thin surface layer[2,3] owing to slow hydrogen diffusion.

(G. G. LIBOWITZ, A. J. MAELAND)

1. E. A. Gulbransen, K. F. Andrew, *Trans. A.I.M.E.*, *185*, 741 (1949).
2. P. Millenbach, M. Givon, *J. Less-Common Met.*, *87*, 179 (1982).
3. E. Brauer, R. Gruner, F. Rauch, *Ber. Bunsenges. Phys. Chem.*, *87*, 341 (1983).

1.12.5.2. Involving Zirconium.

The preparation of ZrH_2 follows the procedure as in §1.12.5.1 for TiH_2. In the $Zr-H_2$ system the oxide layer reduces[1] the rate markedly below 250°C. Sintering Ni to Zr catalyzes H_2 absorption and compensates for the effect of the oxide layer, improving the rate below 250°C; this may be important for getter applications[1].

Cathodic charging[2] also is used to prepare ZrH_2. Hydride formation again is limited to a thin surface layer as in the formation of TiH_2 by the same method.

(G. G. LIBOWITZ, A. J. MAELAND)

1. G. Kuus, W. Martens, *J. Less-Common Met.*, *75*, 111 (1980).
2. E. Brauer, R. Gruner, F. Rauch, *Ber. Bunsenges. Phys. Chem.*, *87*, 341 (1983).

1.12.5.3. Involving Hafnium.

The procedure for HfH_2 is the same as for TiH_2 (see §1.12.5.1). The optimum T, however, is higher than for TiH_2 or ZrH_2. To obtain reasonable rates, T should be[1] at least $\geq 500°C$.

The hydride also has been made in thin surface layers by cathodic charging[2].

(G. G. LIBOWITZ, A. J. MAELAND)

1. R. K. Edwards, E. Veleckis, *J. Phys. Chem.*, *66*, 1657 (1962).
2. B. Streb, E. Brauer, *Z. Metallkd.*, *74*, 680 (1983).

1.12.6. with Group VA Transition Metals.

The nonstoichiometric monohydrides formed by V, Nb and Ta have structures that are determined by T and the H content[1,2]. Order–disorder transitions involving the H atoms lead to structural complexity; in the ordered phases the bcc metal lattice distorts to tetragonal, orthorhombic or monoclinic. In addition, V and Nb, but not Ta, at > 0.1 MPa H_2 form dihydrides that have the fluorite structure. Hence, the phase diagrams of these elements with H_2 are complex.

The group VA metals do not react with H_2 at RT because an oxide layer on the surface of the metal[3-7] prevents or inhibits the catalytic dissociation of H_2 and the subsequent solution in the metal. Even in the absence of an oxide layer, however, there is an intrinsic barrier to H_2 absorption because of a strongly bound surface state of hydrogen[8,9]. Activation procedures consequently are required before the reaction with H_2 proceeds at an appreciable rate, e.g., heating to 300–500°C and utilizing absorption–desorption cycles or heating to high T in ultrahigh vacuum to dissolve[7] the surface film before H_2 is admitted in situ.

The rate of H_2 absorption by the group VA metals is increased by thin overlayers[8-12] of Pd or Pt; Ni and Fe are less effective, and Cu and Ag have no appreciable effect, at least not with[13] Ta. The solution of small amounts of a second metal yields alloys that react rapidly with H_2 at RT without activation[14]. The second metal should have a radius $\geq 5\%$ smaller than that of the solvent group VA metal, e.g., Co, Fe and Ni.

The reaction of Nb with H_2 occurs readily with $LaNi_5$, which acts as a catalyst[15]; e.g., Nb powder mixed with activated $LaNi_5$ (50 : 1) absorbs H_2 to give[15] NbH_2 at $\geq 22°C$.

Treating the metals or the monohydrides with 10% HF yields[16,17] NbH_2 and VH_2. Atomic hydrogen is generated from the metal with HF and subsequently is absorbed by the metal or the monohydride. Atomic H generated by other sources, such as the thermal dissociation of H_2 on heated W filaments[18], also may be used, but the hydrogen concentration is difficult to regulate.

Electrochemical charging also yields hydrides of the group VA metals[19], e.g., NbH_2 is formed[19] when Nb metal is made the cathode in an electrolyte solution such as H_3PO_4 or H_2SO_4; however, chemical dissolution of the metal occurs and the hydrogen concentration in the sample is difficult to control.

(G. G. LIBOWITZ, A. J. MAELAND)

1. H. Asano, M. Hirabayashi, Z. Phys. Chem. N.F. (Frankfurt-am-Main), 114, 1 (1979).
2. M. Hirabayashi, H. Asano, in Metal Hydrides, G. Bambakidis, ed., Plenum Press, New York, 1981, p. 53.
3. S. M. Ko, L. D. Schmidt, Surface Sci., 42, 508 (1974).
4. L. Johnson, M. J. Dresser, E. E. Donaldson, J. Vac. Sci. Technol., 9, 857 (1972).
5. S. M. Ko, L. D. Schmidt, Surface Sci., 47, 557 (1975).
6. A. J. Pryde, C. G. Titcomb, Trans. Faraday Soc., 65, 2758 (1969).
7. E. Fromm, H. Uchida, J. Less-Common Met., 66, 77 (1979).
8. M. A Pick, M. G. Greene, J. Less-Common Met., 73, 89 (1980).
9. M. A. Pick, Phys. Rev., B, 24, 4287 (1981).
10. N. Boes, H. Züchner, Z. Naturforsch, Teil A, 31, 754 (1976).
11. M. A. Pick, J. W. Davenport, M. Strongin, G. J. Dienes, Phys. Rev. Lett., 43, 286 (1979).
12. T. Schober, A. Carl, J. Less-Common Met., 63, P53 (1979).
13. K. Nakamura, H. Uchida, E. Fromm, J. Less-Common Met., 80, P19 (1981).
14. A. J. Maeland, G. G. Libowitz, J. F. Lynch, G. Rak, J. Less-Common Met., 104, 133 (1984).
15. D. H. W. Carstens, J. D. Farr, J. Inorg. Nucl. Chem., 36, 461 (1974).
16. G. Brauer, H. Müller, Angew. Chem., 70, 53 (1958).
17. A. J. Maeland, T. R. P. Gibb Jr., D. P. Schumacher, J. Am. Chem. Soc., 83, 3728 (1961).
18. W. A. Oates, T. B. Flanagan, Can. J. Chem., 53, 694 (1975).
19. G. Brauer, H. Müller, J. Inorg. Nucl. Chem., 17, 102 (1961).

1.12.7. with Groups VIA, VIIA and VIII Transition Metals

Except for Pd, the groups VIA, VIIA and VIII metals do not form hydrides under ordinary conditions of T and H_2 pressure. If, however, the thermodynamic activity of hydrogen is increased by going to high P, these metals do react. The required pressures are > 10 MPa and special high-P equipment is therefore necessary for the synthesis of these hydrides.

(G.G. LIBOWITZ, A.J. MAELAND)

1.12.7.1. Involving Chromium and Molybdenum.

Hydrides of the bcc group VIA elements Cr and Mo, but not W, are prepared by high-P techniques[1-6]. Chromium hydride is synthesized[1] directly from Cr with H_2 by keeping a thin Cr foil at 150°C under 2200 MPa for 4 h. The H/Cr atom ratio is 0.93 and the metal lattice structure hcp[1,3,5,6]. The hydride, therefore, is equivalent to the hexagonal hydride phase prepared earlier[8] by electrodeposition of Cr from H_2SO_4. A dihydride phase (fcc) also results from this procedure, but this phase is not prepared by the direct reaction. The formation pressure (plateau pressure observed in absorption) is[3,4] 1.74–1.82 GPa at 150°C, whereas the decomposition pressure, i.e., plateau pressure observed in desorption, at the same T is 0.32 GPa. This hysteresis is observed frequently in metal–hydrogen systems.

Nearly stoichiometric MoH is prepared[2] at 350°C and high H_2 pressure; the formation pressure[2,6] at this T is 23.5 GPa. X-Ray diffraction analysis[2], performed at $-145°C$ because of the instability of the hydride, shows an hcp Mo lattice. The volume expansion on formation of the hydride phase ($\Delta V_{Mo} = V_{MoH} - V_{Mo}$) is approximately the same as in the Cr system, 20% vs. 19%, respectively[2]. An fcc dideuteride, MoD_2, forms in thin films during bombardment of Mo with D_2^+ ions[7]. This phase is the counterpart of the fcc CrH_2 phase prepared electrochemically[8].

(G.G. LIBOWITZ, A.J. MAELAND)

1. B. Baranowski, K. Bojarski, *Rocz. Chem.*, 46, 525 (1972); *Chem. Abstr.*, 77, 13,347 (1972).
2. I. T. Belash, V. E. Antonov, E. G. Ponyatovskii, *Proc. Acad. Sci. USSR (Engl. Transl.)*, 235, 665 (1977).
3. B. Baranowski, *Ber. Bunsenges. Phys. Chem.*, 76, 714 (1972); review.
4. B. Baranowski, in *Hydrogen in Metals*, Vol. II, G. Alefeld, J. Völk, eds., Springer-Verlag, Berlin, 1978. p. 157; review.
5. B. Baranowski, in *Metal Hydrides*, G. Bambakidis, ed., Plenum Press, New York, 1981, p. 193; review.
6. E. G. Ponyatovskii, V. E. Antonov, I. T. Belash, *Inorg. Mater. (Engl. Transl.)*, 14, 1227 (1978); review.
7. V. Kh. Alimov, A. E. Gorodetskii, A. P. Zakharov, V. M. Sharapov, *Proc. Acad. Sci. USSR (Engl. Transl.)*, 241, 595 (1978).
8. C. A. Snavely, *Trans. Electrochem. Soc.*, 92, 552 (1947).

1.12.7.2. Involving Manganese and Technetium.

Treating Mn with H_2 at 2.2 GPa and RT gives no evidence of hydride formation[1]. However, at 300°C and 1.4 GPa, MnH forms[1]. The need for the higher T arises from kinetic barriers (surface contamination), which must be overcome before hydrogen absorption takes place[1,2]. Hydride formation is confirmed at[3,4] 1.8 GPa H_2 and 350°C. These pressures are higher than the absorption plateau pressures[2]; e.g., at 448°C, the plateau pressure is only 0.56 GPa and at 729°C it is 1.04 GPa. The maximum hydrogen content, H/Mn, is ca. 0.8, and hydride formation changes the metal-lattice structure from cubic to hexagonal[1-5]. Despite the lower absorption pressures observed in the formation of MnH as compared to CrH and the similarity in their structures, Mn does not form a hydride phase during cathodic codeposition or cathodic saturation with H_2, whereas Cr does[2].

The preparation of TcH occurs[6] at 300°C and 1.9 GPa H_2 with the composition $TcH_{0.73}$. Both $TcH_{0.73}$ and the hydrogen-free metal have an hcp metal lattice, which makes it difficult, without extensive x-ray work showing the dependence of lattice parameters on composition, to determine whether $TcH_{0.73}$ is a hydride phase or just a solid solution of hydrogen in Tc. However, from the behavior of the electrical resistance of TcH as a function of composition, it can be deducted that a hydride phase is formed[6] below 300°C.

There are no reports of hydride formation in the $Re-H_2$ system.

(G.G. LIBOWITZ, A.J. MAELAND)

1. M. Krukowski, B. Baranowski, *Rocz. Chem.*, 49, 1183 (1975); *Chem. Abstr.*, 83, 150,966 (1975).
2. B. Baranowski, in *Hydrogen in Metals*, Vol II, G. Alefeld, J. Völk, eds., Springer-Verlag, Berlin, 1978, p. 157; review.
3. E. G. Ponyatovskii, I. T. Belash, *Proc. Acad. Sci. USSR (Engl. Transl.)*, 224, 570 (1975).
4. E. G. Ponyatovskii, V. E. Antonov, I. T. Belash, *Inorg. Mater. (Engl. Transl.)*, 14, 1227 (1978); review.
5. B. Baranowski, *Z. Phys. Chem. N. F. (Frankfurt-am-Main)*, 114, 59 (1979); review.
6. V. I. Spitsyn, E. G. Ponyatovskii, V. E. Antonov, I. T. Belash, O. A. Balakhovskii, *Proc. Acad. Sci. USSR (Engl. Trans.)*, 247, 723 (1979).

1.12.7.3. Involving Nickel, Rhodium and Palladium.

Nearly stoichiometric NiH is prepared[1] at RT by exposing Ni foil to > 1.0 GPa H_2. The formation pressure[2,3] at RT is actually ca. 0.6 GPa. The decomposition pressure of the hydride at RT is, in contrast, 0.34 GPa, indicating considerable hysteresis in this system[2-7]. During hydride formation the Ni lattice remains fcc but undergoes[8] a sudden 18% increase in volume[8].

Nickel hydride also can be prepared[9] by cathodic charging of Ni in aq H_2SO_4 in the presence of a promoter (e.g., As); this delays the recombination of H atoms into molecules at the Ni surface, enhancing the concentration of active hydrogen[2].

Rhodium hydride is prepared[10] at 250°C and > 4.0 GPa H_2 (the formation pressure)[6,8]. Maximum concentration of hydrogen in the hydride is $H/Rh = 0.65$ at 6.0 GPa. The hydride has an fcc metal lattice like Rh itself, but with a lattice parameter[10] ca. 6% larger than that of pure Rh.

Palladium hydride can be prepared without high H_2 pressures; e.g., at 30°C the equilibrium absorption pressure[11] is ca. 2.40 kPa; at 160°C, 203 kPa and even at 250°C it is only 1.11×10^3 kPa. The reaction is slow at RT, however, unless activated (Pd black) material is used[11]. Heating to 100–200°C speeds up the reaction when H_2 pressures greater than the dissociation pressure of hydride are maintained[11]. The hydride is formed by cathodic charging in aqueous electrolytes as well as by immersing Pd in aq H_2SO_4 and bubbling H_2 gas through[11-13]. The composition of the hydride at atmospheric pressure of H_2 and RT is[11,12] ca. $PdH_{0.7}$. The metal lattice in the hydride has the same structure as in the metal phase, i.e., fcc, but expanded by 10.8%. Near stoichiometric PdH can be prepared at RT by increasing the H_2 pressure[3] to 1.2 GPa or by electrolytic charging[11,12].

Hydrogen is soluble in Co at high pressures[8,14]; at 6.5 GPa and 225°C the equilibrium composition is $CoH_{0.51}$. However, the solubility varies smoothly with pressure, indicating a continuous solid solution rather than hydride formation[8,14].

(G.G. LIBOWITZ, A.J. MAELAND)

1. B. Baranowski, R. Wisniewski, *Bull. Acad. Polon. Sci.*, 14, 273 (1966); *Chem. Abstr.*, 65, 14,810 (1966).
2. B. Baranowski, *Ber. Bunsenger. Phys. Chem.*, 76, 714 (1972); review.
3. B. Baranowski, in *Hydrogen in Metals*, Vol. II, G. Alefeld, E. Völk, eds., Springer-Verlag, Berlin, 1978, p. 157; review.
4. B. Baranowski, in *Metal Hydrides*, G. Bambakidis, ed., Plenum Press, New York, 1981, p. 193; review.
5. T. Skoskiewicz, *Phys. Status Solidi, A*, 6, 29 (1971).
6. B. Baranowski, *Z. Phys. Chem. N.F. (Frankfurt-am-Main)*, 114, 59 (1979); review.
7. B. Baranowski, K. Bochenska, *Rocz. Chem.*, 38, 1419 (1964); *Chem. Abstr.*, 62, 9825 (1965).

8. E. G. Ponyatovskii, V. E. Antonov, I. T. Belash, *Inorg. Mater. (Engl. Trans.)*, *14*, 1227 (1978); review.
9. B. Baranowski, M. Smialowski, *J. Phys. Chem. Solids*, *12*, 206 (1959).
10. V. E. Antonov, I. T. Belash, V. F. Degtyareva, E. G. Ponyatovskii, *Proc. Acad. Sci. USSR (Engl. Transl.)*, *239*, 222 (1978).
11. F. A. Lewis, *The Palladium-Hydrogen System*, Academic Press, London, 1967.
12. B. Siegel, G. G Libowitz, in *Metal Hydrides*, W. M. Mueller, J. P. Blackledge, G. G. Libowitz, eds., Academic Press, New York, 1968, p. 545.
13. T. B. Flanagan, F. A. Lewis, *J. Phys. Chem.*, *29*, 1417 (1958).
14. I. T. Belash, V. E. Antonov, E. G. Ponyatovskii, *Proc. Acad. Sci. USSR (Engl. Transl.)*, *235*, 128 (1977).

1.12.8. to Form Ternary Hydrides

1.12.8.1. from Intermetallics

The formation of hydrides of intermetallic compounds by direct reaction with H_2 is carried out at lower T than formation of binary hydrides, and more rapidly. However, disproportionation of the intermetallic compound may occur at elevated T.

The known hydrides of intermetallics can be classified in a few groups, and the discussion in this section is in terms of those groups.

Most binary intermetallic compounds that form hydrides contain one hydride-forming element (e.g., group IVA, VA, VIA, rare earth or actinide) and one which does not form hydrides under normal conditions (but may under ultra-high P, e.g., group VIII); see §1.12.7. Because bonds to the hydride-forming elements are stronger, hydrogen atoms are found in lattice sites in which the hydride-forming elements are nearest neighbors.

(G. G. LIBOWITZ, A. J. MAELAND)

1.12.8.1.1. Giving AM₅ Compounds.

The AM_5 intermetallic compounds have the hexagonal $CaCu_5$ structure. For the compounds that form hydrides, A is usually a rare-earth element (or Ca or Th) and M is either Ni or Co (but may also be Fe, Pt or Cu). These hydrides are listed in Table 1 along with the minimum H_2 pressure needed to form the hydride and the approximate hydrogen content of the hydride, x.

Because of hysteresis effects[16] the pressure for formation of the hydride may be higher than the dissociation pressure shown in Fig. 1, §1.12.1. The values of x shown in Table 1 are the approximate hydrogen content of the hydride when it is formed initially, but they usually increase with H_2 pressure. At high pressure (150 MPa), the maximum possible hydrogen content in these intermetallic compounds is nine H atoms per formula unit[17].

The pressure necessary to form the hydride may be changed by partial substitution of one of the components by other metals; e.g., the replacement of 20% of the La in $LaNi_5$ with Nd doubles the formation pressure[3]; the pressure may also be raised[3] by replacing some of the Ni and Pd. Conversely, substitutions, such as Al for Ni, decrease the hydride formation pressure[18].

The intermetallic compound is usually prepared by arc melting the elements in the correct proportions under Ar. To insure homogeneity, the alloy is remelted several times. Samples also are prepared by induction melting.

TABLE 1. HYDRIDES OF AM$_5$ INTERMETALLICS

Intermetallic	x[a]	P (MPa)[b]	Ref.
YNi$_5$	1	30	1
	3.5	100	
YCo$_5$	2.8	3	2
LaNi$_5$	6.7	0.13	3
LaCo$_5$	3.2	0.004	4
	4.4	0.02	
LaCu$_5$	2.5	0.024	5
LaPt$_5$	ca. 1.5	20	1
	ca. 2.5	62	
	4	105	
CeNi$_5$	6	4.8	7
CeCo$_5$	2.7	0.16	8
CeFe$_5$	ca. 4	0.075	9
PrNi$_5$	ca. 6	0.8	6
PrCo$_5$	ca. 3		
		ca. 0.07	10
NdNi$_5$	6	1.3	6
NdCo$_5$	3.5	0.08	11
SmNi$_5$	ca. 4	3	6
SmCo$_5$	2.6	0.5	12
EuNi$_5$	>2		13
GdNi$_5$	3	1.2	6
GdCo$_5$	2.5	ca. 0.3	11
YbNi$_5$	2.9	ca. 1.2	6
ThCo$_5$	4.6	ca. 5	14, 2
CaNi$_5$	1	0.0025	15
	4.5	0.056	
	6.5	2.7	

[a] x = number of H atoms per formula unit.
[b] P = maximum H$_2$ pressure needed to form hydride.

These intermetallic compounds react with H$_2$ at RT, provided that the pressure is high enough[19], but there is often an induction period ranging from seconds to days depending on previous treatment of the alloy and time of exposure to air[20]. Freshly prepared samples not exposed to air usually react in seconds because of the catalytic action[21,22] of Ni and Co on Fe. The intermetallic compound is oxidized at its surface to form the rare-earth oxide (e.g., La$_2$O$_3$) and free metallic Ni (or Co or Fe), which acts as a catalyst to dissociate H$_2$. Impurity gases, such as CO, O$_2$ and H$_2$O, decrease the rates of hydride formation and can poison the alloy for reaction[23-25] with H$_2$.

The formation of most hydrides of intermetallic compounds is metastable with respect to disproportionation of the intermetallic compound; e.g., the free energy of:

$$\text{LaNi}_5 + \text{H}_2 \rightleftharpoons \text{LaH}_2 + 5\text{ Ni} \qquad \text{(a)}$$

is more negative than that for:

$$\text{LaNi}_5 + 3\text{ H}_2 \rightleftharpoons \text{LaNi}_5\text{H}_6 \qquad \text{(b)}$$

However, the hydride of the intermetallic compound forms because it is kinetically favored. Reaction (a) requires rearrangement of metal atoms, which is unlikely to occur at low T, whereas reaction (b) involves little motion of metal atoms. However, at elevated T, reaction (a) is more likely to occur; e.g., $LaNi_5H_6$ degrades[26] at 300°C, and $CaNi_5H_x$ disproportionates[15] at even lower T.

The H atoms in the AM_5 compounds occupy two types of T_d sites in the lattice; in one hydrogen is coordinated to two A atoms and two M atoms, and in the second to one A and three M atoms[27]. However, the bonds must occur between the A and the H atoms because the M atoms do not form hydrides.

(G. G. LIBOWITZ, A. J. MAELAND)

1. T. Takeshita, K. A. Gschneidner, J. F. Lakner, *J. Less-Common Met., 78*, 43 (1981).
2. T. Takeshita, W. E. Wallace, R. S. Craig, *Inorg. Chem., 13*, 2282 (1974).
3. H. H. VanMal, K. H. J. Buschow, A. R. Miedema, *J. Less-Common Met., 35*, 65 (1974).
4. F. A. Kuijpers, B. O. Loopstra, *J. Phys. Chem. Solids, 35*, 301 (1974).
5. J. Shinar, D. Shaltiel, D. Davidov, A. Grayevsky, *J. Less-Common Met., 60*, 209 (1978).
6. J. L. Anderson, T. C. Wallace, A. L. Bowman, C. L. Radosevich, M. L. Courtney, USAEC Report No. LA-5320-MS (Los Alamos Scientific Laboratory), July, 1973; *Chem. Abstr., 80,* 125,151 (1974).
7. C. E. Lundin, F. E. Lynch, AFOSR Report No. F44620-74-C-002 (Denver Research Institute), 1976.
8. F. A. Kuijpers, *J. Less-Common Met., 27*, 27 (1972).
9. C. E. Lundin, F. E. Lynch, AFOSR Report No. TR-75-1482 (Denver Research Institute), 1975.
10. J. Clinton, H. Bittner, H. Oesterreicher, *J. Less-Common Met., 41*, 187 (1975).
11. F. A. Kuijpers, *Ber. Bunsenges. Phys. Chem., 76*, 1220 (1972).
12. J. S. Raichlen, R. H. Doremus, *J. Appl. Phys., 42*, 3166 (1971).
13. F. W. Oliver, K. W. West, R. L. Cohen, K. H. J. Buschow, *J. Phys., F: Met. Phys., 8*, 701 (1978).
14. K. H. J. Buschow, H. H. Van Mal, A. R. Miedema, *J. Less-Common Met., 42*, 163 (1975).
15. G. D. Sandrock, J. J. Murray, M. L. Post, J. B. Taylor, *Mater. Res. Bull., 17*, 887 (1982).
16. G. G. Libowitz, *The Solid State Chemistry of Binary Metal Hydrides*, W. A. Benjamin, New York, 1965, p. 83.
17. J. F. Lakner, F. S. Uribe, S. A. Steward, *J. Less-Common Met., 72*, 87 (1980).
18. M. H. Mendelsohn, D. M. Gruen, A. E. Dwight, *Nature (London), 269*, 45 (1977).
19. J. H. N. van Vucht, F. A. Kuijpers, H. C. A. M. Bruning, *Philips Res. Rep., 25*, 133 (1970); *Chem. Abstr., 73,* 28,184 (1970).
20. H. H. Van Mal, Ph.D. Thesis, Technical Univ., Delft, Neth., 1976.
21. H. C. Siegmann, L. Schlapbach, C. R. Brundle, *Phys. Rev. Lett., 40,* 972 (1978).
22. W. E. Wallace, R. F. Karlicek, H. Imamura, *J. Phys. Chem., 83,* 1708 (1979).
23. G. D. Sandrock, P. D. Goodell, *J. Less-Common Met., 73*, 161 (1980).
24. P. D. Goodell, *J. Less-Common Met., 89*, 45 (1983).
25. F. G. Eisenberg, P. D. Goodell, *J. Less-Common Met., 89*, 55 (1983).
26. R. L. Cohen, K. W. West, J. H. Wernick, *J. Less-Common Met., 73*, 273 (1980).
27. P. Fischer, A. Furrer, G. Busch, L. Schlapbach, *Helv. Phys. Acta, 50*, 421 (1977).

1.12.8.1.2. Giving AB_2 Compounds.

The AB_2 intermetallics discussed here have either a cubic $MgCu_2$-type or a hexagonal $MgZn_2$-type structure where A has a metallic radius larger than that of B, ideally by a factor of 1.225. The radius ratios of A to B vary from ca. 1.05 to 1.68, but the radii contract or expand on compound formation to approach the ideal value[1] of 1.225.

The unit cell of the cubic structure has 136 T_d interstices, or 17 per formula unit. Many of these interstices are available for occupancy by hydrogen. However, the hydrogen content in these hydrides is generally under seven atoms per formula unit, as seen in Table 1. The maximum number of H atoms per formula unit is given by the

TABLE 1. HYDRIDES OF AB_2 INTERMETALLICS

Compound	x^a	P (kPa)^b	Ref.
$CaAl_2$	2 (0.1)	ND	2
$ScMn_2$	3.8 (3)	ca. 10^{-3}	3, 4
$ScFe_2$	3.1 (3)	6×10^{-1}	3, 4
$ScCo_2$	2.2 (0.1)	ND	5
$ScNi_2$	2.0 (0.1)	ND	5
$TiBe_2$	ca. 3 (>15)	ND	6
$TiCr_2$	2 (7)	ND	7
$TiMn_{1.5}$	2.6 (5)	7×10^2	8
YMg_2	3.2 (3)	ND	3
YMn_2	3.4 (0.1–5)	ND	9
YFe_2	4.3 (1.1)	6×10^1	10
YCo_2	4.2 (0.1–5)	5	9, 11
YNi_2	3.6 (0.1–5)	ND	9
ZrV_2	5.3 (1.2)	ca. 10^{-2}	4, 11
$ZrCr_2$	4 (6.1)	ca. 2×10^{-1}	11
$ZrMn_2$	3.6 (0.8)	10^{-1}	11
$ZrTa_2$	2.4 (NR)	ND	12
$LaMg_2$	6.4 (3)	ND	2
$LaNi_2$	4.6 (3–5)	ND	13
$LaRu_2$	4.5 (0.34)	6×10^{-4}	4, 11
$LaRh_2$	4.9 (6)	5×10^2 ($x \approx 1$)	11, 14
		1 ($x \approx 2$)	
$CeMg_2$	6.2 (3)	ND	3
$CeFe_2$	4 (NR)	ND	15
$CeCo_2$	4.1 (1)	ND	16
$CeNi_2$	4 (4)	ND	17
$CeRu_2$	5.2 (1)	ND	11
$PrCo_2$	4.0 (6)	ND	18
$PrNi_2$	4.4 (6)	ND	18
$NdMg_2$	4 (2.8)	ND	2
$NdFe_2$	3.0 (6)	ND	18
$NdNi_2$	3.7 (6)	ND	18
$NdRu_2$	5.5 (NR)	4×10^{-3}	4
$SmMg_2$	3 (0.2)	ND	2
$SmMn_2$	4.2 (3)	ND	3
$SmFe_2$	2.9 (6)	ND	18
$SmCo_2$	3.2 (3)	ND	3
$SmNi_2$	3.8 (6)	ND	18
$SmRu_2$	4.6 (5)	3×10^{-2}	4, 19
$EuNi_2$	>2 (17)	ND	20
$EuRh_2$	3 (NR)	ND	21
$GdMn_2$	3 (1.5–2)	6×10^{-7}	14
$GdFe_2$	4.7 (1.1)	20	10
		2.3×10^2 ($x \approx 2$)	
$GdCo_2$	4.5 (6.1)	2×10^1 ($x \approx 3$)	11, 14
		6×10^1 ($x \approx 3.5$)	
$GdNi_2$	4.35 (6)	3.5×10^{-7}	14
$GdRu_2$	4.5 (0.34)	ND	11, 22
$GdRh_2$	4.9 (6)	ND	11

1.12.8. to Form Ternary Hydrides
1.12.8.1. from Intermetallics
1.12.8.1.2. Giving AB₂ Compounds.

253

TABLE 1. HYDRIDES OF AB_2 INTERMETALLICS (CONTINUED)

Compound	x^a	P (kPa)^b	Ref.
$TbCo_2$	3.2 (6)	ND	18
$TbNi_2$	3.5 (6)	ND	18
$DyMn_2$	3.4 (15)	ND	23
$DyFe_2$	7.7 (14)	5×10^{-4} ($x = 1.5$)	24, 25
		2×10^{-3} ($x = 2.0$)	
		3 ($x = 3.4$)	
		100 ($x = 4.3$)	
		10^3 ($x = 5$)	
$DyCo_2$	3.8 (6)	ND	18
$DyNi_2$	3.6 (6)	ND	18
$DyRu_2$	3.1 (1)	ND	26
$HoFe_2$	4.5 (3.6)	ND	27
$HoCo_2$	3.8 (6)	ND	18
$HoNi_2$	3.8 (6)	ND	18
$HoRu_2$	4.2 (NR)	2.6×10^{-1}	4, 28
$ErMn_2$	4.6 (6.5)	1.5	29
$ErFe_2$	4.1 (13)	5×10^{-3} ($x = 1.4$)	25, 30
		5×10^{-1} ($x \approx 2.3$)	
		2 ($x \approx 3.2$)	
		700 ($x \approx 4.0$)	
$ErCo_2$	3.65 (3.6)	ND	31
$ErNi_2$	3.6 (6)	ND	18
$ErRu_2$	NR	8.5×10^1	4
$TmFe_2$	4.6 (3.6)	6.8×10^{-3} ($x = 1.6$)	27, 32
		5.3 ($x = 3.4$)	
		$> 1.3 \times 10^3$ ($x = 4.2$)	
$YbNi_2$	3.1 (1)	ND	26
$LuFe_2$	4 (NR)	ND	33
HfV_2	4.5 (3.4)	5×10^{-5}	4, 34
$ThRu_2$	5 (5.9)	ND	11

^{a} NR, not reported.
^{b} ND, not determined.

value of x in column 2. The numbers in parenthesis represent the H_2 pressure (in MPa) at which the x values are determined.

The third column gives the dissociation pressures of the hydride phase at RT. These are the minimum pressures required to form the hydride. For some cases (e.g., $GdCo_2$) there is more than one hydride phase. These hydrides are usually more stable than the hydrides of the AM_5 compounds (see §1.12.8.1.1); i.e., the dissociation pressures are much lower but, as with the AM_5 compounds, the stability and maximum hydrogen content can be modified by changing the stoichiometry or by alloying with other elements. For example, $TiMn_2$ only absorbs ca. 0.1 H atoms per formula unit at the stoichiometric composition; however, as the composition becomes Mn deficient, the hydrogen content increases to 2.6 H atoms per formula unit[8].

The effect of additional alloying elements can be illustrated with $ZrMn_2$, the hydride of which has a dissociation pressure of 0.1 kPa (see Table 1). The addition of 0.8

mol Fe increases[35] the dissociation pressure to 30 kPa, whereas 0.8 mol Co raises it to 400 kPa.

The intermetallics themselves are synthesized by melting together the elements in the correct proportions either by arc melting under an inert atmosphere (usually Ar gas) or by induction melting. The alloys are remelted several times to make them more homogeneous. To insure homogeneity the samples are then annealed at 600–950°C for 24 h to 2 weeks.

Although some intermetallics react directly with H_2 with no activation, some activation procedure is required for most; e.g., for the rare-earth Ni, Co and Fe compounds[4] the alloy is evacuated and heated at 70°C for 1.5 h and at 250°C for 1.5 h. After it has cooled to RT the sample alloy is contacted with 5–10 MPa of H_2 and becomes fully hydrided in 15 min to 3 h. Pulverizing the sample may be necessary before it can react with H_2; e.g., $TiMn_{1.5}$ is pulverized to 5–20 mesh (0.8 to 4 mm), evacuated at RT for 1.2 h and then contacted[8] with H_2 at 4.5 MPa. Likewise, $CeCo_2$ that is crushed to -20 to $+60$ mesh (0.25 to 0.8 mm) in a glove box forms the hydride after evacuation and contact with H_2 at 1 MPa and RT.

The mechanism of hydride formation of $ZrMn_2$ and $TiMn_{1.5}$ is related to segregation and preferential oxidation of the Mn on the surface[36] and precipitation of metallic Zr and Ti at the subsurface below the oxidized Mn. The Zr and Ti metallic precipitates catalyze the $H_2 \rightleftharpoons 2\ H$ reaction. In addition, the Zr and Ti precipitates themselves may form hydrides and spill over atomic H to the intermetallic compound lying below.

Because the AB_2 hydrides have a greater tendency to disproportionate than the AB_5 compounds, T must be kept sufficiently low during preparation of the hydride to prevent disproportionation. For example, $LaNi_2H_{4.5}$ is prepared by hydriding[37] $LaNi_2$ at 25°C and 10 MPa of H_2; however, at 100°C decomposition to LaH_3 occurs. Heating $LaNi_2H_{4.6}$ to 400°C gives[38]:

$$LaNi_2H_{4.6} \rightarrow LaH_2 + LaNi_nH_x\ (n > 2) \tag{a}$$

The heat of hydride formation can cause local excursions in T that disproportionate the hydride: in attempts to prepare the hydride of $GdFe_2$ by exposing to H_2 at 1.3 MPa, the resulting hydride contains only 3.4 H atoms per formula unit, and the hydrogen could not be desorbed[10]. Therefore, local heating resulting from rapid hydrogen absorption gives:

$$GdFe_2 + H_2 \rightarrow GdH_2 + 2\ Fe \tag{b}$$

More gentle hydrogenation at 0.2 MPa and 20°C (after activation) and then increasing the pressure to 1.3 MPa forms the hydride $GdFe_2H_{4.73}$.

The AB_2 intermetallics (vide supra) have three types of interstices that may be available for occupation by H atoms[2]. In all three, the H atoms would be T_d coordinated. There are 12 sites per formula unit in which the coordination is to two A and two B atoms, four sites with coordination to one A and three B atoms, and one with coordination to four B atoms. Because the A atoms are normally the hydride-forming elements, the H atoms should prefer the (2 A + 2 B) sites, for this would give maximum bonding to the A atoms; in the ZrB_2 compounds (B = Cr, Mn, V) the (2 Zr + 2 B) sites are the first to be occupied[39–41] and at higher concentrations of hydrogen the (1 Zr + 3 B) sites also become occupied. In ZrV_2 the number of atoms in the (1 Zr + 3 V) sites exceeds that in the (2 Zr + 2 V) sites[40,42]. However, V is also a hydride former[43], although it does not form as strong a bond with hydrogen as does Zr. There is no evidence for occupation of the 4 B sites by hydrogen.

1. J. H. Wernick, in *Intermetallic Compounds*, J. H. Westbrook, ed., Wiley, New York, 1967, p. 197.
2. D. Shaltiel, *J. Less-Common Met.*, 62, 407 (1978).
3. M. E. Kost, M. V. Raevskaya, A. L. Shilov, E. I. Yaropolova, V. I. Mikheeva, *Russ. J. Inorg. Chem. (Engl. Transl.)*, 24, 1803 (1979).
4. A. L. Shilov, L. N. Padurets, M. E. Kost, *Russ. J. Phys. Chem. (Engl. Transl.)*, 57, 555 (1983).
5. V. V. Burnasheva, A. V. Ivanov, V. A. Yartys, K. N. Semenenko, *Inorg. Mater. (Engl. Transl.)* 17, 704 (1981).
6. A. J. Maeland, G. G. Libowitz, *J. Less-Common Met.*, 89, 197 (1983).
7. I. Jacob, A. Stern, A. Moran, D. Shaltiel, D. Davidov, *J. Less-Common Met.*, 73, 369 (1980).
8. T. Gamo, Y. Moriwaki, N. Yanagihara, T. Yamashita, T. Iwaki, in *Hydrogen Energy Progress*, T. N. Veziroglu, ed., Pergamon Press, Oxford, 1981, p. 2127.
9. H. H. VanMal, K. H. J. Buschow, A. R. Miedema, *J. Less-Common Met.*, 49, 473 (1976).
10. H. A. Kierstead, *J. Less-Common Met.*, 86, L1 (1982).
11. D. Shaltiel, I. Jacob, D. Davidov, *J. Less-Common Met.*, 53, 117 (1977).
12. H. W. Newkirk, USERDA Report UCRL-52110 (Lawrence Livermore Laboratory), Aug., 1976; *Chem. Abstr.*, 87, 120,446 (1977).
13. V. I. Mikheeva, M. E. Kost, A. L. Shilov, *Russ. J. Inorg. Chem. (Engl. Transl.)* 23, 657 (1978).
14. I. Jacob, D. Shaltiel, *J. Less-Common Met.*, 65, 117 (1979).
15. K. H. J. Bushow, *Solid State Commun.*, 19, 421 (1976).
16. R. A. Guidotti, G. B. Atkinson, M. M. Wong, *J. Less-Common Met.*, 52, 13 (1977).
17. R. H. Van Essen, K. H. J. Buschow, *J. Less-Common Met.*, 70, 189 (1980).
18. V. V. Burnasheva, A. V. Ivanov, K. N. Semenenko, *Inorg. Mater. (Engl. Transl.)* 14, 1017 (1978).
19. A. L. Shilov, E. I. Yaropolova, M. V. Raevskaya, M. E. Kost, *Russ. J. Inorg. Chem. (Engl. Transl.)* 23, 1871 (1978).
20. F. W. Oliver, K. W. West, R. L. Cohen, K. H. J. Buschow, *J. Phys. F: Met. Phys.*, 8, 701 (1978).
21. R. L. Cohen, K. W. West, K. H. J. Buschow, *Solid State Commun.*, 25, 293 (1978).
22. S. K. Malik, W. E. Wallace, *Solid State Commun.*, 24, 283 (1977).
23. R. L. Cohen, K. W. West, F. Oliver, K. H. J. Buschov, *Phys. Rev., B*, 21, 941 (1980).
24. F. Pourarian, W. E. Wallace, A. Elatter, J. F. Lakner, *J. Less-Common Met.*, 74, 161 (1980).
25. H. A. Kierstead, *J. Less-Common Met.*, 70, 199 (1980).
26. H. Oesterreicher, K. Ensslen, E. Bucher, *Appl. Phys.*, 22, 303 (1980).
27. D. M. Gualtieri, K. S. V. L. Narasimhan, W. E. Wallace, *A.I.P. Conf. Proc.*, 34, 219 (1976); *Chem. Abstr.*, 86, 49,910 (1977).
28. A. L. Shilov, M. E. Kost, *Russ. J. Inorg. Chem. (Engl. Transl.)*, 26, 163 (1981).
29. P. J. Viccaro, G. K. Shenoy, D. Niarchos, B. D. Dunlap, *J. Less-Common Met.*, 73, 265 (1980).
30. T. B. Flanagan, N. B. Mason, G. E. Biehl, *J. Less-Common Met.*, 91, 107 (1983).
31. D. M. Gualtieri, W. E. Wallace, *J. Less-Common Met.*, 55, 53 (1977).
32. H. A. Kierstead, *J. Less-Common Met.*, 85, 213 (1982).
33. K. H. J. Buschow, P. H. Smit, R. M. van Essen, *J. Magn. Magnet. Mater.*, 15–18, 1261 (1980).
34. P. Duffer, D. M. Gualtieri, V. U. S. Rao, *Phys. Rev. Lett.*, 37, 1410 (1976).
35. F. Pourarian, V. K. Sinha, W. E. Wallace, *J. Less-Common Met.*, 96, 237 (1984).
36. L. Schlapbach, *J. Less-Common Met.*, 89, 37 (1983).
37. H. Oesterreicher, J. Clinton, H. Bittner, *Mater. Res. Bull.*, 11, 1241 (1976).
38. M. E. Kost, A. L. Shilov, *Inorg. Mater. (Engl. Transl.)*, 14, 1270 (1978).
39. J. J. Didisheim, K. Yvon, D. Shaltiel, P. Fischer, *Solid State Commun.*, 31, 47 (1979).
40. J. J. Didisheim, K. Yvon, P. Fischer, D. Shaltiel, *J. Less-Common Met.*, 73, 355 (1980).

41. D. Fruchart, A. Rouault, C. B. Shoemaker, D. P. Shoemaker, *J. Less-Common Met.*, *73*, 363 (1980).
42. J. J. Didisheim, K. Yvon, D. Shaltiel, P. Fischer, P. Bujard, E. Walker, *Solid State Commun.*, *32* 1087 (1979).
43. G. G. Libowitz, *Solid State Chemistry of Binary Metal Hydrides*, W. A. Benjamin, New York, 1965, pp. 66, 76.

1.12.8.1.3. Giving Hydrides of Other Intermetallics.

Although the AM_5 $CaCu_5$-type and AB_2 ($MgCu_2$ and $MgZn_2$ types) compound hydrides are most extensively covered, other hydrides of intermetallics are known (see Table 1). The formation of these hydrides is similar to those of the AM_5 and AB_2 hydrides.

The AM_3 intermetallics ($CeNi_3$ type) can be considered derivatives of the AM_5 and AB_2 compounds[7]. The AM_3 structure is obtained by stacking one unit of AM_5 onto one unit of AB_2 (see §1.12.8.1.2). Consequently, the sites available for occupation by H atoms are the same as those in the AM_5 and AB_2 intermetallics. Similarly, the A_2M_7 (Cu_2Ni_7-type) structure may be formed by two layers of AM_5 and one layer of AB_2. The equilibrium dissociation pressures (i.e., minimum H_2 pressure necessary to form the hydride at RT) are ca. 0.01–50 kPa for the AM_3 and 0.1–1000 kPa for the A_2M_7 hydrides.

The A_6M_{23} intermetallics have a complex fcc structure with 116 atoms (four formula units) per unit cell[8]. Although there are many interstices available for occupation by hydrogen, the preferred sites are those that are coordinated by one or more atoms of the metal that forms a stable hydride (i.e., the A atom)[9]. Dissociation pressures range[10,11] 0.4–300 kPa.

Hydrides are made from 1:1 intermetallics of groups V and VIII transition metals. For example, TiFe, TiCo and TiNi have the cubic CsCl structure, but on hydriding the first two form monohydrides and dihydrides that are orthorhombic and monoclinic, respectively, although they may be viewed as distorted CsCl structures. No hydride phase is formed from TiNi at RT (or above), but a solid solution of maximum composition $TiNiH_{1.47}$ forms[12]. The ZrCo and HfCo intermetallics also have the CsCl structure but transform to the orthorhombic CrB structure on formation of a monohydride

TABLE 1. HYDRIDES OF INTERMETALLIC COMPOUNDS

Intermetallic general formula	Examples of hydrides	Ref.
AM_3	$CeNi_3H_3$, $YFe_3H_{4.8}$	1
(A = rare earth, Y, Th;	$LuCo_3H_{3.5}$	1
M = Fe, Co, Ni)		2
A_2M_7	$Ce_2Co_7H_6$, $Y_2Ni_7H_2$	1
(A = rare earth, Y, Th;	$Th_2Fe_7H_{6.1}$	1
M = Fe, Co, Ni)		3
A_6M_{23}	$Y_6Fe_{23}H_{22.5}$	1
(A = rare earth, Y, Th;	$Th_6Mn_{23}H_{30}$	4
M = Fe, Ni)		
AB	TiFeH, $TiFeH_2$	5
(A = Ti, Zr, Hf;	ZrNiH, $ZrNiH_3$	6
B = Fe, Co, Ni)		

or trihydride, whereas ZrNi and HfNi already have the CrB structure before they form monohydrides and dihydrides of the same structure. Dissociation pressures for hydrides in this group range from 2×10^{-6} kPa for ZrCo monohydride[13] to 1000 kPa for TiFe dihydride[5].

Most of these alloys require activation to react with H_2; e.g., HfCo and HfNi are heated at 50°C for a few hours at 4 MPa of H_2 to effect complete absorption[14] of H_2 and TiCo is activated by heating at 325°C for 20 min at 2 MPa of H_2. The alloy is then cooled to RT and more H_2 added. Hydrogen is then desorbed from the sample and then reabsorbed according to the above procedure. After a few such absorption–desorption cycles the sample can be hydrided fully[15]. It may be necessary to activate[16] ZrNi by heating in vacuum at 800°C, introducing H_2 into the system after cooling and then reheating under H_2 to 400–450°C. However, it is not necessary to activate this alloy if it is kept under Ar after buffing and cleaning with emory paper[6]. Also ZrCo requires no activation as it reacts[13] with H_2 at 200°C under a pressure of only 3 kPa.

Because of its low cost and appropriate dissociation pressures, the hydrides of TiFe (along with $LaNi_5$) are considered for H_2 storage, but TiFe that has been exposed to air does not react with H_2 unless it is activated by granulating to 10 mesh (2 mm), heating to 400°C (in H_2 or vacuum) followed by cooling to RT and applying H_2 at ca. 6 MPa. If the sample is in ingot form, several hydriding and dehydriding cycles are required. The alloy is dehydrided by heating to 200°C while outgassing. Activation[17,18] also occurs on substituting a few % Mn, Cr, Co, Ni or Nb for Fe. The mechanism of activation of TiFe is still unknown[19]. Two mechanisms are possible. One involves the formation of oxides on the surface, primarily Fe_2Ti_4O, that either catalyze the dissociation of H_2 or absorb and transmit hydrogen to the underlying TiFe. The other[20] involves segregation of Ti at the surface, oxidation to TiO_2, and reduction of the Fe (by H2) to metallic Fe. Dissociation of the H_2 occurs on the Fe particles that are formed. Evidence[21] for the first mechanism is the appearance of Ti_4Fe_2O on the surface of a TiFe sample which undergoes heat treatment simulating the activation procedure under 0.1-Pa dynamic pressure of O_2. Evidence for the second mechanism is the observation[19,22] of TiO_2 and Fe clusters on the surface on reaction of TiFe with O_2. Discrepancies in the effects of activation on the TiFe samples may result from differences in preparation conditions[19].

The interstices occupied by hydrogen in the hydrides of the AB compounds are those in which the coordination is predominantly to hydride-forming elements. In $ZrNiH_3$[23] and $ZrCoH_3$[24] the H atoms are in two sites, T_d sites coordinated by three Zr and one Ni atom, and five-coordinated sites of four Zr and one Ni. In the monohydride of TiFe, the coordination is[25] distorted O_h to four Ti and two Ni. However, in all three cases, the H atoms are 9–25% closer to the non-hydride-forming element (Ni or Co) than to the hydride-forming element (Zr or Ti). In the dihydride of FeTi there are four types of distorted O_h sites[26]; in three the coordination is similar to that in the monohydride. However, in the fourth site, the H atoms are coordinated to four Ni and two Ti. The last type of site is difficult to fill (because of weak H—Fe bonding) and explains why stoichiometric $TiFeH_2$ cannot be attained. In the monohydride of ZrNi, the H atoms are T_d coordinated to four Zr atoms[27].

Other hydrides of intermetallic compounds include the A_2M_7 (A = rare earth, M = Ni, Co)[28] or the $MoSi_2$ structure intermetallics (where Mo = Cu, Pd; Si = Ti, Zr, Hf)[29], etc.[30]. The formation of these hydrides does not differ from that of those already covered; however, because of their technological value, the formation of Mg_2Ni hydrides is discussed.

Pulverizing the alloy to -25 mesh (< 0.7 mm) (in a dry box), loading into a high-pressure reactor and exposing to H_2 at ca. 2.4 MPa at 350°C yields[31] Mg_2NiH_4. The alloy reacts slowly at ca. 2 MPa and 325°C. The kinetics improve after several hydriding–dehydriding cycles so that reaction occurs at 200°C and 1.4 MPa. The intermetallic compound is hexagonal, but there are two forms of the hydride: a high-T antifluorite cubic form[32] (> 240°C) and a low-T form which is probably monoclinic[33] (< 236°C). As in the AB hydrides, the H atom coordination is predominantly to the hydride-forming elements, four Mg and one Ni atom[32]. However, the Ni–H distance, 0.147 nm, is closer than the Mg–H distance, 0.230 nm.

The dissociation pressure of Mg_2NiH_4 at RT is[31] ca. 800 Pa.

(G. G. LIBOWITZ, A. J. MAELAND)

1. R. H. Van Essen, K. H. J. Bushow, *J. Less-Common Met.*, *70*, 189 (1980).
2. H. A. Kierstead, *J. Less-Common Met.*, *96*, 133 (1984).
3. K. H. J. Buschow, H. H. VanMal, A. R. Miedema, *J. Less-Common Met.*, *42*, 163 (1975).
4. S. K. Malik, T. Takeshita, W. E. Wallace, *Solid State Commun.*, *23*, 599 (1977).
5. J. J. Reilly, R. H. Wiswall, *Inorg. Chem.*, *13*,218 (1974).
6. G. G. Libowitz, H. F. Hayes, T. R. P. Gibb, *J. Phys. Chem.*, *62*, 76 (1958).
7. B. D. Dunlap, P. J. Viccaro, G. K. Shenoy, *J. Less-Common Met.*, *74*, 75, 1980.
8. J. V. Florio, R. E. Rundle, A. I. Snow, *Acta. Crystallogr.* 5, 449 (1952).
9. D. G. Westlake, *Scripta Metall.*, *16*, 1049 (1982).
10. H. K. Smith, W. E. Wallace, R. S. Craig, *J. Less-Common Met.*, *94*, 89 (1983).
11. S. K. Malik, G. T. Bayer, E. B. Boltich, W. E. Wallace, *J. Less-Common Met.*, *98*, 109 (1984).
12. R. Burch, N. B. Mason, *Z. Phys. Chem. N.F.* (*Frankfurt-am-Main*), 116, 185 (1979).
13. S. J. C. Irvine, I. R. Harris, *J. Less-Common Met.*, *74*, 33 (1980).
14. R. M. Van Essen, K. H. J. Buschow, *J. Less-Common Met.*, *64*, 277 (1979).
15. R. Burch, N. B. Mason, *J. Chem. Soc., Faraday Trans.*, 1, 75, 561 (1979).
16. L. N. Padurets, A. A. Chertkov, V. I. Mikheeva, *Inorg. Mater.* (*Engl. Transl.*), 14, 1267 (1978).
17. M. H. Mintz, S. Voknin, S. Biderman, Z. Hadari, *J. Appl. Phys.*, *52*, 463 (1981).
18. T. Sasai, K. Oku, H. Konno, K. Onouwe, S. Kashu, *J. Less-Common Met.*, *89*, 281 (1983).
19. H. Züchner, G. Kirch, *J. Less-Common Met.*, *99*, 143 (1984).
20. L. Schlapbach, T. Riesterer, *Appl. Phys.*, *A32*, 169 (1983).
21. A. Venkert, M. Talianker, M. P. Dariel, *Mater. Lett.*, *2*, 45 (1983).
22. H. Züchner, U. Bilitewski, G. Kirch, *J. Less-Common Met.*, *101*, 441 (1984).
23. S. W. Peterson, V. N. Sadana, W. L. Korst, *J. Phys.* (*Paris*), 25, 451 (1964).
24. A. V. Irodova, V. A. Somenkov, S. Sh. Shil'shtein, L. N. Padurets, A. A. Chertkov, *Sov. Phys.-Crystallogr.*, 23, 591 (1978).
25. P. Thompson, M. A. Pick, F. Reidinger, L. M. Corliss, J. M. Hastings, J. J. Reilly, *J. Phys. F: Met. Phys.*, 8, L75 (1978).
26. P. Thompson, J. J. Reilly, F. Reidinger, J. M. Hastings, L. M. Corliss, *J. Phys. F: Met. Phys.*, 9, L62 (1979).
27. D. G. Westlake, H. Shaked, P. R. Mason, B. R. McCart, M. H. Mueller, *J. Less-Common Met.*, *88*, 17 (1982).
28. K. H. J. Buschow, in *Handbook on the Physis and Chemistry of Rare Earths* Vol. 6, K. A. Gschneidner, L. Eyring, eds., North-Holland, Amsterdam, 1984, p. 1.
29. A. J. Maeland, G. G. Libowitz, *J. Less-Common Met.*, *74*, 295 (1980).
30. K. H. J. Buschow, P. C. P. Bouton, A. R. Miedema, *Rep. Prog. Phys.*, *45*, 937 (1982).
31. J. J. Reilly, R. H. Wiswall, *Inorg. Chem.*, *7*, 2254 (1968).
32. J. Schefer, P. Fischer, W. Hälg, F. Stucki, L. Schlapbach, J. J. Didisheim, K. Yvon, A. F. Andresen, *J. Less-Common Met.*, *74*, 65 (1980).
33. A. F. Andresen, *J. Less-Common Met.*, *88*, 1 (1982).

1.12.8.2. from Metal–Nonmetal Systems

Hydrogen reacts with metal borides, carbides, silicides, nitrides, phosphides, oxides, sulfides, and halides to form a solid solution of hydrogen in the compound with distortion of the original lattice or with ordering of the H atoms in the lattice. New structures may be formed. Characteristic of all interactions of H with MX_n compounds (where M is the metal and X the nonmetal) are[1]:

1. Only metallic compounds take up hydrogen.
2. The metal must form a hydride.
3. The H atoms preferentially occupy those sites most distant from the X atoms in the lattice. Therefore, only M—H bonds, and no X—H bonds, are formed.

(G. G. LIBOWITZ, J. A. MAELAND)

1.12.8.2.1. Involving Carbides.

Hydrogen forms solid solutions with most carbides. In the hexagonal M_2C structures the H atoms occupy the T_d sites, and in the fcc MC lattices they occupy O_h sites not occupied by C atoms. The solid solutions of hydrogen in these carbides are prepared by heating in H_2; e.g., in the carbides V_2C and Nb_2C, solns containing ≤ 0.1 mol fraction of hydrogen are prepared[2] by introducing H_2 to the carbide at high T, cooling, reheating to high T and evacuating, and then repeating the procedure several times.

With Th carbides, two carbohydride phases, Th_2CH_2 and Th_3CH_4, are obtained[3] by heating the carbide in H_2 gas at 0.1 MPa and 850°C. Pressure–composition isotherms show that these are definite phases, not solid solutions. The Th_2CH_2 is hexagonal and the Th_3CH_4 monoclinic (probably distorted hexagonal). These are stable compounds; extrapolation of the pressure–composition–temperature data to RT indicates dissociation pressures of ca. 10^{-21} and 10^{-14} Pa, respectively. However, the hexagonal Th_2CH_2 transforms[4] to a cubic phase at 380°C.

Carbohydrides of Yb are prepared[5] by reacting YbH_2 with graphite for 2–10 h at 900°C under 50 kPa of H_2. Two carbohydride phases are obtained, a hexagonal $YbC_{0.5}H$ and a cubic $YbCH_{0.5}$.

(G. G. LIBOWITZ, J. A. MAELAND)

1. S. Rundqvist, R. Tellgren, Y. Andersson, *J. Less-Common Met., 101*, 145 (1984).
2. H. F. Franzen, A. S. Khan, D. T. Peterson, *J. Solid State Chem., 19*, 81 (1976).
3. D. T. Peterson, J. Rexer, *J. Inorg. Nucl. Chem., 24*, 519 (1962).
4. M. Makovec, Z. Ban. *J. Less-Common Met., 22*, 383 (1970).
5. J. M. Haschke, *Inorg. Chem., 14*, 779 (1975).

1.12.8.2.2. Involving Oxides and Sulfides.

Hydrogen forms solid solutions with several metallic oxides, although the solubility is usually less than in the pure metal. This includes ZrO [1], Ti_4M_2O (where M = Ni, Fe, Co, Cu)[2-5], and Zr_2PdO [6].

The compound Zr_3V_3O absorbs hydrogen up to the composition $Zr_3V_3OH_{5.5}$ by heating[7] granules (<1 mm) of the alloy to 300°C in 0.1–0.2 MPa of H_2 for 0.5–2 h.

The hydrogen molybdenum and tungsten bronzes H_xMoO_3 and H_xWO_3, where x = 0.6, may be prepared by electrochemical reduction of WO_3 or MoO_3. They violate the condition that the metal must be a hydride former, but they cannot be prepared by direct reaction of the oxide with H_2, and they differ from other compounds because the hydrogen prefers to bond with oxygen rather than the metal.

The sufides form solid solutions with no change in structure; Ta_6SH_2 and $Nb_{21}S_8H_5$ are prepared[9] by outgassing the sulfides at 850–900°C for 4 h, introducing H_2, cooling to 400°C and holding this T overnight. The sulfides are then reheated to higher T and evacuated. This procedure is repeated several times until the maximum hydrogen content is obtained. Thermodynamic data indicate that hydrogen bonds with the metal and not with sulfur; sulfur and hydrogen may be in competition for interactions with the metal atoms.

(G. G. LIBOWITZ, A. J. MAELAND)

1. R. K. Edwards, P. Levesque, *J. Am. Chem. Soc., 77,* 1312 (1955).
2. M. H. Mintz, Z. Hadari, M. P. Dariel, *J. Less-Common Met., 63,* 181 (1979).
3. M. H. Mintz, Z. Hadari, M. P. Dariel, *J. Less-Common Met., 74,* 287 (1980).
4. K. Hiebl, E. Tuscher, H. Bittner, *Monatsh. Chem., 110,* 869 (1979).
5. E. Tuscher, H. Bittner, *Monatsh. Chem., 111,* 1229 (1980).
6. A. J. Maeland, *J. Less-Common Met., 89,* 173 (1983).
7. M. H. Mendelsohn, in *Proc. 4th World Hydrogen Energy Conf.,* T. N. Veziroglu, W. D. Van Vorst, J. H. Kelley, eds., Pergamon Press, Oxford, 1982.
8. R. H. Jarman, P. G. Dickens, *J. Electrochem. Soc., 129,* 2276 (1982).
9. H. F. Franzen, A. S. Khan, D. T. Peterson, *J. Solid State Chem., 17,* 283 (1976).

1.12.8.2.3. Involving Nitrides and Phosphides.

Hydrogen forms solid solutions with phosphides with no structural change. Whereas Pd_6P reacts readily with H_2 at RT or below[1], Pd_3P_{1-x} only dissolves hydrogen when it is nonstoichiometric[2]; i.e., solubility increases with increasing value of x.

Hydrogen also forms solid solutions with nitrides. However, in some Ti nitrohydrides ($TiN_{1-x}H_{0.15}$, where $0.16 \leq x \leq 0.26$), ordering of both the N and H atoms occurs at lower T, the N atoms over O_h and the H atoms over the T_d interstices[3]. Ordering of N in the lattice results from the presence of H atoms. The nitrohydrides are prepared by high-T synthesis of $TiH_{0.15}$ with N_2.

In the fcc compound[4] $ThN_{1-x}H_y$, for $x = 0.31$ the maximum value of y is 2.5. The compound is stable; dissociation pressure is ca. 0.1 Pa.

(G. G. LIBOWITZ, A. J. MAELAND)

1. T. B. Flanagan, B. S. Bowerman, *J. Chem. Soc., Faraday Trans., I, 79,* 1605 (1983).
2. T. B. Flanagan, G. E. Biehl, J. D. Clewley, S. Rundqvist, Y. Andersson, *J. Chem. Soc., Faraday Trans., 1, 76,* 196 (1980).
3. I. Khiderov, I. Karimov, V. T. Ém. V. É. Loryan, I. P. Borovinskaya, M. M. Antonova, *Inorg. Mater. (Engl. Transl.),* 7, 1055 (1980).
4. D. T. Peterson, S. D. Nelson, *J. Less-Common Met., 80,* 221 (1981).

1.12.8.2.4. Involving Halides.

Many metal-rich transition-metal halides absorb hydrogen reversibly[1-3]. The halohydrides are prepared by exposing the halide to H_2 at 6–240 kPa and RT–300°C. Faster formation rates are obtained at higher P, but several hours are required to reach maximum hydrogen content. Hemihydrides and monohydrides, $ZrXH_{0.5}$ and $ZrXH$, form[1] from ZrCl and ZrBr. Dissociation pressures at RT are ca. 10^{-10} Pa for $ZrClH_{0.5}$ and $ZrBrH_{0.5}$ and 10^{-9} Pa for ZrClH and ZrBrH. The H atoms occupy sites in which they are T_d coordinated to four metal atoms[4]. High H_2 pressure may cause disproportionation, e.g.:

$$2 ZrXH \rightarrow ZrH_2 + ZrX_2 \qquad (a)$$

Two hydride phases, $ThI_2H_{0.7}$ and $ThI_2H_{1.7}$, with dissociation pressures at RT of ca. 10^{-4} Pa form[2] from ThI_2. Other halohydrides prepared include $LaI_2H_{0.75}$, $CeI_2H_{0.82}$,

1.12.8. to Form Ternary Hydrides
1.12.8.2. from Metal–Nonmetal Systems
1.12.8.2.4. Involving Halides.

261

$PrI_2H_{1.23}$, $ScI_{2.17}H_{0.64}$, $TiI_{2.5}H_{0.37}$, $Mo_6Cl_{12}H_{0.82}$, $Mo_6Cl_{12}H_{0.66}$, $ScCl_{1.5}H_{0.68}$, $GdCl_{1.5}H_{0.90}$ and $Nb_6I_{11}H_{1.33}$. The H atoms in $Nb_6I_{11}H_{1.33}$ are[5] at the centers of Nb O_h.

(G. G. LIBOWITZ, A. J. MAELAND)

1. A. W. Struss, J. D. Corbett, *Inorg. Chem.*, *16*, 360 (1977).
2. A. W. Struss, J. D. Corbett, *Inorg. Chem.*, *20*, 965 (1978).
3. H. Imoto, J. D. Corbett, *Inorg. Chem.*, *20*, 630 (1981).
4. H. S. Marek, J. D. Corbett, R. L. Daake, *J. Less-Common Met.*, *89*, 243 (1983).
5. A. Simon, *Z. Anorg. Allg. Chem.*, *355*, 311 (1967).

Abbreviations

abs	absolute
a.c.	alternating current
Ac	acetyl, CH_3CO
acac	acetylacetonate anion
acacH	acetylacetone, $CH_3C(O)CH_2C(O)CH_3$
ads	adsorbed
AIBN	2,2'-azobis(isobutyronitrile), $2,2'-[(CH_3)_2CCN]_2N_2$
Alk	alkyl
am	amine
amt	amount
Am	amyl, C_5H_{11}
amu	atomic mass unit
anhyd	anhydrous
aq	aqueous
Ar	aryl
asym	asymmetrical, asymmetric
at	atom (not atomic, except in atomic weight)
atm	atmosphere (not atmospheric)
av	average
bcc	body-centered cubic
BD	butadiene
bipy	2,2'-bipyridyl
bipyH	protonated 2,2'-bipyridyl
bp	boiling point
Bu	butyl, C_4H_9
Bz	benzyl, $C_6H_5CH_2$
ca.	circa, about, approximately
catal	catalyst (not catalyzing, catalysis, catalyzed, etc.)
CDT	cyclododecatriene
Ch.	chapter
COD	cyclooctadiene
conc	concentrated (not concentration)
const.	constant
COT	cyclooctatriene
Cp	cyclopentadienyl, C_5H_5
CPE	controlled-potential electrolysis
cpm	counts per minute
CT	charge-transfer
CV	cyclic voltammetry
CVD	chemical vapor deposition
CW	continuous wave
d	day, days
DABIP	N,N'-diisopropyl-1,4-diazabutadiene
DBA	dibenzylideneacetone
d.c.	direct current
DDT	dichlorodiphenyltrichloroethane, 1,1,1'-trichloro-2,2-bis-(4-chlorophenyl)ethane
dec	decomposed

DED	1,1-bis(ethoxycarbonyl)ethene-2,2-dithiolate, $[[(H_5C_2OC(O)]_2C=CS_2]^{2-}$
depe	1,2-bis(diphenylphosphino)ethene, $(C_6H_5)_2PCH=CHP(C_6H_5)_2$
diars	1,2-bis(dimethylarsino)benzene, o-phenylenebis(dimethylarsine), $1,2-(CH_3)_2AsC_6H_4As(CH_3)_2$
dien	diethylenetriamine, $[H_2N(CH_2)_2]_2NH$
diglyme	diethyleneglycol dimethylether, $CH_3O(CH_2CH_2O)_2CH_3$
dil	dilute
diop	2,3-O-isopropylidene-2,3-dihydroxy-1,4-bis(diphenylphos-phino)butane, $(C_6H_5)_2PCH_2CH[OCH(CH_3)=CH_2]CH$ $[OCH(CH_3)=CH_2]CH_2P(C_6H_5)_2$
diphos	1,2-bis(diphenylphosphino)benzene, $1,2-(C_6H_5)_2PC_6H_4P(C_6H_5)_2$
Div.	division
dme	dropping mercury electrode
DME	1,2-dimethoxyethane, glyme, $CH_3O(CH_2)_2OCH_3$
DMF	N,N-dimethylformamide, $HC(O)N(CH_3)_2$
DMG	dimethylglyoxime, $CH_3C(=NOH)C(=NOH)CH_3$
DMP	1,2-dimethoxybenzene, $1,2-(CH_3O)_2C_6H_4$
dmpe	1,2-bis(dimethylphosphino)ethane, $(CH_3)_2P(CH_2)_2P(CH_3)_2$
DMSO	dimethylsulfoxide, $(CH_3)_2SO$
dpam	bis(diphenylarsino)methane, $[(C_6H_5)_2As]_2CH_2$
dpic	dipicolinate ion
DPP	differential pulse polarography
dppb	1,4-bis(diphenylphosphino)butane, $1,4-(C_6H_5)_2P(CH_2)_4P(C_6H_5)_2$
dppe	1,2-bis(diphenylphosphino)ethane, $1,2-(C_6H_5)_2P(CH_2)_2P(C_6H_5)_2$
dppm	bis(diphenylphosphino)methane, $[(C_6H_5)_2P]_2CH_2$
dppp	1,3-bis(diphenylphosphino)propane, $1,3-(C_6H_5)_2P(CH_2)_3P(C_6H_5)_2$
dptpe	1,2-bis(di-p-tolylphosphino)ethane, $1,2-(4-CH_3C_6H_4)_2P(CH_2)_2P-(C_6H_4CH_3-4)_2$
DTA	differential thermal analysis
DTBQ	3,5-di-t-butyl-o-benzoquinone
DTH	1,6-dithiahexane, butane-1,4-dithiol, $1,4-HS(CH_2)_4SH$
DTS	dithiosquarate
ed.	edition, editor
eds.	editors
EDTA	ethylenediaminetetraacetic acid, $[HOC(O)]_2N(CH_2)_2N[C(O)OH]_2$
e.g.	exempli gratia, for example
emf	electromotive force
en	ethylenediamine, $H_2N(CH_2)_2NH_2$
enH	protonated ethylenediamine
EPR	electron paramagnetic resonance
equimol	equimolar
equiv	equivalent
EPR	electron paramagnetic resonance
Eq.	equation
ERF	effective reduction factor
ES	excited state
ESR	electron-spin resonance
esu	electrostatic unit
Et	ethyl, CH_2CH_2
etc.	et cetera, and so forth

Et$_2$O	diethyl ether, $(C_2H_5)_2O$
EtOH	ethanol, C_2H_5OH
et seq.	et sequentes, and the following
eu	entropy unit
fac	facial
fcc	face-centered cubic
ff.	following
Fig.	figure
Fl	fluorenyl
fp	freezing point
g	gas
g-at	gram-atom
glyme	1,2-dimethoxyethane, $CH_3O(CH_2)_2OCH_3$
graph	graphite
GS	ground state
h	hour, hours
Hex	hexyl
hmde	hanging mercury drop electrode
HMPA	hexamethylphosphoramide, $[(CH_3)_2N]_3PO$
HOMO	highest occupied molecular orbital
i.e.	id est, that is
Im	imidazole
inter alia	among other things
IR	infrared
irrev	irreversible
ISC	intersystem crossing
isn	isonicotinamide
l	liquid
L	ligand
LC	ligand centered
LF	ligand field
LFER	linear free-energy relationship
liq	liquid
LMCT	ligand-to-metal charge transfer
Ln	lanthanides, rare earths
LSV	linear-scan voltammetry
LUMO	lowest unoccupied molecular orbital
m	meta
max	maximum
M	metal
MC	metal centered
Me	methyl, CH_3
Men	menthyl
MeOH	methanol, CH_3OH
mer	meridional; the repeating unit of an oligomer or polymer
mhp	2-hydroxy-6-methylpyridine, 2-HO, 6-$CH_3C_5H_3N$
min	minimum, minute, minutes
MLCT	metal-to-ligand charge transfer
MO	molecular orbital
mol	molar
mp	melting point
MV	methyl viologen, 1,1'-dimethyl-4,4'-bipyridinium dichloride
n.a.	not available

napy	naphthyridine
NBD	norbornadiene, [2.2.1]bicyclohepta-2,5-diene
neg	negative
nhe	normal hydrogen electrode
NMR	nuclear magnetic resonance
No.	number
np	tris-[2-(diphenylphosphino)ethyl]amine, $N[CH_2CH_2P(C_6H_5)_2]_3$
Np	naphthyl
NPP	normal pulse polarography
NQR	nuclear quadrupole resonance
NTA	nitrilotriacetate
o	ortho
obs	observed
Oct	octyl
OF	oxidation factor
O_h	octahedral
Oq	oxyquinolate
p	para
p.	page
P	pressure
Pat.	patent
pet.	petroleum
Ph	phenyl, C_6H_5
phen	1,10-phenanthroline
Ph_2PPy	2-(diphenylphosphino)pyridine, $2\text{-}(C_6H_5)_2PC_5H_4N$
pip	piperidine, $C_5H_{10}N$
PMDT	pentamethyldiethylenetriamine, $(CH_3)_2N(CH_2)_2N(CH_3)(CH_2)_2N(CH_3)_2$
PMR	proton magnetic resonance
pn	propylene-1,3-diamine, $1,3\text{-}H_2NCH_2CH_2CH_2NH_2$
pos	positive
pp.	pages
ppb	parts per billion
ppm	parts per million
ppn	bis(diphenylphosphino)amine, $[(C_6H_5)_2P]_2NH$
ppt	precipitate
Pr	propyl, C_3H_7
PSS	photostationary state
PVC	poly(vinyl chloride)
PY	pyridine, C_5H_5N
pyr	pyrazine
PZE	potential of zero charge
rac	racemic mixture, racemate
R	organic group; universal gas constant
RDE	rotated disk electrode
RE	rare earths, lanthanides
ref.	reference
rev	reversible
rf	radiofrequency
RF	reduction factor
rh	rhombohedral
rms	root mean square
rpm	revolutions per minute

RT	room temperature
s	second, seconds; solid
sce	saturated calomel electrode
SCE	standard calomel electrode
sec	secondary
Sep	sepulcrate, 1,3,6,8,10,13,16,19-octaazabicyclo[6.6.6]eicosane
soln	solution
solv	solvated
sp	specific
STP	standard temperature and pressure
subl	sublimes
Suppl.	supplement
sym	symmetrical, symmetric
t	time; tertiary
T	temperature
T_d	tetrahedral
TCNE	tetracyanoethylene
TEA	tetraethylammonium ion, $[(C_2H_5)_4N]^+$
terpy	2,2'2''-terpyridyl
tetraphos	$Ph_2PCH_2CH_2PPhCH_2CH_2PPhCH_2CH_2PPh_2$
TGA	thermogravimetric analysis
THF	tetrahydrofuran
THP	tetrahydropyran
THT	tetrahydrothiophene
TLC	thin-layer chromatography
TMED	N,N,N',N'-tetramethylethylenediamine, $(CH_3)_2N(CH_2)_2N(CH_3)_2$
TMPH	2,2,6,6-tetramethylpiperidine, 2,2,6,6-$(CH_3)_4C_5H_6N$
Tos	tosyl, tolylsulfonyl, 4-$CH_3C_6H_4SO_2$
TPA	tetraphenylarsonium ion, $[(C_6H_5)_4As]^+$
triars	bis-[2-(dimethylarsino)phenyl]methylarsine, $[2-(CH_3)_2AsC_6H_4]_2AsCH_3$
triphos	1,1,1-tris(diphenylphosphinomethyl)ethane, $[C_6H_5)_2PCH_2]_3CCH_3$
trien	triethylenetetraamine, $H_2N(CH_2)_2NH(CH_2)_2NH(CH_2)_2NH_2$
UV	ultraviolet
v	vicinal
Vi	(E)-[2-$(CH_3)_2NCH_2C_6H_4]C=C(CH_3)C_6H_4CH_3$-4
viz.	videlical, that is to say, namely
vol., Vol.	volume
VPE	vapor-phase epitaxy
vs.	versus
wk.	week
wt	weight
X	halogen or pseudohalogen
xs	excess
yr.	year
§	section

Author Index

The entries of this index were derived directly by computer program from the lists of references. The accuracy of the references was the sole responsibility of the authors. No editorial check, except for format and journal title abbreviation, was applied. Consequently, errors occurring in authors' names in the references will recur in this index.

Each entry in the index refers to the appropriate section number.

Nainan, K. C.
1.9.5.2
Naish, P. J.
1.6.3.1.3
1.6.4.1.4
Nakajima, M.
1.5.3.2.2
Nakamura, K.
1.12.6
Naldini, L.
1.9.5.1
Namba, S.
1.6.2.1.1
Nametkin, N. S.
1.6.3.1.3
1.6.4.1.5
Nanda, R. K.
1.10.7.1
Narasimhan, K. S. V. L.
1.12.8.1.2
Naslain, R.
1.7.2
Näumann, F.
1.10.6.2.1
Nawich, H.
1.8.2.2
Negishi, E.
1.9.5.1
Neilson, R. H.
1.5.3.1.3
Neldini, L.
1.9.1
Nelson, S. D.
1.12.8.2.3
Nelson, W. J. H.
1.10.6.2.4
Nemirovskaya, I. B.
1.10.6.1.1
Neumann, F. K.
1.8.2.1
Neumann, S. M.
1.10.7.2
Neumann, W. P.
1.6.4.1.4
1.6.4.4.1
1.6.4.4.2
1.6.4.5.1
1.6.4.5.2
1.6.4.5.3
1.6.5.5
Neumauer, H.
1.9.5.1
Newkirk, A. E.
1.7.2
Newkirk, H. W.
1.12.8.1.2
Newmann, W. P.
1.6.3.4.3

Niarchos, D.
1.12.8.1.2
Nibler, J. W.
1.9.5.2
Nicholls, D.
1.7.2
Nickl, J.
1.7.2
Niebylski, L. M.
1.8.3.1
Niecke, E.
1.5.3.2.2
1.5.5.2.1
Niedenzu, K.
1.5.3.1.3
Nielsen, E.
1.7.5.1
Nielson, L.
1.5.3.2.2
1.5.3.3.3
Nilzbach, K. E.
1.8.3.1
1.8.3.2
Nixon, J. F.
1.5.3.2.2
Noack, M.
1.5.3.1.2
Nohr, R. S.
1.10.5.6.1
Noltes, J. G.
1.5.3.1.3
Norman, A. D.
1.5.2.2
1.5.2.3
1.5.2.4
1.5.3.1.3
1.5.3.2.1
1.5.3.2.2
1.5.3.3.1
1.5.3.3.3
1.5.3.4
1.5.3.5
1.5.4.1.4
1.5.4.2.1
1.5.4.2.2
1.5.4.2.4
1.5.4.3
1.5.5.3.1
1.5.5.3.2
1.5.5.4
1.5.6.3
1.5.6.4
1.5.7.1.2
1.6.2.2
1.6.2.4
1.6.3.2.2
1.6.3.3.1
1.6.3.3.2

1.6.3.4.1
1.6.4.1.4
1.6.4.2.1
1.6.4.2.2
1.6.4.2.3
1.6.4.3.3
1.6.5.2.1
1.6.5.2.3
1.6.5.3.1
1.6.5.4.1
1.6.7.1.2
1.6.7.1.3
1.6.7.1.4
1.7.7.2
Norton, J. R.
1.6.4.1.5
1.10.6.2.1
1.10.6.2.2
1.10.6.2.3
1.10.6.2.4
1.6.4.1.5
Norton, M. C.
1.10.5.3
Nosyrev, S. A.
1.5.5.4
Nöth, H.
1.5.5.5
1.7.3.2
1.7.4.4
1.7.5.1
1.7.5.2
1.9.5.2
1.10.8.3
Novotnah, G. C.
1.10.4.1.4
Novotny, J.
1.8.2.1
Nowak, E. J.
1.12.4.1
Nutton, A.
1.10.8.1
Nyholm , R. S.
1.5.3.3.1

O
Oates, W. A.
1.12.6
O'Brien, D. H.
1.5.3.3.3
Ochrymiek, S. B.
1.5.5.2.3
Ochsler, B.
1.5.5.2.2
1.6.5.4.3
Odell K. J.
1.10.5.3
Odiaka, T. L.
1.10.5.6.1

Compound Index

This index lists individual, fully specified compositions of matter that are mentioned in the text. It is an index of empirical formulas, ordered according to the following system: the elements within a given formula occur in alphabetical sequence except for C, or C and H if present, which always come first. The formulas are ordered alphanumerically without exception.

The index is augmented by successively permuted versions of all empirical formulas. As an example, $C_3H_3AlO_9$ will appear as such and, at the appropriate positions in the alphanumeric sequence, as $H_3AlO_9*C_3$, $AlO_9*C_3H_3$ and $O_9*C_3H_3Al$. The asterisk identifies a permuted formula and allows the original formula to be reconstructed by shifting to the front the elements that follow the asterisk.

Whenever an empirical formula does not show how the elements are combined in groups, it is followed by a linearized structural formula, which reveals the connectivity of the compound(s) underlying the empirical formula and serves to distinguish substances which are identical in composition but differ in the arrangement of elements.

The nonpermuted empirical formulas are followed by keywords. They describe the context in which the compounds represented by the empirical formulas are discussed. Section numbers direct the reader to relevant positions in the book.

AcCl$_3$
Formation: 1.12.4.2
Preparation of Ac metal: 1.12.4.2

AcH$_2$
Formation: 1.12.4.2

AcH$_3$O$_3$
Ac(OH)$_3$
Formation of AcCl$_3$: 1.12.4.2

Ag
Oxidative addition of methane: 1.10.5.3

Ag[AlH$_4$]
Formation and decomposition: 1.9.5.1

Ag[BH$_4$]
Formation and decomposition: 1.9.5.1

AgBO$_2$*C$_2$H$_6$

AgBO$_2$P$_3$*C$_{42}$H$_{47}$
AgBO$_2$P$_3$*C$_{56}$H$_{51}$
AgB$_3$P$_2$*C$_{36}$H$_{38}$
AgB$_3$P$_3$*C$_{54}$H$_{53}$

Ag[GaH$_4$]
Formation and decomposition: 1.9.5.1

AgH
Formation: 1.9.2

AgPd
Pd–Ag
Reaction with H$_2$: 1.9.1

Ag$_2$*C$_2$

Ag$_2$Ca
CaAg$_2$
Reaction with H$_2$: 1.9.2

309

Ag₂CaH

Ag$_2$CaH

CaAg$_2$H

Formation: 1.9.2

Al

Catalyst for reaction of H$_2$with Mg metal: 1.8.3.2

Reduction of alkali-metal plumbates: 1.6.3.5

Reduction of triorganochlorostannanes: 1.6.3.4.1

AlAs$_4$H$_8$Li

Li[Al(AsH$_2$)$_4$]

Hydrolysis to AsH$_3$: 1.5.3.3.3

AlBH$_4$Na

Na[AlBH$_4$]

Reduction of (C$_2$H$_5$O)$_4$Si: 1.6.5.2.2

AlB$_3$H$_{12}$

Al(BH$_4$)$_3$

Formation: 1.7.3.2, 1.7.5.1, 1.7.5.2

Reaction with (C$_2$H$_5$O)$_4$Si: 1.6.5.2.2

Reaction with C$_2$H$_4$: 1.6.5.1.4

Reaction with chlorosilanes: 1.6.5.2.1

Reduction of (C$_2$H$_5$O)$_3$CH: 1.6.5.1.2

AlBr$_3$

Catalyst in DBr exchange with C$_6$H$_6$: 1.6.7.2.1

Al*C$_3$H$_9$

Al*C$_4$H$_{11}$

Al*C$_6$H$_{15}$

Al*C$_7$H$_{17}$

Al*C$_8$H$_{19}$

Al*C$_{12}$H$_{27}$

Al*C$_{16}$H$_{27}$

Al*C$_{19}$H$_{39}$

AlClLi*C$_8$H$_9$

AlCl$_3$

Catalysis of DBr exchange with C$_6$H$_6$: 1.6.7.2.1

Catalysis of NaH–SiCl$_4$ reductions: 1.6.4.2.1

Catalysis of redistribution reactions: 1.6.4.3.1

Solvent for reaction of Al and H$_2$ with SiCl$_4$: 1.6.6.2

AlCl$_3$H$_4$InLi

LiInCl$_3$(AlH$_4$)

Formation: 1.7.5.2

AlCl$_4$Li

LiAlCl$_4$

Product of reduction of methylchlorobismuthines by Li[AlH$_4$]: 1.5.7.1.5

AlCsH$_4$

CsAlH$_4$

Formation: 1.7.2

AlCuH$_4$

CuAlH$_4$

Formation and decomposition: 1.9.5.1

AlD*C$_4$H$_{10}$

AlD*C$_8$H$_{18}$

AlD$_3$

Reduction of C$_2$H$_5$CHO: 1.6.7.1.1

AlD$_4$Li

Li[AlD$_4$]

Formation: 1.7.7.2

Reduction of C$_2$H$_5$Br: 1.6.7.1.1

Reduction of [(C$_6$H$_5$)$_3$Ge]$_2$O: 1.6.7.1.3

Reduction of C$_6$H$_5$CON(C$_2$H$_5$)$_2$: 1.6.7.1.1

Reduction of C$_6$H$_5$NO$_2$: 1.5.7.1.1

Reduction of C$_6$H$_5$P(O)Cl$_2$: 1.5.7.1.2

Reduction of (CH$_3$)$_2$P(O)OC$_2$H$_5$: 1.5.7.1.2

Reduction of PCl$_3$: 1.5.7.1.2

Reduction of (i-C$_3$H$_7$O)$_3$SiC$_2$H$_5$: 1.6.7.1.2

Reduction of SiH$_3$Cl: 1.6.7.1.2

Reduction of (SiCl$_3$)$_2$O: 1.6.7.1.2

Reduction of Si$_2$Cl$_6$: 1.6.7.1.2

Reduction of SnCl$_4$: 1.6.7.1.4

Reduction of chlorobismuthines: 1.5.7.2.1, 1.5.7.1.4

Reduction of chlorophosphines: 1.5.7.1.2

Reduction of chlorostibines: 1.5.7.1.4

Reduction of halogermanes: 1.6.7.1.3

Reduction of haloplumbanes: 1.6.7.1.5

Reduction of halosilanes: 1.6.7.1.2

Reduction of halostannanes: 1.6.7.1.4

AlH$_3$

Formation: 1.7.2

Formation as the ether adduct: 1.7.5.2

Reducing action: 1.7.3.2

Reduction of Si$_2$Cl$_6$: 1.6.4.2.1

Reduction of C$_2$H$_5$CHO: 1.6.4.1.2

AlH$_3$I$_2$Zn

ZnI$_2$(AlH$_3$)

Formation: 1.9.5.2

AlH$_4$*Ag

AlH$_4$K

K[AlH$_4$]

Formation: 1.7.4.2

AlH$_4$Li

Li[AlH$_4$]

Cleavage of (C$_2$H$_5$)$_3$GeP(C$_6$H$_5$)$_2$: 1.5.5.2.3

Cleavage of C$_6$H$_5$(CH$_3$)NP(O)CH$_3$: 1.5.5.2.3

Cleavage of [(CH$_3$)$_3$Si]$_2$NPSi(CH$_3$)$_3$: 1.5.5.2.3

CH$_3$Br$_3$Ge
  CH$_3$GeBr$_3$
    Reduction by Na[BH$_4$]: 1.6.5.3.1
CH$_3$Cl
    Reaction with Cu–Si at high T: 1.6.6.2
    Reaction with elemental Si and H$_2$:
      1.6.2.2
CH$_3$Cl$_2$P
  CH$_3$PCl$_2$
    Hydrolysis to form CH$_3$PH(O)OH:
      1.5.6.2
    Reaction with CH$_3$OH: 1.5.3.2.3
    Reaction with LiH: 1.5.4.2.1
    Reaction with HF: 1.5.3.2.3
CH$_3$Cl$_2$Sb
  CH$_3$SbCl$_2$
    Reduction by Li[AlH$_4$]: 1.5.5.4
CH$_3$Cl$_3$Ge
  CH$_3$GeCl$_3$
    Reduction by Li[(t-C$_4$H$_9$O)$_3$AlH]:
      1.6.5.3.1
    Reduction by Li[AlH$_4$]: 1.6.6.3
CH$_3$Cl$_3$Si
  CH$_3$SiCl$_3$
    Reduction by Li[AlH$_4$]: 1.6.5.2.1
CH$_3$Cl$_3$Sn
  CH$_3$SnCl$_3$
    Reduction by (C$_2$H$_5$)$_2$AlH: 1.6.4.4.1
    Reduction with Li[AlH$_4$]: 1.6.5.4.1
CH$_3$D
    Formation: 1.6.7.1.1
CH$_3$F$_3$P$_2$
  CF$_3$P(H)PH$_2$
    Formation: 1.5.4.2.4
CH$_3$IZn
  CH$_3$ZnI
    Reduction with NaH: 1.9.4.1
CH$_3$KO$_2$Sn
  K[CH$_3$SnO$_2$]
    Reduction by Na[BH$_4$]: 1.6.5.4.2
CH$_3$Li
  CH$_3$Li
    Hydrolysis: 1.6.2.5
CH$_4$
  CH$_4$
    Formation: 1.6.2.1.1, 1.6.2.1.2, 1.6.2.2,
      1.6.2.5, 1.6.3.1.3, 1.6.4.1.2, 1.6.4.1.5,
      1.6.5.1.2, 1.6.5.1.3, 1.6.6.1
    Impurity in H$_2$: 1.8.2
    Protonation in gas phase: 1.6.3.1.3
CH$_4$AlLiO
  Li[H$_2$AlCH$_2$O]
    Formation: 1.6.5.1.2

CH$_4$AsCl
  CH$_3$AsHCl
    Formation: 1.5.4.3
CH$_4$AsF$_3$P$_2$
  CF$_3$As(PH$_2$)$_2$
    Decomposition to CF$_3$AsH$_2$: 1.5.4.3
CH$_4$Cl$_2$Si
  CH$_3$SiCl$_2$H
    Exchange with SiCl$_3$D in presence of
      H$_2$PtCl$_6$: 1.6.7.2.2
    Formation: 1.6.2.2
    Industrial formation: 1.6.6.2
    Reaction with C$_6$H$_5$C(CH$_3$)=CH$_2$:
      1.6.4.1.4
CH$_4$Cl$_4$N$_3$P$_3$
  (NPCl$_2$)$_2$NP(H)CH$_3$
    Formation: 1.5.3.2.3
CH$_4$F$_2$Ge
  CH$_3$GeF$_2$H
    Formation: 1.6.4.3.1
CH$_4$F$_3$P
  CH$_3$P(F)$_3$H
    Formation: 1.5.3.2.3
CH$_4$F$_3$P$_3$
  CF$_3$P(PH$_2$)$_2$
    Disproportionation to (PH)$_n$ and CF$_3$*
      P(H)PH$_2$: 1.5.4.2.4
CH$_4$GeO$_2$
  H$_3$GeCO$_2$H
    Disproportionation: 1.6.4.3.3
CH$_4$N$_2$O
  (H$_2$N)$_2$CO
    Thermolysis to NH$_3$: 1.5.4.1.3
CH$_4$N$_2$O$_2$
  H$_2$NNHCO$_2$H
    Thermolysis to N$_2$H$_4$: 1.5.4.1.3
CH$_4$O
  CH$_3$OH
    Cleavage of group-IVB phosphines:
      1.5.3.2.3
    Formation: 1.6.2.1.2
    Formation in CO reaction with H$_2$ us-
      ing a Zn–Cr$_2$O$_3$ catalyst: 1.6.6.1
    Formation in CO reaction with H$_2$ us-
      ing a Pd–Ca$_2$O$_3$ catalyst: 1.6.6.1
    Reaction with Mn(CO)$_3$[P(C$_6$H$_5$)$_3$]
      CO$_2$NH$_2$: 1.5.3.1.2
    Reaction with Cl$_3$PNP(O)Cl$_2$: 1.5.3.1.3
    Reaction with (CH$_3$)$_3$SiN[C(CH$_3$)$_3$]P*
      (S)C$_4$H$_9$-t: 1.5.3.1.3
    Reaction with (CH$_3$)$_3$SiN[Si(CH$_3$)$_3$]P*
      (S)NC$_4$H$_9$-t: 1.5.3.1.3
    Reaction with [(CH$_3$)$_3$Si]$_2$NC$_6$H$_5$:
      1.5.3.1.2

Reaction with $(C_6H_5)_3SiSi(C_6H_5)_2CH_3$:
1.6.4.2.3
Reaction with $(C_6H_5)_2SnH_2$: 1.6.3.4.3
Reaction with $(h^5-C_5H_5)_2TiN_2MgCl$:
1.5.3.1.3
Reaction with $W_2[N(CH_3)_2]_6$: 1.5.3.1.2
Reaction with $(CH_3)_2Cd$: 1.6.3.1.3

CH$_4$OSi
$(CH_3SiHO)_n$
Reduction of $[(n-C_4H_9)_3Sn]_2O$: 1.6.6.4
Reduction of stannyl oxides: 1.6.4.4.2
Reaction with $(CH_3)_2Cd$: 1.6.3.1.3

CH$_4$O$_2$P
CH_3PO_2H
Formation: 1.5.3.2.3

CH$_4$Th$_3$
Th_3CH_4
Formation: 1.12.8.2.2

CH$_5$As
CH_3AsH_2
Formation: 1.5.4.3

CH$_5$Bi
CH_3BiH_2
Disproportionation: 1.5.5.5
Formation: 1.5.5.5, 1.5.7.2.1

CH$_5$FGe
CH_3GeH_2F
Disproportionation: 1.6.4.3.1

CH$_5$N
CH_3NH_2
Formation: 1.5.3.1.3, 1.6.2.1.1
Protonation in H_2O: 1.5.3.1.1
Reaction with P_4 to form
CH_3PH_2: 1.5.6.2
Relative basicity: 1.5.3.1.3

CH$_5$O$_2$P
$CH_3PH(O)OH$
Formation: 1.5.3.2.3
Industrial synthesis: 1.5.6.2
Thermal disproportionation: 1.5.6.2

CH$_5$P
CH_3PH_2
Formation: 1.5.4.2.4, 1.5.5.2.1, 1.5.5.2.2
Synthesis from $CH_3PH(O)OH$: 1.5.6.2
Synthesis from P_4 with CH_3NH_2 in
presence of carbon: 1.5.6.2

CH$_5$Sb
CH_3SbH_2
Formation: 1.5.5.4

CH$_6$Ge
CH_3GeH_3
Formation: 1.6.4.3.1, 1.6.4.3.3, 1.6.5.3.1
Industrial formation: 1.6.6.3

CH$_6$N$_2$O$_2$
$[NH_4][H_2NCO_2]$
Thermolysis to NH_3: 1.5.4.1.3

CH$_6$P$_2$
$H_2C(PH_2)_2$
Formation: 1.5.5.2.1

CH$_6$Si
CH_3SiH_3
Formation: 1.6.5.2.1
Reaction with $[CH_3Si]^+$: 1.6.4.2.3
Reaction with Si atoms in matrix: 1.6.2.2
Reaction with Si_2H_6: 1.6.4.2.3

CH$_6$Sn
CH_3SnH_3
Formation: 1.6.4.4.1, 1.6.4.4.3, 1.6.5.4.1,
1.6.5.4.2

CH$_7$AsGe
$CH_3GeH_2AsH_2$
Cleavage by protonic acids: 1.5.3.3.3

CH$_7$AsSi
$CH_3SiH_2AsH_2$
Cleavage by protonic acids: 1.5.3.3.3

CH$_7$BZn
CH_3ZnBH_4
Formation: 1.9.5.2

CH$_7$GeP
$CH_3Ge(PH_2)H_2$
Formation: 1.6.4.3.3

CH$_8$B$_2$
$CH_3B_2H_5$
Formation: 1.7.3.2

CH$_8$GeP$_2$
$CH_3Ge(PH_2)_2H$
Disproportionation: 1.6.4.3.3

CH$_8$Si$_2$
$CH_3Si_2H_5$
Formation: 1.6.4.2.3

CH$_{13}$B$_{10}$Na
$Na[B_{10}H_{12}CH]$
Formation: 1.6.4.1.3

CH$_{15}$B$_{10}$N
$B_{10}H_{12}CNH_3$
Formation: 1.5.3.1.3

CH$_{44}$Cl$_2$GeP$_2$Pt
$[(C_6H_5)_2PCH_2CH_2P(C_6H_5)_2]Pt^*$
$Cl_2(H)Ge(C_6H_5)_3$
Thermolysis: 1.6.4.3.3

CIN
ICN
Reaction with atomic H: 1.6.2.1.1

CKN
KCN
Reaction with $(C_6H_5)_3PAuGe(C_6H_5)_3$
in CH_3OH: 1.6.3.3.3

CNNaO
  NaNCO
    Reaction with protonic acids: 1.5.3.1.3
CNb$_2$
  Nb$_2$C
    Reaction with H$_2$: 1.12.8.2.2
CO
    Poisoning of hydriding reaction:
      1.12.8.1.1
    Reaction with Li[AlH$_4$]: 1.6.5.1.2
    Reaction with (h^5-C$_5$H$_5$)$_2$ZrHCl:
      1.6.4.1.2
    Reaction with H$_2$: 1.6.2.1.2
    Reaction with H$_2$ and [HRu$_3$(CO)$_{10}$*
      [Si(C$_2$H$_5$)$_3$]$_2$]$^-$: 1.6.2.2
    Reaction with H$_2$ and n-C$_4$H$_9$CH=C*
      H$_2$: 1.6.2.1.2
    Reaction with H$_2$ and alkenes: 1.6.6.1
    Reaction with H$_2$ over catalysts to pro-
      duce hydrocarbons: 1.6.6.1
    Reaction with H$_2$ to form CH$_3$OH:
      1.6.6.1
    Reaction with H$_2$ to produce
      (CH$_2$OH)$_2$: 1.6.6.1
    Reduction by CaH$_2$: 1.6.4.1.2
    Reduction by KH: 1.6.4.1.2
    Surface impurity: 1.12.1
CO$_2$
    Electrolytic reduction to [HCO$_2$]$^-$:
      1.6.2.5
    Reaction with Na[BH$_4$]: 1.6.5.1.2
    Reaction with group-VIA, metal-car-
      bonyl hydrides: 1.6.4.1.2
    Reaction with CaH$_2$: 1.6.4.1.2
    Reaction with H$_2$ over Ni: 1.6.2.1.2
    Reaction with alkali-metal hydrides:
      1.6.4.1.2
    Reduction at Hg cathode: 1.6.2.5
    Reduction by Li[BH$_4$]: 1.6.5.1.2
    Reduction by CO$_2$: 1.6.5.1.2
CV$_2$
  V$_2$C
    Reaction with H$_2$: 1.12.8.2.2
C$_2$
    Reactions with hydrogen atoms:
      1.6.2.1.1
C$_2$Ag$_2$
  Ag$_2$C$_2$
    Hydrolysis: 1.6.2.5
C$_2$AsF$_6$I
  (CF$_3$)$_2$AsI
    Reaction with HI and Hg: 1.5.3 3.3

    Reduction by Cu–Zn alloy in acid:
      1.5.3.3.1
C$_2$Au$_2$
  Au$_2$C$_2$
    Hydrolysis: 1.6.2.5
C$_2$CaN$_2$
  Ca(CN)$_2$
    Reaction with H$_2$: 1.5.2.1.2
C$_2$Cl$_2$F$_5$N
  C$_2$F$_5$NCl$_2$
    Reaction with (CH$_3$)$_3$SiH: 1.5.4.1.2
C$_2$Cl$_6$N$_2$P$_2$
  (Cl$_3$CPN)$_2$
    Reaction with HCO$_2$H: 1.5.3.1.3
C$_2$Cs$_2$
  Cs$_2$C$_2$
    Hydrolysis: 1.6.2.5
C$_2$Cu$_2$
  Cu$_2$C$_2$
    Hydrolysis: 1.6.2.5
C$_2$D$_2$
    Formation: 1.6.7.2.1
C$_2$D$_4$
    Formation: 1.6.7.2.1
C$_2$F$_4$
    Reaction with h^5-C$_5$H$_5$Fe(CO)$_2$H:
      1.6.4.1.4
C$_2$F$_4$
    Reaction with (CH$_3$)$_2$SiH$_2$: 1.6.4.1.4
C$_2$F$_6$
    Formation: 1.6.4.1.5
C$_2$F$_6$IP
  (CF$_3$)$_2$PI
    Reaction with HI and Hg: 1.5.3.2.3
C$_2$F$_6$P$_2$
  (CF$_3$P)$_2$
    Formation of d,l- and meso: 1.5.3.2.3
C$_2$F$_7$P
  (CF$_3$)$_2$PF
    Reaction with K[HF$_2$]: 1.5.3.2.3
C$_2$HAsF$_6$
  (CF$_3$)$_2$AsH
    Formation: 1.5.3.3.1, 1.5.3.3.3, 1.5.4.3
    Thermolysis: 1.6.4.1.5
C$_2$HCl$_7$NP
  (Cl$_3$C)$_2$PClNH
    Reaction with HF: 1.5.3.1.3
C$_2$HF$_3$O$_2$
  CF$_3$CO$_2$H
    Reaction with [h^5-(CH$_3$)$_5$C$_5$]$_2$Ni:
      1.6.3.1.3
    Reaction with [(CH$_3$)$_3$Si]$_4$N$_4$: 1.5.3.1.3

C₂HF₄N
  CF₃CFNH
    Formation: 1.5.4.1.2
C₂HF₆P
  (CF₃)₂PH
    Formation: 1.5.3.2.3
    Reaction with Fe(CO)₅: 1.10.5.2
    Reaction with Fe₃(CO)₁₂: 1.10.5.2
    Reaction with (CH₃)₂SiAs(CH₃)₂:
      1.5.4.2.3
C₂HF₉KP
  K[(CF₃)₂P(H)F₃]
    Formation: 1.5.3.2.3
C₂H₂
    Exchange with basic D₂O: 1.6.7.2.1
    Formation: 1.6.2.1.1, 1.6.2.5, 1.6.5.1.3,
      1.6.6.1
    Impurity in H₂: 1.8.2
C₂H₂As₂F₆
  (CF₃AsH)₂
    Formation in d,l- and meso forms:
      1.5.3.3.3
C₂H₂Cl₇FNP
  (CCl₃)₂P(F)ClNH₂
    Formation: 1.5.3.1.3
C₂H₂F₆P₂
  (CF₃P)₂H₂
    Formation: 1.5.3.2.3
C₂H₃Cl
  CH₂CHCl
    Reaction with (C₆H₅)₃SiH: 1.6.4.1.1,
      1.6.7.1.1
C₂H₃ClO
  CH₃COCl
    Reduction by Li[AlH₄]: 1.6.5.1.2
C₂H₃D
  CH₂CHD
    Formation: 1.6.7.1.1
C₂H₄
  CH₂=CH₂
    Dimerization reaction: 1.6.4.1.4
    Exchange with D₂SO₄: 1.6.7.2.1
    Formation: 1.6.2.1.1, 1.6.2.5, 1.6.3.1.3,
      1.6.5.1.1, 1.6.5.1.2
    Formation in hydrocarbon cracking:
      1.6.6.1
    Reaction with Al(BH₄)₃: 1.6.5.1.4
    Reaction with B₂H₆: 1.6.6.1
    Reaction with PtH(Cl)[P(C₂H₅)₃]₂:
      1.6.4.1.4
    Reaction with H₂O in presence of [Pd*
      Cl₄]²⁻–CuCl₂: 1.6.6.1

    Reaction with HCN: 1.6.6.1
C₂H₄AlLiO₄
  LiAl(OCH₂O)₂
    Formation: 1.6.5.1.2
C₂H₄BLiO₄
  LiBO(OCH₃)O₂CH
    Formation: 1.6.5.1.2
C₂H₄Cl₄Si
  CH₃CHClSiCl₃
    Reduction by LiH: 1.6.4.2.1
C₂H₄N₂O₆S
  O₂S(NHCO₂H)₂
    Thermolysis to O₂S(NH₂)₂: 1.5.4.1.3
C₂H₄O
  CH₃CHO
    Industrial formation: 1.6.6.1
C₂H₄O₂
  CH₃CO₂H
    Reaction with BH₃ in THF: 1.6.4.1.2
    Reaction with [(C₂H₅)₃Ge]₂Cd:
      1.6.3.3.3
C₂H₅Br
    Reduction by Li[AlD₄]: 1.6.7.1.1
    Reduction by Zn: 1.6.2.5
    Reduction with Li[AlH₄]: 1.6.5.1.1
C₂H₅BrMg
  C₂H₅MgBr
    Pyrolysis to MgH₂: 1.8.3.2
C₂H₅ClMg
  C₂H₅MgCl
    Reaction with (n-C₄H₉)₃SnH and D₂O
      1.6.7.1.4
C₂H₅Cl₂P
  C₂H₅PCl₂
    Hydrolysis to form C₂H₅PH(O)OH:
      1.5.6.2
    Reaction with HF: 1.5.3.2.3
C₂H₅Cl₂Sb
  C₂H₅SbCl₂
    Reduction by Li[AlH₄]: 1.5.5.4
C₂H₅Cl₃Ge
  C₂H₅GeCl₃
    Reduction with Li[AlH₄]: 1.6.5.3.1
C₂H₅Cl₃Si
  C₂H₅SiCl₃
    Reduction by Li[AlH₄]: 1.6.5.2.1
C₂H₅Cl₃Sn
  C₂H₅SnCl₃
    Reduction by (C₂H₅)₂AlH: 1.6.4.4.1
C₂H₅D
    Formation: 1.6.7.1.1

C₂H₆NNa

$Na[(CH_3)_2N]$

Reaction with DCl: 1.5.3.1.3

C₂H₆O

CH_3CH_2OH

Formation: 1.6.4.1.2, 1.6.5.1.2

Reaction with $(C_6H_5)_3PPt(Cl)_2CNC_6*$ H_5: 1.5.3.1.2

Reaction with $[(CH_3)_2N]_3P$: 1.5.3.1.2

Reaction with $C_6H_5PCl_2$: 1.5.6.2

Reaction with $Cl_3PNP(O)Cl_2$: 1.5.3.1.3

Reaction with $W_2[N(CH_3)_2]_6$: 1.5.3.1.2

C₂H₆O₂

$HOCH_2CH_2OH$

Formation in $CO-H_2$ reaction: 1.6.2.1.2

Industrial formation: 1.6.6.1

C₂H₆Si

$(CH_3)_2Si$

Reaction with CH_3OH: 1.6.3.2.3

Reaction with H_2O: 1.6.3.2.1

Reaction with alcohols: 1.6.3.2.3

Reaction with amines: 1.6.3.2.3

C₂H₆Sn

$(CH_3)_2Sn$

Reaction with HCl or $[NH_4][HF_2]$: 1.6.3.4.3

C₂H₆Zn

$(CH_3)_2Zn$

Reduction to $K[ZnH(CH_3)_2]$ by KH: 1.9.4.1

C₂H₇As

$(CH_3)_2AsH$

Formation: 1.5.3.3.1, 1.5.3.3.3, 1.5.6.3

Reaction with $(CF_3)_4P_2$: 1.5.3.2.3

Reaction with $CF_3C \equiv CCF_3$: 1.6.4.1.4

C₂H₇AsO

$(CH_3)_2AsOH$

Reduction by Zn in HCl: 1.5.3.3.1

C₂H₇AsO₂

$(CH_3)_2As(O)OH$

Reduction by Zn in HCl: 1.5.3.3.1, 1.5.6.3

C₂H₇BO₂

$(CH_3O)_2BH$

Redistribution: 1.7.5.1

Formation: 1.7.5.1

C₂H₇B₅

$B_5C_2H_7$

Formation: 1.6.2.1.1

C₂H₇Bi

$(CH_3)_2BiH$

Formation: 1.5.5.5, 1.5.7.2.1

C₂H₇ClGe

$(CH_3)_2GeClH$

Formation: 1.6.3.3.3

$(CH_3)_2GeHCl$

Formation: 1.6.4.3.1

C₂H₇ClO₂Si

$SiHCl(OCH_3)_2$

Reduction by $Li[BH_4]$: 1.6.5.2.2

C₂H₇ClSi

$CH_3CHClSiH_3$

Formation: 1.6.4.2.1

$(CH_3)_2SiHCl$

Formation: 1.6.4.2.3, 1.6.4.3.1

C₂H₇ClSn

$(CH_3)_2SnHCl$

Formation: 1.6.3.4.3

C₂H₇D₂PSi

$(CH_3)_2SiDPHD$

Formation from $SiH_3PH_2-(CH_3)_2SiD_2$ exchange: 1.5.4.2.4

C₂H₇FGe

$C_2H_5GeH_2F$

Redistribution: 1.6.4.3.1

C₂H₇KZn

$KZnHMe_2$

Formation: 1.9.4.2

C₂H₇N

$(CH_3)_2NH$

Formation: 1.5.3.1.2, 1.5.3.1.3, 1.5.4.1.3

Protonation in H_2O: 1.5.3.1.1

Relative basicity: 1.5.3.1.3

$C_2H_5NH_2$

Formation: 1.5.3.1.3, 1.5.4.1.3

Industrial synthesis: 1.5.6.1

Reaction with $(CH_3)_2AsN(CH_3)_2$: 1.5.3.1.3

C₂H₇OP

$(CH_3)_2P(O)H$

Formation: 1.5.3.2.1

C₂H₇O₂P

$(CH_3O)_2PH$

Formation: 1.5.4.2.2

$CH_3PH(O)OCH_3$

Formation: 1.5.3.2.3

$C_2H_5PH(O)OH$

Industrial synthesis: 1.5.6.2

Thermal disproportionation: 1.5.6.2

C₂H₇O₃P

$(CH_3O)_2P(O)H$

Formation: 1.5.3.2.2

C₂H₇P

$(CH_3)_2PH$

Cleavage of $(CH_3)_2PN(CH_3)_2$: 1.5.4.1.3

Formation: 1.5.3.2.3, 1.5.4.2.3, 1.5.5.2.3

Reaction with $(CF_3)_4As_2$: 1.5.4.3

$C_2H_5PH_2$

Formation: 1.5.3.2.3, 1.5.5.2.2, 1.5.6.2

C₂H₇PS
  (CH₃)₂P(S)H
    Conversion to (CH₃)₂P(O)H: 1.5.3.2.1
    Formation: 1.5.3.2.1
C₂H₇PSi
  CH₃H₂SiPH₂
    Redistribution in presence of
    Li[OC₂H₅]: 1.5.4.2.4
C₂H₇Sb
  (CH₃)₂SbH
    Formation: 1.5.5.4
  C₂H₅SbH₂
    Formation: 1.5.5.4
    Reaction with (CH₃)₂AsCl: 1.5.4.3
C₂H₇Si
  (CH₃)₃SiH
    Reagent for trapping GeH₂: 1.6.4.3.3
C₂H₈AlNaO
  Na[AlH₃OC₂H₅]
    Formation: 1.7.4.2
C₂H₈BP
  [BH₂P(CH₃)₂]ₓ
    Formation: 1.5.4.2.4
C₂H₈B₁₀O₂
  B₁₀H₈(CO)₂
    Reaction with Li[AlH₄]: 1.6.5.1.2
C₂H₈ClP
  [(CH₃)₂PH₂]Cl
    Formation: 1.5.3.2.1
C₂H₈Ge
  (CH₃)₂GeH₂
    Exchange with halogermanes: 1.6.4.3.1
    Formation: 1.6.3.3.1, 1.6.4.3.1, 1.6.4.3.3,
    1.6.5.3.1, 1.6.5.3.3
  C₂H₅GeH₃
    Formation: 1.6.4.3.1, 1.6.5.3.1
C₂H₈N₂
  H₂C(NH₂)C(NH₂)H₂
    Formation: 1.5.5.1
C₂H₈OSi
  C₂H₅OSiH₃
    Formation: 1.6.5.2.2
C₂H₈O₂Si
  (CH₃O)₂SiH₂
    Formation: 1.6.5.2.2
C₂H₈P₂
  H₂P(CH₂)₂PH₂
    Formation: 1.5.5.2.2
C₂H₈Pb
  (CH₃)₂PbH₂
    Formation: 1.6.5.5
C₂H₈Si
  (CH₃)₂SiH₂
    Formation: 1.6.4.2.1, 1.6.4.2.2, 1.6.5.2.1,
    1.6.5.2.2, 1.6.5.2.3

Reaction with Si atoms in matrix: 1.6.2.2
Reaction with C₂F₄: 1.6.4.1.4
Reduction of GeCl₄: 1.6.4.3.1
C₂H₅SiH₃
    Formation: 1.6.4.2.2, 1.6.5.2.1
C₂H₈Sn
  (CH₃)₂SnH₂
    Formation: 1.6.4.4.1, 1.6.4.4.2, 1.6.4.4.3,
    1.6.5.4.1
    Reaction with (n-C₄H₉)₂Sn: 1.6.4.4.3
  C₂H₅SnH₃
    Formation: 1.6.4.4.3
C₂H₉F₅NP
  [(CH₃)₂NH₂]PF₅H
    Formation: 1.5.3.2.3
C₂H₉GeP
  (CH₃)₂Ge(PH₂)H
    Redistribution: 1.6.4.3.3
C₂H₉NPS₃
  [(CH₃)₂NH₂][HP(S)S₂]₂
    Formation: 1.5.3.2.3
C₂H₉NaZn₂
  Na[Zn₂H₃(CH₃)₂]
    Formation and thermolysis:1.9.4.1,
    1.9.4.2
C₂H₁₀AlLiZn
  Li[H₂AlH₂ZnMe₂]
    Formation: 1.9.5.2
C₂H₁₀BGa
  (CH₃)₂GaBH₄
    Formation: 1.7.3.2, 1.7.5.2
C₂H₁₀BP
  (CH₃)₂PHBH₃
    Formation: 1.5.4.2.4
C₂H₁₀GeP₂
  (CH₃)₂Ge(PH₂)₂
    Formation: 1.6.4.3.3
    Redistribution to [(CH₃)₂Ge]₆P₄ and
    PH₃: 1.5.4.2.4
C₂H₁₀Ge₂
  C₂H₅Ge₂H₅
    Formation: 1.6.4.3.3
C₂H₁₀Si₂
  (CH₃)₂Si₂H₄
    Formation: 1.6.2.2
C₂H₁₁PSi₂
  (CH₃SiH₂)₂PH
    Formation: 1.5.4.2.4
C₂H₁₂Ge₃
  C₂H₅Ge₃H₇
    Formation: 1.6.4.3.3
C₂H₁₃B₅
  B₅H₈C₂H₅
    Formation: 1.6.4.1.4

C₃H₁₁AsGe

$(CH_3)_3GeAsH_2$
  Cleavage by protonic acids: 1.5.3.3.3

C₃H₁₁AsSi

$(CH_3)_3SiAsH_2$
  Cleavage by protonic acids: 1.5.3.3.3

C₃H₁₂BGeK

$K[H_3GeB(CH_3)_3]$
  Reaction with aq acid: 1.6.3.3.1

C₃H₁₂BN

$(CH_3)_3NBH_3$
  Formation: 1.7.2
  Reaction with HF: 1.5.3.1.3

C₃H₁₂Si₂

$CH_3SiH_2CH_2SiH_2CH_3$
  Formation: 1.6.4.2.3
$(CH_3)_3Si_2H_3$
  Formation: 1.6.2.2

C₃H₁₃BPb

$[(CH_3)_3Pb][BH_4]$
  Reaction with CH_3OH: 1.6.4.5.2

C₃H₁₅BGaN

$(CH_3)_3NGaH_2(BH_4)$
  Formation: 1.7.5.2

C₃Mg₂

$Mg_2[C_3]$
  Hydrolysis: 1.6.2.5

C₃O₂

  Reaction with hydrogen atoms: 1.6.2.1.1

C₄

  Reaction with hydrogen atoms: 1.6.2.1.1

C₄As₂F₁₂

$(CF_3)_4As_2$
  Reaction with $(CH_3)_2PH$: 1.5.4.3

C₄D₈

$CD_2CDCD_2CD_3$
  Formation: 1.6.7.2.1

C₄F₆

CF_3CCCF_3
  Reaction with $(CH_3)_2AsH$: 1.6.4.1.4

C₄F₁₂P₄

$(CF_3P)_4$
  Hydrolysis to $CF_3P(O)(OH)H$: 1.5.3.2.1
  Hydrolysis to CF_3PH_2: 1.5.3.2.1
  Reaction with H_2O: 1.5.3.2.3

C₄HCoO₄

$HCo(CO)_4$
  Catalysis of CO reaction with H_2:
  1.6.6.1
  Equilibrium acidity: 1.10.6.2.2
  Formation: 1.10.2, 1.10.3.2

Generation by protonation of a metal
anion: 1.10.6.2.1

C₄HFeKO₄

$K[HFe(CO)_4]$
  Protonation at metal: 1.10.6.2.1

C₄HN₃

$(CN)_3CH$
  Formation: 1.6.2.5, 1.6.3.1.3

C₄H₂FeO₄

$H_2Fe(CO)_4$
  Equilibrium acidity: 1.10.6.2.2
  Generation by protonation of a metal
  anion: 1.10.6.2.1

C₄H₂O₄Os

$H_2Os(CO)_4$
  Equilibrium acidity: 1.10.6.2.2
  Formation: 1.10.3.1
  Generation by protonation of a metal
  anion: 1.10.6.2.1
  Rate of formation by protonation of
  anion: 1.10.6.2.3

C₄H₂O₄Ru

$H_2Ru(CO)_4$
  Formation: 1.10.3.2

C₄H₄

  Formation: 1.6.2.5
  Formation from C_4 and H_2: 1.6.2.1.1

C₄H₄AlLiO₈

$LiAl(O_2CH)_4$
  Formation: 1.6.5.1.2

C₄H₄BrNO₂

$\overline{C(O)CH_2CH_2C}(O)NBr$
  Reaction with HBr: 1.5.3.1.3

C₄H₄O

  Reaction with H_2 over Ni: 1.6.6.1

C₄H₅NO₂

$\overline{C(O)CH_2CH_2C}(O)NH$
  Formation: 1.5.3.1.3

C₄H₆

$CH_2CH=CHCH_2$
  Reaction with $(CH_3)_3SiH$: 1.6.4.1.4
$CH_3CHC\equiv CH_2$
  Formation from C_4 and hydrogen:
  1.6.2.1.1
$CH_3C\equiv CCH_3$
  Reaction with H_2 over Pd–C: 1.6.2.1.2
  Reaction with Na in liq NH_3: 1.6.3.1.2
  Reaction with H_2 in presence of Pt:
  1.6.6.1
  Reaction with H_2O in presence of acid
  catalyst: 1.6.6.1

C$_6$BF$_4$MnO$_6$
  [Mn(CO)$_6$][BF$_4$]
    Reaction with H$_2$O: 1.10.9
C$_6$CrO$_6$
  Cr(CO)$_6$
    Reaction with [OH]$^-$: 1.10.9
C$_6$D$_6$
    Formation: 1.6.7.2.1
C$_6$F$_3$MnO$_8$
  [Mn(CO)$_5$][O$_3$SCF$_3$]
    Formation by protonation of HMn(C*
    O)$_5$: 1.10.6.2.1
C$_6$HF$_5$
  C$_6$F$_5$H
    Formation: 1.6.3.1.3
C$_6$HO$_6$V
  HV(CO)$_6$
    Equilibrium acidity: 1.10.6.2.2
C$_6$H$_4$BrO$_5$Re
  Br(OC)$_4$ReC(OH)CH$_3$
    Formation by protonation of an acyl
    anion: 1.10.6.2.1
C$_6$H$_4$F$_7$N$_2$P
  [p-FC$_6$H$_4$N$_2$][PF$_6$]
    Reaction with h^5-C$_5$H$_5$WH$_2$: 1.5.4.1.2
C$_6$H$_5$BiBr$_2$
  C$_6$H$_5$BiBr$_2$
    Reduction by Li[AlH$_4$]: 1.5.5.5
C$_6$H$_5$Cl$_2$OP
  C$_6$H$_5$P(O)Cl$_2$
    Reduction by Li[AlH$_4$]: 1.5.5.2.1
    Reaction with Li[AlD$_4$]: 1.5.7.1.2
    Reaction with HN[CH$_2$CH(CH$_3$)OH]$_2$:
    1.5.3.2.3
C$_6$H$_5$Cl$_2$P
  C$_6$H$_5$PCl$_2$
    Hydrolysis to form C$_6$H$_5$PH(O)OH:
    1.5.6.2
    Reaction with C$_2$H$_5$OH: 1.5.6.2
    Reaction with LiH: 1.5.4.2.1
    Reaction with HF: 1.5.3.2.3
    Reaction with diols: 1.5.3.2.3
    Reduction by (C$_6$H$_5$)$_2$SiH$_2$ or HSiCl$_3$:
    1.5.4.2.2
C$_6$H$_5$Cl$_3$Ge
  C$_6$H$_5$GeCl$_3$
    Reduction by LiAlH$_4$: 1.6.6.3
C$_6$H$_5$Cl$_3$Si
  C$_6$H$_5$SiCl$_3$
    Reduction by Li[AlH$_4$]: 1.6.5.2.1, 1.6.6.2
    Reduction by LiH: 1.6.6.2
C$_6$H$_5$Cl$_3$Sn
  C$_6$H$_5$SnCl$_3$
    Reduction by (C$_2$H$_5$)$_2$AlH: 1.6.4.4.1
    Reduction by Na[BH$_4$]: 1.6.5.4.2

C$_6$H$_5$D
    Formation: 1.6.7.1.1
C$_6$H$_5$D$_2$N
  C$_6$H$_5$ND$_2$
    Formation: 1.5.7.1.1
C$_6$H$_5$D$_2$P
  C$_6$H$_5$PD$_2$
    Formation: 1.5.7.1.2
C$_6$H$_5$F$_2$P
  C$_6$H$_5$PF$_2$
    Alcoholysis: 1.5.3.2.3
    Hydrolysis: 1.5.3.2.3
C$_6$H$_5$F$_6$N$_2$P
  C$_6$H$_5$N$_2$[PF$_6$]
    Reaction with h^5-C$_5$H$_5$WH$_2$: 1.5.4.1.2
C$_6$H$_5$I
  C$_6$H$_5$I
    Reduction by NaH: 1.6.4.1.1
C$_6$H$_5$I$_2$Sb
  C$_6$H$_5$SbI$_2$
    Reduction by Li[BH$_4$]: 1.5.5.4
C$_6$H$_5$NO
    Reaction with H$_2$: 1.5.2.1.2
    Reaction with D$_2$ over Pt: 1.5.7.1.1
C$_6$H$_5$NO$_2$
    Industrial conversion to C$_6$H$_5$NH$_2$:
    1.5.6.1
    Reaction with H$_2$: 1.5.2.1.2
    Reduction by Li[AlD$_4$]: 1.5.7.1.1
    Reduction by CaH$_2$: 1.5.4.1.1
    Reduction by carbon in steam: 1.5.6.1
    Reduction with D$_2$ over Pt: 1.5.7.1.1
C$_6$H$_5$Na$_2$P
  Na$_2$[C$_6$H$_5$P]
    Reaction with D$_2$O: 1.5.7.1.2
C$_6$H$_5$O$_5$Rh
  Rh(CO)$_2$CH$_3$COCH$_2$CO$_2$
    Catalyst in formation of (CH$_2$OH)$_2$:
    1.6.2.1.2
C$_6$H$_6$
  C$_6$H$_6$
    Exchange with D$_2$, DF or DBr: 1.6.7.2.1
    Protonation in strong-acid media:
    1.6.3.1.3
    Reaction with AsCl$_3$: 1.5.4.3,1.5.6.3
    Reaction with H$_2$: 1.6.2.1.2
  CH$_3$C≡CC≡CCH$_3$
    Reaction with R$_2$SnH$_2$:1.6.4.1.4
C$_6$H$_6$B$_2$Fe$_2$O$_6$
  Fe$_2$(B$_2$H$_6$)(CO)$_6$
    Thermolysis: 1.10.5.2
C$_6$H$_6$FOP
  C$_6$H$_5$P(O)FH
    Disproportionation to C$_6$H$_5$PF$_2$ and
    C$_6$H$_5$PF$_3$H: 1.5.3.2.3

Formation: 1.5.3.2.3

C₆H₆F₂Ge
C₆H₅GeF₂H
Formation: 1.6.4.3.1

C₆H₆NO₂Re
h⁵-C₅H₅Re(CO)(NO)H
Formation: 1.10.9

C₆H₆O₂
1,2-(HO)₂C₆H₄
Reaction with C₆H₄O₂P(Cl)NSO₂C₆*
H₄Cl: 1.5.3.1.3

C₆H₇As
C₆H₅AsH₂
Formation: 1.5.3.3.1, 1.5.4.3, 1.5.5.3.1,
1.5.6.3

C₆H₇AsF₆
(CH₃)₂AsC(CF₃)=C(CF₃)H
Formation: 1.6.4.1.4

C₆H₇AsO₂
C₆H₅As(OH)₂
Reduction by Zn amalgam in acid:
1.5.3.3.1

C₆H₇N
C₆H₅NH₂
Formation: 1.5.2.1.2, 1.5.3.1.2, 1.5.3.1.3,
1.5.4.1.1, 1.5.4.1.3
Industrial synthesis: 1.5.6.1
Protonation in H₂O: 1.5.3.1.1
Reaction with [(C₂H₅)₂N]₃P: 1.5.3.1.3

C₆H₇O₂P
C₆H₅PH(O)OH
Industrial synthesis: 1.5.6.2
Thermal disproportionation: 1.5.6.2

C₆H₇O₃P
C₆H₅P(O)(OH)₂
Reduction by LiAlH₄: 1.5.5.2.2

C₆H₇P
C₆H₅PH₂
Formation: 1.5.3.2.3, 1.5.5.2.1, 1.5.5.2.2
Reaction with (C₆H₅P)₅: 1.5.4.2.4
Synthesis from C₆H₅PH(O)OH: 1.5.6.2

C₆H₇Sb
C₆H₅SbH₂
Formation: 1.5.5.4

C₆H₈Ge
C₆H₅GeH₃
Disproportionation at 200°C: 1.6.4.3.3
Industrial synthesis: 1.6.6.3

C₆H₈NP
o-NH₂C₆H₄PH₂
Formation: 1.5.5.2.2

C₆H₈Si
C₆H₅SiH₃
Formation: 1.6.5.2.1
Industrial synthesis: 1.6.6.2

C₆H₈Sn
C₆H₅SnH₃
Formation: 1.6.4.4.3, 1.6.5.4.2

C₆H₉B
(CH₂=CH)₃B
Reaction with HCl: 1.6.3.1.3

C₆H₉Cl₅NSb
[NH₄][C₆H₅SbCl₅]
Reduction by LiAlH₄: 1.5.5.4

C₆H₉F₆NSn
(CH₃)₃SnNC(CF₃)₂
Reaction with PtHCl[P(C₆H₅)₃]₂:
1.5.4.1.2

C₆H₉MoNO₄P
h⁵-Mo(CO)₂P(OCH₂CH₂)₂NH
Formation: 1.5.3.1.3

C₆H₉O₃P
(CH₃CO)₃P
Reaction with CH₃OH: 1.5.3.2.3

C₆H₁₀
Formation: 1.6.4.1.2
Reaction with [(CH₃)₂Si]₆: 1.6.4.2.3

C₆H₁₀N₂O₄
(C₂H₅CO₂N)₂
Formation: 1.5.3.1.3

C₆H₁₀Pb
(C₂H₃)₃PbH
Formation: 1.6.4.5.3

C₆H₁₁BrMg
BrMgCH₂CH(C₂H₅)CH=CH₂
Thermal rearrangement: 1.6.4.1.5

C₆H₁₁ClMg
C₆H₁₁MgCl
Reaction with (C₆H₅)₃SiCl: 1.6.4.2.1

C₆H₁₁NO₂
Dehydrogenation-reduction: 1.5.2.1.2

C₆H₁₂
(CH₃)₃CCH=CH₂
Formation: 1.6.2.1.2
Reaction with HCl: 1.6.6.1
n-C₄H₉CHCH₂
Reaction with CO and H₂: 1.6.2.1.2

C₆H₁₂NO₃P
P(OCH₂CH₂)₃N
Protonation by [R₃O][BF₄]: 1.5.3.2.3

C₆H₁₃BF₄NO₃P
[HP(OCH₂CH₂)₃N]BF₄
Formation: 1.5.3.2.3

C₆H₁₃ClMoN₃O₆
[Mo(CH₂CH₂NH)₃(NO)₂H][ClO₄]
Equilibrium acidity at metal: 1.10.6.1.2

C₆H₁₃Cl₃Si
C₆H₁₃SiCl₃
Reduction by Li[AlH₄]: 1.6.6.2

C₈H₁₁N₂OP
C₆H₄O(NH)PN(CH₃)₂
Reaction with diols: 1.5.3.2.3

C₈H₁₁O₂P
C₆H₅PH(O)OC₂H₅
Synthesis from C₆H₅PCl₂ and
C₂H₅OH: 1.5.6.2

C₈H₁₁O₃P
C₆H₅CH(OH)CH₂PH(O)OH
Industrial synthesis: 1.5.6.2

C₈H₁₁P
(C₂H₅)C₆H₅PH
Formation: 1.5.3.2.3

C₈H₁₂CrF₉P₃
h⁵-C₈H₁₁CrH(PF₃)₃
Formation: 1.10.2

C₈H₁₂GeO₈Sn
(CH₃CO₂)₃SnGeCO₂CH₃
Reduction by LiAlH₄: 1.6.5.4.2

C₈H₁₂Mo₂O₈
Mo₂(O₂CMe)₄
Reaction with HCl: 1.10.5.1
Reaction with H₂ and PMe₃: 1.10.3.2

C₈H₁₂Pb
(CH=CH₂)₄Pb
Reaction with B₂H₆: 1.6.4.5.3

C₈H₁₃Co
HCo(C₄H₆-1,3)₂
Formation: 1.10.2

C₈H₁₆
(CH₃)₄C₄H₄
Formation: 1.6.2.5

C₈H₁₆Si
C₆H₉(CH₃)₂SiH
Formation: 1.6.4.2.3

C₈H₁₇Br
CH₃(CH₂)₆CH₂Br
Reduction by (KCuH)ₙ: 1.6.5.1.1

C₈H₁₇N₄P
[(CH₂)₂]₄N₄PH
Formation: 1.5.3.2.3

C₈H₁₈
CH₃(CH₂)₆CH₃
Formation: 1.6.5.1.1
(CH₃)₂CHCH₂C(CH₃)₃
Formation: 1.6.6.1

C₈H₁₈AlD
(i-C₄H₉)₂AlD
Formation: 1.7.4.2

C₈H₁₈Be
[(CH₃)₃C]₂Be
Pyrolysis to form BeH₂: 1.8.3.1

C₈H₁₈ClOP
(C₄H₉)₂P(O)Cl
Reduction by HSiCl₃: 1.5.6.2
(n-C₄H₉)₂P(O)Cl
Reduction by HSiCl₃: 1.5.4.2.2

C₈H₁₈ClSb
(n-C₄H₉)₂SbCl
Reduction by LiAlH₄: 1.5.5.4

C₈H₁₈Cl₂Pb
(n-C₄H₉)₂PbCl₂
Reduction by LiAlH₄: 1.6.5.5

C₈H₁₈Cl₂Sn₂
(n-C₄H₉SnCl)₂
Reduction by LiAlH₄: 1.6.5.4.1

C₈H₁₈KP₃
K[(t-C₄H₉P)₂P]
Protonolysis to (t-C₄H₉P)₂PH: 1.5.3.2.3

C₈H₁₈K₂P₂
K₂(t-C₄H₉P)₂]
Protonolysis to (t-C₄H₉P)₂H₂: 1.5.3.2.3

C₈H₁₈OSn
(n-C₄H₉)₂SnO
Reaction with (n-C₄H₉)₂Sn(Cl)H:
1.6.4.4.2

C₈H₁₈Si
(CH₃)₂Si[(CH₃)₂C]₂
Reaction with (C₂H₅)₃SiH: 1.6.4.2.3
(C₂H₅)₃SiCH=CH₂
Reaction with (C₂H₅O)₂P(S)SH:
1.6.4.1.4

C₈H₁₈Sn
(n-C₄H₉)₂Sn
Reaction with (CH₃)₂SnH₂: 1.6.4.4.3

C₈H₁₈Zn
(s-C₄H₉)₂Zn
Reduction to K₂ZnH₄ by KH: 1.9.4.1

C₈H₁₉Al
(C₄H₉)₂AlH
Reaction with [(CH₃)₃Si]₂O: 1.6.4.2.2
Reduction of aminostannanes: 1.6.4.4.3
(i-C₄H₉)₂AlH
Formation: 1.7.3.3, 1.7.4.2
Reaction with GeO₂: 1.6.4.3.2
Reduction of C₆H₅CH=CH₂: 1.6.4.1.4
(n-C₄H₉)₂AlH
Reduction of aminostannanes: 1.6.4.4.3
(t-C₄H₉)₂AlH
Reduction of acyl metal complexes:
1.6.4.1.2

C₈H₁₉As
(n-C₄H₉)₂AsH
Formation: 1.5.5.3.1

C$_9$H$_2$O$_9$Os$_3$S
  Os$_3$H$_2$(S)(CO)$_9$
    Formation: 1.10.5.6.2
C$_9$H$_3$Cl$_9$O$_9$Os$_3$Si$_3$
  Os$_3$(μ-H)$_3$(SiCl$_3$)$_3$(CO)$_9$
    Reaction with HSiCl$_3$: 1.10.5.4
C$_9$H$_5$NbO$_4$
  h^5-C$_5$H$_5$Nb(CO)$_4$
    Protonation at metal: 1.10.6.1.1
C$_9$H$_5$O$_4$V
  h^5-C$_5$H$_5$V(CO)$_4$
    Reaction with borohydride: 1.10.6.3
C$_9$H$_6$AsCrF$_6$PO$_5$
  (CF$_3$)$_2$PAs(CH$_3$)$_2$Cr(CO)$_5$
    Reaction with (CH$_3$)$_3$SnH: 1.5.4.3
C$_9$H$_6$AsF$_6$MoO$_5$PO$_5$
  (CF$_3$)$_2$PAs(CH$_3$)$_2$Mo(CO)$_5$
    Reaction with (CH$_3$)$_3$SnH: 1.5.4.3
C$_9$H$_6$As$_2$CrF$_6$O$_5$
  (CH$_3$)$_2$AsAs(CF$_3$)$_2$Cr(CO)$_5$
    Reaction with (CH$_3$)$_3$SnH: 1.5.4.3
C$_9$H$_6$As$_2$F$_6$MoO$_5$
  (CH$_3$)$_2$AsAs(CF$_3$)$_2$Mo(CO)$_5$
    Reaction with (CH$_3$)$_3$SnH: 1.5.4.3
C$_9$H$_6$CrF$_6$O$_5$P$_2$
  (CH$_3$)$_2$PP(CF$_3$)$_2$Cr(CO)$_5$
    Reaction with (CH$_3$)$_3$SnH: 1.5.3.2.3
C$_9$H$_6$CrO$_3$
  h^6-C$_6$H$_6$Cr(CO)$_3$
    Exchange with C$_2$H$_5$OD in presence of
    [OD]$^-$: 1.6.7.2.1
C$_9$H$_6$F$_4$FeO$_2$
  h^5-C$_5$H$_5$Fe(CO)$_2$C$_2$F$_4$H
    Formation: 1.6.4.1.4
C$_9$H$_6$F$_6$MoO$_5$P$_2$
  (CH$_3$)$_2$PP(CF$_3$)$_2$Mo(CO)$_5$
    Reaction with (CH$_3$)$_3$SnH: 1.5.3.2.3
C$_9$H$_7$F$_8$OP
  C$_6$H$_5$P[OCH(CF$_3$)$_2$]F$_2$H
    Formation: 1.5.3.2.3
C$_9$H$_7$MnO$_3$
  C$_5$H$_4$CH$_3$-h^{5}Mn(CO)$_3$
    Reaction with R(+)-SiHMePh(1-
    C$_{10}$H$_7$): 1.10.5.4
  h^5-C$_5$H$_5$CH$_3$Mn(CO)$_3$
    Formation: 1.6.4.1.4
C$_9$H$_9$O$_3$Re
  h^5-C$_5$H$_5$(OC)$_2$Re=C(OH)CH$_3$
    Formation by protonation of an acyl
    anion: 1.10.6.2.1
C$_9$H$_{10}$
  C$_6$H$_5$C(CH$_3$)=CH$_2$
    Reaction with HMn(CO)$_5$: 1.6.4.1.4
    Reaction with CH$_3$SiCl$_2$H: 1.6.4.1.4

  C$_6$H$_5$CH$_2$CH=CH$_2$
    Electrochemical reduction: 1.6.3.1.3
C$_9$H$_{11}$CoN$_2$O$_2$
  h^5-C$_5$H$_5$Co(ON)$_2$C$_2$(CH$_3$)$_2$
    Reduction by Li[AlH$_4$]: 1.5.5.1
C$_9$H$_{12}$
  C$_6$H$_5$CH(CH$_3$)$_2$
    Formation: 1.6.4.1.4
  C$_6$H$_7$CH$_2$CHCH$_2$
    Formation: 1.6.3.1.3
C$_9$H$_{12}$Cl$_2$Si
  C$_6$H$_5$C(CH$_3$)HCH$_2$SiCl$_2$H
    Formation: 1.6.4.1.4
C$_9$H$_{12}$CrO$_5$P$_2$
  (CH$_3$)$_2$PP(CH$_3$)$_2$Cr(CO)$_5$
    Reaction with (CH$_3$)SnH: 1.5.3.2.3
C$_9$H$_{12}$FeO
  h^5-C$_5$H$_5$Fe(CO)CH(CH$_3$)$_2$
    Formation: 1.6.5.1.4
C$_9$H$_{12}$MoO$_5$P$_2$
  (CH$_3$)$_2$PP(CH$_3$)$_2$Mo(CO)$_5$
    Reaction with (CH$_3$)$_3$SnH: 1.5.3.2.3
C$_9$H$_{15}$AsGe
  (CH$_3$)$_3$GeAsH(C$_6$H$_5$)
    Redistribution: 1.5.4.3
C$_9$H$_{15}$AsSi
  (CH$_3$)$_3$SiAsH(C$_6$H$_5$)
    Redistribution: 1.5.4.3
C$_9$H$_{15}$PSi
  (CH$_3$)$_3$SiP(H)C$_6$H$_5$
    Cleavage by CH$_3$OH: 1.5.3.2.3
    Formation: 1.5.3.2.3
C$_9$H$_{16}$BF$_4$MoN$_3$O$_3$
  [Mo(CH$_2$CH$_2$NH)$_3$(CO)$_3$H][BF$_4$]
    Equilibrium acidity at metal: 1.10.6.1.2
C$_9$H$_{18}$BD$_3$
  (CH$_3$CH$_2$CHD)$_3$B
    Formation: 1.6.7.1.1
C$_9$H$_{19}$N$_4$P
  [(CH$_2$)$_2$]$_3$(CH$_2$)$_3$N$_4$PH
    Formation: 1.5.3.2.3
C$_9$H$_{21}$B
  (CH$_3$CH$_2$CH$_2$)$_3$B
    Formation: 1.6.4.1.4
C$_9$H$_{21}$BO$_3$
  B[(CH$_3$)$_2$CHO]$_3$
    Reduction: 1.7.3.2
  B(OC$_3$H$_7$-i)$_3$
    Reduction with NaAlH$_4$: 1.7.5.1
  B(OC$_3$H$_7$)$_3$
    Formation: 1.6.4.1.2
    Reaction with H$_2$O: 1.6.4.1.2
    Reduction: 1.7.3.2

C$_{12}$D$_4$O$_{12}$Ru$_4$
  D$_4$Ru$_4$(CO)$_{12}$
    Formation: 1.10.4.3.3
C$_{12}$F$_{10}$Zn
  (C$_6$F$_5$)$_2$Zn
    Forms Na$_2$[ZnH(C$_6$F$_5$)$_2$]$_2$ with NaH:
    1.9.4.1
C$_{12}$FeO$_{12}$Ru$_2$
  FeRu$_2$(CO)$_{12}$
    Reaction with H$_2$: 1.10.4.3.3
C$_{12}$Fe$_3$O$_{12}$
  Fe$_3$(CO)$_{12}$
    Reaction with alkanethiols: 1.10.5.6.2
C$_{12}$HCo$_3$FeO$_{12}$
  HFeCo$_3$(CO)$_{12}$
    Formation by protonation of metal
    core: 1.10.6.2.4
C$_{12}$HCo$_3$O$_{12}$Os
  HOsCo$_3$(CO)$_{12}$
    Formation by protonation of metal
    core: 1.10.6.2.4
C$_{12}$HF$_6$O$_{12}$Os$_3$P
  [HOs$_3$(CO)$_{12}$][PF$_6$]
    Formation by protonation of a metal-
    –metal bond: 1.10.6.1.4
C$_{12}$HFe$_4$NO$_{12}$
  HFe$_4$N(CO)$_{12}$
    Formation by protonation of metal
    core: 1.10.6.2.4
C$_{12}$HMnO$_{12}$Os$_2$
  HMnOs$_2$(CO)$_{12}$
    Formation by protonation of metal
    core: 1.10.6.2.4
C$_{12}$H$_2$Co$_2$O$_{12}$Ru$_2$
  H$_2$Ru$_2$Co$_2$(CO)$_{12}$
    Formation: 1.10.4.3.3
C$_{12}$H$_2$D$_2$O$_{12}$Ru$_4$
  H$_2$D$_2$Ru$_4$(CO)$_{12}$
    Formation: 1.10.4.3.3
C$_{12}$H$_2$F$_3$NO$_{10}$Os$_3$
  HOs$_3$(CO)$_{10}$NCHCF$_3$
    Reaction with H$_2$: 1.5.2.1.2
C$_{12}$H$_2$O$_{11}$Os$_3$
  Os$_3$(CO)$_{11}$CH$_2$
    Formation: 1.6.3.1.3
C$_{12}$H$_3$BFe$_4$O$_{12}$
  Fe$_4$H(BH$_2$)(CO)$_{12}$
    Formation: 1.10.5.2
C$_{12}$H$_3$BO$_{12}$Ru$_4$
  Ru$_4$H(BH$_2$)(CO)$_{12}$
    Formation: 1.10.5.2

C$_{12}$H$_3$CoO$_{12}$Ru$_3$
  H$_3$CoRu$_3$(CO)$_{12}$
    Formation: 1.10.4.3.3
C$_{12}$H$_3$Co$_3$O$_{10}$
  Co$_3$(CO)$_9$CCOCH$_3$
    Reaction with (C$_2$H$_5$)$_3$SiH: 1.6.4.1.2
C$_{12}$H$_3$O$_{12}$Re$_3$
  H$_3$Re$_3$(CO)$_{12}$
    Formation: 1.10.9
  Re$_3$H$_3$(CO)$_{12}$
    Formation: 1.10.3.2
C$_{12}$H$_4$F$_3$NO$_{10}$Os$_3$
  HOs$_3$(CO)$_{10}$HNCH$_2$CF$_3$
    Formation: 1.5.2.1.2
C$_{12}$H$_4$FeO$_{12}$Os$_3$
  H$_4$FeOs$_3$(CO)$_{12}$
    Formation: 1.10.4.3.4
C$_{12}$H$_4$FeO$_{12}$Ru$_3$
  H$_4$FeRu$_3$(CO)$_{12}$
    Equilibrium acidity and rate of forma-
    tion by protonation of an anion:
    1.10.6.2.4
    Formation: 1.10.3.2, 1.10.4.3.3
C$_{12}$H$_4$Fe$_3$O$_{11}$
  HFe$_3$(μ-COCH$_3$)(CO)$_{10}$
    Reaction with H$_2$ and SbPh$_3$: 1.10.3.2
    Reaction with H$_2$: 1.10.4.3.4
C$_{12}$H$_4$O$_{11}$Os$_3$
  HOs$_3$(μ-COCH$_3$)(CO)$_{10}$
    Reaction with H$_2$: 1.10.3.2, 1.10.4.3.4
C$_{12}$H$_4$O$_{11}$Ru$_3$
  HRu$_3$(μ-COCH$_3$)(CO)$_{10}$
    Reaction with H$_2$: 1.10.3.2, 1.10.4.3.3
C$_{12}$H$_4$O$_{12}$Os$_4$
  H$_4$Os$_4$(CO)$_{12}$
    Equilibrium acidity and rate of forma-
    tion by protonation of an anion:
    1.10.6.2.4
    Formation: 1.10.3.2
C$_{12}$H$_4$O$_{12}$Re$_4$
  H$_4$Re$_4$(CO)$_{12}$
    Formation: 1.10.3.2
C$_{12}$H$_4$O$_{12}$Ru$_4$
  H$_4$Ru$_4$(CO)$_{12}$
    Equilibrium acidity and rate of forma-
    tion by protonation of an anion:
    1.10.6.2.4
    Formation: 1.10.6.2.4, 1.10.3.2, 1.10.4.3.3,
    1.10.9

C$_{12}$H$_4$O$_{16}$Re$_4$
Re(OH)$_4$(CO)$_{12}$
Formation: 1.10.5.6.1

C$_{12}$H$_5$Co$_3$O$_9$
Co$_3$(CO)$_9$CCH$_2$CH$_3$
Formation: 1.6.4.1.2

C$_{12}$H$_6$O$_{10}$Os$_3$S
HOs$_3$(CO)$_{10}$SC$_2$H$_5$
Formation: 1.10.5.6.2

C$_{12}$H$_6$O$_{10}$Ru$_3$S
HRu$_3$(SC$_2$H$_5$)(CO)$_{10}$
Formation: 1.10.5.6.2

C$_{12}$H$_8$Cl$_2$NO$_4$PS
C$_6$H$_4$O$_2$P(Cl)NSO$_2$C$_6$H$_4$Cl
Reaction with 1,2-(HO)$_2$C$_6$H$_4$: 1.5.3.1.3

C$_{12}$H$_9$N$_3$
C$_6$H$_5$C$_6$H$_4$N$_3$
Reaction with H$_2$: 1.5.2.1.2

C$_{12}$H$_{10}$BiCl
(C$_6$H$_5$)$_2$BiCl
Reduction by Li[AlH$_4$]: 1.5.5.5

C$_{12}$H$_{10}$Br$_2$Ge
(C$_6$H$_5$)$_2$GeBr$_2$
Reduction by Li[AlH$_4$]: 1.6.5.3.1

C$_{12}$H$_{10}$ClOP
(C$_6$H$_5$)$_2$P(O)Cl
Reduction by Li[AlH$_4$]: 1.5.5.2.1

C$_{12}$H$_{10}$ClP
(C$_6$H$_5$)$_2$PCl
Hydrolysis: 1.5.6.2
Reduction by Na[(CH$_3$OCH$_2$CH$_2$O)$_2$*
AlH$_2$]: 1.5.5.2.1
Reduction by metals: 1.5.6.2
Reduction by (C$_6$H$_5$)$_2$SiH$_2$ or HSiCl$_3$:
1.5.4.2.2

C$_{12}$H$_{10}$ClSb
(C$_6$H$_5$)$_2$SbCl
Reduction by Li[BH$_4$]: 1.5.5.4

C$_{12}$H$_{10}$Cl$_2$Ge
(C$_6$H$_5$)$_2$GeCl$_2$
Electrochemical reduction in H$_2$O:
1.6.3.3.1
Redistribution with germanes: 1.6.4.3.1

C$_{12}$H$_{10}$Cl$_2$Si
(C$_6$H$_5$)$_2$SiCl$_2$
Reduction by Li[AlH$_4$]: 1.6.5.2.1

C$_{12}$H$_{10}$Cl$_2$Sn
(C$_6$H$_5$)$_2$SnCl$_2$
Electrolytic reduction to (C$_6$H$_5$)$_2$SnH$_2$:
1.6.3.4.1
Reduction by Li[AlH$_4$]: 1.6.5.4.1

Reduction by (C$_2$H$_5$)$_2$AlH: 1.6.4.4.1

C$_{12}$H$_{10}$Cl$_3$P
(C$_6$H$_5$)$_2$PCl$_3$
Reduction by Li[AlH$_4$]: 1.5.5.2.1

C$_{12}$H$_{10}$DP
(C$_6$H$_5$)$_2$PD
Formation: 1.5.7.1.2

C$_{12}$H$_{10}$D$_2$Si
(C$_6$H$_5$)$_2$SiD$_2$
Reduction of halophosphines: 1.5.7.1.2

C$_{12}$H$_{10}$LiP
(C$_6$H$_5$)$_2$PLi
Reaction with D$_2$O: 1.5.7.1.2

C$_{12}$H$_{10}$LiSb
(C$_6$H$_5$)$_2$SbLi
Reaction with H$_2$O: 1.5.3.4

C$_{12}$H$_{10}$NaP
Na[((C$_6$H$_5$)P]
Protonation to (C$_6$H$_5$)$_2$PH: 1.5.3.2.2

C$_{12}$H$_{10}$Ni$_2$O$_2$
Ni$_2$(C$_5$H$_5$)$_2$(CO)$_2$
Reaction with H$_2$ and Os$_3$(CO)$_{12}$:
1.10.3.2

C$_{12}$H$_{10}$O$_2$Ti
(h^5-C$_5$H$_5$)$_2$Ti(CO)$_2$
Reaction with H$_2$: 1.6.2.1.2

C$_{12}$H$_{10}$O$_2$Zr
(h^5-C$_5$H$_5$)$_2$Zr(CO)$_2$
Reaction with H$_2$: 1.6.2.1.2

C$_{12}$H$_{11}$As
(C$_6$H$_5$)$_2$AsH
Formation: 1.5.3.3.3, 1.5.4.3, 1.5.5.3.1,
1.5.6.3

C$_{12}$H$_{11}$BrSn
(C$_6$H$_5$)$_2$SnH(Br)
Formation: 1.6.4.4.1

C$_{12}$H$_{11}$ClGe
(C$_6$H$_5$)$_2$GeHCl
Formation: 1.6.4.3.1

C$_{12}$H$_{11}$ClSi
(C$_6$H$_5$)$_2$SiHCl
Formation: 1.6.4.2.3, 1.6.4.3.1

C$_{12}$H$_{11}$ClSn
(C$_6$H$_5$)$_2$SnHCl
Formation: 1.6.3.4.3, 1.6.4.4.1

C$_{12}$H$_{11}$FSn
(C$_6$H$_5$)$_2$SnHF
Formation: 1.6.4.4.1

C$_{12}$H$_{11}$N
C$_6$H$_5$C$_6$H$_4$NH$_2$
Formation: 1.5.2.1.2, 1.9.4.2

C$_{12}$H$_{11}$NaZn
Na[ZnH(C$_6$H$_5$)$_2$]
From NaZnH(C$_6$H$_5$)$_2$ with (C$_6$H$_5$)$_2$Zn:
1.9.4.2

C$_{12}$H$_{11}$OP
(C$_6$H$_5$)$_2$P(O)H
Formation: 1.5.3.2.3
Industrial formation: 1.5.6.2

C$_{12}$H$_{11}$O$_2$P
(C$_6$H$_5$)$_2$P(O)OH
Reduction by LiAlH$_4$: 1.5.5.2.2

C$_{12}$H$_{11}$O$_3$P
(C$_6$H$_5$O)$_2$P(O)H
Formation: 1.5.3.2.2

C$_{12}$H$_{11}$P
(C$_6$H$_5$)$_2$PH
Formation: 1.5.3.2.1, 1.5.3.2.2, 1.5.3.2.3,
1.5.5.2.1, 1.5.5.2.2, 1.5.5.2.3
Industrial formation: 1.5.6.2

C$_{12}$H$_{11}$Sb
(C$_6$H$_5$)$_2$SbH
Formation: 1.5.5.4

C$_{12}$H$_{12}$CrO$_3$
h^6-C$_6$H$_3$Me$_3$Cr(CO)$_3$
Protonation at metal: 1.10.6.1.1

C$_{12}$H$_{12}$Ge
(C$_6$H$_5$)$_2$GeH$_2$
Exchange with halogermanes: 1.6.4.3.1
Formation: 1.6.3.3.1, 1.6.4.3.3, 1.6.5.3.1

C$_{12}$H$_{12}$MoO$_2$
h^5-C$_5$H$_5$Mo(CO)$_2$C$_5$H$_7$-h^3
Formation: 1.10.5.3

C$_{12}$H$_{12}$Os$_3$
Os$_3$(CO)$_{12}$
Reaction with H$_2$: 1.10.3.2

C$_{12}$H$_{12}$P$_2$
(C$_6$H$_5$PH)$_2$
Formation: 1.5.4.2.4

C$_{12}$H$_{12}$Si
(C$_6$H$_5$)$_2$SiH$_2$
Formation: 1.6.4.2.3, 1.6.5.2.1, 1.6.5.2.2
Reaction with C$_5$H$_{10}$P(O)OH: 1.5.4.2.2
Reduction of GeCl$_4$: 1.6.4.3.1
Reduction of chlorophosphines: 1.5.4.2.2

C$_{12}$H$_{12}$Sn
(C$_6$H$_5$)$_2$SnH$_2$
Exchange reaction with dihalostan-
nanes: 1.6.4.4.1
Formation: 1.6.3.4.1, 1.6.4.4.1, 1.6.4.4.2,
1.6.4.4.3, 1.6.5.4.1
Reaction with (C$_2$H$_5$)$_3$SnN(C$_6$H$_5$)COH:
1.5.4.1.3

Reaction with CH$_3$OH: 1.6.3.4.3

C$_{12}$H$_{13}$N$_2$PS
(C$_6$H$_5$)$_2$P(S)NHNH$_2$
Redistribution: 1.5.4.1.3

C$_{12}$H$_{14}$W
C$_6$H$_5$CH$_3$WHC$_5$H$_5$-h^5
Formation: 1.10.2

C$_{12}$H$_{15}$F$_6$N$_2$PW
[(h^5-C$_5$H$_5$)W(H$_2$NNC$_6$H$_5$CH$_3$-p)][PF$_6$]
Formation: 1.5.4.1.2

C$_{12}$H$_{15}$IrO$_2$
Ir(C$_5$Me$_5$-h^5)(CO)$_2$
Reaction with cyclohexane: 1.10.5.3

C$_{12}$H$_{15}$KO$_2$Ru
K[h^5-C$_5$(CH$_3$)$_5$Ru(CO)$_2$]
Protonation at metal: 1.10.6.2.1

C$_{12}$H$_{15}$O$_2$Rh
h^5-C$_5$(CH$_3$)$_5$Rh(CO)$_2$
Reaction with acid to give a dinuclear
complex: 1.10.6.1.4

C$_{12}$H$_{15}$O$_3$P
C$_6$H$_5$CCP(O)(OC$_2$H$_5$)$_2$
Reduction of LiAlH$_4$: 1.5.5.2.2

C$_{12}$H$_{15}$O$_4$P
CH$_3$(C$_6$H$_5$)P(H)C(CO$_2$CH$_3$)$_2$
Tautomeric exchange with CH$_3$(C$_6$H$_5$)*
PCH(CO$_2$CH$_3$)$_2$: 1.5.3.2.3
CH$_3$(C$_6$H$_5$)P(H)(CCO$_2$CH$_3$)$_2$
Tautomeric exchange with CH$_3$(C$_6$H$_5$)*
P(H)C(CO$_2$CH$_3$)$_2$: 1.5.3.2.3

C$_{12}$H$_{16}$BF$_4$IrO$_2$
[h^5-C$_5$(CH$_3$)$_5$Ir(CO)$_2$H][BF$_4$]
Formation by metal protonation:
1.10.6.1.1

C$_{12}$H$_{16}$NO$_3$P
N[CH$_2$C(CH$_3$)O]$_2$PH(OC$_6$H$_5$)
Formation in d,l- and meso-forms:
1.5.3.2.3

C$_{12}$H$_{16}$O$_2$Os
[h^5-C$_5$(CH$_3$)$_5$]Os(CO)$_2$H
Reaction with H$_2$: 1.10.4.3.4

C$_{12}$H$_{16}$O$_2$Ru
h^5-C$_5$(CH$_3$)$_5$Ru(CO)$_2$H
Formation by protonation of a metal
anion: 1.10.6.2.1

C$_{12}$H$_{16}$Ti
(h^5-C$_5$H$_5$)$_2$Ti(CH$_3$)$_2$
Reaction with DCl: 1.6.7.1.1

C$_{12}$H$_{17}$Ta
(h^5-C$_5$H$_4$CH$_3$)$_2$TaH$_3$
Formation: 1.10.7.2

C₁₅H₁₈Ta
H₃Ta(h⁵-C₅H₅)₃
  Catalyst in D₂ exchange with C₆H₆:
  1.6.7.2.1

C₁₅H₁₉O₃Re
[h⁵-(CH₃)₆C₆H] Re(CO)₃
  Formation: 1.6.5.1.4

C₁₅H₂₁CoO₆
Co(acac)₃
  Reduction with (i-Bu)₂AlOEt: 1.10.8.2
  Reduction with (i-Bu)₃Al: 1.10.8.2

C₁₅H₂₁MoO₆
Mo(acac)₃
  Reaction with AlEt₃: 1.10.8.2

C₁₅H₂₁Ni₃P₃
[h⁵-(C₅H₅)NiPH₂]₃
  Formation: 1.5.3.2.3

C₁₅H₂₇O₂PSn
(C₂H₅)₃SnCH₂P(O)(C₆H₅)OC₂H₅
  Reduction with LiAlH₄: 1.6.5.4.3

C₁₅H₃₃PPt
h³-C₃H₅PtHP(CMe₃)₃
  Formation: 1.10.9

C₁₅H₃₇ClP₂Pt
trans-(n-Pr)PtCl(PEt₃)₂
  Formation: 1.10.8.4

C₁₅H₄₅FeO₁₅P₅
Fe[P(OMe)₃]₅
  Protonation at metal: 1.10.6.1.1

C₁₅H₄₆F₆FeO₁₅P₆
[HFe[P(OMe)₃]₅][PF₆]
  Formation by metal protonation:
  1.10.6.1.1

C₁₅H₄₆N₃PSi₅
{[(CH₃)₃Si]₂N}₂P(H)NSi(CH₃)₃
  Formation: 1.5.3.2.3

C₁₆H₇NO₁₀Os₃
Os₃H(NHPh)(CO)₁₀
  Formation: 1.10.5.5.1

C₁₆H₉NO₁₁Ru₃
Ru₃(CO)₁₁[CN(C₄H₉-t)]
  Reaction with H₂: 1.10.3.2, 1.10.4.3.3

C₁₆H₁₀Cr₂O₆
Cr₂(C₅H₅)₂(CO)₆
  Reaction with H₂: 1.10.3.2

C₁₆H₁₀O₆W₂
[h⁵-C₅H₅W(CO)₃]₂
  Protonation at metal–metal bond:
  1.10.6.1.4

C₁₆H₁₁F₆O₆PW₂
[[h⁵-C₅H₅W(CO)₃]₂H][PF₆]
  Formation by protonation of a metal-
  -metal bond: 1.10.6.1.4

C₁₆H₁₂MnNO₁₂Os₂
[Me₄N][MnOs₂(CO)₁₂]
  Protonation at metal core: 1.11.6.2.4

C₁₆H₁₃NO₁₁Ru₄
H₄Ru₄(CO)₁₁[CN(C₄H₉-t)]
  Formation: 1.10.4.3.3

C₁₆H₁₄O₃Ru₂
[Ru₂(CO)₃(h⁵-C₅H₅)₂(h³-C₃H₄)]
  Reaction with HBF₄: 1.6.3.1.3

C₁₆H₁₅BF₄O₃Ru₂
[Ru₂(CO)₃(n⁵-C₅H₅)₂C(CH₃)CH₂][BF₄]
  Formation: 1.6.3.1.3

C₁₆H₁₅IrO₂
h⁵-C₅Me₅Ir(CO)₂
  Protonation at metal: 1.10.6.1.1

C₁₆H₂₀NP
(C₂H₅)₂PN(C₆H₅)₂
  Reaction with H₂S: 1.5.3.2.3

C₁₆H₂₀O₂Si
(C₆H₅)₂Si(OC₂H₅)₂
  Reduction by Li[AlH₄]: 1.6.5.2.2

C₁₆H₂₂Cl₄P₂W
WCl₄(PMe₂Ph)₂
  Reaction with Na[BH₄]: 1.10.9

C₁₆H₂₂Ru
Ru(C₈H₁₂)(C₈H₁₀)
  Reaction with H₂: 1.10.4.3.3

C₁₆H₂₄Cl₂Ir₂
[Ir(C₈H₁₂)Cl]₂
  Reaction with EtOH and PEt₂Ph: 1.10.8
  Reaction with MeMgI: 1.10.8.4
  Reaction with i-PrMgBr: 1.10.8.4

C₁₆H₂₄Cl₄Ni₂
{[(CH₃)₄C₄]NiCl₂}₂
  Reduction with Zn: 1.6.2.5

C₁₆H₂₄Pt
Pt(C₈H₁₂)₂
  Reaction with phosphines: 1.10.4.1.2

C₁₆H₂₆Cl₄Ir₂
[IrHCl₂(C₈H₁₂)]₂
  Formation: 1.10.8

C₁₆H₂₇Al
(i-C₄H₉)₂AlCH₂CH₂C₆H₅
  Formation: 1.6.4.1.4

C₁₆H₂₇PPt
Pt(C₂H₄)₂[P(C₆H₅)(C₃H₇-i)₂]
  Reaction with H₂: 1.10.4.3.4

C₁₆H₂₈O₄PSi
CH₃(C₆H₅)CHC[OSi(CH₃)₃]*
P(O)(OC₂H₅)₂
  Formation: 1.5.3.2.2

$C_{16}H_{29}P_2Re$
  $H_7Re[P(CH_3)_2C_6H_5]_2$
    Formation: 1.10.4.3.2
$C_{16}H_{32}ClIrSi$
  h^5-$C_5Me_5IrH_2(SiEt_3)Cl$
    Formation: 1.10.5.4
$C_{16}H_{33}ClP_2Ru$
  h^5-$C_5Me_5Ru(PMe_3)_2Cl$
    Reaction with organomagnesium halide
    reagents: 1.10.8.2
$C_{16}H_{33}CoP_2$
  h^5-$C_5Me_5Co[P(Me)_3]_2$
    Protonation at metal: 1.10.6.1.1
$C_{16}H_{34}CoF_6P_3$
  $[h^5$-$C_5Me_5Co(PMe_3)_2H][PF_6]$
    Formation: 1.10.6.1.1
$C_{16}H_{34}GeO_2$
  $(n$-$C_4H_9)_3GeCH_2CH_2CO_2CH_3$
    Formation: 1.6.4.1.4
$C_{16}H_{34}P_2Ru$
  h^5-$C_5Me_5Ru(PMe_3)_2H$
    Formation from $Me_2CHMgCl$ and
    Me_3CMgCl: 1.10.8.2
$C_{16}H_{35}PPt$
  $Pt(C_2H_4)_2[P(C_4H_9$-$t)_3]$
    Reaction with H_2: 1.10.4.3.4
$C_{16}H_{36}Ir_2N_6O_2P_2S_2$
  $\{Ir(\mu$-SC_4H_9-$t)(CO)P(NCH_3)_3\}_2$
    Reaction with H_2: 1.10.4.1.1
$C_{16}H_{36}Ir_2O_2P_2S_2$
  $\{Ir(\mu$-S-t-$C_4H_9)(CO)P(CH_3)_3\}_2$
    Reaction with H_2: 1.10.4.1.1
$C_{16}H_{36}Ir_2O_8P_2S_2$
  $\{Ir(\mu$-S-t-$C_4H_9)(CO)P(OCH_3)_3\}_2$
    Reaction with H_2: 1.10.4.1.1
$C_{16}H_{36}P_2S_2$
  $(n$-$C_4H_9)_2P(S)P(S)(C_4H_9$-$n)_2$
    Hydrolysis to $(n$-$C_4H_9)_2P(S)H$: 1.5.3.2.1
$C_{16}H_{36}ZrO_4$
  $Zr(OCMe_3)_4$
    Reaction with Et_2AlH and C_8H_8:
    1.10.8.2
$C_{16}H_{38}Ir_2N_6O_2P_2S_2$
  $\{HIr(\mu$-SC_4H_9-$t)(CO)P(NCH_3)_3\}_2$
    Formation: 1.10.4.1.1
$C_{16}H_{38}Ir_2O_2P_2S_2$
  $\{HIr(\mu$-S-t-$C_4H_9)(CO)P(CH_3)_3\}_2$
    ·Formation: 1.10.4.1.1
$C_{16}H_{38}Ir_2O_8P_2S_2$
  $\{HIr(\mu$-S-t-$C_4H_9)(CO)P(OCH_3)_3\}_2$
    Formation: 1.10.4.1.1

$C_{16}H_{38}Sn_2$
  $[(n$-$C_4H_9)_2SnH]_2$
    Formation: 1.6.5.4.1
$C_{16}H_{40}N_4U$
  $U[N(C_2H_5)_2]_4$
    Reaction with C_5H_6: 1.5.3.1.3
$C_{16}H_{53}B_{10}Cl_2CuN_2$
  $[Et_4N]_2(Cl_2CuB_{10}H_{13})$
    Formation as solvate: 1.9.5.1
$C_{17}H_5Co_3O_{10}$
  $Co_3(CO)_9CCOC_6H_5$
    Reaction with $(C_2H_5)_3SiH$ in CF_3CO_2*
    H: 1.6.4.1.2
$C_{17}H_7Co_3O_9$
  $Co_3(CO)_9CCH_2C_6H_5$
    Formation: 1.6.4.1.2
$C_{17}H_{12}NO_9Os_3$
  $H_2Os_3(CO)_9(CH_3NCH_2C_6H_5)$
    Formation: 1.5.4.1.2
$C_{17}H_{16}O_5Ru_2$
  $Ru_2(CO)_3(\mu$-$CHCO_2C_2H_5)(C_5H_5)_2$
    Reaction with H_2: 1.10.4.3.3
$C_{17}H_{20}NP$
  $C_6H_5NC(t$-$C_4H_9)P(C_6H_5)H$
    Formation: 1.5.3.2.2
$C_{17}H_{21}NO_9Re_2$
  $[Et_4N][Re_2(CO)_9H]$
    Formation: 1.10.7.2
$C_{17}H_{24}Zr$
  $(h^5$-$C_5H_5)_2Zr(H)CH_2C_6H_{11}$
    Reaction with H_2: 1.6.2.1.2
$C_{17}H_{38}BF_4OPPt$
  $[h^3$-$C_3H_5PtP(CMe_3)_3(Me_2O)][BF_4]$
    Reaction with borohydride: 1.10.9
$C_{18}H_2O_{18}Os_6$
  $H_2Os_6(CO)_{18}$
    Formation by diprotonation of dianion:
    1.10.6.2.4
$C_{18}H_7Fe_4NO_{13}$
  $[pyH][HFe_4(CO)_{13}]$
    Formation by protonation of metal
    core: 1.10.6.2.4
$C_{18}H_{13}ClNO_6PS$
  $(C_6H_4O_2)_2PNHSO_2C_6H_4Cl$
    Formation: 1.5.3.1.3
$C_{18}H_{15}As$
  $(C_6H_5)_3As$
    Protonation by strong acids: 1.5.3.3.2
    Reaction with H_2 over Ni: 1.6.2.1.2
$C_{18}H_{15}BO_2$
  $C_6H_5B(OC_6H_5)_2$
    Reduction to $C_6H_5BH_2$: 1.7.5.1

$C_{26}H_{26}Cl_4MoP_2$
  $MoCl_4[P(C_6H_5)_2CH_3]_2$
  Reaction with $Na[BH_4]$: 1.10.9
$C_{26}H_{28}BCuP_2$
  $[(C_6H_5)_2PCH_2]_2CuH_2BH_2$
  Formation: 1.9.5.1
$C_{26}H_{28}O_{10}Os_3$
  $Os_3(CO)_{10}(C_8H_{14})_2$
  Reaction with HCO_2H: 1.10.5.6.1
  Reaction with aniline: 1.10.5.5.1
$C_{26}H_{29}ClN_4O_4Rh$
  $ClRh(dmgH)_2(C_6H_5)_3$
  Reaction with $Na[BH_4]$: 1.10.9
$C_{26}H_{30}ClIrP_2$
  $IrCl(PPh_3)_2$
  Reaction with $1,2-C_2B_{10}H_{12}$: 1.11.5.2
$C_{26}H_{30}N_4O_4PRh$
  $HRh(dmgH)_2PPh_3$
  Formation: 1.10.9
$C_{26}H_{30}N_4O_4Rh$
  $HRh(dmgH)_2(C_6H_5)_3$
  Equilibrium acidity: 1.10.6.2.2
$C_{26}H_{33}NPSi$
  $(CH_3)_3SiNC_6H_5C(t-C_4H_9)P(C_6H_5)_2$
  Methanolysis: 1.5.3.2.2
$C_{26}H_{33}P_2Re$
  $H_7Re[P(CH_3)(C_6H_5)_2]_2$
  Formation: 1.10.4.3.2
$C_{26}H_{40}Cr_2N_2O_{10}$
  $[(C_2H_5)_4N]_2[Cr_2(CO)_{10}]$
  Protonation on metal–metal bond:
  1.10.6.2.4
$C_{26}H_{41}FeO_2P$
  $h^5-C_5H_5[(C_6H_{11})_3P](OC)FeC(O)CH_3$
  Protonation of acyl oxygen: 1.10.6.1.1
$C_{26}H_{42}BF_4FeO_2P$
  $[h^5-C_5H_5[(C_6H_{11})_3P](OC)Fe=C^*$
  $(OH)CH_3][BF_4]$
  Formation by acyl protonation:
  1.10.6.1.1
$C_{26}H_{55}As_2F_3O_2Pt$
  $HPt(OCOCF_3)[As(C_4H_9-t)_3]_2$
  Formation: 1.10.5.6.1
$C_{26}H_{64}P_3Rh$
  $HRh[P(C_3H_7-i)_3]_3$
  Reaction with CO_2 and H_2O: 1.10.5.6.1
$C_{27}H_{16}F_9IrN_4O_9S_3$
  $[Ir(bipy)_2(OSO_2CF_3)_2][CF_3SO_3]$
  Reaction with $PPh_3$and $HOCH_2CH_2$-
  OH: 1.10.8.1
$C_{27}H_{24}D_3IrOP_2$
  $D_3Ir(CO)[(C_6H_5)_2P(CH_2)_2P(C_6H_5)_2]$
  Formation: 1.10.4.1.1

$C_{27}H_{25}IrOP_2$
  $HIr(CO)[(C_6H_5)_2P(CH_2)_2P(C_6H_5)_2]$
  Reaction with D_2: 1.10.4.1.1
  Reaction with H_2: 1.10.4.1.1
$C_{27}H_{26}ClMoP$
  $MoCl(h^3-C_3H_5)P(C_6H_5)_3C_6H_6-h^6$
  Reaction with $NaBH_4$: 1.10.9
$C_{27}H_{26}Cl_2NOPPt$
  $[(C_6H_5)_3P]$ $Pt(Cl)_2C(OC_2H_5)NHC_6H_5$
  Formation: 1.5.3.1.2
$C_{27}H_{27}IrOP_2$
  $H_3Ir(CO)[(C_6H_5)_2P(CH_2)_2P(C_6H_5)_2]$
  Formation: 1.10.4.1.1
$C_{27}H_{28}P_2Pt$
  $HPtCH_3[(C_6H_5)_2PCH_2CH_2P(C_6H_5)_2]$
  Formation from methyl cation by treat-
  ment with formate ion: 1.10.9
$C_{27}H_{37}NO_{11}Ru_3$
  $[(C_4H_9)_4N][HRu_3(\mu-CO)(CO)_{10}]$
  Protonation on bridging carbonyl oxy-
  gen: 1.10.6.2.4
$C_{27}H_{39}Br_2N_2P_3W$
  $WBr_2[N_2C(CH_3)_2][P(CH_3)_2C_6H_5]_3$
  Reduction by $LiAlH_4$: 1.5.5.1
$C_{27}H_{44}P_2Pt$
  $H_2Pt[P(C_3H_7-i)_3]_3$
  Formation: 1.10.4.3.4
$C_{27}H_{58}P_2PtSn$
  trans-$PtH(SnPh_3)[P(C_3H_7-i)_3]$
  Formation: 1.10.5.4
$C_{27}H_{63}P_3Pd$
  $Pd[P(C_3H_7-i)_3]_3$
  Reaction with HO_2CCF_3: 1.10.5.6.1
$C_{27}H_{63}P_3Pt$
  $Pt[P(CH_3)(C_4H_9-t)_2]_3$
  Formation: 1.10.4.1.2
  $Pt[P(C_3H_7-i)_3]_3$
  Reaction with H_2: 1.10.4.3.4
$C_{27}H_{64}P_3Rh$
  $HRh\{P[CH(CH_3)_2]_3\}_3$
  Protonation at metal: 1.10.6.1.1
  $HRh[P(C_3H_7-i)_3]_3$
  Reaction with H_2: 1.10.4.3.4
$C_{28}H_{21}IrNO_2P$
  $Ir(C_9N_1OH_6)(CO)P(C_6H_5)_3$
  Reaction with HCl: 1.10.5.1
$C_{28}H_{22}$
  $C_6H_5CHC(C_6H_5)C(C_6H_5)CH(C_6H_5)$
  Formation: 1.6.2.1.2, 1.6.5.1.4
$C_{28}H_{22}ClIrNO_2P$
  $HIrCl(C_9NOH_6)(CO)P(C_6H_5)_3$
  Formation: 1.10.5.1

$C_{42}H_{72}Cl_3OsP_3$
OsCl$_3$[P(C$_4$H$_9$-n)$_2$C$_6$H$_5$]$_3$
Reaction with N$_2$H$_4$: 1.10.8.2

$C_{42}H_{72}NiOP_2$
HNi(OPh)[P(C$_6$H$_{11}$)$_3$]$_2$
Formation: 1.10.5.6.1

$C_{43}H_{30}Fe_5N_6O_{13}$
[Fe(py)$_6$][Fe$_4$(CO)$_{13}$]
Protonation at metal core: 1.10.6.2.4

$C_{43}H_{36}ClNOP_2Pt$
PtCl(NHCOC$_6$H$_5$)[P(C$_6$H$_5$)$_3$]$_2$
Formation: 1.5.4.1.2

$C_{43}H_{36}FeP_2$
(h^5-C$_5$H$_5$)Fe[(C$_6$H$_5$)$_3$P]$_2$C≡CH
Reaction with HCl: 1.6.3.1.3

$C_{43}H_{37}ClFeP_2$
{(h^5-C$_5$H$_5$)Fe[(C$_6$H$_5$)$_3$P]$_2$CCH$_2$}Cl
Formation: 1.6.3.1.3

$C_{43}H_{37}Cl_3NP_2Re$
Re[P(C$_6$H$_5$)$_3$]$_2$(NC$_6$H$_4$CH$_3$)Cl$_3$
Reaction with various basic alcohols:
1.10.8.1

$C_{43}H_{38}Cl_2NP_2Re$
Re(PPh$_3$)$_2$NC$_6$H$_4$CH$_3$(H)Cl$_2$
Formation from basic isopropanol:
1.10.8.1

$C_{43}H_{39}BNO_2P_2V$
[[(C$_6$H$_5$)$_3$P]$_2$N][h^5-C$_5$H$_5$V(CO)$_2$BH$_4$]
Formation: 1.10.9

$C_{43}H_{42}ClIrOP_2$
Ir(CO)Cl[P(CH$_2$C$_6$H$_5$)$_3$]$_2$
Enthalpy for reaction with H$_2$: 1.10.4.1.1
Ir(CO)Cl[P(C$_6$H$_4$CH$_3$-p)$_3$]$_2$
Enthalpy for reaction with H$_2$: 1.10.4.1.1

$C_{44}H_{31}Fe_2NO_8P_2$
[[P(C$_6$H$_5$)$_3$]$_2$N][HFe$_2$(CO)$_8$]
Formation by protonation of a metal-
-metal bond: 1.10.6.2.4

$C_{44}H_{36}As_4O_6Ru_3$
(μ-H)$_2$Ru$_3$(CO)$_6$[(μ-As(C$_6$H-
$_5$)CH$_2$As(C$_6$H$_5$)$_2$]$_2$
Formation: 1.10.4.3.3

$C_{44}H_{36}NO_3V$
[[P(C$_6$H$_5$)$_3$N]$_2$][h^5-C$_5$H$_5$V(CO)$_3$H]
Formation: 1.10.9

$C_{44}H_{42}BF_4IrP_2$
[Ir(C$_8$H$_{12}$)[P(C$_6$H$_5$)$_3$]$_2$][BF$_4$]
Reaction with H$_2$: 1.10.4.3.4

$C_{44}H_{43}IrP_2$
HIr(C$_8$H$_{12}$)[P(C$_6$H$_5$)$_3$]$_2$
Formation: 1.10.8.4

$C_{44}H_{48}BF_4IrO_2P_2$
[H$_2$Ir(C$_4$H$_8$O)$_2$[P(C$_6$H$_5$)$_3$]$_2$][BF$_4$]
Formation: 1.10.4.3.4

$C_{44}H_{52}BF_4IrO_2P_2$
[H$_2$Ir(C$_4$H$_{10}$O)$_2$[P(C$_6$H$_5$)$_3$]$_2$][BF$_4$]
Formation: 1.10.4.3.4

$C_{45}H_{36}Fe_3O_8Sb_2$
H$_3$Fe$_3$(μ_3-COMe)(CO)$_7$(SbPh$_3$)$_2$
Formation: 1.10.3.2

$C_{45}H_{45}B_{10}IrOP_2$
Ir[P(C$_6$H$_5$)$_3$]$_2$(CO)-7-C$_6$H$_5$-1,7-B$_{10}$C$_2$H$_{10}$
Reaction with RCN: 1.10.4.1.1
Reaction with H$_2$: 1.10.4.1.1

$C_{45}H_{45}IrP_2$
Ir(C$_8$H$_{12}$)[P(C$_6$H$_5$)$_3$]$_2$CH$_3$
Formation: 1.10.8.4

$C_{45}H_{47}B_{10}IrOP_2$
H$_2$Ir[P(C$_6$H$_5$)$_3$]$_2$(CO)-7-C$_6$H$_5$-1,7-
B$_{10}$C$_2$H$_{10}$
Formation: 1.10.4.1.1

$C_{46}H_{48}Ir_2O_2P_2S_2$
{Ir(μ-SC$_4$H$_9$-t)(CO)P(C$_6$H$_5$)$_3$}$_2$
Reaction with H$_2$: 1.10.4.1.1

$C_{46}H_{49}ClN_2P_2Rh$
RhCl[P(C$_6$H$_4$CH$_3$-p)$_3$]$_2$C$_4$H$_7$N$_2$
Reaction with H$_2$: 1.10.4.1.3

$C_{46}H_{50}ClP_2RhS$
RhCl[P(p-CH$_3$C$_6$H$_4$)$_3$]$_2$C$_4$H$_8$S
Reaction with H$_2$: 1.10.4.1.3

$C_{46}H_{50}Ir_2O_2P_2S_2$
{HIr(μ-SC$_4$H$_9$-t)(CO)P(C$_6$H$_5$)$_3$}$_2$
Formation: 1.10.4.1.1

$C_{46}H_{51}ClN_2P_2Rh$
H$_2$RhCl[P(C$_6$H$_4$CH$_3$-p)$_3$]$_2$(C$_4$H$_7$N$_2$)
Formation: 1.10.4.1.3

$C_{46}H_{52}ClP_2RhS$
H$_2$RhCl[P(C$_6$H$_4$CH$_3$-p)$_3$]$_2$C$_4$H$_8$S
Formation: 1.10.4.1.3

$C_{47}H_{47}ClNP_2Rh$
RhCl[P(C$_6$H$_4$CH$_3$-p)$_3$]$_2$(C$_5$H$_5$N)
Reaction with H$_2$: 1.10.4.1.3

$C_{47}H_{49}ClNP_2Rh$
H$_2$RhCl[P(C$_6$H$_4$CH$_3$-p)$_3$]$_2$(C$_5$H$_5$N)
Formation: 1.10.4.1.3

$C_{48}H_{40}Si_4$
[(C$_6$H$_5$)$_2$Si]$_4$
Reaction with Li[AlH$_4$]: 1.6.5.2.1,
1.6.5.2.3

$C_{48}H_{60}GeP_2Pd$
[(C$_2$H$_5$)$_3$P]$_2$Pd[Ge(C$_6$H$_5$)$_3$]$_2$
Reaction with H$_2$: 1.6.2.3

$C_{48}H_{60}Ge_2P_2Pd$
[(C$_2$H$_5$)$_3$P]$_2$Pd[Ge(C$_6$H$_5$)$_3$]$_2$
Reaction with HCl: 1.6.3.3.3

$C_{48}H_{60}Ge_2P_2Pt$
[(C$_2$H$_5$)$_3$P]$_2$Pt[Ge(C$_6$H$_5$)$_3$]$_2$
Reaction with H$_2$: 1.6.2.3

C$_{52}$H$_{56}$MoP$_4$
MoH$_4$[P(C$_6$H$_5$)$_2$CH$_3$]$_4$
Formation: 1.10.9

C$_{52}$H$_{56}$P$_4$W
WH$_4$[P(C$_6$H$_5$)$_2$CH$_3$]$_4$
Formation from LiEt$_3$BH: 1.10.7.2
Protonation at metal: 1.10.6.1.1

C$_{52}$H$_{57}$F$_6$P$_5$W
[WH$_5$[P(C$_6$H$_5$)$_2$CH$_3$]$_4$][PF$_6$]
Formation by metal protonation:
1.10.6.1.1

C$_{52}$H$_{90}$P$_2$Pt$_2$Si$_2$
{Pt(μ-H)[Si(CH$_3$)$_2$C$_6$H$_5$]P(C$_6$H$_{11}$)$_3$}$_2$
Thermolysis: 1.6.4.2.3

C$_{53}$H$_{40}$FeIrO$_5$P$_3$
(CO)$_3$[P(C$_6$H$_5$)$_3$FeIr[μ-
P(C$_6$H$_5$)$_2$](CO)$_2$[P(C$_6$H$_5$)$_3$]
Formation and reaction with H$_2$:
1.10.4.1.3

C$_{53}$H$_{42}$FeIrO$_5$P$_3$
(CO)$_3$[P(C$_6$H$_5$)$_3$]FeIr[μ-
P(C$_6$H$_5$)$_2$]H$_2$(CO)$_2$[P(C$_6$H$_5$)$_3$]
Formation: 1.10.4.1.3

C$_{53}$H$_{51}$IMoN$_2$P$_4$
MoIN$_2$(CH$_3$)[(C$_6$H$_5$)$_2$P(CH$_2$)$_2$P*
(C$_6$H$_5$)$_2$]$_2$
Protonolysis by HBF$_4$: 1.5.3.1.1

C$_{53}$H$_{52}$BF$_4$IMoN$_2$P$_4$
[MoIN$_2$H(CH$_3$)[(C$_6$H$_5$)$_2$P(CH$_2$)$_2$P*
(C$_6$H$_5$)$_2$]$_2$][BF$_4$]
Formation: 1.5.3.1.1

C$_{53}$H$_{52}$BrClN$_2$P$_4$W
{trans-
WBr(N$_2$HCH$_3$)[(C$_6$H$_5$)$_2$P(CH$_2$)$_2$P*
(C$_6$H$_5$)$_2$]$_2$}Cl
Formation: 1.5.3.1.3

C$_{54}$H$_{31}$NO$_{18}$P$_2$Ru$_6$
[[(C$_6$H$_5$)$_3$P]$_2$N][HRu$_6$(CO)$_{18}$]
Formation by protonation of metal
core: 1.10.6.2.4

C$_{54}$H$_{45}$ClCuP$_3$
CuCl[P(C$_6$H$_5$)$_3$]$_3$
Reaction with LiAlH$_4$: 1.10.7.2

C$_{54}$H$_{45}$ClIrP$_3$
IrCl[P(C$_6$H$_5$)$_3$]$_3$
Cyclometallation: 1.10.5.3

C$_{54}$H$_{45}$ClP$_3$Rh
RhCl[P(C$_6$H$_5$)$_3$]$_3$
Catalysis of CO reaction with H$_2$:
1.6.6.1
Reaction with H$_2$: 1.10.4.1.3
Reaction with H$_2$S: 1.10.5.6.2

C$_{54}$H$_{45}$ClRh
RhCl[P(C$_6$H$_5$)$_3$]$_3$
Reduction with (i-C$_3$H$_7$)$_3$Al: 1.10.8.2

C$_{54}$H$_{45}$Cl$_2$P$_3$Ru
RuCl$_2$[P(C$_6$H$_5$)$_3$]$_3$
Reaction with methanolic base: 1.10.8

C$_{54}$H$_{45}$Ge$_4$Li
{[(C$_6$H$_5$)$_3$Ge]$_3$Ge}Li
Hydrolysis: 1.6.3.3.3

C$_{54}$H$_{45}$IrNOP$_3$
Ir(NO)[P(C$_6$H$_5$)$_3$]$_3$
Reaction with HCl: 1.5.3.1.3

C$_{54}$H$_{45}$P$_3$Pt
Pt[P(C$_6$H$_5$)$_3$]$_3$
Protonation at metal: 1.10.6.1.1
Reaction with OPHPh$_2$: 1.10.5.2
Reaction with HS(CH$_2$)$_2$SMe: 1.10.5.6.2
Reaction with HS(CH$_2$)$_2$S(CH$_2$)$_3$SMe:
1.10.5.6.2
Reaction with H$_2$S: 1.10.5.6.2
Reaction with HCN: 1.10.5.3

C$_{54}$H$_{46}$ClP$_3$Ru
RuHCl[P(C$_6$H$_5$)$_3$]$_3$
Reaction with AlEt$_3$: 1.10.8.2

C$_{54}$H$_{46}$CoN$_2$P$_3$
HCo(N$_2$)[P(C$_6$H$_5$)$_3$]$_3$
Formation: 1.10.8.2

C$_{54}$H$_{46}$Ge$_4$
[(C$_6$H$_5$)$_3$Ge]$_3$GeH
Formation: 1.6.3.3.3

C$_{54}$H$_{46}$NOP$_3$Ru
HRu(NO)[P(C$_6$H$_5$)$_3$]$_3$
Formation: 1.10.8

C$_{54}$H$_{46}$P$_3$Rh
HRh[P(C$_6$H$_5$)$_3$]$_3$
Formation: 1.10.8.2

C$_{54}$H$_{47}$ClP$_3$Rh
H$_2$RhCl[P(C$_6$H$_5$)$_3$]$_3$
Formation: 1.10.4.1.3

C$_{54}$H$_{47}$N$_2$P$_3$Ru
H$_2$Ru(N$_2$)[P(C$_6$H$_5$)$_3$]$_3$
Formation: 1.10.8.2

C$_{54}$H$_{48}$ClIrP$_4$
[Ir[(C$_6$H$_5$)$_2$P(CH$_2$)$_2$P(C$_6$H$_5$)$_2$]$_2$]Cl
Reaction with B$_{10}$H$_{14}$: 1.10.5.1

C$_{54}$H$_{48}$CoP$_3$
CoH$_3$[P(C$_6$H$_5$)$_3$]$_3$
Formation: 1.10.9

C$_{54}$H$_{48}$IrP$_3$
H$_3$Ir[P(C$_6$H$_5$)$_3$]$_3$
Formation: 1.10.9
Reaction with H$_2$: 1.10.4.3.4

$C_{58}H_{47}F_6O_6P_3Pt_6$
   $[Pt(PPh_3)_3H][(CF_3CO_3)_2H]$
      Formation by metal protonation:
      1.10.6.1.1
$C_{60}H_{50}Si_5$
   $[(C_6H_5)_2Si]_5$
      Formation: 1.6.5.2.1
$C_{60}H_{64}MoN_4P_4$
   $Mo(N_2)_2[(m-$
   $CH_3C_6H_4)_2P(CH_2)_2P(CH_3C_6H_4-m)_2]_2$
      Reaction with H_2: 1.10.4.3.2
   $Mo(N_2)_2[(p-$
   $CH_3C_6H_4)_2P(CH_2)_2P(C_6H_4CH_3-$
   $m)_2]_2P(m-CH_3C_6H_4)_2]_2$
      Reaction with H_2: 1.10.4.3.2
$C_{60}H_{68}MoP_4$
   $H_4Mo[(m-$
   $CH_3C_6H_4)_2P(CH_2)_2P(C_6H_4CH_3-m)_2]_2$
      Formation: 1.10.4.3.2
   $H_4Mo[(p-$
   $CH_3C_6H_4)_2P(CH_2)_2P(C_6H_4CH_3-p)_2]_2$
      Formation: 1.10.4.3.2
$C_{60}H_{90}Cl_4P_6Ru_2$
   $[Ru_2Cl_3[P(C_2H_5)_2C_6H_5]_6]Cl$
      Reaction with ethanolic KOH: 1.10.8
$C_{62}H_{64}BF_4O_4P_4Rh$
   $[Rh[(+)-C_{31}H_{32}O_2P_2)]_2][BF_4]$
      Reaction with H_2: 1.10.4.2.1
$C_{62}H_{66}BF_4O_4P_4Rh$
   $[H_2Rh[(+)-C_{31}H_{32}O_2P_2]_2][BF_4]$
      Formation: 1.10.4.2.1
$C_{63}H_{63}ClP_3Rh$
   $RhCl[P(C_6H_4CH_3-p)_3]_3$
      Reaction with H_2: 1.10.4.1.3
$C_{63}H_{65}ClP_3Rh$
   $H_2RhCl[P(C_6H_4CH_3-p)_3]_3$
      Formation: 1.10.4.1.3
$C_{64}H_{82}BIrP_4$
   $[IrH_2[P(C_2H_5)_2C_6H_5]_4][BPh_4]$
      Formation: 1.11.0.8
$C_{66}H_{55}DGeNiP_3$
   $[(C_6H_5)_3P]_3NiGe(C_6H_5)_2D$
      Formation: 1.6.7.2.3
$C_{66}H_{56}GeNiP_3$
   $[(C_6H_5)_3P]_3NiGe(C_6H_5)_2H$
      Exchange with D_2O: 1.6.7.2.3
$C_{72}H_{60}P_4Pt$
   $Pt[P(C_6H_5)_3]_4$
      Reaction with HCN: 1.10.5.3
      Reaction with phthalimide: 1.10.5.5.1
      Reaction with saccharin: 1.10.5.5.1
      Reaction with succinimide: 1.10.5.5.1

$C_{72}H_{61}CoO_{12}P_4$
   $HCo[P(OC_6H_5)_3]_4$
      Formation: 1.10.9
$C_{72}H_{61}P_4Rh$
   $HRh[P(C_6H_5)_3]_4$
      Formation: 1.10.8, 1.10.9
$C_{72}H_{62}Sn_6$
   $H[(C_6H_5)_2Sn]_6H$
      Formation: 1.6.3.4.3
$C_{72}H_{63}P_4Re$
   $H_3Re[P(C_6H_5)_3]_4$
      Reaction with H_2: 1.10.4.3.2
$C_{72}H_{64}B_2Cu_2F_4P_4$
   $[[(C_6H_5)_3P]_2CuH_2BH-$
   $_2Cu[P(C_6H_5)_3]_2][BF_4]$
      Formation: 1.9.5.1
$C_{72}H_{70}B_{10}Cu_2P_4$
   $[[(C_6H_5)_3P]_2Cu]_2B_{10}H_{10}$
      Formation: 1.9.5.1
$C_{75}H_{68}P_6Ru_3$
   $\{H_2Ru[(C_6H_5)_2PCH_2P(C_6H_5)_2]_2\}_3$
      Formation: 1.10.4.3.3
$C_{76}H_{64}P_4W$
   $WH_4(PPh_3)_4$
      Formation: 1.10.7.2
$C_{84}H_{84}Cl_2Rh_2P_4$
   $\{RhCl[P(C_6H_4CH_3-p)_3]_2\}_2$
      Reaction with H_2: 1.10.4.1.3
$C_{90}H_{60}N_2O_{18}P_4Ru_6$
   $[[(C_6H_5)_3P]_2N]_2[Ru_6(CO)_{18}]$
      Protonation on metal core: 1.10.6.2.4
Ca
   Ca
      Reaction with H_2: 1.8.3, 1.8.3.3
$Ca*Ag_2$
$Ca*Al_2$
$CaCu_5$
   $CaCu_5$
      Structure type: 1.12.8.1.1
CaF_2
   CaF_2
      Structure type: 1.12.5
CaGe
   CaGe
      Protonolysis: 1.5.3.2.1, 1.6.3.3.1
$CaH*Ag_2$
CaH_2
   CaH_2
      Formation: 1.8.3.3
      Reaction with $C_6H_5NO_2$ in presence of
      $PtCl_2$ catalyst: 1.5.4.1.1
      Reaction with CO: 1.6.4.1.2

ClH$_8$*B$_5$

ClH$_8$Tl*B$_2$

ClIrNO$_2$P*C$_{28}$H$_{22}$

ClIrNP$_2$*C$_{37}$H$_{31}$

ClIrO*C$_{37}$H$_{30}$As$_2$

ClIrOP*C$_{19}$H$_{15}$

ClIrOP$_2$*C$_7$H$_{18}$

ClIrOP$_2$*C$_{13}$H$_{30}$

ClIrOP$_2$*C$_{19}$H$_{42}$

ClIrOP$_2$*C$_{33}$H$_{38}$

ClIrOP$_2$*C$_{37}$H$_{30}$

ClIrOP$_2$*C$_{37}$H$_{32}$

ClIrOP$_2$*C$_{37}$H$_{66}$

ClIrOP$_2$*C$_{43}$H$_{42}$

ClIrOP$_4$*C$_{13}$H$_{32}$

ClIrO$_7$P$_2$*C$_{37}$H$_{30}$

ClIrO$_7$P$_2$*C$_{39}$H$_{30}$

ClIrP$_2$*C$_{21}$H$_{46}$

ClIrP$_2$*C$_{22}$H$_{48}$

ClIrP$_2$*C$_{26}$H$_{30}$

ClIrP$_2$*C$_{37}$H$_{30}$

ClIrP$_3$*C$_{12}$H$_{51}$B$_{30}$

ClIrP$_3$*C$_{36}$H$_{37}$

ClIrP$_3$*C$_{36}$H$_{39}$

ClIrP$_3$*C$_{54}$H$_{45}$

ClIrP$_4$*C$_{54}$H$_{48}$

ClIrSi*C$_{16}$H$_{32}$

ClLi*C$_8$H$_9$Al

ClMg*C$_2$H$_5$

ClMg*C$_3$H$_7$

ClMg*C$_6$H$_{11}$

ClMgN$_2$Ti*C$_{10}$H$_{10}$

ClMgN$_2$Ti$_2$*C$_{20}$H$_{20}$

ClMnO$_2$Si*C$_{20}$H$_{18}$

ClMo*C$_{33}$H$_{25}$

ClMoN$_3$O$_6$*C$_6$H$_{13}$

ClMoO$_3$*C$_8$H$_5$

ClMoP*C$_{27}$H$_{26}$

ClN*C$_6$H$_{16}$

ClNOP$_2$Pt*C$_{19}$H$_{36}$

ClNOP$_2$Pt*C$_{43}$H$_{36}$

ClNO$_2$OsP$_2$*C$_{37}$H$_{30}$

ClNO$_3$*C$_7$H$_4$

ClNO$_6$PS*C$_{18}$H$_{13}$

ClNP*C$_{18}$H$_{17}$

ClNP$_2$Pt*C$_{18}$H$_{36}$

ClNP$_2$Pt*C$_{42}$H$_{36}$

ClNP$_2$Rh*C$_{47}$H$_{47}$

ClNP$_2$Rh*C$_{47}$H$_{49}$

ClNSi*C$_4$H$_{12}$

ClN$_2$O$_2$W*C$_5$H$_5$

ClN$_2$P*C$_4$H$_{12}$

ClN$_2$P$_2$Pt*C$_{18}$H$_{35}$

ClN$_2$P$_2$Rh*C$_{46}$H$_{49}$

ClN$_2$P$_2$Rh*C$_{46}$H$_{51}$

ClN$_2$P$_3$W*C$_{24}$H$_{36}$Br$_2$

ClN$_2$P$_4$W*C$_{53}$H$_{52}$Br

ClN$_3$P$_3$*C$_{25}$H$_{24}$Au

ClN$_3$Si$_6$Th*C$_{18}$H$_{54}$

ClN$_4$O$_4$Rh*C$_{26}$H$_{29}$

ClNaO$_2$

  NaClO$_2$

    Reaction with phosphorus to form

    Na$_2$[H$_2$P$_2$O$_4$]$_2$: 1.5.6.2

ClNiP$_2$*C$_{36}$H$_{67}$

ClO*CAu

ClO*C$_2$H$_3$

ClOOsP$_3$*C$_{55}$H$_{46}$

ClOP*C$_5$H$_{10}$

ClOP*C$_8$H$_{18}$

ClOP*C$_{10}$H$_{10}$

ClOP*C$_{10}$H$_{14}$

ClOP*C$_{12}$H$_{10}$

ClOP*C$_{14}$H$_{14}$

ClOP$_2$Rh*C$_{37}$H$_{30}$

ClOP$_3$Ru*C$_{31}$H$_{46}$

ClOZr*C$_{11}$H$_{11}$

ClO$_2$*C$_2$H$_6$B

ClO$_2$P*C$_2$H$_6$

ClO$_2$P$_2$Pt*C$_{13}$H$_{31}$

ClO$_2$Si*C$_2$H$_7$

ClO$_2$Si*C$_4$H$_{11}$

ClO$_7$Os$_2$P$_2$*C$_{39}$H$_{31}$

ClP*C$_2$H$_6$

ClP*C$_2$H$_8$

ClP*C$_{12}$H$_{10}$

ClP*C$_{13}$H$_{10}$

ClP*C$_{19}$H$_{18}$

ClPRhSi*C$_{36}$H$_{31}$

ClP$_2$IrS*C$_{36}$H$_{32}$

ClP$_2$Pt*C$_{12}$H$_{31}$

ClP$_2$Pt*C$_{14}$H$_{35}$

ClP$_2$Pt*C$_{15}$H$_{37}$

ClP$_2$Pt*C$_{28}$H$_{47}$

ClP$_2$Pt*C$_{36}$H$_{31}$

ClP$_2$Rh*C$_{21}$H$_{44}$

ClP$_2$Rh*C$_{21}$H$_{46}$

ClP$_2$RhS*C$_{36}$H$_{32}$

ClP$_2$RhS*C$_{46}$H$_{50}$

ClP$_2$RhS*C$_{46}$H$_{52}$

ClP$_2$Ru*C$_{16}$H$_{33}$

ClP$_2$Ru*C$_{41}$H$_{35}$

ClP$_3$Rh*C$_{54}$H$_{45}$

ClP$_3$Rh*C$_{54}$H$_{47}$

Cl₆Ga₂
Ga₂Cl₆
Reduction by (CH₃)₃SiH: 1.7.3.2
Cl₆GeSi
Cl₃GeSiCl₃
Reduction by LiAlH₄: 1.6.5.3.1
Cl₆H₂Pt
H₂PtCl₆
Catalyst in SiCl₃D exchange with CH₃*
Cl₂SiH: 1.6.7.2.2
Cl₆H₁₂IrNa₂O₆
Na₂IrCl₆·6 H₂O
Reaction with NaBH₄: 1.10.9
Reaction with alcoholic KOH and
PPh₃: 1.10.8
Cl₆H₁₂Na₂O₆Os
Na₂OsCl₆·6 H₂O
Reaction with alcoholic PPh₃: 1.10.8
Cl₆IrH₂
H₂[IrCl₆]
Reaction with EtOH and diene: 1.10.8
Cl₆IrNa₂
Na₂[IrCl₆]
Reaction with Na[BH₄]: 1.10.9
Cl₆N₂P₂*C₂
Cl₆N₃P₃
(NPCl₂)₃
Conversion to (NPCl₂)₂NP(H)CH₃:
1.5.3.2.3
Conversion to (NPCl₂)₂NP(H)C₂H₅:
1.5.3.2.3
Conversion to (NPCl₂)₂NP(H)C₃H₇-n:
1.5.3.2.3
Cl₆OSi₂
(SiCl₃)₂O
Reduction by Li[AlD₄]: 1.6.7.1.2
Reduction by Li[AlH₄]: 1.6.5.2.2
Cl₆P₄Ta₂*C₁₂H₃₈
Cl₆Si₂
Si₂Cl₆
Reduction by Li[AlD₄]: 1.6.7.1.2
Reduction by Li[AlH₄]: 1.6.5.2.1
Reduction by AlH₃: 1.6.4.2.1
Cl₇FNP*C₂H₂
Cl₇NP*C₂H
Cl₈Si₃
Si₃Cl₈
Reduction by Li[AlH₄]: 1.6.5.2.1
Cl₉O₉Os₃Si₃*C₉H₃
Cl₁₀Si₄
Si₄Cl₁₀
Reduction by Li[AlH₄]: 1.6.5.2.1

Cl₁₂Si₅
Si₅Cl₁₂
Reduction by Li[AlH₄]: 1.6.5.2.1
Co
Catalysis of CO reaction with H₂:
1.6.6.1
Oxidative addition of CH₄: 1.10.5.3
Reaction with PF₃ and H₂: 1.10.2
Reaction with CO and H₂: 1.10.2
Co*C₈H₁₃
Co*C₁₀H₁₅
Co*Cl₂
CoF₆P₃*C₁₁H₂₄
CoF₆P₃*C₁₆H₃₄
CoF₈P₃*C₁₄H₃₂B₂
CoF₁₂HP₄
CoH(PF₃)₄
Formation: 1.10.2, 1.10.3.2
CoH₂N₂O₇
Co[NO₃]₂H₂O
Reaction with Na[BH₄]: 1.10.9
CoHf
HfCo
Reaction with H₂: 1.12.8.1.3
Structure: 1.12.8.1.3
CoKO₁₃Ru₃*C₁₃
CoN₂O₂*C₇H₇
CoN₂O₂*C₉H₁₁
CoN₂O₂*C₁₁H₁₅
CoN₂P₃*C₅₄H₄₆
CoN₄O₄*C₂₀H₄₁Cl
CoN₄O₄*C₂₀H₄₂
CoN₄O₄P*C₂₀H₄₂
CoO₃P*C₂₁H₁₆
CoO₄*C₄H
CoO₆*C₁₅H₂₁
CoO₆P*C₂₁H₁₆
CoO₁₂P₄*C₃₆H₈₅
CoO₁₂P₄*C₇₂H₆₁
CoO₁₂Ru₃*C₁₂H₃
CoO₁₃Ru₃*C₁₃H
CoP₂*C₁₁H₂₃
CoP₂*C₁₆H₃₃
CoP₃*C₅₄H₄₈
CoP₄*C₃₂H₄₅
CoTi
TiCo
Reaction with H₂: 1.12.8.1.3
CoZr
ZrCo
Structure: 1.12.8.1.3
Co₂*C₈O₈

H$_2$ *cont.*

Reaction with $CH_3C_6H_5$ over $Pt–SiO_2$: 1.6.6.1

Reaction with $[(C_2H_5)_3P]_2Pt[Ge(C_6^*H_5)_3]_2$: 1.6.2.3

Reaction with $Pt[Sn(CH_3)_3]_2[P(C_6^*H_5)_3]_2$: 1.6.2.4

Reaction with CO and $[HRu(CO-)_{10}[Si(C_2H_5)_3]_2]^-$: 1.6.2.2

Reaction with SiO_2 in presence of Al and $AlCl_3$: 1.6.2.2

Reaction with $SiCl_4$: 1.6.2.2

Reaction with $SiCl_4$ and Al in $AlCl_3$: 1.6.6.2

Reaction with $(C_6H_5)_3SiSi(C_6H_5)_3$ over Cr: 1.6.2.2

Reaction with $(h^5-C_5H_5)_2Zr(CO)_2$: 1.6.2.1.2

Reaction with $(h^5-C_5H_5)_2Zr(H)CH_2C_6^*H_{11}$: 1.6.2.1.2

Reaction with $C_6H_5CH(NC_6H_5)$: 1.5.2.1.2

Reaction with $C_6H_5NO_2$: 1.5.2.1.2

Reaction with $CH_3CO(CH_3O)C_6H_3C^*H(CO_2H)NHCOCH_3$: 1.6.2.1.2

Reaction with C_6H_5NO: 1.5.2.1.2

Reaction with N-nitrosoamine: 1.5.2.1.2

Reaction with $Ca(CN)_2$: 1.5.2.1.2

Reaction with N_2: 1.5.2.1.1

Reaction with N_2 to form NH_3: 1.5.6.1

Reaction with CO and $n-C_4H_9CH=C^*H_2$: 1.6.2.1.2

Reaction with CO to form CH_3OH: 1.6.6.1

Reaction with $[C_2H_5]^+$ in gas phase: 1.6.2.1.2

Reaction with C_4H_4O in presence of Ni: 1.6.6.1

Reaction with CO_2 over Ni: 1.6.2.1.2

Reaction with CO: 1.6.2.1.2

Reaction with CO and olefins: 1.6.6.1

Reaction with $(CH_3)_2CO$: 1.6.2.1.2

Reaction with $CH_3C\equiv CCH_3$: 1.6.2.1.2, 1.6.6.1

Reaction with $CH_2CHCH_2CH_2CH_3$: 1.6.6.1

Reaction with C_6H_6: 1.6.2.1.2

Reaction with C_6H_5CN: 1.6.2.1.2

Reaction with H_2 to produce $(CH_2OH)_2$: 1.6.6.1

Reaction with alkenes: 1.6.6.1

Reaction with alkynes: 1.6.6.1

Reaction with arenes: 1.6.6.1

Reaction with azide: 1.5.2.1.2

Reaction with carbon at high temperatures: 1.6.2.1.1

Reaction with carbon atoms from C_3O_2: 1.6.2.1.1

Reaction with distannanes: 1.6.2.4

Reaction with esters: 1.6.6.1

Reaction with imine: 1.5.2.1.2

Reaction with organic substrates in presence of catalysts: 1.6.2.1.2

Reaction with oxime: 1.5.2.1.2

Reaction with selenophosphinic acid: 1.5.2.2

Reaction with thiophosphinic acid: 1.5.2.2

Reaction with transition metals: 1.9.1

Reduce CuI in liq NH_3: 1.9.5.2

Reduction of $(h^5-C_5H_5)_2Ti(CO)_2$: 1.6.2.1.2

Reduction of C_5H_5N over Pt: 1.5.6.1

H_2^*Ac

H_2^*AsD

$H_2AsF_3^*C$

$H_2As_2F_6^*C_2$

$H_2^*B_{10}D_{12}$

H_2^*Ba

H_2^*Be

$H_2^*Br_2Ge$

$H_2^*C_2$

H_2^*Ca

H_2^*Cd

H_2^*Ce

$H_2^*Cl_2Ge$

$H_2^*Cl_6Ir$

$H_2Cl_7FNP^*C_2$

$H_2Co_2O_{12}Ru_2^*C_{12}$

H_2^*Cr

$H_2D_2O_{12}Ru_4^*C_{12}$

H_2^*Dy

H_2^*Er

$H_2F_3N^*C$

$H_2F_3NO_{10}Os_3^*C_{12}$

$H_2F_3O_2P^*C$

$H_2F_3P^*C$

$H_2F_6P_2^*C_2$

$H_2F_9P_3^*C_3$

$H_2F_{12}Fe_2O_6P_2^*C_{10}$

$H_2F_{20}Na_2Zn_2^*C_{24}$

$H_2FeO_4^*C_4$

$H_2FeO_{13}Os_3^*C_{13}$

$H_2FeO_{13}Ru_3^*C_{13}$

$H_2Fe_5O_{12}^*C_{13}$

H₂O *cont.*

Reaction with $(CF_3)_4P_2$: 1.5.3.2.3

Reaction with $(CH_3)_3SiN_3$: 1.5.3.1.3

Reaction with $(CH_3)_2Si$: 1.6.3.2.1

Reaction with $(C_2H_5)_3SnB[N(CH_3)_2]_2$: 1.6.3.4.1

Reaction with $(C_2H_5)_3SnC_6H_5$: 1.6.2.5

Reaction with $[(C_2H_5)_3SnN(CO_2C_2*H_5)]_2$: 1.5.3.1.3

Reaction with $[(C_6H_5)_3Sn]_2Mg$: 1.6.3.4.1

Reaction with CaH_2, SrH_2, BaH_2: 1.8.3.3

Reaction with ZnC_2 to form C_2H_2: 1.6.6.1

Reaction with $CH_3ZrCl(C_5H_5-h^5)_2$: 1.6.3.1.3

Reaction with $[CN]^-$, $[(CN)_3C]^-$ or $[(NO_2)_3C]^-$: 1.6.2.5

Reaction with carbon atoms: 1.6.2.1.1

Reaction with elemental boron: 1.7.2

Reaction with metals to form hydrides: 1.12.4.1

H₂O*C

H₂O*Ge

H₂O₂*C

H₂O₂*Ge

H₂O₂V

V(OH)₂

Reduction of aq N_2 to N_2H_4: 1.5.3.1.1

H₂O₄Os*C₄

H₂O₄P₂*Ba

H₂O₄Ru*C₄

H₂O₄S

H_2SO_4

Electrolyte cathodic charging of Ti: 1.12.5

Electrolyte for electrodeposition of CrH and CrH_2: 1.11.2.7

Electrolyte in cathodic charging of Ni: 1.12.7

Electrolyte in cathodic charging of group-VA metals: 1.12.6

Hydrolysis of Mg_2Sn: 1.6.3.4.1

Preparation of PdH: 1.12.7

Protonation of Mg_2Si: 1.6.6.2

Protonation of amines: 1.5.3.1.2

Reactions with amides: 1.5.3.1.3

Reaction with $M(N_2)_2[P(CH_3)_2C_6H_5]_4(M=Mo, W)$: 1.5.3.1.2

Reaction with $M(N_2)_2[PCH_3(C_6H_5)_2]_4(M=Mo, W)$: 1.5.3.1.2

H₂O₈Os₂*C₈

H₂O₉Os₃S*C₉

H₂O₁₀Os₃*C₁₀

H₂O₁₁Os₃*C₁₂

H₂O₁₁Ru₃*C₁₁

H₂O₁₂Os₃*C₁₁

H₂O₁₃Ru₄*C₁₃

H₂O₁₈Os₆*C₁₈

H₂OsP₄*F₁₂

H₂P*Cs

H₂P*CsF₄

H₂P*F₃

H₂P₄*F₁₂Fe

H₂P₄Ru*F₁₂

H₂Pb₂

Pb_2H_2

Formation: 1.6.3.5

H₂Pr

PrH_2

Formation: 1.12.4.1

H₂Pt*Cl₆

H₂S

Cleavage of group-IVB–As bonds: 1.5.3.3.3

Impurity in H_2: 1.8.2

Reaction with $[(CH_3)_2N]_3P$: 1.5.3.2.3

Reaction with dialkylaminodialkyl-phosphines: 1.5.3.2.3

H₂SO₄

Reaction with $CH_3CH=CH_2$: 1.6.6.1

H₂STa₆

Ta_6SH_2

Preparation: 1.12.8.2.3

H₂Sc

ScH_2

Formation: 1.12.4.1

H₂Si*AsF₃

H₂Sm

SmH_2

Formation: 1.12.4.1

H₂Sn

$(SnH_2)_n$

Formation: 1.6.3.4.2

H₂Sr

SrH_2

Formation: 1.8.3.3

H₂Tb

TbH_2

Formation: 1.12.4.1

H₂Th

ThH_2

Formation: 1.12.4.2

H₂Th*C

$H_8Li*AlAs_4$
$H_8MoO_3*C_{11}$
$H_8NO_2Re*C_7$
$H_8NO_3Re*C_7$
H_8NP*C_6
$H_8N_2*C_2$
$H_8NiO_9Ru_3*C_{14}$
H_8O*C_3
H_8O*C_4
H_8O*C_7
H_8O*C_8
H_8OSi*C_2
$H_8O_2*C_4$
$H_8O_2Si*C_2$
$H_8O_4S*C_3$
$H_8P_2*C_2$
H_8Pb*C_2
H_8Si*C_2
H_8Si*C_3
H_8Si*C_6
H_8Si*Ge_2
H_8Si_2*C
H_8Si_2*Ge
H_8Si_3
　Si_3H_8
　　Formation: 1.6.2.2, 1.6.3.2.1, 1.6.4.2.3,
　　1.6.5.2.1
　　Formation from SiH_4 in silent electric
　　discharge: 1.5.4.2.2
　　Formation in $H-SiH_4$ reaction: 1.6.2.2
　　Pyrolysis in presence of Ge_2H_6:
　　1.6.4.2.3, 1.6.4.3.3
H_8Sn*C_2
H_8Sn*C_6
H_8Tl*B_2Cl
H_8Zn*B_2
H_9Al*C_3
$H_9AlClLi*C_8$
H_9As*C_7
H_9*AsGe_3
H_9AsO*C_7
$H_9AsO_2*C_7$
$H_9AsO_4*C_7$
H_9B*C_6
$H_9BF_7N_2O_2Re*C_{14}$
$H_9BO_3*C_3$
H_9*B_2Ga
H_9*B_4D
H_9*B_5
$H_9B_5FeO_3*C_3$
$H_9BrGe*C_3$
$H_9BrMg*C_4$

$H_9BrSn*C_3$
$H_9ClGe*C_3$
$H_9ClPb*C_3$
$H_9ClSi*C_3$
$H_9ClSn*C_3$
$H_9Cl_2P*C_4$
$H_9Cl_2Sb*C_4$
$H_9Cl_3Sn*C_4$
$H_9Cl_5NSb*C_6$
H_9D*C_4
$H_9D_2NSi*C_3$
$H_9F_2OP*C_7$
$H_9F_5NP*C_2$
$H_9F_6NSn*C_6$
$H_9FeNO_2*C_{10}$
H_9Ga*C_3
$H_9GeN_3*C_3$
H_9GeP*C_2
H_9In*C_3
H_9K*B_4
H_9K*B_6
$H_9K*B_9D_5$
H_9Li*C_4
$H_9MoNO_4P*C_6$
H_9N*C_7
H_9N*C_8
H_9NO*C_8
$H_9NO_2*C_4$
$H_9NO_{10}Ru_3*C_{15}$
$H_9NO_{11}Ru_3*C_{16}$
$H_9NPS_3*C_2$
H_9NSi_3
　$(H_3Si)_3N$
　　Disproportionation in presence of B_5H_9:
　　1.6.4.2.3
　　Disproportionation in presence of NH_3:
　　1.6.3.2.2, 1.6.4.2.3
$H_9N_3*C_{12}$
$H_9N_3Si*C_3$
$H_9NaSn*C_3$
$H_9NaZn_2*C_2$
H_9Na_2Re
　$Na_2[ReH_9]$
　　Formation from Na and EtOH: 1.10.8.1
H_9OP*C_7
$H_9O_2P*C_3$
$H_9O_2P*C_7$
$H_9O_3P*C_3$
$H_9O_3P*C_6$
$H_9O_3Re*C_9$
H_9P*C_7
H_9P*C_8

$H_{33}F_6O_3PRh_2*C_{20}$
$H_{33}IrOP_2*C_{37}$
$H_{33}IrO_4P_2*C_{40}$
$H_{33}LiPb*C_{18}$
$H_{33}NPSi*C_{26}$
$H_{33}NSn*C_{14}$
$H_{33}N_3Sn*C_{13}$
$H_{33}PPt*C_{15}$
$H_{33}P_2Re*C_{26}$
$H_{33}P_3W*C_9$
$H_{34}BCuP_2*C_{36}$
$H_{34}B_{12}O_2Zn*C_8$
$H_{34}CoF_6P_3*C_{16}$
$H_{34}Ge*C_{18}$
$H_{34}Ge*C_{20}$
$H_{34}GeO_2*C_{16}$
$H_{34}OPtSP_2*C_{38}$
$H_{34}OZr*C_{21}$
$H_{34}O_2P_2Ru*C_{36}$
$H_{34}P_2Pt*C_{14}$
$H_{34}P_2Pt*C_{22}$
$H_{34}P_2Ru*C_{16}$
$H_{34}Pb*C_{18}$
$H_{35}B_{10}IrNOP*C_{29}$
$H_{35}B_{10}IrNOP*C_{34}$
$H_{35}Br_2MoN_2P_3*C_{24}$
$H_{35}Br_2N_2P_3W*C_{24}$
$H_{35}ClN_2P_2Pt*C_{18}$
$H_{35}ClP_2Pt*C_{14}$
$H_{35}ClP_2Ru*C_{41}$
$H_{35}IrP_2*C_{36}$
$H_{35}N_3Sn*C_{14}$
$H_{35}PPt*C_{16}$
$H_{35}PRu*C_{30}$
$H_{36}As_4O_6Ru_3*C_{44}$
$H_{36}BF_4IrO_2P_2*C_{36}$
$H_{36}B_2CuN_2P_2*C_{38}$
$H_{36}Br_2ClN_2P_3W*C_{24}$
$H_{36}ClNOP_2Pt*C_{19}$
$H_{36}ClNOP_2Pt*C_{43}$
$H_{36}ClNP_2Pt*C_{18}$
$H_{36}ClNP_2Pt*C_{42}$
$H_{36}ClP_3Si_4*C_{12}$
$H_{36}Cl_2N_2P_2Pt*C_{18}$
$H_{36}Co_7Fe_2O_{30}*C_{42}$
$H_{36}F_6IrP_5*C_{12}$
$H_{36}F_6P_2Ru*C_{30}$
$H_{36}FeP_2*C_{43}$
$H_{36}FeP_4*C_{12}$
$H_{36}Fe_3O_8Sb_2*C_{45}$
$H_{36}Ge_6P_4*C_{12}$
$H_{36}Ir_2N_6O_2P_2S_2*C_{16}$

$H_{36}Ir_2O_2P_2S_2*C_{16}$
$H_{36}Ir_2O_8P_2S_2*C_{16}$
$H_{36}Li_2Si_6*C_{12}$
$H_{36}NO_3V*C_{44}$
$H_{36}N_2Si_2Sn*C_{14}$
$H_{36}N_2Si_4Sn*C_{12}$
$H_{36}N_4Si_4*C_{12}$
$H_{36}N_6W_2*C_{12}$
$H_{36}P_2Pt*C_{18}$
$H_{36}P_2Ru*C_{41}$
$H_{36}P_2S_2*C_{16}$
$H_{36}P_4Pd*C_{12}$
$H_{36}P_4Si_4*C_{12}$
$H_{36}Si_6*C_{12}$
$H_{36}ZrO_4*C_{16}$
$H_{37}BCuP_3*C_{24}$
$H_{37}B_{10}IrNOP*C_{34}$
$H_{37}B_{10}IrOP_2*C_{29}$
$H_{37}ClFeP_2*C_{43}$
$H_{37}ClIrP_3*C_{36}$
$H_{37}ClP_2Pt*C_{15}$
$H_{37}ClSi_6*C_{12}$
$H_{37}Cl_3NP_2Re*C_{43}$
$H_{37}NO_{11}Ru_3*C_{27}$
$H_{37}OP_2Rh*C_{40}$
$H_{37}O_2P_3Pt*C_{42}$
$H_{37}P_2Re*C_{20}$
$H_{37}P_2Re*C_{36}$
$H_{38}AgB_3P_2*C_{36}$
$H_{38}BF_4IrN_2P_2*C_{40}$
$H_{38}BF_4OPPt*C_{17}$
$H_{38}B_3CuP_2*C_{36}$
$H_{38}B_3IrOP_2*C_{37}$
$H_{38}B_5CuP_2*C_{36}$
$H_{38}ClIrOP_2*C_{33}$
$H_{38}Cl_2NP_2Re*C_{43}$
$H_{38}Cl_6P_4Ta_2*C_{12}$
$H_{38}F_6IrOP_5*C_{13}$
$H_{38}F_6P_3Ir*C_{34}$
$H_{38}Ir_2N_6O_2P_2S_2*C_{16}$
$H_{38}Ir_2O_2P_2S_2*C_{16}$
$H_{38}Ir_2O_8P_2S_2*C_{16}$
$H_{38}MoP_2*C_{42}$
$H_{38}N_2Sn*C_{19}$
$H_{38}OsP_4*C_{12}$
$H_{38}P_3Re*C_{24}$
$H_{38}P_4Si_2*C_{30}$
$H_{38}Si_4Sn*C_{14}$
$H_{38}Si_6*C_{12}$
$H_{38}Sn_2*C_{16}$
$H_{39}Al*C_{19}$
$H_{39}BF_4OP_2Pt*C_{19}$

Yb

Reaction with 1-hexyne: 1.10.5.3

Zn

Oxidative addition of CH$_4$: 1.10.5.3
Reduction of (CH$_3$)$_2$AsOH: 1.5.3.3.1
Reduction of (CH$_3$)$_2$As(O)OH: 1.5.3.3.1,
1.5.6.3
Reduction of C$_6$H$_5$(CH$_3$)AsO$_2$H:
1.5.3.3.1
Reduction of p-CH$_3$OC$_6$H$_4$AsO(OH)$_2$:
1.5.3.3.1
Reduction of Ge(IV) species in H$_2$O:
1.6.3.3.1
Reduction of [(CH$_3$)$_2$GeS]$_2$ in aq acid:
1.6.3.3.1
Reduction of {[(CH$_3$)$_4$C$_4$]NiCl$_2$}$_2$:
1.6.2.5
Reduction of C$_6$H$_5$CONH$_2$: 1.6.3.1.3
Reduction of alkyl halides: 1.6.2.5
Reduction of aq P$_4$ to form PH$_3$: 1.5.6.2
Reduction of aq Sb^{3+} to SbH$_3$: 1.5.3.4
Reduction of germyl halides: 1.6.3.3.1
Use in industrial reduction of nitro
compounds: 1.5.6.1

Subject Index

This index supplements the compound index and the table of contents by providing access to the text by way of methods, techniques, reaction conditions, properties, effects and other phenomena. Reactions of specific bonds and compound classes are noted when they are not accessed by the heading of the section in which they appear.

For multiple entries, additional keywords indicate contexts and thereby avoid the retrieval of information that is irrelevant to the user's need.

Section numbers are used to direct the reader to those positions in the volume where substantial information is to be found.

Periodic T

Period	Group IA	Group IIA	Group IIIA	Group IVA	Group VA	Group VIA	Group VIIA	
1 1s	1 H							
2 2s2p	3 Li	4 Be						
3 3s3p	11 Na	12 Mg						
4 4s3d 4p	19 K	20 Ca	21 Sc	22 Ti	23 V	24 Cr	25 Mn	F
5 5s4d 5p	37 Rb	38 Sr	39 Y	40 Zr	41 Nb	42 Mo	43 Tc	F
6 6s (4f) 5d 6p	55 Cs	56 Ba	57* La	72 Hf	73 Ta	74 W	75 Re	C
7 7s (5f) 6d	87 Fr	88 Ra	89** Ac	104	105	106	107	1

*Lanthanide series 4f	58 Ce	59 Pr	60 Nd	61 Pm	62 Sm	

**Actinide series 5f	90 Th	91 Pa	92 U	93 Np	94 Pu	A

Adapted from F.A. Cotton, G. Wilkinson, *Advanced Inorgan*